Modern Concepts of Biological Engineering

Modern Concepts of Biological Engineering

Edited by Suzy Hill

SYRAWOOD
PUBLISHING HOUSE

New York

Published by Syrawood Publishing House,
750 Third Avenue, 9th Floor,
New York, NY 10017, USA
www.syrawoodpublishinghouse.com

Modern Concepts of Biological Engineering
Edited by Suzy Hill

International Standard Book Number: 978-1-68286-453-1 (Hardback)

Cataloging-in-publication Data

Modern concepts of biological engineering / edited by Suzy Hill.
    p. cm.
Includes bibliographical references and index.
ISBN 978-1-68286-453-1
1. Bioengineering. 2. Tissue engineering. I. Hill, Suzy.
TA164 .M64 2017
660.6--dc23

Printed in the United States of America.

TABLE OF CONTENTS

Preface..IX

Chapter 1   **Preparation of Laponite Bioceramics for Potential Bone Tissue Engineering Applications** 1
Chuanshun Wang, Shige Wang, Kai Li, Yaping Ju, Jipeng Li, Yongxing Zhang, Jinhua Li, Xuanyong Liu, Xiangyang Shi, Qinghua Zhao

Chapter 2   **Polymer-Ceramic Spiral Structured Scaffolds for Bone Tissue Engineering: Effect of Hydroxyapatite Composition on Human Fetal Osteoblasts** 12
Xiaojun Zhang, Wei Chang, Paul Lee, Yuhao Wang, Min Yang, Jun Li, Sangamesh G. Kumbar, Xiaojun Yu

Chapter 3   **Electrospun Poly(ester-Urethane)- and Poly(ester-Urethane-Urea) Fleeces as Promising Tissue Engineering Scaffolds for Adipose-Derived Stem Cells** 22
Alfred Gugerell, Johanna Kober, Thorsten Laube, Torsten Walter, Sylvia Nürnberger, Elke Grönniger, Simone Brönneke, Ralf Wyrwa, Matthias Schnabelrauch, Maike Keck

Chapter 4   **The Effect of Gamma Irradiation on the Biological Properties of Intervertebral Disc Allografts: *In Vitro* and *In Vivo* Studies in a Beagle Model**........................36
Yu Ding, Dike Ruan, Keith D. K. Luk, Qing He, Chaofeng Wang

Chapter 5   **Evaluation of Physical and Mechanical Properties of Porous Poly (Ethylene Glycol)-co-(L-Lactic Acid) Hydrogels during Degradation** 46
Yu-Chieh Chiu, Sevi Kocagöz, Jeffery C. Larson, Eric M. Brey

Chapter 6   **A Cost-Minimization Analysis of Tissue-Engineered Constructs for Corneal Endothelial Transplantation** 57
Tien-En Tan, Gary S. L. Peh, Benjamin L. George, Howard Y. Cajucom-Uy, Di Dong, Eric A. Finkelstein, Jodhbir S. Mehta

Chapter 7   **Dermal Substitutes Support the Growth of Human Skin-Derived Mesenchymal Stromal Cells: Potential Tool for Skin Regeneration** 66
Talita da Silva Jeremias, Rafaela Grecco Machado, Silvia Beatriz Coutinho Visoni, Maurício José Pereima, Dilmar Francisco Leonardi, Andrea Gonçalves Trentin

Chapter 8   **Fabrication, Characterization and Cellular Compatibility of Poly (Hydroxy Alkanoate) Composite Nanofibrous Scaffolds for Nerve Tissue Engineering**...74
Elahe Masaeli, Mohammad Morshed, Mohammad Hossein Nasr-Esfahani, Saeid Sadri, Janneke Hilderink, Aart van Apeldoorn, Clemens A. van Blitterswijk, Lorenzo Moroni

Chapter 9   **Regional Variations in the Cellular, Biochemical, and Biomechanical Characteristics of Rabbit Annulus Fibrosus** 87
Jun Li, Chen Liu, Qianping Guo, Huilin Yang, Bin Li

Chapter 10   **Comparison of Decellularization Protocols for Preparing a Decellularized Porcine Annulus Fibrosus Scaffold**..98
Haiwei Xu, Baoshan Xu, Qiang Yang, Xiulan Li, Xinlong Ma, Qun Xia, Yang Zhang, Chunqiu Zhang, Yaohong Wu, Yuanyuan Zhang

Chapter 11   **Tissue Engineered Skin Substitutes Created by Laser-Assisted Bioprinting Form Skin-Like Structures in the Dorsal Skin Fold Chamber in Mice**..........................111
Stefanie Michael, Heiko Sorg, Claas-Tido Peck, Lothar Koch, Andrea Deiwick, Boris Chichkov, Peter M. Vogt, Kerstin Reimers

Chapter 12   **Tissue Engineering Bone using Autologous Progenitor Cells in the Peritoneum**................123
Jinhui Shen, Ashwin Nair, Ramesh Saxena, Cheng Cheng Zhang, Joseph Borrelli Jr., Liping Tang

Chapter 13   **Altering the Architecture of Tissue Engineered Hypertrophic Cartilaginous Grafts Facilitates Vascularisation and Accelerates Mineralisation**..............................130
Eamon J. Sheehy, Tatiana Vinardell, Mary E. Toner, Conor T. Buckley, Daniel J. Kelly

Chapter 14   **Induced Collagen Cross-Links Enhance Cartilage Integration**..............................140
Aristos A. Athens, Eleftherios A. Makris, Jerry C. Hu

Chapter 15   **Modulating Gradients in Regulatory Signals within Mesenchymal Stem Cell Seeded Hydrogels: A Novel Strategy to Engineer Zonal Articular Cartilage**........146
Stephen D. Thorpe, Thomas Nagel, Simon F. Carroll, Daniel J. Kelly

Chapter 16   **Noninvasive Quantification of *In Vitro* Osteoblastic Differentiation in 3D Engineered Tissue Constructs using Spectral Ultrasound Imaging**.......................159
Madhu Sudhan Reddy Gudur, Rameshwar R. Rao, Alexis W. Peterson, David J. Caldwell, Jan P. Stegemann, Cheri X. Deng

Chapter 17   **Similar Properties of Chondrocytes from Osteoarthritis Joints and Mesenchymal Stem Cells from Healthy Donors for Tissue Engineering of Articular Cartilage**.................169
Amilton M. Fernandes, Sarah R. Herlofsen, Tommy A. Karlsen, Axel M. Küchler, Yngvar Fløisand, Jan E. Brinchmann

Chapter 18   **Functional Characterization of Detergent-Decellularized Equine Tendon Extracellular Matrix for Tissue Engineering Applications**...183
Daniel W. Youngstrom, Jennifer G. Barrett, Rod R. Jose, David L. Kaplan

Chapter 19   **3D Non-Woven Polyvinylidene Fluoride Scaffolds: Fibre Cross Section and Texturizing Patterns have Impact on Growth of Mesenchymal Stromal Cells**..................192
Anne Schellenberg, Robin Ross, Giulio Abagnale, Sylvia Joussen, Philipp Schuster, Annahit Arshi, Norbert Pallua, Stefan Jockenhoevel, Thomas Gries, Wolfgang Wagner

Chapter 20   **Tissue Engineering in Animal Models for Urinary Diversion**...............................201
Marije Sloff, Rob de Vries, Paul Geutjes, Joanna IntHout, Merel Ritskes-Hoitinga, Egbert Oosterwijk, Wout Feitz

Chapter 21   **A Glycosaminoglycan Based, Modular Tissue Scaffold System for Rapid Assembly of Perfusable, High Cell Density, Engineered Tissues** 211
Ramkumar Tiruvannamalai-Annamalai, David Randall Armant, Howard W. T. Matthew

Permissions

List of Contributors

Index

PREFACE

This book elucidates the concepts and innovative models around prospective developments with respect to biological engineering. This field of study is a branch of engineering focusing on the application of concepts and methods of biology to life sciences. While understanding the long-term perspectives of the topics, the book makes an effort in highlighting their impact as a modern tool for the growth of the discipline. It aims to highlight the varied aspects of biological engineering through the use of extensive examples. This book is a complete source of knowledge on the present status of this important field. Engineers, biologists, researchers, experts, professionals and students will benefit alike from this book.

This book unites the global concepts and researches in an organized manner for a comprehensive understanding of the subject. It is a ripe text for all researchers, students, scientists or anyone else who is interested in acquiring a better knowledge of this dynamic field.

I extend my sincere thanks to the contributors for such eloquent research chapters. Finally, I thank my family for being a source of support and help.

Editor

Preparation of Laponite Bioceramics for Potential Bone Tissue Engineering Applications

Chuanshun Wang[1,9], Shige Wang[2,9], Kai Li[1], Yaping Ju[1], Jipeng Li[1], Yongxing Zhang[1], Jinhua Li[3], Xuanyong Liu[3], Xiangyang Shi[2,4]*, Qinghua Zhao[1]*

1 Department of Orthopaedics, Shanghai First People's Hospital, School of Medicine, Shanghai Jiao Tong University, Shanghai, P. R. China, 2 State Key Laboratory for Modification of Chemical Fibers and Polymer Materials, College of Materials Science and Engineering, Donghua University, Shanghai, P. R. China, 3 State Key Laboratory of High Performance Ceramics and Superfine Microstructure, Shanghai Institute of Ceramics, Chinese Academy of Sciences, Shanghai, P. R. China, 4 College of Chemistry, Chemical Engineering and Biotechnology, Donghua University, Shanghai, P. R. China

Abstract

We report a facile approach to preparing laponite (LAP) bioceramics via sintering LAP powder compacts for bone tissue engineering applications. The sintering behavior and mechanical properties of LAP compacts under different temperatures, heating rates, and soaking times were investigated. We show that LAP bioceramic with a smooth and porous surface can be formed at 800°C with a heating rate of 5°C/h for 6 h under air. The formed LAP bioceramic was systematically characterized via different methods. Our results reveal that the LAP bioceramic possesses an excellent surface hydrophilicity and serum absorption capacity, and good cytocompatibility and hemocompatibility as demonstrated by resazurin reduction assay of rat mesenchymal stem cells (rMSCs) and hemolytic assay of pig red blood cells, respectively. The potential bone tissue engineering applicability of LAP bioceramic was explored by studying the surface mineralization behavior via soaking in simulated body fluid (SBF), as well as the surface cellular response of rMSCs. Our results suggest that LAP bioceramic is able to induce hydroxyapatite deposition on its surface when soaked in SBF and rMSCs can proliferate well on the LAP bioceramic surface. Most strikingly, alkaline phosphatase activity together with alizarin red staining results reveal that the produced LAP bioceramic is able to induce osteoblast differentiation of rMSCs in growth medium without any inducing factors. Finally, in vivo animal implantation, acute systemic toxicity test and hematoxylin and eosin (H&E)-staining data demonstrate that the prepared LAP bioceramic displays an excellent biosafety and is able to heal the bone defect. Findings from this study suggest that the developed LAP bioceramic holds a great promise for treating bone defects in bone tissue engineering.

Editor: Xiaohua Liu, Texas A&M University Baylor College of Dentistry, United States of America

Funding: This work was financially supported by the Medicine and Engineering Joint Foundation of Shanghai Jiao Tong University (No. YG2011MS28), the High-Tech Research and Development Program of China (2012AA030309), and the program for Professor of Special Appointment (Eastern Scholar) at Shanghai Institutions of Higher Learning. The funders had no role in study design, data collection and analysis, decision to publish, or preparation of the manuscript.

Competing Interests: The authors have declared that no competing interests exist.

* Email: xshi@dhu.edu.cn (XS); sawboneszhao@163.com (QZ)

⑨ These authors contributed equally to this work.

Introduction

Bone defects arising from trauma, tumor or bone-related diseases are causing more social issues due to the lack of ideal bone tissue substitutes [1]. Since a Dutch surgeon first used a piece of a dog's skull to repair a soldier's cranium in the 17th century [2], repair of bone defect effectively using substitutes such as autografts and allografts has been of great importance. However, both of the traditional autografts and allografts are not the best candidates, since autografts may face to the donor shortage and donor site morbidity, whereas allografts may suffer the risk of disease transmission and immune response [3]. With the advances of tissue engineering and regenerative medicine [2,4], a majority of damage to any tissue or organ is expected to be solved in clinic [5]. One of the most important issues in tissue engineering and regenerative medicine is to develop various artificial 3-dimensional scaffolds with appropriate physical and/or chemical properties that can closely mimic the natural extracellular matrix [6]. These scaffolds should not bring any immune or other adverse responses

after implantation, should be porous in nature with high surface area to volume ratio to facilitate cell attachment, proliferation, and differentiation so that new tissue can be easily formed, and should be biodegradable so that they do not require any additional surgical procedures to be removed out of body [7].

Beyond polymer scaffolds, inorganic bioceramic materials have been received a great deal of attention for uses as implantation and/or fixation biomaterials [8]. Since the late 1960s, bioceramic has been used as alternatives to metals in order to increase the biocompatibility of the implants [9]. Bioceramic could be composed of several elements including alumina, zirconia, carbon, silica-contained compounds, and some other chemical ingredients [8]. Till now, bioceramic including bioactive glasses [10–12], sintered hydroxyapatite (HA) [13,14], glass ceramics [10,15], and composite materials [16,17] have been intensively studied due to their compatibility with living bone through interfacial formation of a HA interface layer [18]. Taking the biocompatibility and biodegradability into account, bioceramics have been chosen as a promising candidate for potential bone tissue engineering applications. Silicate bioceramics have received significant attention in the past several years due to the fact that they can efficiently stimulate the proliferation, differentiation, and osteogenic gene expression of tissue cells as well as the regeneration of bone tissue by release of Si-containing ionic products [19–21] and their special surface composition renders them the ability to be used as a template for the formation of artificial bone tissue [22,23].

Laponite $(Na^+_{0.7}(Si_8Mg_{5.5}Li_{0.3})O_{20}(OH)_4)^-_{0.7}$, LAP) is a kind of synthetic silica clay material that can be degraded into nontoxic products under physiological conditions [24,25]. LAP is biocompatible and has been used as a drug carrier because its interlayer space can be used to encapsulate drug molecules with high retention capacity [24–27]. In our previous study, we fabricated poly(lactic-co-glycolic acid) (PLGA) nanofibers incorporated with LAP nanodisks for osteogenic differentiation of human mesenchymal stem cells (hMSCs) [28]. We show that the incorporated LAP is beneficial to promote the cell adhesion and proliferation when compared with pure PLGA nanofibers. More strikingly, the doped LAP within the PLGA nanofibers is able to induce the osteoblast differentiation of hMSCs in growth medium without any inducing factors, such as dexamethasone [28], which is likely ascribed to the fact that the ionic Si and Mg can be released from LAP.

In this study, we prepared LAP ceramic by sintering LAP powder compact at 1200°C for 6 h for potential bone tissue engineering applications. The sintering behavior, mechanical properties, and other physical properties including line shrinkage, relative density, and contact angle of LAP bioceramic under different temperatures, heating rates, and soaking time periods were investigated. The surface morphology of the LAP ceramic was observed using scanning electron microscopy (SEM). The hemocompatibility of the LAP ceramic was investigated via hemolysis assay, while the cytocompatibility of the material was evaluated via resazurin reduction assay as well as SEM observation of the morphologies of rat MSCs (rMSCs) cultured onto the LAP bioceramic. The potential bone tissue engineering applicability of LAP bioceramic was explored by studying its surface mineralization behavior via soaking in simulated body fluid (SBF), as well as the osteogenic differentiation of rMSC cultured onto the material. Finally, the *in vivo* bone defect repair ability and biosafety were studied using a pig model. To our knowledge, this is the first report concerning the preparation of LAP bioceramics for bone tissues engineering applications.

Materials and Methods

Materials

LAP with a diameter of 50 nm and a thickness of 7 nm was purchased from Zhejiang Institute of Geologic and Mineral Resources Co., Ltd. (Hangzhou, China). Eagle's Minimal Essential Medium (α-MEM), Dulbecco's Modified Eagle's Medium (DMEM), fetal bovine serum (FBS), phosphate buffer saline (PBS), penicillin, and streptomycin were purchased from Gibco (Carlsbad, CA). β-Glycerophosphate (β-GP), ascorbic acid, resazurin, p-nitrophenyl phosphate, and p-nitrophenol standard were from Sigma (St. Louis, MO). Reporter Lysis Buffer and Picogreen DNA quantification kit were from Molecular Probes, Inc. (Eugene, OR). rMSC and heparin-stabilized pig blood was kindly provided by Shanghai First People's Hospital (Shanghai, China). All other chemicals were from Sinopharm Chemical Reagent Co., Ltd (Shanghai, China) and used as received. Water used in all experiments was purified using a Milli-Q Plus 185 water

purification system (Millipore, Bedford, MA) with resistivity higher than 18 MΩ·cm.

Preparation of LAP bioceramic

LAP bioceramic was produced by uniaxial pressing of 0.45 g LAP powder which was placed in a mold with a diameter of 14 mm under 10 MPa and sintering in a roasting furnace (P300, Nabertherm, German) at different temperatures, heating rates, and sintering time periods (Table 1). Finally, the formed LAP bioceramic was cooled down to room temperature and stored in a desiccator before use.

Characterization

The surface morphology of various LAP bioceramic was observed using SEM (JEOL JSM-5600LV, Japan) with an accelerating voltage of 10 kV. All samples were sputter coated with gold films with a thickness of 10 nm before observation. The diameter of each sample before and after sintering was measured using a micrometer, and the line shrinkage was calculated by dividing the diameter difference before and after sintering by the diameter of the original LAP compacts. The relative density of each sample after sintering was determined by dividing the apparent density measured in water using the Archimedean technique by the density of LAP powder (2.60 g/cm³) [29]. The surface hydrophilicity of LAP bioceramic was evaluated via water contact angle test using a contact angle goniometer (DSA-30, Kruss, Germany). Before analysis, 1 μL water was dropped onto the surface of each sample at the randomly selected areas in triplicate. The contact angle was recorded when the droplet was stable at ambient temperature and humidity. The mechanical property of LAP bioceramic was studied via nanoindentation experiments using a nano indenter (Agilent, Nano Indenter G200, Santa Clara, CA). A diamond Berkovich indenter with a tip radius of 20 nm was used. The constant value of Poisson ratio was 0.25, the vibration frequency of indenter was 45 Hz, and the maximum indentation depth was 4000 nm. The sintering behavior LAP compacts under different temperatures, heating rates, and soaking time periods were comparatively investigated by X-ray diffraction (XRD) using a Rigaku D/max-2550 PC XRD system (Rigaku Co., Tokyo, Japan) using Cu Kα radiation with a wavelength of 1.54 Å at 40 kV and 200 mA.

Hemolysis assay

The hemocompatibility of the formed LAP bioceramic was examined via hemolysis assay according to our previous study [28]. Briefly, pig red blood cells (pRBCs) were obtained by removing the serum via centrifugation (5000 rpm, 3 min) and washing with PBS for 3 times. The obtained pRBCs were 10 times diluted with PBS. Each sample was placed in the individual well of a 24-well tissue culture plate, and 2 mL of the diluted pRBCs suspension was added. Another two wells containing 0.4 mL of the diluted pRBCs and 1.6 mL of water and PBS solution were set as positive and negative control, respectively. The plate was then incubated at 37°C for 2 h, and the supernatant was centrifuged (10000 rpm, 1 min) and the absorbance of the supernatant related to hemoglobin was recorded using a Lambda 25 UV-Vis spectrophotometer (Perkin Elmer, Waltham, MA) at 541 nm. The hemolytic percentage (HP) can be calculated using the following equation [30],

$$HP(\%) = \frac{D_t - D_{nc}}{D_{pc} - D_{nc}} \times 100\% \quad\quad (1)$$

Table 1. Physicochemical parameters of LAP compact and LAP-bioceramic (all data are given as mean ± SD, n = 3).

Sample ID	1*	2	3	4	5
Sintering temperature (°C)	-	600	800	800	800
Heating rate (°C/h)	-	5	5	10	5
Sintering time (h)	-	6	2	6	6
Line shrinkage (%)	-	3.62±0.37	5.99±0.65	7.94±0.65	8.80±1.34
Relative density (%)	-	72.7±1.8	86.5±2.2	87.7±3.1	96.2±2.4
Contact angle (°)	-	27.62±2.33	29.64±0.96	25.29±3.22	18.37±1.24

where D_t is the absorbance of the test samples; D_{pc} and D_{nc} are the absorbances of the positive and negative controls, respectively.

Biomineralization

The formed LAP bioceramic (Sample #5, Table 1) was immersed into a 1.5-times concentrated simulated body fluid (SBF) at 37°C up to 7 days, and the SBF solution was changed every 24 h [31]. The LAP bioceramic was removed from the SBF solution after 7 day incubation, gently rinsed with water, and air-dried at room temperature. The formation of HA onto the LAP surface was confirmed using SEM and energy-dispersive spectroscopy (EDS, IE300X, Oxford, U.K.) attached to the SEM equipment.

Serum adsorption onto LAP bioceramic

The serum adsorption onto the surface of LAP ceramic was quantified according to procedures described in our previous study [28]. Briefly, LAP bioceramic exposed by UV light for 2 h was fixed in a 24-well tissue culture plate (TCP). After that, 1 mL FBS (10%, in PBS) solution was added to each well and incubated for 24 h at 37°C. TCP without LAP bioceramic was set as control. The concentration of FBS before and after adsorption was quantified using a Lambda 25 UV-Vis spectrophotometer at 280 nm based on the FBS calibration curve at the same wavelength. The adsorbed FBS on the surface of LAP bioceramic was also observed by SEM with an accelerating voltage of 10 kV.

rMSC culture and seeding

rMSCs (passage 2) were cultured in 25 cm² tissue culture flasks with 5 mL complete medium (α-MEM supplemented with 10% FBS, 1% ascorbic acid solution (5 mg/mL in PBS), and 1% β-GP solution (1 M in PBS), 100 U/mL penicillin, and 100 U/mL streptomycin) in a humidified incubator with 5% CO_2 at 37°C. The culture medium was replaced every 3 days and cells were 1:3 passaged when reaching a confluence of 80–90%. Before cell seeding, LAP was sterilized after exposure under UV light for 2 h. TCPs were set as control. rMSCs (passage 3) were seeded at a density of 2×10^4 cells per well with 1 mL α-MEM per well and incubated at 37°C and 5% CO_2. The medium was replaced every 3 days.

Metabolic activity of rMSCs

The metabolic activity of rMSCs cultured onto LAP bioceramic was evaluated using resazurin reduction assay. At each predetermined time point, medium was replaced with 900 μL complete α-MEM and 100 μL resazurin solution (1 mg/mL in PBS). Then the plate was incubated for another 4 h, and the fluorescence intensity in proportion to the viability of the cells was measured by a BioTek Synergy 2 multilabel plate reader ($\lambda_{ex} = 530$ nm, $\lambda_{em} = 590$ nm).

The morphology rMSCs cultured onto the LAP bioceramic surface after 14 days was observed by SEM with an accelerating voltage of 10 kV. Before observation, cell samples were rinsed 3 times with PBS solution to remove non-adherent cells, and then fixed with 2.5 wt% glutaraldehyde at 4°C for 2 h, followed by dehydrating through a series of gradient ethanol solutions of 30%, 50%, 70%, 80%, 90%, 95%, and 100%. After air dried overnight, samples were sputter coated with a 10 nm thick gold film before SEM observation.

Alkaline phosphatase activity and DNA content assay

After 14 day culture, rMSCs cultured in 24-well plate were rinsed with PBS for 3 times. After that, 200 μL Reporter Lysis Buffer was added to each well to lyse cells according to the manufacturer's instruction. The cell lysates were stored at −20°C before analysis. For the alkaline phosphatase (ALP) activity assay, 20 μL of the cell lysates was mixed with 200 μL of ALP substrate and incubated for 1 h at 37°C in the dark. Thereafter, 10 μL of 0.02 M NaOH was added to each well to stop the hydrolysis. For comparison, 220 μL of ALP substrate mixed with 10 μL of 0.02 M NaOH in triplicate was used as a blank control. The absorbance was read at 405 nm and the ALP content was calculated from a standard calibration curve.

The DNA content of each cell sample was quantified using Picogreen DNA kit. Briefly, 20 μL cell lysates was mixed with 80 μL Tris-EDTA buffer and transferred to a clean 96-well plate. Then, 100 μL Picogreen working reagent was added to each well and incubated at room temperature for 5 min. The fluorescence intensity was immediately monitored by a BioTek Synergy 2 multilabel plate reader ($\lambda_{ex} = 485$ nm, $\lambda_{em} = 538$ nm). The DNA content was calculated from a standard calibration curve.

Alizarin red staining

Before histochemical assay, cells were first fixed with 3.7% formaldehyde in PBS solution for 2 h at 4°C and then rinsed with water for 3 times to remove all traces of formaldehyde. The fixed cells were first covered with Alizarin red S solution (1%, in water with a pH range of 6.3–6.4 adjusted using 0.28% NH_4OH) for 2 min, washed with water and acidic ethanol (1 part of concentrated HCl to 10000 parts of ethanol 95%), and then observed using Leica DM IL LED inverted phase contrast microscope with a magnification of 200× for each sample.

In vivo biosafety evaluation

All animal studies (including the acquisition of rMSCs and heparin-stabilized pig blood as mentioned in "Materials" section) were approved by the Animal Ethics Committee of Shanghai Jiao Tong University School of Medicine (project number 2012008) according to Regulations for the Administration of Affairs

Concerning Experimental Animals (approved by the State Council of the People's Republic of China) and Guide for the Care and Use of Laboratory Animals (Department of Laboratory Science, Shanghai Jiao Tong University School of Medicine, laboratory animal usage license number SYXK 2008-0050, certificated by Shanghai Committee of Science and Technology). Female SD rats (180–205 g) were obtained from Shanghai SLAC Laboratory Animal Co., Ltd. (Shanghai, China). LAP bioceramic was soaked in saline in a concentration of 2 g/mL under 37°C for 3 days, and the biomaterial extract was then sterilized through a filter (Millipore, 0.22 μm) and stored at 4°C. *In vivo* biosafety of LAP bioceramic was evaluated using acute systemic toxicity test and intramuscular stimulation test, respectively. For the acute systemic toxicity test, six female SD rats (divided into two groups, three for experimental group and three for control group) were chosen for the acute systemic toxicity test. Rats in experimental group were intraperitoneally injected with the prepared extract with a dose of 50 mL/Kg according to body weight, while the control group was treated with saline in the same manner. The general toxic effects including appetite, breathe, movement, body temperature, body weight, and survival rate were monitored daily during the first week. For the intramuscular stimulation test, another female SD rat was used and the hair on the back was removed. Then 12 dots on the back were injected with 100 μL saline (dots 1–6, negative controls), 100 μL alcohol (dots 10–12, positive controls, 5% v/v), and 100 μL extract (dot 7–9), respectively. Erythema, edema, and necrosis of skin around the injection region were monitored immediately and at 1, 2, and 3 days post injection.

Histological analyses were performed to further evaluate *in vivo* biosafety of LAP bioceramic extract. Briefly, female SD rat treated with saline or LAP bioceramic extract were euthanized after 14 days, major organs including the heart, liver, spleen, lung, and kidney were harvested and fixed with 10% neutral buffered formalin. Then, the above organs were embedded in paraffin, sectioned into slices with a thickness of 8 μm, and stained with hematoxylin and eosin (H&E). Each stained slide was examined using a Leica DM IL LED inverted phase contrast microscope with a magnification of 100×.

Animal experiments

Two mature male pigs (Bama miniature swine, 25 kg, 10 months old) were used. Each pig was anesthetized using a mixture of ketamine hydrochloride (60 mg/kg body weight) and xylazine (6 mg/kg body weight). When the animals were in supine position, a rectangular bone defect with a dimension of 1.0 cm×0.2 cm×0.5 cm (length×width×depth) was prepared on the right (control group) and left front (experimental group) leg diaphysis using a orthopaedic bone drills, respectively. Note that this is a non-weight bearing model of assessment. The defect of the experimental groups was implanted with LAP bioceramic while the control group was not implanted with any additional materials. After implantation, penicillin was injected intramuscularly to avoid wound infections. Radiographs of the implant region for each animal immediately after surgery and 24 weeks after euthanized were obtained by X-ray (GE OEC 9900 Elite) under standardized conditions: 60 kV, 4 mA, 4.5 seconds exposure time, 65 cm film-radiation beam distance. Each radiograph was calibrated at the same grey scale and the radiographs were converted to digitalized images using a digital camera.

Statistical analysis

One way ANOVA statistical analysis was carried out to assess the significance of the experimental data. 0.05 was selected as the significance level, the results were indicated with (*) for $p < 0.05$, (**) for $p < 0.01$, and (***) for $p < 0.001$, respectively.

Results and Discussion

Preparation and characterization of LAP bioceramic

In our previous study, we have shown that LAP-doped PLGA nanofibers are able to induce the osteoblast differentiation of hMSCs in growth medium without any inducing factors [28], which may be due to the released Si ions from LAP. Inspired by this, we hypothesize that a scaffold produced from LAP powder without any organic components may also induce the osteoblast differentiation. To prove our hypothesis, in this present study, we prepared LAP bioceramic under a high temperature sintering process, which is expected not to compromise the biocompatibility of LAP [32]. The *in vitro* and *in vivo* performances of the formed LAP bioceramic to act as an artificial scaffold for bone tissue engineering were evaluated.

As shown in Table 1, heating rate, sintering time, and sintering temperature are main factors that may have immediate influence on the formed LAP bioceramic. We first comparatively studied the synergistic relationship between these factors and the properties of the formed LAP bioceramic. Apparently, compared to LAP compact (Sample #1), the chosen sintering temperature (not lower than 600°C) can sufficiently render the LAP compact with a ceramic prototype. With the increase of the temperature (Sample #5 versus Sample #2), and sintering time (Sample #5 versus Sample #4), the line shrinkage ($p < 0.01$) and relative density ($p < 0.05$) significantly increased, implying a densification process of the LAP compacts. The densification may lead to a smooth surface, thus a decreased water contact angle. With the increase of heating rate, the densification of LAP compacts may be slowed down, and the line shrinkage and relative density decrease, while water contact angle increases ($p < 0.01$, Sample #5 versus Sample #2). This shows a contrary variation tendency with sintering temperature and time.

Figure 1 shows the morphology of LAP compacts and bioceramics under different sintering conditions. Although the LAP compact shows a regular smooth surface, the poor water stability limits its applications. The sintering process does not seem to alter the relatively smooth surface when compared to the LAP compact before sintering (Figures 1b, 1d, and 1e), except the case shown in Figure 1c, where cracks are observed on the surface of LAP bioceramic. It appears that in this case, while the sintering temperature is sufficient to convert the LAP compact to ceramic prototype, however, the sintering time is too short to lead to a smooth surfaced biocermaic. It is notable that some regular small holes can be detected in Figure 1d. With such a rough surface, the attachment and proliferation of stem cells may be facilitated. We then compared the mechanical property of the LAP bioceramic via an instrumented nanoindentation test (Figure 2). The hardness-displacement (Figure 2a) and modulus-displacement (Figure 2b) curves of LAP compact and LAP bioceramics show that both hardness and modulus have an apparent dip in the depth range of less than 20 nm, then increase gradually and level off. Apparently, the final hardness and modulus increase with sintering temperature and sintering time. Figures 2c and 2d show the harmonic contact stiffness-displacement and load-displacement curves. The nearly identical relationship between maximum harmonic contact stiffness and maximum load as a function of sintering temperature and sintering time can be found. Therefore, a optimized hardness, modulus, stiffness, and load can be obtained at a sintering temperature of 800°C, sintering time of 6 h, and heating rate of

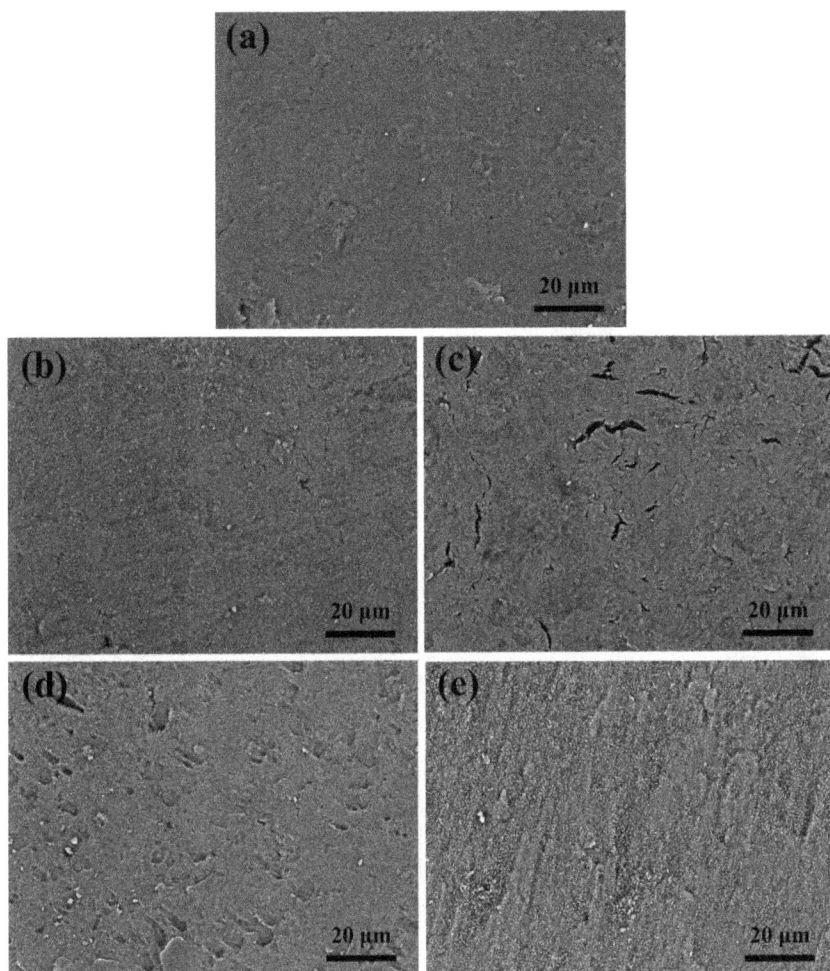

Figure 1. The SEM surface morphologies of the LAP compact. The LAP compact (a) before and (b–e) after sintered. (b) Sample #2, (c) Sample #3, (d) Sample #5, and (e) Sample #4. See Table 1 for sample information.

5°C/min. LAP ceramics prepared under the optimized conditions was selected for subsequent studies.

XRD was used to characterize the crystalline phase change of the LAP compacts before and after sintering (Figure 3). It is obvious that all of the featured LAP peaks exist at a low sintering temperature (600°C). At a high sintering temperature, besides the existing main peaks related to LAP, some new peaks belonging to sodium mica (JCPD: 46-0740) and enstatite (JCPD: 19-0768) emerge in the pattern, illustrating that part of the LAP has been changed into new crystalline phases under a high temperature. Our results indicate that the high-temperature sintering process enables the generation of LAP bioceramic composed of the crystals of LAP, sodium mica, and enstatite.

Hemocompatibility assay

Hemocompatibility has been considered as one of the key issues for an ideal material to be used in tissue engineering applications, especially when the designed scaffold materials are required to contact blood [30]. Like our previous study [28], we evaluated the hemocompatibility of LAP bioceramic via hemolysis assay *in vitro* (Figure 4). Obviously, the pRBCs were totally damaged after exposed to water (Figure 4b, a positive control). In contrast, similar to the pRBCs exposed to PBS solution utilized as a negative control, no visible hemolysis phenomenon was observed after exposure of pRBCs to PBS solution containing LAP bioceramics (p<0.01, Sample #2, #3, #4, and #5, versus sample #1, respectively, as shown in Figure 4a). The hemolytic effects of each sample were quantified by recording the absorbance of the supernatant at 541 nm, which is in proportion to the hemoglobin concentration (Figure 4a). It is found that LAP bioceramics formed under different sintering temperatures have hemolysis percentages all lower than 5% ($2.1\pm0.3\%$, $1.5\pm0.3\%$, $2.6\pm0.1\%$, and $2.3\pm0.3\%$ for Sample #2, #3, #4, and #5, respectively). This indicates that LAP bioceramic possesses good hemocompatibility [33], which is essential for their applications in bone tissue engineering. It is worth noting that the LAP compact without sintering shows a slight hemolysis effect (hemolysis percentage of $8.3\pm0.2\%$, Figure 4a; pRBCs was partially destroyed, Figure 4b, tube 1), which may be due to the fact that the degradation products destroyed part of the pRBCs. Our results suggest that a sintering process enables the LAP with improved hemocompatibility, and degradation products of in vivo inserted LAP implant at a relatively high concentration may generate certain degree of hemolysis.

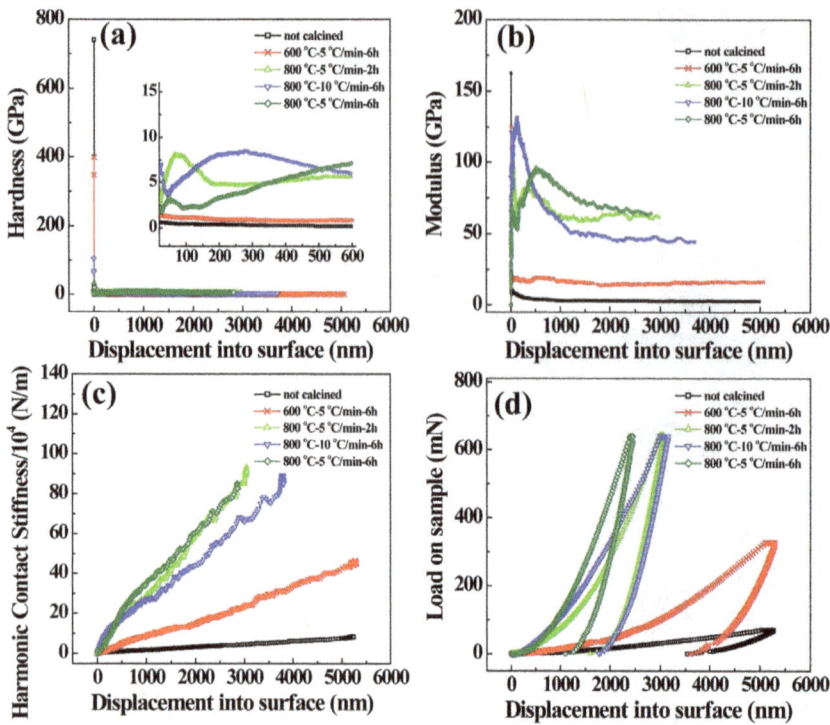

Figure 2. Mechanical properties of the LAP bioceramics. (a) Hardness–displacement, (b) elastic modulus–displacement, (c) harmonic contact stiffness-displacement, and (d) load-displacement curves of LAP compact before and after sintered under different conditions. Inset in (a) shows the hardness change in the range of 20–600 nm.

HA formation and serum adsorption onto LAP bioceramic

Bone-like HA plays an essential role in the formation, growth, and maintenance of the tissue-biomaterial interface [34]. Bioceramic was reported to be compatible with living bone through interfacial formation of HA interface layer [18]. We then

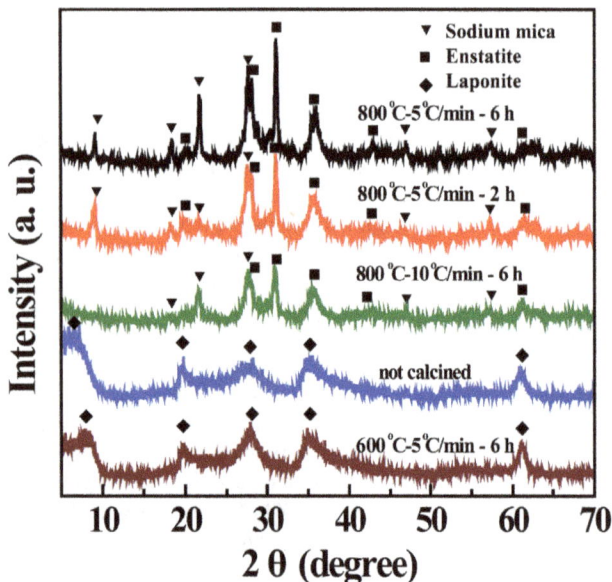

Figure 3. XRD patterns of LAP compact before and after sintering under different conditions.

evaluated the HA formation ability by soaking LAP bioceramic into SBF. Figures 5a and 5c illustrate the surface morphology of LAP ceramic before and after soaking in SBF solution for 7 days, respectively. Obviously, the HA deposits could be found on the LAP bioceremic surface after 7 day incubation in SBF. The mineralized HA shows a particulate morphology, with a diameter of nearly 1 μm. Figures 5b and 5d show the EDS analysis of the LAP ceramic before and after soaking in SBF solution for 7 days, respectively. It can be seen that before soaking in SBF, the element of Si, Mg, Na, and O belonging to the LAP itself can be detected. After mineralization for 7 days in SBF, besides the above elements associated with LAP, Ca and P with a molar ratio of 1.57 similar to the stoichiometric molar ratio of HA can be found, further confirming that LAP biocermic possesses the apatite formation ability, which is essential for its bone tissue engineering applications.

An ideal material for bone tissue engineering applications should also have the ability to absorb protein onto its surface, thus providing enough nutrition for cell growth and migration [35]. Based on this point, we then explored the serum (FBS) adsorption capacity of LAP bioceramic (Figures 6a–c). As shown in Figure 6a, the LAP bioceramic can absorb similar amount of FBS when compared to TCP after 24 h incubation (9.1 ± 0.9 mg/well versus 8.4 ± 0.2 mg/well, $p > 0.05$). The adsorbed FBS was further confirmed by SEM (Figures 6b–c), where solid-state FBS is attached onto the surfaces of both TCP and LAP bioceramic. Since TCP has been already coated with other materials to enhance cell attachment and proliferation, it is reasonable to conclude that the protein adsorption onto LAP bioceramic surface may induce enhanced cellular response, similar to our previous reports [28,36].

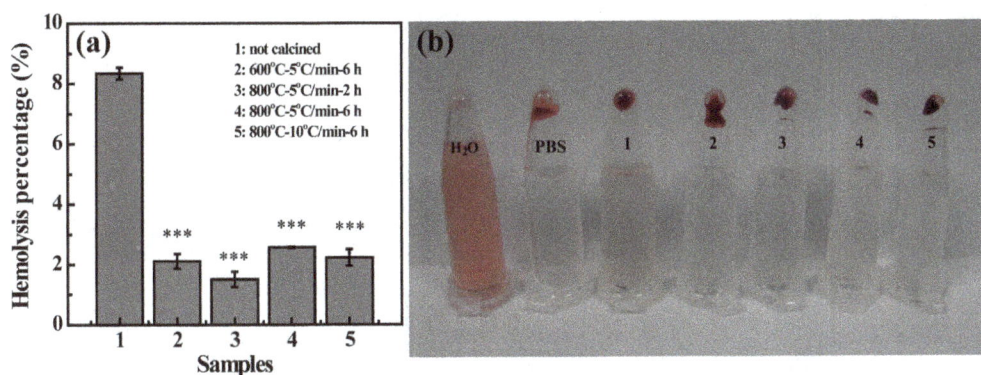

Figure 4. Hemolytic assay data of LAP bioceramics. (a) Hemolytic percentage (%) of pRBCs after treatment with LAP before and after sintered under different conditions for 2 h (mean $\pm$ SD, n = 3). (b) shows the photograph of rRBC suspensions after treatment with different LAPs (shown in (a)), followed by centrifugation.

Cytocompatibility assay of LAP bioceramic

With the good hemocompatibility and excellent protein adsorption capacity, we next explored the potential to use the LAP bioceramic as a scaffold for proliferation and osteogenic differentiation of rMSCs, which is important in bone tissue engineering applications. We first analyzed the metabolic activity of the rMSCs cultured onto LAP bioceramic. Figure 6d shows the resazurin reduction assay data of rMSCs at different time points. Apparently, during the first 3 days, the rMSCs grow slowly. During day 3 to day 14, rMSCs experience a typical exponential phase growth [37] and the metabolic activity is enhanced rapidly likely due to autocrine secretion of extracellular matrix (ECM) from cells. Importantly, metabolic activity of rMSCs cultured onto

both TCP and LAP bioceramic does not show any significant difference at each time point ($p > 0.05$), implying the good cytocompatibility of the prepared LAP bioceramic.

The morphology of rMSCs seeded onto LAP bioceramic after 14 day culture was observed by SEM (Figures 6e–f). Clearly, rMSCs could adhere onto the LAP bioceramic tightly, confirming that the porous surface morphology is favorable for cell adhesion and proliferation, in agreement with the metabolic activity assay results. A higher magnification SEM image clearly reveals the cell pseudopodia structure (Figure 6f). The rMSCs grown onto the porous LAP bioceramic have filopodia extended and migrated onto the surface to form structured 3D cell-scaffold network, suggesting that the bioactive LAP bioceramic can promote the cell

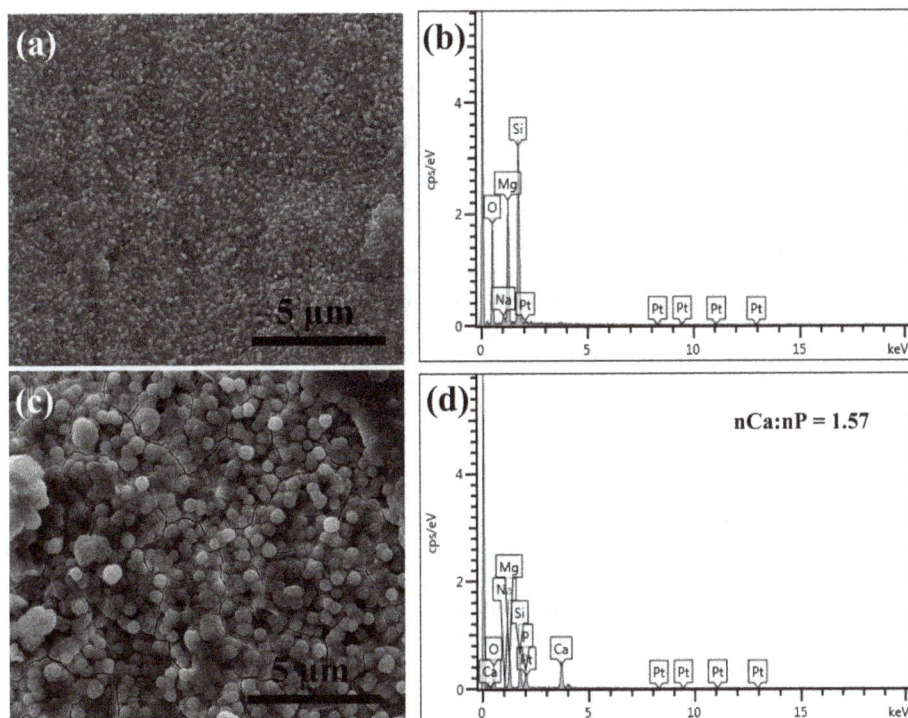

Figure 5. Biomineralization onto the surface of LAP ceramic. (a) and (c) show the SEM surface morphology of LAP ceramic before and after soaking in SBF solution for 7 days, respectively. (b) and (d) show the EDS analysis of the LAP ceramic before and after soaking in SBF solution for 7 days, respectively.

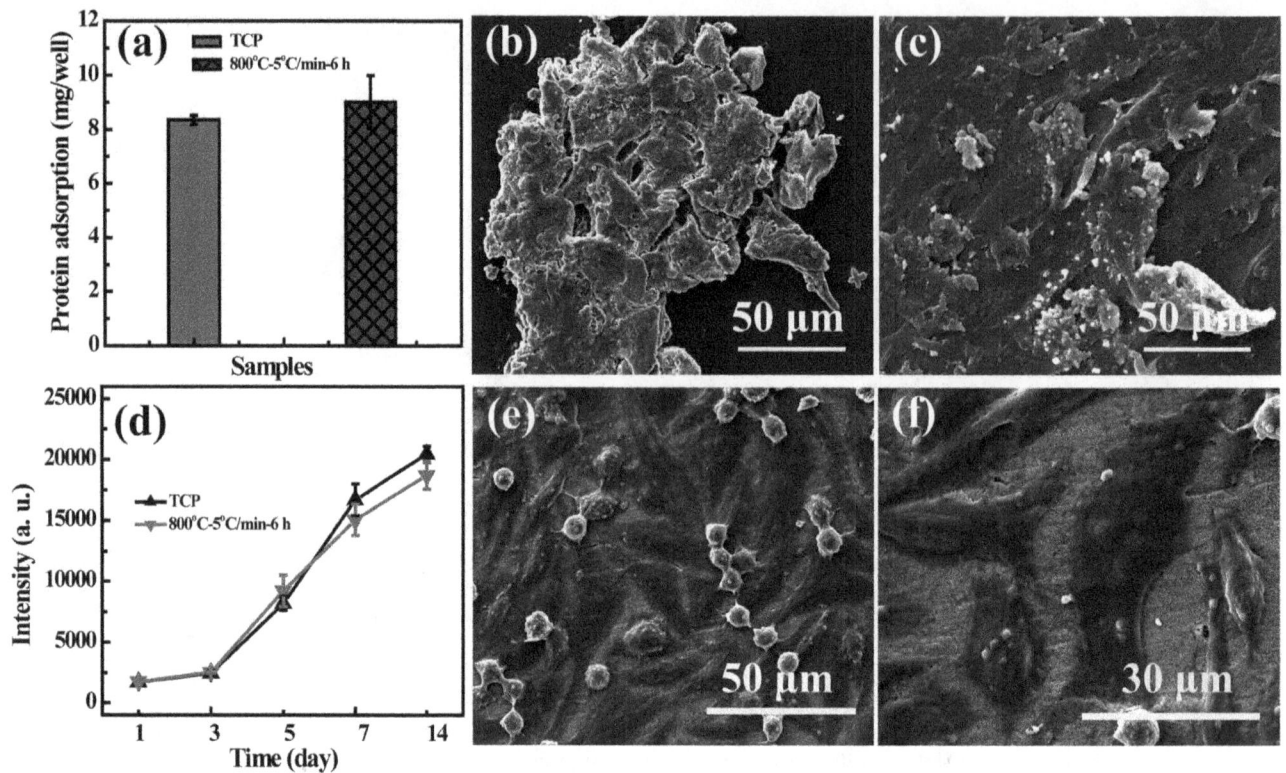

Figure 6. Protein adsorption and metabolic activity assay of rMSCs. (a) The adsorption of protein onto TCP and LAP ceramic (mean ± S.D., n = 3). (b) and (c) show the SEM micrographs of the TCP and LAP ceramic with protein adsorption, respectively. (d) shows the metabolic activity assay of rMSCs cultured onto TCP and LAP ceramic (mean ± S.D., n = 3). (e) shows the micrograph of rMSCs proliferated onto the LAP bioceramic for 14 days. (f) is the magnified image of (e).

attachment and proliferation. Overall, the sintered LAP bioceramic with porous surface structure, good hemocompatibility, excellent protein adsorption capacity, sufficient mechanical durability, and good cytocompatibility may serve as a porous scaffolding material for bone tissue engineering applications.

In vivo biosafety of LAP bioceramic

We nest explored the *in vivo* biosafety of LAP bioceramic through the use of acute systemic toxicity test and intramuscular stimulation test, respectively. As shown in Figures 7a and 7d, after intraperitoneal injection of the extract, no obvious body temperature and body weight alteration can be detected between rats treated with saline and LAP bioceramic extract within the first week (p>0.05, rats treated with saline versus rats treated with LAP at each time points). And no death and other toxic symptoms such as appetite reduction, breathing difficulty and mobility impairments were shown for either saline or the extract groups (data not shown). Intramuscular stimulation results (Figures 7b, 7c, 7e, and 7f) illustrate that there is no obvious erythema, edema and necrosis for normal rat skin treated with the extract (dots 7–9, Figures 7b, 7c, 7e, and 7f), similar to that treated with saline (dots 1–6, Figures 7b, 7c, 7e, and 7f) at all monitored time points. In sharp contrast, the rat skin treated with alcohol (dots 10–12, Figures 7b, 7c, 7e, and 7f) displayed obvious skin damage at the injected area. H&E staining was further used to assess the in vivo biosafty of LAP bioceramic, as shown in Figure 8. Compared with rat treated with saline (control group), all of the studied organs from mice treated with LAP bioceramic extract showed no appreciable abnormality or noticeable damage, further suggesting a good in vivo

biocompatibility of LAP bioceramic. Our results suggest that LAP bioceramic has no irritation to normal skin of rats, thus possessing an excellent *in vivo* biosafety.

In vitro and in vivo bone formation ability of LAP bioceramic

We finally explored the ability of LAP bioceramic in regulating the osteogenic differentiation of rMSCs. ALP enzyme is an important marker of osteogenesis which is usually used to monitor the osteogenic differentiation of osteoblasts. It is highly active in osteoblasts involved in the early initiation of mineralization of newly formed bone tissue [38]. We then measured the ALP activity of rMSCs on day 14, and the results (normalized for the total DNA content) are shown in Figure 9a. It is clear that rMSCs cultured onto LAP bioceramic show a significantly higher ALP activity (p<0.05) than those cultured onto TCP in growth medium without any inducing factors on day 14. To further qualitatively confirm the osteoblastic differentiation of rMSCs cultured onto LAP bioceramic, alizarin red staining was performed. Alizarin red S is an ionic dye which tends to bind with the calcium deposition associated with the osteoblastic differentiation and generates a visible red complex. As shown in the inset of Figure 9a, only rMSCs cultured onto LAP bioceramic show an obvious red color, implying the formation of calcium deposit during the osteoblastic differentiation, further confirming that LAP bioceramic can induce osteogenic differentiation of rMSCs in growth medium without any inducing factors.

In vivo implantation experiment showed that no apparent wound infection was identified during the healing period or at the time of

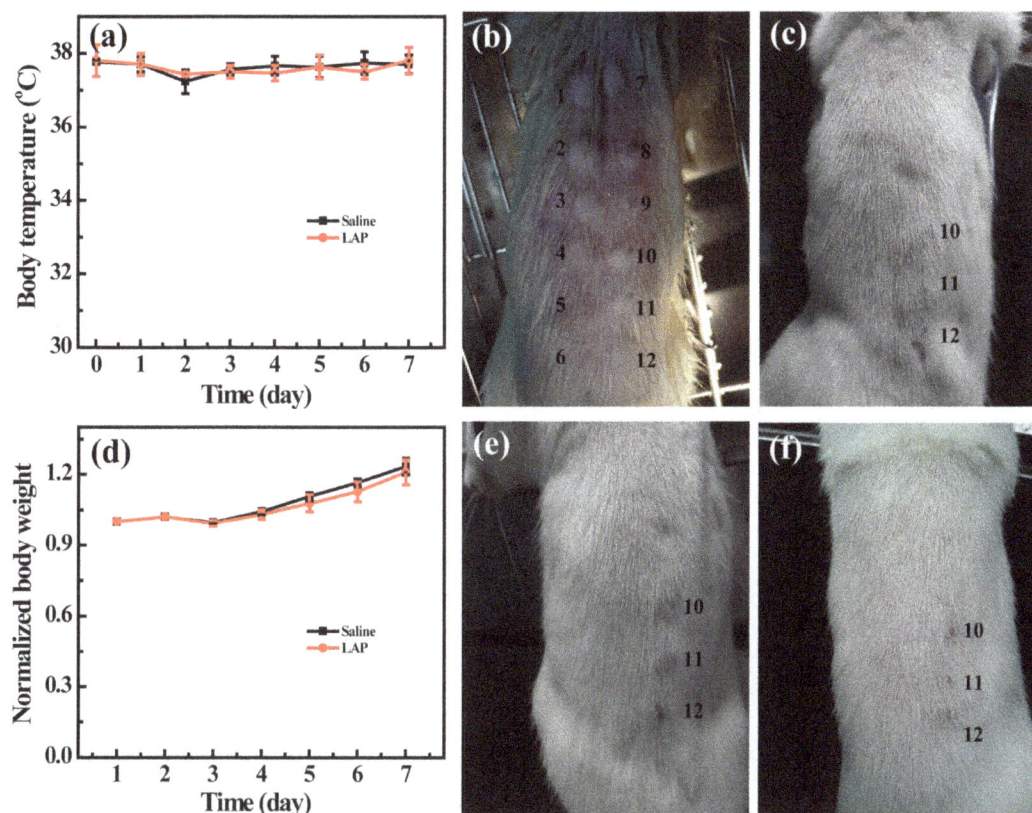

Figure 7. Biosafety test of LAP ceramic extract. (a) Body temperature and normalized body temperature curves of SD rat treated with saline or LAP ceramic extract. (b), (c), (e), and (f) show the results of acute toxicity on normal skin of SD rat immediately after injection or 1 d, 2 d or 3 d post injection of saline (dots 1–6), LAP ceramic extracts (dots 7–9), and alcohol (dots 10–12), respectively.

retrieval and all of the pigs survived after the implantation procedures (data not shown). The bone defects of experimental groups was healed (Figure 9e) while control group was not, illustrating that the defect size is large enough to prove our hypothesis, in agreement with the literature [39]. LAP bioceramic was not distinguishable from surrounding bone and a good union at the LAP bioceramic-implanted interface was observed 24 weeks post-surgery except for a trace of residuals (Figure 9e and Figure S1 in Supporting Information, pointed by an arrow head),

implying an excellent ability to use LAP bioceramic to induce the bone formation. Moreover, the defects exhibited some resistance on palpation, further indicating that LAP bioceramic is able to induce bone regeneration [40,41].

The bone formation was then evaluated using X-ray analysis. Figures 9c and 9f display the representative radiographs of control leg and treated leg 24 weeks after the implantation. Obviously, the residual LAP bioceramic can be clearly detected under the X-ray (Figure 9f). However, the density of the implant region in which

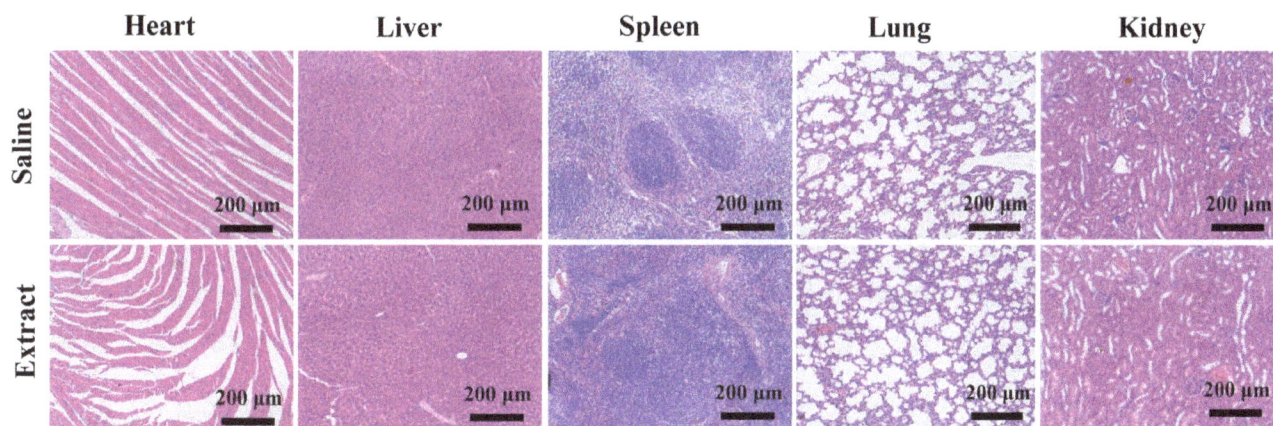

Figure 8. Histological examination. H&E-stained tissue sections of major organs, including the heart, liver, spleen, lung, and kidney from mice treated with saline or LAP ceramic extract for 14 days.

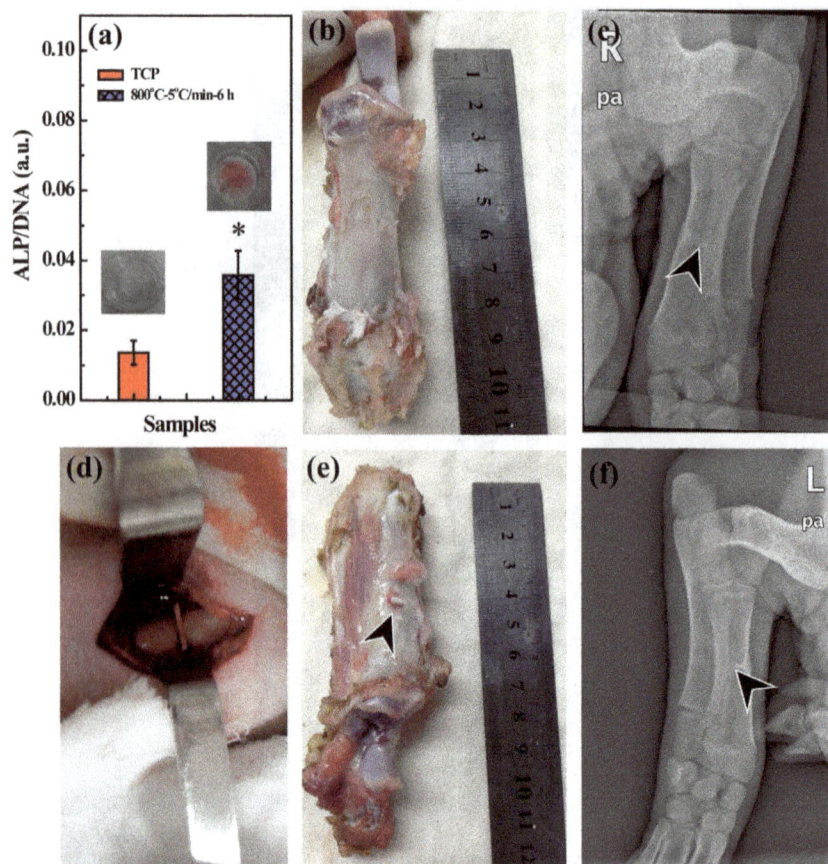

Figure 9. *In vitro* osteogenic differentiation of rMSCs and *in vivo* bone repair evaluation. (a) ALP activity (normalized for the DNA content, nmol of transformed substrate per unit of time and per mass of DNA) of rMSCs cultured onto different substrates in growth medium after 14 days culture. Insert of (a) shows the picture of alizarin red staining of rMSCs cultured onto TCP (left) and LAP ceramic (right) in growth medium without any inducing factors on day 14. (b–f) show the macroscopic appearance of bone defects (A 2-mm bone defect was created in the middle of the tibia, which was implanted with laponite ceramic as shown in (d)). (b) and (e) show the macroscopic appearance of defects without and with implantation for 24 weeks, a trace of laponite ceramic residual is observed in (e) as pointed by an arrow. (c) and (f) show the radiographic images of bone defects without and with LAP implantation after 24 weeks.

the LAP bioceramic was degraded is much higher than that of the control leg (as marked by the arrow head), suggesting the strong new bone tissue deposition capacity of the LAP bioceramic. This may be due to the fact that LAP bioceramic is able to induce the bone formation *in vivo*, in accordance with the *in vitro* bone formation assay data (Figure 9a). Overall, our data demonstrated the excellent implant integration and bone reconstruction ability of LAP bioceramic.

Conclusion

In summary, we report a facile approach to preparing LAP bioceramic via sintering LAP powder compacts for bone tissue engineering applications. The sintering behavior and mechanical properties of LAP compacts under different temperatures, heating rates, and soaking time periods were comparatively investigated. We show that LAP bioceramic with a smooth surface and relatively regular holey structures can be formed at 800°C for 6 h with a heating rate of 5°C/h under air. The formed LAP bioceramic possesses an excellent surface hydrophilicity and protein adsorption behavior. Besides, the LAP bioceramic is quite cytocompatible and hemocompatible, able to induce the HA

deposition onto its surface when soaked in SBF, and enables well proliferation of rMSCs on its surface. Most strikingly, the produced LAP bioceramic is able to induce the osteoblast differentiation of rMSCs in growth medium without any inducing factors. With the good bone healing effect demonstrated by *in vivo* experiments and excellent biosafety, the prepared LAP bioceramic holds a great promise for treating bone defects or other applications in bone tissue engineering.

Author Contributions

Conceived and designed the experiments: XS QZ. Performed the experiments: CW SW KL YJ Jipeng Li YZ Jinhua Li XL. Analyzed the data: CW SW KL YJ Jipeng Li YZ Jinhua Li XL XS QZ. Contributed reagents/materials/analysis tools: CW SW Jinhua Li XL. Wrote the paper: CW SW XS QZ.

References

1. Murugan R, Ramakrishna S. (2004) Bioresorbable composite bone paste using polysaccharide based nano hydroxyapatite. Biomaterials 25: 3829–3835.
2. Zaidman N, Bosnakovski D. (2012) Advancing with ceramic biocomposites for bone graft implants. Recent Patents on Regenerative Medicine 2: 65–72.
3. Swetha M, Sahithi K, Moorthi A, Srinivasan N, Ramasamy K, et al. (2010) Biocomposites containing natural polymers and hydroxyapatite for bone tissue engineering. Int J Biol Macromol 47: 1–4.
4. Barbieri D, Yuan H, Luo X, Farè S, Grijpma DW, et al. (2013) Influence of polymer molecular weight in osteoinductive composites for bone tissue regeneration. Acta Biomater 9.
5. Cao Z, Wen J, Yao J, Chen X, Ni Y, et al. (2013) Facile fabrication of the porous 3-dimensional regenerated silk fibroin scaffolds. Mater Sci Eng C-Mater Biol Appl 33: 3522–3529.
6. Qi R, Guo R, Shen M, Cao X, Zhang L, et al. (2010) Electrospun poly (lactic-co-glycolic acid)/halloysite nanotube composite nanofibers for drug encapsulation and sustained release. J Mater Chem 20: 10622–10629.
7. Sowmya S, Bumgardener JD, Chennazhi KP, Nair SV, Jayakumar R. (2013) Role of nanostructured biopolymers and bioceramics in enamel, dentin and periodontal tissue regeneration. Prog Polym Sci 38.
8. Vallet-Regí M. (2001) Ceramics for medical applications. J Chem Soc, Dalton Trans: 97–108.
9. Dorozhkin SV. (2010) Bioceramics of calcium orthophosphates. Biomaterials 31: 1465–1485.
10. Hoppe A, Güldal NS, Boccaccini AR. (2011) A review of the biological response to ionic dissolution products from bioactive glasses and glass-ceramics. Biomaterials 32: 2757–2774.
11. Gentleman E, Fredholm YC, Jell G, Lotfibakhshaiesh N, O'Donnell MD, et al. (2010) The effects of strontium-substituted bioactive glasses on osteoblasts and osteoclasts in vitro. Biomaterials 31: 3949–3956.
12. Wu C, Fan W, Chang J. (2013) Functional mesoporous bioactive glass nanospheres: synthesis, high loading efficiency, controllable delivery of doxorubicin and inhibitory effect on bone cancer cells. J Mater Chem B 1: 2710–2718.
13. Rossi AL, Barreto IC, Maciel WQ, Rosa FP, Rocha-Leão MH, et al. (2012) Ultrastructure of regenerated bone mineral surrounding hydroxyapatite-alginate composite and sintered hydroxyapatite. Bone 50: 301–310.
14. Kumar A, Webster TJ, Biswas K, Basu B. (2013) Flow cytometry analysis of human fetal osteoblast fate processes on spark plasma sintered hydroxyapatite-titanium biocomposites. J Biomed Mater Res Part A 101: 2925–2938.
15. Magallanes-Perdomo M, De Aza AH, Mateus AY, Teixeira S, Monteiro FJ, et al. (2013) In vitro study of the proliferation and growth of human bone marrow cells on apatite-wollastonite-2M glass ceramics. Acta Biomater 6: 2254–2263.
16. Chung C-J, Long H-Y. (2011) Systematic strontium substitution in hydroxyapatite coatings on titanium via micro-arc treatment and their osteoblast/osteoclast responses. Acta Biomater 7: 4081–4087.
17. Huang J, Best SM, Bonfield W, Buckland T. (2011) Development and characterization of titanium-containing hydroxyapatite for medical applications. Acta Biomater 6: 241–249.
18. Hench LL, Splinter RJ, Allen WC, Greenlee TK. (1971) Bonding mechanisms at the interface of ceramic prosthetic materials. J Biomed Mater Res Part A 5: 117–141.
19. Wu C-T, Chang J. (2013) Silicate bioceramics for bone tissue regeneration. J Inorg Mater 28: 29–39.
20. Wu C, Han P, Xu M, Zhang X, Zhou Y, et al. (2013) Nagelschmidtite bioceramics with osteostimulation properties: material chemistry activating osteogenic genes and WNT signalling pathway of human bone marrow stromal cells. J Mater Chem B 1: 876–885.
21. Huang Y, Jin X, Zhang X, Sun H, Tu J, et al. (2009) In vitro and in vivo evaluation of akermanite bioceramics for bone regeneration. Biomaterials 30: 5041–5048.
22. Xynos ID, Hukkanen MVJ, Batten JJ, Buttery LD, Hench LL, et al. (2000) Bioglass® 45S5 stimulates osteoblast turnover and enhances bone formation in vitro: implications and applications for bone tissue engineering. Calcified Tissue International 67: 321–329.
23. Loty C, Sautier JM, Tan MT, Oboeuf M, Jallot E, et al. (2001) Bioactive glass stimulates in vitro osteoblast differentiation and creates a favorable template for bone tissue formation. Journal of Bone and Mineral Research 16: 231–239.
24. Wang S, Zheng F, Huang Y, Fang Y, SHen M, et al. (2012) Encapsulation of amoxicillin within laponite-doped poly(lactic-co-glycolic acid) nanofibers: preparation, characterization and antibacterial activity. ACS Appl Mater Interfaces 4: 6393–6401.
25. Wang S, Wu Y, Guo R, Huang Y, Wen S, et al. (2013) Laponite nanodisks as an efficient platform for doxorubicin delivery to cancer cells. Langmuir 29: 5030–5036.
26. Jung H, Kim HM, Choy YB, Hwang SJ, Choy JH. (2008) Itraconazole-laponite: kinetics and mechanism of drug release. Appl Clay Sci 40: 99–107.
27. Viseras C, Cerezo P, Sanchez R, Salcedo I, Aguzzi C. (2010) Current challenges in clay minerals for drug delivery. Appl Clay Sci 48: 291–295.
28. Wang S, Castro R, An X, Song C, Luo Y, et al. (2012) Electrospun laponite-doped poly (lactic-co-glycolic acid) nanofibers for osteogenic differentiation of human mesenchymal stem cells. J Mater Chem 22: 23357–23367.
29. Jabbari-Farouji S, Tanaka H, Wegdam GH, Bonn D. (2008) Multiple nonergodic disordered states in Laponite suspensions:a phase diagram. Phys Rev E 78: 061405.
30. Meng ZX, Zheng W, Li L, Zheng YF. (2010) Fabrication and characterization of three-dimensional nanofiber membrane of PCL-MWCNTs by electrospinning. Mater Sci Eng C-Mater Biol Appl 30: 1014–1021.
31. Tas AC, Bhaduri SB. (2004) Rapid coating of Ti6Al4V at room temperature with a calcium phosphate solution similar to 10× simulated body fluid. J Mater Res 19: 2742–2749.
32. Wu C, Chang J. (2006) A novel akermanite bioceramic: preparation and characteristics. J Biomater Appl 21: 119–129.
33. Wang QZ, Chen XG, Li ZX, Wang S, Liu CS, et al. (2008) Preparation and blood coagulation evaluation of chitosan microspheres. J Mater Sci - Mater Med 19: 1371–1377.
34. Wu C, Chang J, Wang J, Ni S, Zhai W. (2005) Preparation and characteristics of a calcium magnesium silicate (bredigite) bioactive ceramic. Biomaterials 26: 2925–2931.
35. Qi R, Cao X, Shen M, Guo R, Yu J, et al. (2012) Biocompatibility of electrospun halloysite nanotube-doped poly (lactic-co-glycolic acid) composite nanofibers. J Biomater Sci, Polym Ed 23: 299–313.
36. Liao H, Qi R, Shen M, Cao X, Guo R, et al. (2011) Improved cellular response on multiwalled carbon nanotube-incorporated electrospun polyvinyl alcohol/chitosan nanofibrous scaffolds. Colloid Surf B-Biointerfaces 84: 528–535.
37. Fernandes LF, Costa MA, Fernandes MH, Tomás H. (2009) Osteoblastic behavior of human bone marrow cells cultured over adsorbed collagen layer, over surface of collagen gels, and inside collagen gels. Connect Tissue Res 50: 336–346.
38. da Silva HM, Mateescu M, Damia C, Champion E, Soares G, et al. (2010) Importance of dynamic culture for evaluating osteoblast activity on dense silicon-substituted hydroxyapatite. Colloid Surf B-Biointerfaces 80: 138–144.
39. Schmitz JP, Hollinger JO. (1986) The critical size defect as an experimental model for craniomandibulofacial nonunions. Clin Orthop Rel Res 205: 299–308.
40. Marcacci M, Kon E, Zaffagnini S, Giardino R, Rocca M, et al. (1999) Reconstruction of extensive long-bone defects in sheep using porous hydroxyapatite sponges. Calcif Tissue Int 64: 83–90.
41. Yoon E, Dhar S, Chun DE, Gharibjanian NA, Evans GRD. (2007) In vivo osteogenic potential of human adipose-derived stem cells/poly lactide-co-glycolic acid constructs for bone regeneration in a rat critical-sized calvarial defect model. Tissue Eng 13: 619–627.

Polymer-Ceramic Spiral Structured Scaffolds for Bone Tissue Engineering: Effect of Hydroxyapatite Composition on Human Fetal Osteoblasts

Xiaojun Zhang[1,2], Wei Chang[1], Paul Lee[1], Yuhao Wang[1], Min Yang[1], Jun Li[1¤], Sangamesh G. Kumbar[3], Xiaojun Yu[1]*

1 Department of Chemistry, Chemical Biology, and Biomedical Engineering, Stevens Institute of Technology, Hoboken, New Jersey, United States of America, 2 Department of Physics and Mathematics, School of Biomedical Engineering, Fourth Military Medical University, Xi'an, Shaanxi, People's Republic of China, 3 Department of Orthopaedic Surgery, University of Connecticut Health Center, Farmington, Connecticut, United States of America

Abstract

For successful bone tissue engineering, a scaffold needs to be osteoconductive, porous, and biodegradable, thus able to support attachment and proliferation of bone cells and guide bone formation. Recently, hydroxyapatites (HA), a major inorganic component of natural bone, and biodegrade polymers have drawn much attention as bone scaffolds. The present study was designed to investigate whether the bone regenerative properties of nano-HA/polycaprolactone (PCL) spiral scaffolds are augmented in an HA dose dependent manner, thereby establishing a suitable composition as a bone formation material. Nano-HA/PCL spiral scaffolds were prepared with different weight ratios of HA and PCL, while porosity was introduced by a modified salt leaching technique. Human fetal osteoblasts (hFOBs) were cultured on the nano-HA/PCL spiral scaffolds up to 14 days. Cellular responses in terms of cell adhesion, viability, proliferation, differentiation, and the expression of bone-related genes were investigated. These scaffolds supported hFOBs adhesion, viability and proliferation. Cell proliferation trend was quite similar on polymer-ceramic and neat polymer spiral scaffolds on days 1, 7, and 14. However, the significantly increased amount of alkaline phosphatase (ALP) activity and mineralized matrix synthesis was evident on the nano-HA/PCL spiral scaffolds. The HA composition in the scaffolds showed a significant effect on ALP and mineralization. Bone phenotypic markers such as bone sialoprotein (BSP), osteonectin (ON), osteocalcin (OC), and type I collagen (Col-1) were semi-quantitatively estimated by reverse transcriptase polymerase chain reaction analysis. All of these results suggested the osteoconductive characteristics of HA/PCL nanocomposite and cell maturation were HA dose dependent. For instance, HA:PCL = 1:4 group showed significantly higher ALP mineralization and elevated levels of BSP, ON, OC and Col-I expression as compared other lower or higher ceramic ratios. Amongst the different nano-HA/PCL spiral scaffolds, the 1:4 weight ratio of HA and PCL is shown to be the most optimal composition for bone tissue regeneration.

Editor: Mário A. Barbosa, Instituto de Engenharia Biomédica, University of Porto, Portugal

Funding: This study was supported by the Early Career Translational Research Award in Biomedical Engineering from the Wallace H. Coulter Foundation. The funders had no role in study design, data collection and analysis, decision to publish, or preparation of the manuscript.

Competing Interests: The authors have declared that no competing interests exist.

* E-mail: xyu@stevens.edu

¤ Current address: BioFocus LLC., Marlboro, New Jersey, United States of America

Introduction

Bone defects caused by trauma, tumor resection, pathological degeneration, or congenital deformity are one of the majors clinical challenges in orthopedic treatments [1]. Every year, there are around 2.2 million bone graft surgeries in the worldwide [2,3], and this number is increasing annually [4]. Currently, autologous bone grafts remains the gold standard in the treatment of bone defects [5,6]. However, significant issues of autografts such as the need for two surgical procedures for harvesting and implantation, limited supply, and donor site morbidity [7]. Bone allografts, alternatives to autografts, suffer from drawbacks such as cost-expensive, disease transmission, and adverse host immune reaction [4]. These significant caveats have increased the need for synthetic bone graft substitutes to repair bone defects.

Hydroxyapatite (HA) is the major inorganic component of natural bone and has long been used as an orthopedic and dental material [8,9]. HA is well known for its excellent biocompatibility, osteoconductive potential, slow degradation, non-cytotoxicicity, non-inflammatory, as well as non-immunogenic properties [10,11]. Although HA is a constituent of the natural bone, the intrinsic properties of HA, such as difficulty in remodeling, hardness, fragility, and lack of flexibility, make it difficult to be shaped in the specific form required for bone repair and implantation, which limits its application as a load-bearing implant scaffold material [12,13]. To circumvent these drawbacks great efforts are focused on combining HA with polymers, which may not only overcome the mechanical properties of HA but also improve the osteoconductivity properties of the polymers. Some nature polymers, such as collagen [14,15], chitosan [16], chitin [17], alginate [18], and silk [19] are employed for combination

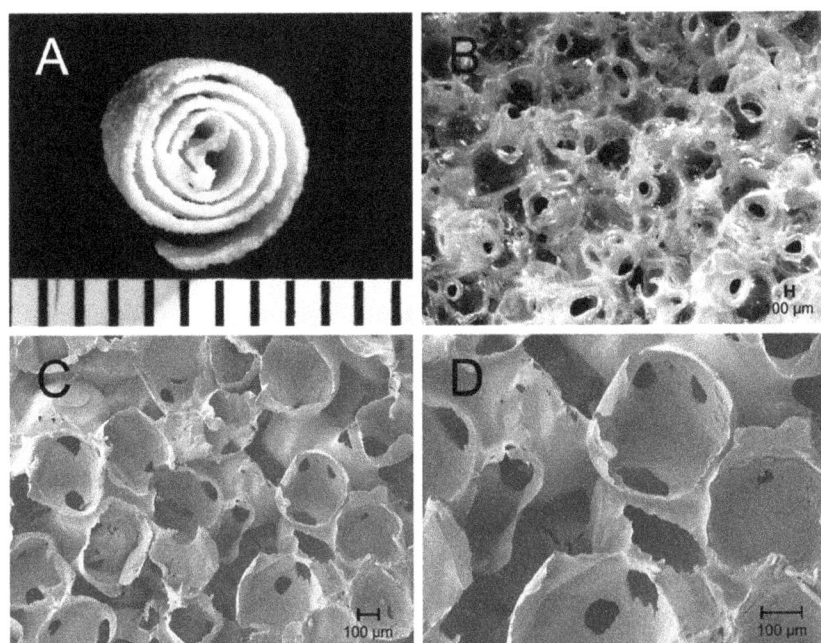

Figure 1. Representative photographs of nano-HA/PCL spiral scaffold. Gross view of the spiral scaffold (A). The morphology of nano-HA/PCL spiral scaffold under stereomicroscopy (B) and scanning electron microscopy (C and D).

with HA to apply in bone tissue engineering. But the problem of using such natural materials is their inferiority in mechanical properties [20]. Therefore, many studies have been reported on combination HA with biodegradable synthetic materials, such as poly (lactic acid) [21], poly (ε-caprolactone) (PCL) [22–24], poly (lactic-co-glycolic acid) [20], Poly (D,L-lactide) [25], to circumvent the drawbacks of natural polymers.

Among the available biodegradable synthetic polymers, PCL polymer-based composites have been focused with more attention than other synthetic polymer composite for bone tissue engineering applications, due to its sustained biodegradability, elastic characteristics, and low inflammatory response [26]. Many studies had focused on the mechanical characters of HA/PCL scaffold as a bone graft substitute. Yu, et al. evaluated the microstructures and mechanical properties of HA/PCL scaffold [27]. Johari, et al. also investigated the mechanical properties of nano-fluoridated HA/PCL scaffolds in order to obtain an optimized composition [28]. While Eosoly, et al. fabricated PCL/HA composites and examined the effect of HA addition on surface roughness, wettability, mechanical behavior and surface morphology and MC-3T3 osteoblast like cells' activity [29]. In regards to the cell responses of different studies, there is conflicting data reported in literature about the effect of HA addition to PCL composites in terms of cell attachment, proliferation and differentiation. Some authors demonstrated that PCL/HA scaffolds improved osteo-blasts differentiation and matrix mineralization [30]. However, Chim et al. showed that the presence of HA in PCL scaffolds has little or no effect on biological response [31]. Furthermore, Causa et al. found that human osteoblasts cultured on different volume ratio HA/PCL scaffolds had similar cell viability, but decreased cell differentiation with increasing the volume of HA in the HA/PCL scaffolds [32].

It is still unclear whether HA/PCL scaffold stimulates its own osteogenic capability in an HA dose dependent manner *in vitro* and whether the ratio of HA and PCL is optimized in previous HA/PCL scaffolds with regard to osteoconductive characteristics as

application in bone tissue engineering. Our group has designed a spiral structured scaffold which can provide thinner scaffold walls for cells to easily grow across and thorough gaps between scaffold walls to ensure sufficient nutrient supply and metabolic waste removal [33,34]. In the present study, spiral scaffolds of various nano-HA/PCL composites were prepared with different weight ratios of HA and PCL. The morphological structures of these spiral scaffolds were analyzed. Moreover, human fetal osteoblasts (hFOBs) cellular responses to the composites were examined in terms of the cellular attachment, proliferation, differentiation, and mineralized matrix deposition as well as the expression of bone-related genes. The aim of this study was designed to establish a suitable composition of nano-HA/PCL as a scaffold of bone graft substitute for clinical applications such as non union bone defect repairs.

Materials and Methods

1. Preparation and characterization of nano-HA/PCL spiral scaffolds

The scaffolds were prepared by a salt leaching method with minor modifications. Briefly, PCL (Sigma-Aldrich, USA) was dissolved in Dichloromethane (DCM) with a ratio of 8% (w/v) by stirring the solution for overnight. Nanosized HA (particle size <200 nm, Sigma-Aldrich, USA) were added to PCL/DCM solution at 1:8, 1:4, and 1:2 of the PCL weight ratios. The resultant mixture was stirred for 24 h at room temperature to increase homogenization. PCL/DCM solution without the addition of nano-HA powder was used as a control. After coating and drying a monolayer of sodium chloride (crystal, Fisher Scientific, USA) particles, 250–400 μm in size, on the bottom of glass petri dish, 7.5 ml of PCL or nano-HA/PCL solutions were added to the glass petri dish. Uniformity was ensured by moving the petri dish around until the polymer solution was evenly distributed. Afterwards DCM was allowed to evaporate in the hood for 3 to 5 min. Subsequently, the same sized sodium chloride was added to

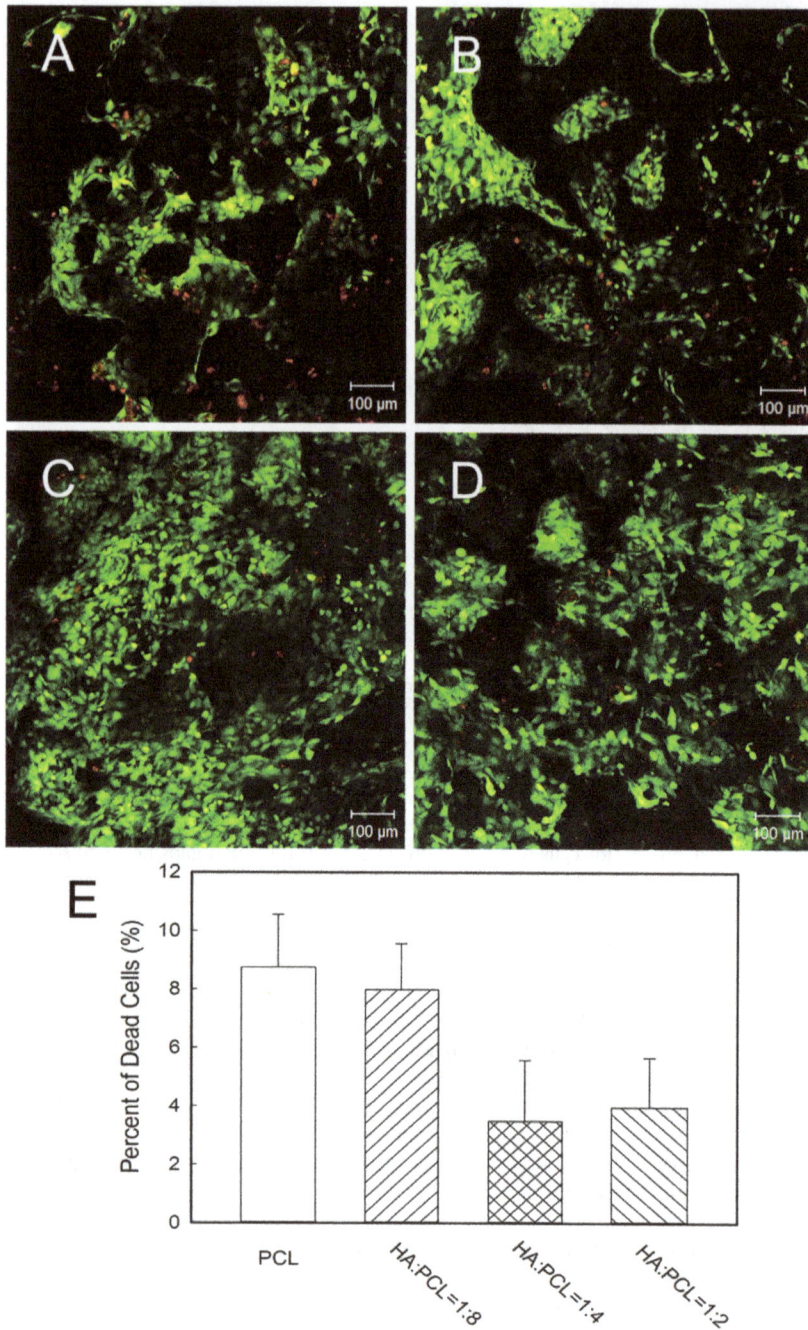

Figure 2. Cell viability cultured on nano-HA/PCL spiral scaffolds using Live/Dead assay. Cells cultured on PCL spiral scaffold (A), HA:PCL = 1:8 nano-HA/PCL spiral scaffold (B), HA:PCL = 1:4 nano-HA/PCL spiral scaffold (C), HA:PCL = 1:2 nano-HA/PCL spiral scaffold (D) for 7 days. (E) Quantitative analysis of the percent of dead cells within each spiral scaffolds. Live cells were stained green, dead cells were stained red and nucleus were stained blue. Data represent the mean ± standard deviation, n = 5.

the surface of PCL or nano-HA/PCL solution to form another salt layer, which was pressed after 2 min. Then, DCM was evaporated under reduced pressure to form a dry layer. Once all DCM is evaporated, overnight salt leaching in DI water, the highly porous nano-HA/PCL or PCL sheets were obtained with a thickness of 0.3–0.4 mm. Each sheet was cut into rectangular strips that measure 45 mm×5 mm. Each strip was then rolled into a spiral shape and held in place with a strip of copper sheeting, which acts as the mold to form a spiral structure. After being incubated at 45°C in an oven for 30 min, the scaffold was immediately

transferred to −80°C for 24 h to immobilize the shape. The copper mold was finally removed. Both stereomicroscopy (Nikon 1500z stereo optical microscope, Japanese) and scanning electron microscopy (SEM, Leo 982 FEG-SEM, Zeiss, Germany) were used to analyze the morphology, the porous structure, and the integration of the scaffolds.

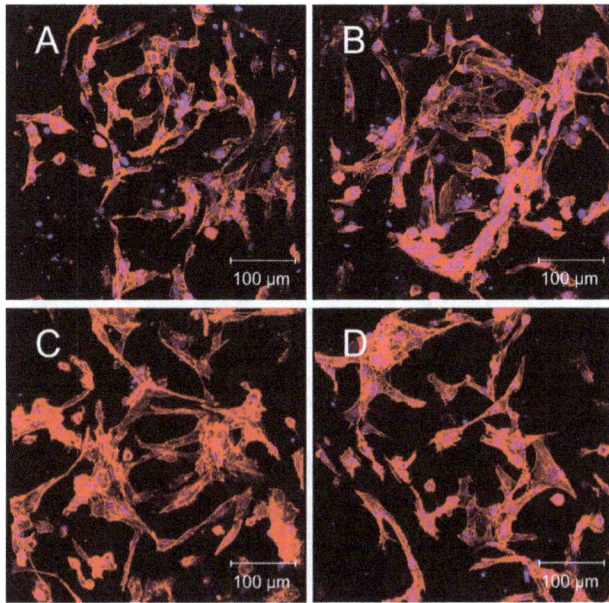

Figure 3. Cytoskeleton structure of osteoblast cultured on nano-HA/PCL spiral scaffolds for 4 days. F-actin staining of cells cultured on PCL spiral scaffold (A), HA:PCL = 1:8 nano-HA/PCL spiral scaffold (B), HA:PCL = 1:4 nano-HA/PCL spiral scaffold (C), HA:PCL = 1:2 nano-HA/PCL spiral scaffold (D). F-actin was stained red and nucleus was stained blue.

2. Cell culture and seeding

HFOB cells (hFOB 1.19, ATCC CRL-11372, USA) were cultured in DMEM (Gibco, USA) supplemented with 10% fetal calf serum and 1% penicillin/streptomycin in a humidified incubator at 37°C with 5% CO_2. The medium was changed every other day.

The scaffolds were sterilized with 70% ethanol for 1 h and then irradiated by UV light for 30 min in PBS. Following rinsing with PBS 3 times, the scaffolds were soaked in cell culture medium for overnight in incubator. In order to uniformly seed cells onto the scaffolds, hFOBs were added into 15 ml tube, which was contained with the spiral scaffolds, at the cell density of 1×10^5 cells per each scaffold and cultured on the rotating shaker at 30 rpm in the humidified incubator at 37°C with 5% CO_2. After 2 h, cell–scaffold constructs were removed from the tube and transferred into 24-well tissue culture plates containing 1 ml of complete media. The medium was changed every day, and the cultures were maintained for 21 days. At the indicated time endpoints, cell–scaffold constructs were removed and characterized for cell viability, proliferation, differentiation, mineralized matrix synthesis, and bone related gene expression, respectively.

3. Cell viability

To determine cell viability within the scaffolds, cell–scaffold constructs were stained using the LIVE/DEAD® Viability/ Cytotoxicity kit for mammalian cells (Invitrogen Life Technologies, USA), according to manufacturer's instructions. Briefly, after 7 days culture, small sections cut from the cell-seeded scaffolds were incubated with 100 µl of the Live/Dead solution containing 4 µM ethidium homodimer-1 (EthD-1) and 2 µM calcein AM at the room temperature for 30 min, then mounted onto a glass slide using SlowFade® Gold Antifade Reagent with DAPI (Invitrogen Life Technologies, USA), and viewed with confocal laser

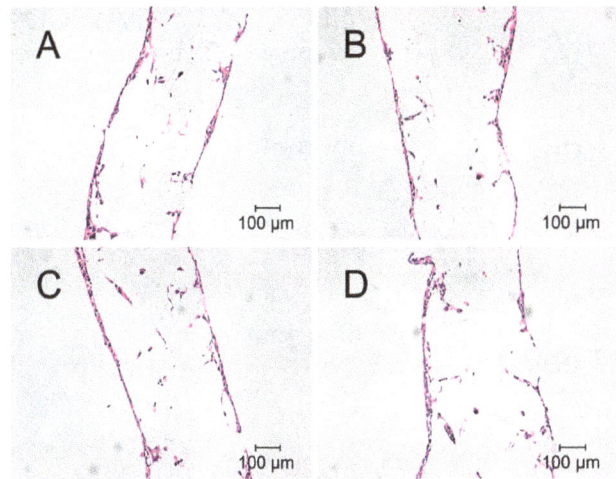

Figure 4. Cells morphology cultured on the nano-HA/PCL spiral scaffolds for 4 days by HE staining. Osteoblasts cultured on PCL spiral scaffold (A), HA:PCL = 1:8 nano-HA/PCL spiral scaffold (B), HA:PCL = 1:4 nano-HA/PCL spiral scaffold (C), HA:PCL = 1:2 nano-HA/ PCL spiral scaffold (D).

microscopy (Zeiss LSM5 Pascal, Germany) with 494 nm (green, Calcein) and 528 nm (red, EthD-1) excitation filters. Images were captured using Zeiss LSM Data Server software. In order to quantitatively analyze, five randomly chosen areas from each sample were captured and the green areas and red areas of each image were recorded.

4. Cell morphology

The 4 day incubation period, cell–scaffold constructs were fixed by 4% paraformaldehyde at 4°C for overnight. Some samples were permeabilized with 0.1% Triton X-100 for 15 min, and then blocked with 1% bovine serum albumin for 30 min. Following that the F-actin cytoskeleton of osteoblasts was stained with Rhodamine Phalloidin (Invitrogen Life Technologies, USA), and the nucleus was stained with DAPI (Invitrogen Life Technologies, USA). The other samples were embedded in paraffin and sectioned at a thickness of 5 µm and then stained with hematoxylin and eosin (H&E).

5. Cell proliferation

Cell proliferation was assessed using the MTS assay (CellTiter96™ AQueous Assay, Promega, USA) and PicoGreen DNA quantification assay (Quant-iT™ PicoGreen® dsDNA Assay Kit, Invitrogen Life Technologies, USA). For MTS assay, cell–scaffold constructs were transferred to new cell culture plates after 1, 7, and 14 days culture. One hundred microliter of pre-warmed MTS solution with 1 ml culture medium was added to each well with continuous culture for 3 h. Two hundred microliter of supernatant from each well was then transferred to a 96-well plate and the absorbance was measured at 490 nm using a SYNERGY HT plate reader (BIO-TEK, USA). Six specimens for each group were tested, and each test was repeated three times.

For PicoGreen DNA quantification assay, cell–scaffold constructs were moved to new 48-well plates, washed with PBS for three times, and treated with 500 µl TE buffer (10 mM Tris-HCl, 1 mM EDTA, pH 7.5) at the indicated time endpoints. Following three freeze–thaw cycles, a volume of 100 µl of supernatant was taken from the samples and added into 100 µl of PicoGreen reagent (diluted 1:200 in TE buffer). Samples were incubated in

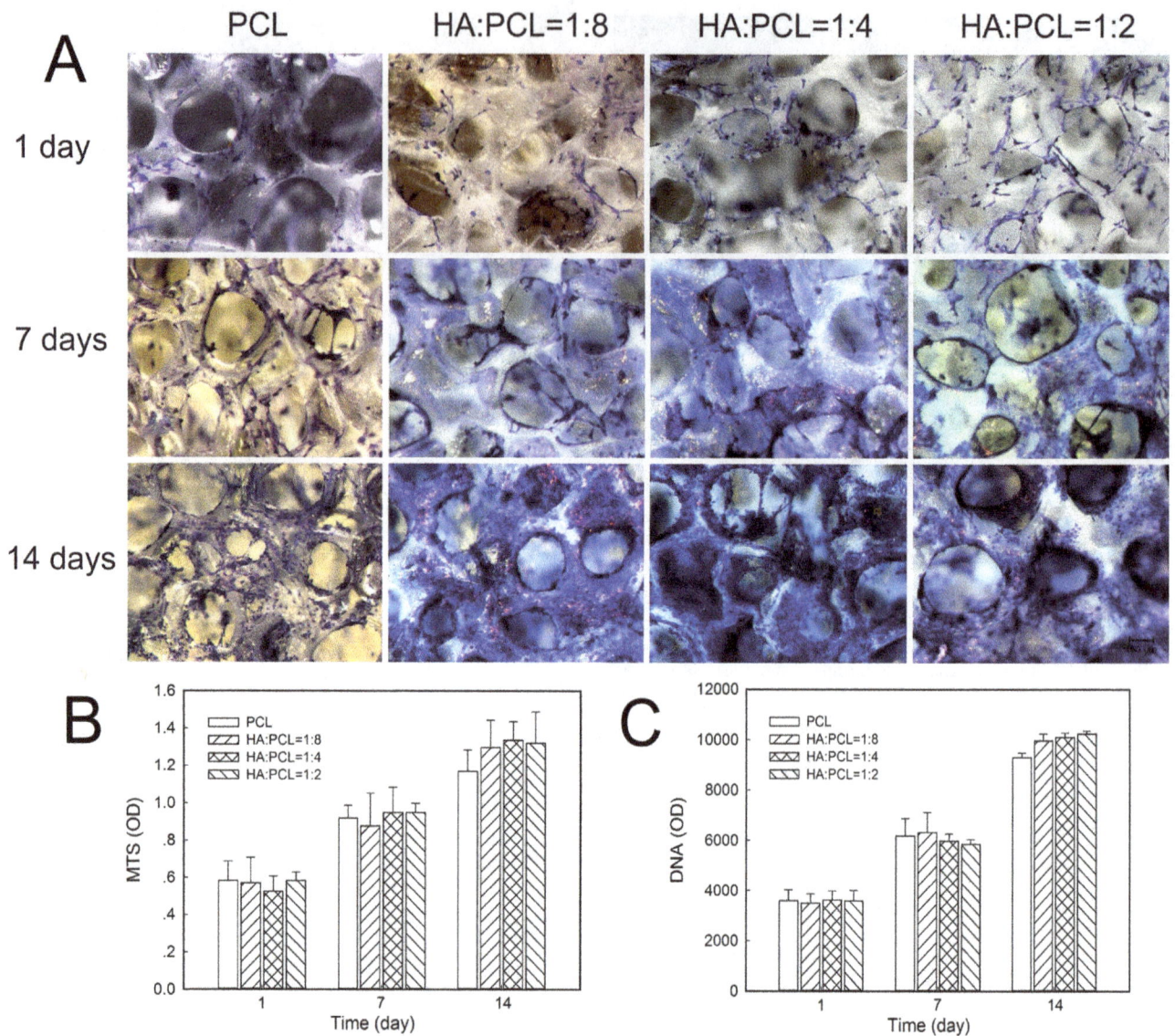

Figure 5. Osteoblasts distribution and proliferation cultured on the nano-HA/PCL spiral scaffolds after 1, 7, and 14 days culture. Cells distribution in the scaffolds evaluated by methylene blue staining (A). Quantitative analysis of cell proliferation measured by MTA assay (B) and PicoGreen DNA quantification assay (C). Data represent the mean ± standard deviation, n = 6.

darkness for 5 min before fluorescence reading at the excitation and emission wavelengths of 485 nm and 520 nm, respectively. To minimize photobleaching, the time used for fluorescence measurement was kept constant for all samples.

6. ALP activity

ALP activity of the cells was measured as an early marker of the maintenance of the osteoblastic phenotype using EnzoLyte pNPP Alkaline Phosphatase Assay Kit (AnaSpec, San Jose, USA) according to the manufacturer's protocol. Briefly, cell–scaffold constructs were homogenized in 500 μl lysis buffer provided in the kit. Lysate was centrifuged for 15 min at 10,000 g at 4°C. Following that, 50 μl supernatant was added to 50 μl of pNPP ALP substrate solution and incubated at 37°C for 60 min. The reaction was then stopped by adding 50 μl of stop solution into each well. The activity of ALP in cell lysates was measured with a microplate reader (SYNERGY HT, BioTek, USA) at 405 nm.

The results were normalized into total cellular protein, which was measured using a Quickstart Bradford Protein assay kit (Bio-Rad, USA) [35]. In short, 150 μl of cell lysate was mixed with 150 μl of working reagent and incubated for 5 min at room temperature. The resulting optical density was read at 595 nm. Six specimens for each group were tested, and each test was repeated three times.

7. Calcium expression

Mineralized matrix synthesis at days 21 was analyzed using an Alizarin Red staining method for calcium deposition [36]. Cell-scaffold constructs were fixed by 4% paraformaldehyde at 4°C for overnight and subsequently stained with 2% Alizarin Red (Sigma-Aldrich, USA) solution for 10 min. To quantify the calcium amount on the scaffold, after washed with acetone and xylene, the red matrix precipitate was solublized in 10% cetylpyridinium chloride (Sigma-Aldrich, USA), and the optical density was read at 562 nm. Nano-HA/PCL spiral scaffolds without seeding hFOBs

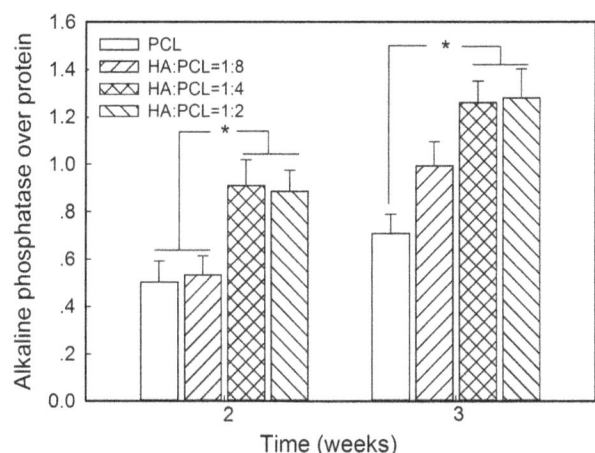

Figure 6. Alkaline phosphatase (ALP) expression on the nano-HA/PCL spiral scaffolds was normalized to protein concentration after 2 and 3 weeks of culture. Data represent the mean ± standard deviation, n = 6. Significant difference between different material groups were denoted as * ($p < 0.05$).

were served as the blank control. Six specimens for each group were tested, and each test was repeated three times.

8. RNA extraction and RT-PCR assay

The expression of osteogenic genes in cell-scaffold constructs were examined by reverse transcription-polymerase chain reaction (RT-PCR). Total RNA was extracted with commercial Trizol Reagent (Invitrogen Life Technologies, USA) from nano-HA/PCL or PCL spiral scaffolds seeded with hFOBs for 21 days. One microgram of total RNA was reverse transcribed with reverse transcriptase (Promega, USA) according to the manufacturer's instructions. All PCR experiments were performed with Taq polymerase (Promega, USA). The primers used as follows: bone sialoprotein (BSP; sense, 5'-AATGAAAACGAAGAAAGCGAA-G-3'; antisense, 5'-ATCATAGCCATCGTAGCCTTGT-3'; 450 bp), osteonectin (ON; sense, 5'-TGGATCTTCTTTCTCCTTT-3'; antisense, 5'-TTCTGCTTCTCAGTCAGA-3'; 569 bp), ALP (sense, 5'-TGGAGCTTCAGAAGCTCAACACCA-3'; antisense, 5'-ATCTCGTTGTCTGAGTACCAGTCC-3'; 454 bp), osteo-calcin (OC; sense, 5'-ATGAGAGCCCTCACACTCCTC-3'; antisense, 5'-GCCGTAGAAGCGCCGATAGGC-3'; 294 bp), type I collagen (Col-1; sense, 5'-GGACACAATGGATTG-CAAGG-3'; antisense, 5'-TAACCACTGCTCCACTCTGG-3'; 461 bp), and β-actin (sense, 5'-GGCATCGTGATGGACTCCG-3'; antisense, 5'-GCTGGAAGGTGGACAGCGA-3'; 613 bp) served as the house-keeping gene control. The cycling conditions consisted of the initial denaturation at 95°C for 5 min, followed by 35 cycles of denaturing at 95°C for 30 s, annealing at 55°C for 1 min, and polymerization at 72°C for 1 min, and then a final 10 min extension at 72°C. The PCR products were separated and visualized on 1.5% agarose gel containing 5 g/L ethidium bromide. In addition, band intensity was normalized to that of β-actin.

9. Statistics

Quantitative data were reported as mean ± standard deviation. The overall statistical significance of any difference resulting from treatment was determined by analysis of variance (ANOVA). If significance was observed, the Student's t test was used to test for

differences between group means. A value of $p < 0.05$ was considered to be statistically significant.

Results

1. Characterization of nano-HA/PCL spiral scaffolds

A representative macrograph and micrograph of the obtained nano-HA/PCL spiral scaffold is depicted in Fig. 1. It can be seen from this figure that the scaffold was porous, and the pores were interconnected. This spiral scaffold contains two types of pores. One is the large pores created by the sale particles that had diameters ranging in size between 250 and 400 μm, and the other is the micropores on the surface of the spherical macropores. The diameter of micropore was measured to be on the average of several 10 s of micrometers. Furthermore, the inclusion of nano-HA powders had little effect on the changes in morphology with different weight proportions from 12.5 to 33.3% since pure PCL scaffolds had the same morphological structures (data not shown).

2. Cell viability on nano-HA/PCL spiral scaffolds

Figure 2 showed nano-HA/PCL spiral scaffolds seeded with hFOBs after performing LIVE/DEAD staining. From the fluorescence signals it could be observed that cells adhered, proliferated, and remained viable after 7 days culture (Fig. 2 A–D). The very low red fluorescence depicted in the images indicated that there were a very low number of dead cells. However, more dead cells could be found in HA:PCL = 1:8 group and PCL group than in HA:PCL = 1:4 group and HA:PCL = 1:2 group (Fig. 2 E).

The morphology of cells on the nano-HA/PCL spiral scaffolds was examined using fluorescence confocal microcopy and HE. After 4 days culture on spiral scaffolds, the cytoskeleton structure were examined using F-actin stain (Fig. 3). It can be observed that the morphology of hFOBs were similar in these four kinds of spiral scaffold. Abundant long actin stress fibers formed in cells were aligned along the interconnected pore of nano-HA/PCL spiral scaffolds. HE staining showed that lots of hFOBs were attached on the surfaces of nano-HA/PCL scaffolds. Furthermore, some cells were spread on the pore of scaffold (Fig. 4).

To further investigate the cell adhesion and distribution within the nano-HA/PCL spiral scaffolds, hFOBs were cultured on the scaffolds for 1, 7, and 14 days and stained with methylene blue (Fig. 5 A). Staining showed cells were not only attached on the surfaces of scaffolds but also spread on the pore of scaffold at all indicated time endpoints. From the staining it can be seen that there were no differences in cellular attachment in four spiral scaffolds in each time group. With the increasing of culture time, the number of cells cultured in nano-HA/PCL spiral scaffolds was dramatically increased and cells were distributed throughout the spiral scaffolds. Furthermore, some cells were aligned together in the pore of scaffolds.

Proliferation of cells on nano-HA/PCLspiral scaffolds were analyzed using MTS assay and DNA assay. The data from the MTS assay (Fig. 5 B) showed that there were no significant differences in cell attachment in all spiral scaffolds at day 1. The results at day 14 indicate that cell proliferation occurred during this period in all the tested scaffolds, as indicated by the higher values of MTS at day 14 when compared to day 1 and day 7. However, the numbers of cells did not appear to make a statistically significant difference among the different ratio nano-HA/PCL spiral scaffolds and PCL spiral scaffolds. The results from the DNA assay were shown in Fig. 5 C. The DNA content on all spiral scaffolds increased gradually until day 14. Furthermore, there were still no significant differences in DNA content amongst the different ratio scaffolds at each time point.

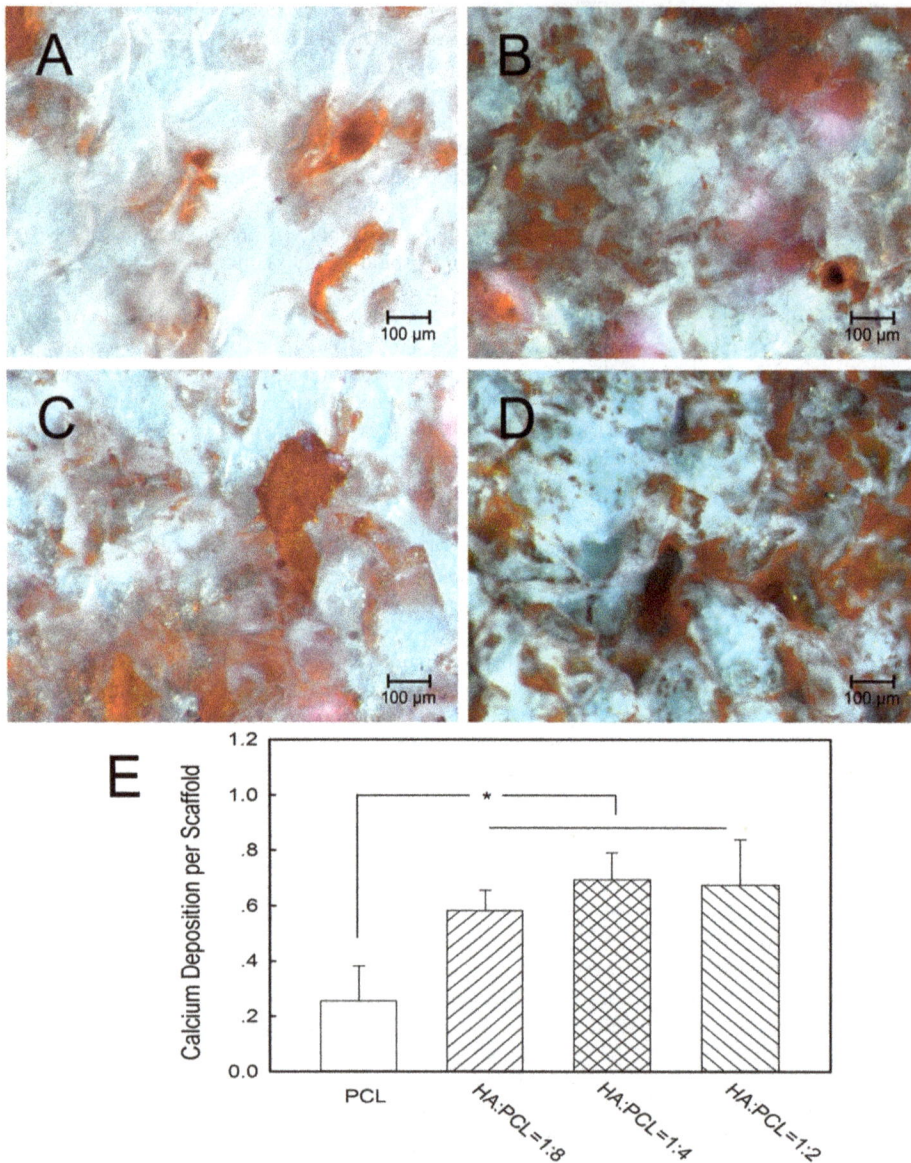

Figure 7. Alizarin S Red staining of calcium deposited on nano-HA/PCL spiral scaffolds. Cells cultured on PCL spiral scaffold (A), HA:PCL = 1:8 nano-HA/PCL spiral scaffold (B), HA:PCL = 1:4 nano-HA/PCL spiral scaffold (C), HA:PCL = 1:2 nano-HA/PCL spiral scaffold (D) for 21 days. (E) Quantitative analysis of the amount of calcium within each spiral scaffolds. Data represent the mean ± standard deviation, n = 6. Significant difference between different material groups were denoted as * ($p<0.05$).

3. Cell differentiation on nano-HA/PCL spiral scaffolds

ALP activity, normalized to protein concentration, was plotted in Fig. 6. After 2 weeks, the expression of ALP activity was significantly higher in the HA:PCL = 1:4 and HA:PCL = 1:2 spiral scaffolds compared with HA:PCL = 1:8 and PCL spiral scaffolds ($p<0.05$). While ALP expression increased in all tested spiral scaffolds after 3 weeks, ALP activity in the HA:PCL = 1:4 and HA:PCL = 1:2 spiral scaffolds continued to be expressed at significantly higher levels compared with those of PCL spiral scaffolds ($p<0.05$). However, no significant differences in the ALP expression were found between HA:PCL = 1:8 spiral scaffolds and HA:PCL = 1:4, or HA:PCL = 1:2 spiral scaffolds in 3 weeks. Furthermore, the levels of ALP expression in the HA:PCL = 1:4 spiral scaffolds were very similar to expression levels in the HA:PCL = 1:2 spiral scaffolds in 2 and 3 weeks.

Calcium assays were performed to assess the mineralized matrix formation on the scaffolds. It could be observed that a mineral matrix formed in the nano-HA/PCL spiral scaffolds (Fig. 7 A–D). Qualitative analysis showed that the average calcium deposition on nano-HA/PCL spiral scaffolds with HA content were significantly higher than that on PCL spiral scaffolds ($p<0.05$, Fig. 6E). The difference in calcium deposition was not statistically significant with varying HA content in the nano-HA/PCL spiral scaffolds ($p>0.05$). However, HA:PCL = 1:4 nano-HA/PCL spiral scaffolds had the highest amount of calcium accumulation.

We further examined the effect of nano-HA/PCL on osteoblastic cell differentiation at the mRNA level using a PCR technique (Fig. 8). After 3 weeks cultured on the nano-HA/PCL spiral scaffolds, BSP and ON mRNAs in the HA:PCL = 1:4 and HA:PCL = 1:2 spiral scaffolds were significantly higher than those in the HA:PCL = 1:8 and PCL spiral scaffolds ($p < 0.05$). Interestingly, the levels of ALP mRNAs in the HA:PCL = 1:4 spiral scaffolds were the highest comparing with other three spiral scaffolds and were significantly higher than expression levels in the HA:PCL = 1:8 and PCL spiral scaffolds ($p < 0.05$). The level of OC mRNAs in the PCL spiral scaffolds were significantly lower than expression levels in the HA:PCL = 1:4 spiral scaffolds ($p < 0.05$). When examining levels of type I collagen in the nano-HA/PCL spiral scaffolds, Col mRNA expression in the nano-HA/PCL spiral scaffolds were the highest comparing to other bone differentiation related gen such as BSP, ON, ALP, and OC. However, there were no significant differences in the Col mRNA expression among all tested spiral scaffolds.

Discussion

Bone tissue engineering is a potentially alternative strategy to repairing bone defects. A critical component of this tissue engineering approach is to develop an osteoconductive, porous, and biodegradable scaffold that may provide a temporary scaffold to guide new tissue in-growth and regeneration. Currently, biodegradable polymer/bioceramic composites have drawn considerable attention as a suitable scaffold material in bone tissue [12,37]. This is because the hybridization of these materials can not only overcome the inflexibility and brittleness of hard ceramic materials, but also improve the osteoconductivity and degradation properties of polymers.

It is well known that the pore size and interconnectivity of scaffold are highly relevant to proper cell migration and proliferation as well as tissue vascularization and diffusion of nutrients and oxygen, which are necessary for bone formation. Previous studies also demonstrated that pore size between 100 and 350 μm is optimum for bone regeneration [38]. Interconnected porosity is also important for maximizing bone ingrowth leading to osteointegration and secure graft fixation. However, scaffolds produced by traditional salt leaching cannot guarantee interconnection of pores due to the low interconnectivity of particles randomly assembled in a polymer dissolved with a solvent [38,39]. Furthermore, it is very difficult to control the pore size of scaffolds during evaporation and agglomeration of salt particles. In the present study, the porous nano-HA/PCL spiral scaffolds with different weight ratio of HA and PCL were fabricated by a modified salt leaching technique. From our SEM results, nano-HA/PCL spiral scaffolds fabricated with the modified salt leaching were composed of lots of macropores, which size was from 250 to 400 μm controlled by the salt size, and micropores in the range of about several 10 s of micrometers as well as formed interconnected porous structure. These microstructure including porosity, pore size and interconnection between pores indicate that nano-HA/PCL spiral scaffolds might be ideal scaffold as bone tissue engineering.

To investigate cell-scaffold interactions, hFOBs were seeded and cultured on the nano-HA/PCL spiral scaffolds. LIVE/DEAD staining showed that hFOBs maintained high cell viability in all spiral scaffolds when cells were cultured on the nano-HA/PCL spiral scaffolds. Moreover, it can be seen that cells were not only attached onto the surface of spiral scaffolds, but also distributed within the whole nano-HA/PCL spiral scaffolds from F-actin staining, HE staining, and methylene blue staining results. From the images it can be observed that the scaffolds with interconnected pores were instrumental in allowing the cells in penetrating throughout the entire scaffold. Considering that previous studies had demonstrated that cell infiltration and distribution within the whole scaffold will greatly affect the overall performance of the

cells/scaffold construct, it can be seen that our porous scaffold offers an optimal biocompatible surface for bone regeneration [40]. Quantity analysis showed that the cell numbers were gradually increased with cultured time, but there were no significant difference in cell proliferation amongst the nano-HA/PCL spiral scaffolds with different weight of HA. This was in agreement with Shor et al. studies where they compared the cell activity between HA/PCL scaffolds contain with 25% weight HA and pure PCL scaffolds and Alamar blue assay showed no statistical difference between two groups [30]. Coincidentally, previous work had also reported that no significant difference between PCL and PCL/HA scaffolds fabricated by fused deposition modeling when using human calvarial osteoblast [31]. Contrarily to what was observed in other studies, porous HA/PCL scaffolds showed a significant increase in osteoblast adhesion and proliferation [22,41]. In addition, Eosoly et al. examined the MC-3T3 mouse calvarial osteoblast cells response of different compositions of PCL/HA scaffolds (0, 15, and 30 weight % HA). Composites with lower HA content (15 weight %) showed the significantly higher cellular proliferation compared to that of higher HA content and pure PCL at the 7 and 14 days [29].

ALP, an early osteogenic marker for differentiation, is important for the construction of bone matrix. Our results showed that nano-HA/PCL scaffolds with higher HA content (HA:PCL = 1:4, 1:2) had a significantly higher expression of ALP compared to scaffolds with lower HA content (HA:PCL = 1:8) or pure PCL scaffolds. Moreover, The release of ALP was the highest for cells grown in the HA:PCL = 1:4 scaffolds group at the different test point. This trend was confirmed by the mRNA expression of ALP after hFOBs were cultured on the nano-HA/PCL scaffolds for 3 weeks, which was reinforced semi-quantitatively by RT-PCR analysis. This observation was in agreement with another study, in which HA steadily increased ALP activity in PCL scaffolds containing with 25% weight HA from days 7 to 21 [30]. Furthermore, there was general consensus of our results with Ngiam et al.'s study whereby there was an enhancement of ALP activity on the PLGA/HA composite scaffolds [20]. Causa et al. examined the Saos-2 cell and human osteoblasts response of different compositions of HA/PCL composites (tested on composition of 13, 20, 32 volume % HA, namely 40, 70, and 132 weight % HA) fabricated by phase inversion and casting technique. The release of ALP was higher for cells grown in the HA/PCL scaffolds contain with 40% (w/w). Furthermore, human osteoblasts on these scaffolds contain with 132% (w/w) HA never reached the ALP levels measured for the other scaffolds [32]. Based on the above results, we can conclude that the ALP activity of osteoblast cultured on the HA/PCL scaffolds increased and then decreased with an increase in percentage of the HA in the HA/PCL scaffolds. Therefore, HA scaffolds induced osteoblastic cell differentiation in a dose dependent manner.

The mineralized matrix formation is a phenotypic marker for a later stage of osteogenic differentiation. The calcified matrix formation in the nano-HA/PCL spiral scaffolds was significantly elevated compared to that of pure PCL spiral scaffolds. Moreover, the formation of mineralized matrix was the highest for cells grown on the HA:PCL = 1:4 scaffolds group. These results were confirmed by the PCR results. The gene expression level of BSP, ON, OC, which are considered as late differentiation markers of the osteoblast phenotype, were significantly upregulated in hFOBs cultured on the nano-HA/PCL spiral scaffolds with higher HA content. All these differentiation markers had the highest expression in the HA:PCL = 1:4 spiral scaffolds group. Therefore, it was further demonstrated that HA scaffolds was capable of inducing cellular differentiation markers in a dose

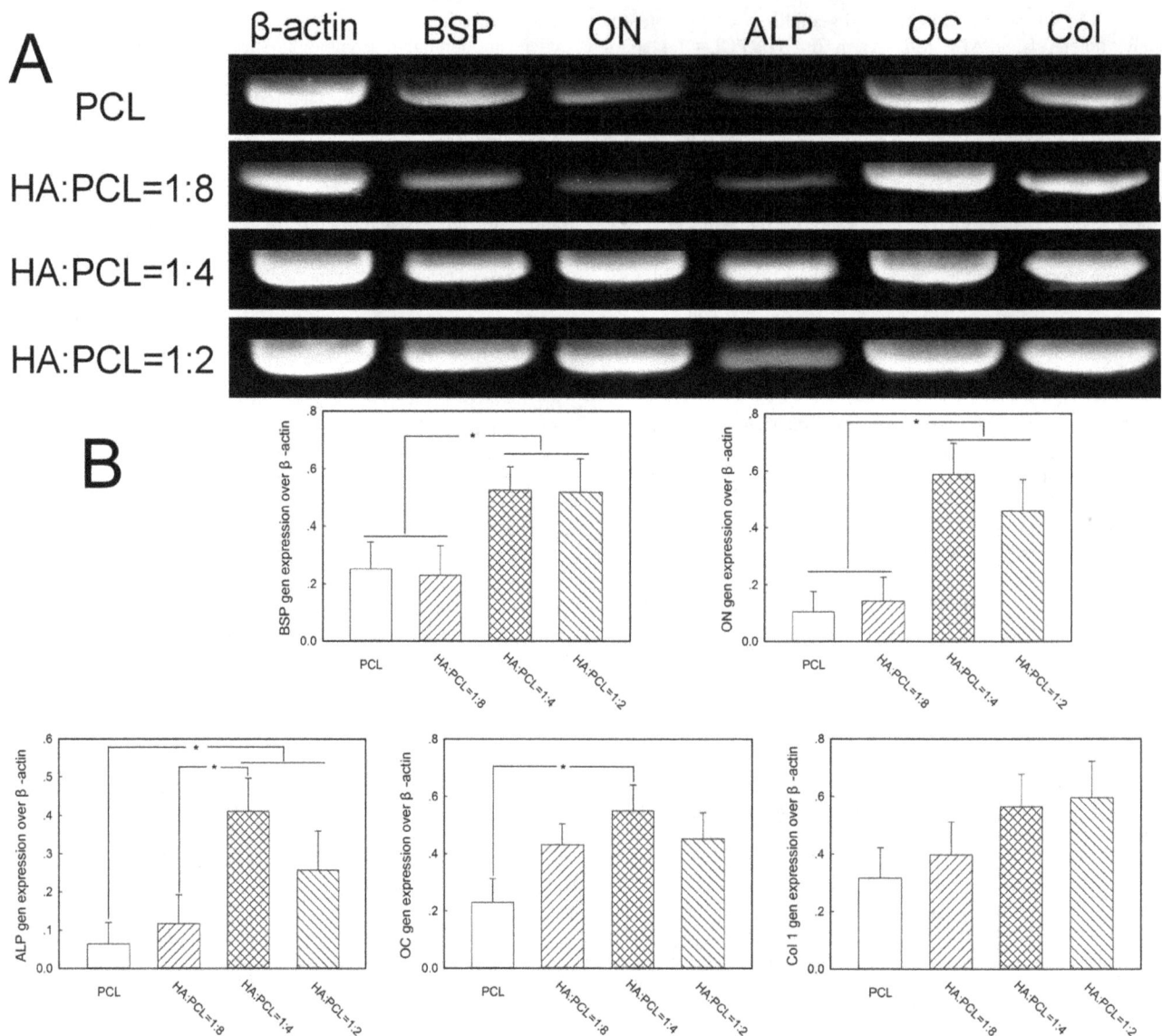

Figure 8. Gene expression of osteogenic markers in nano-HA/PCL spiral scaffolds after 3 weeks of culture. Representative electrophoresis gel (A) and semi-quantitative analysis of gene expression (B). The data presented were normalized with β-actin. Significant difference between different material groups were denoted as * ($p < 0.05$).

dependent manner. Previous studies also demonstrated that another biocermic, octacalcium phosphate, stimulates its own osteogenic capability in an octacalcium phosphate dose–dependent manner in vitro [42,43].

Conclusions

In the present study, we synthesized interconnected porous nano-HA/PCL scaffolds with different weight ratios of HA by a modified salt leaching technique. hFOBs can adhere, infiltrate the interconnected porous of the nano-HA/PCL spiral scaffolds, and also maintain high cellular viability. Although cell proliferation rate were unchanged when the amount of HA were changed in the nano-HA/PCL spiral scaffolds, ALP activity and mineralized matrix formation were significantly promoted with the HA content increase in the nano-HA/PCL spiral scaffolds. Moreover, the highest promotion occurred in the HA:PCL = 1:4 spiral scaffolds

group, instead of lower or higher of HA group. The expression of osteogenic markers, such as ALP, BSP, ON, and OC, were also significantly upregulated in the HA:PCL = 1:4 spiral scaffolds. The results suggested the optimal blend of HA and PCL was around a 1 to 4 ratio by weight.

In future, more in-depth studies, including the investigation into the degradation, the mechanism of cell-nano-HA/PCL spiral scaffolds interaction, and in vivo osteoconductivity properties of these spiral scaffolds, are required to confirm the potential benefits of the nano-HA/PCL spiral scaffolds used for bone regeneration.

Author Contributions

Conceived and designed the experiments: XZ SGK XY. Performed the experiments: XZ WC PL YW MY JL. Analyzed the data: XZ. Wrote the paper: XZ PL XY.

References

1. Jabbarzadeh E, Starnes T, Khan YM, Jiang T, Wirtel AJ, et al. (2008) Induction of angiogenesis in tissue-engineered scaffolds designed for bone repair: a combined gene therapy-cell transplantation approach. Proc Natl Acad Sci USA 105:11099–11104.
2. Saiz E, Zimmermann EA, Lee JS, Wegst UG, Tomsia AP (2013) Perspectives on the role of nanotechnology in bone tissue engineering. Dent Mater 29:103–115.
3. Fu Q, Saiz E, Rahaman MN, Tomsia AP (2011) Bioactive glass scaffolds for bone tissue engineering: state of the art and future perspectives. Mater Sci Eng C Mater Biol Appl 31:1245–1256.
4. Liu X, Rahaman MN, Fu Q (2013) Bone regeneration in strong porous bioactive glass (13–93) scaffolds with an oriented microstructure implanted in rat calvarial defects. Acta Biomater 9:4889–4898.
5. Khan Y, Yaszemski MJ, Mikos AG, Laurencin CT (2008) Tissue engineering of bone: material and matrix considerations. J Bone Joint Surg Am 90:36–42.
6. Yuan H, Fernandes H, Habibovic P, de Boer J, Barradas AM, et al. (2010) Osteoinductive ceramics as a synthetic alternative to autologous bone grafting. Proc Natl Acad Sci USA 107:13614–13619.
7. Hasegawa S, Tamura J, Neo M, Goto K, Shikinami Y, et al. (2005) In vivo evaluation of a porous hydroxyapatite/poly-DL-lactide composite for use as a bone substitute. J Biomed Mater Res A 75:567–579.
8. Jayabalan M, Shalumon KT, Mitha MK, Ganesan K, Epple M (2010) Effect of hydroxyapatite on the biodegradation and biomechanical stability of polyester nanocomposites for orthopaedic applications. Acta Biomater 6:763–775.
9. Xie C, Lu H, Li W, Chen FM, Zhao YM (2012) The use of calcium phosphate-based biomaterials in implant dentistry. J Mater Sci Mater Med 23:853–862.
10. Murugan R, Ramakrishna S (2004) Bioresorbable composite bone paste using polysaccharide based nano hydroxyapatite. Biomaterials 25:3829–3835.
11. Swetha M, Sahithi K, Moorthi A, Srinivasan N, Ramasamy K, et al. (2010) Biocomposites containing natural polymers and hydroxyapatite for bone tissue engineering. Int J Biol Macromol 47:1–4.
12. Sun F, Zhou H, Lee J (2011) Various preparation methods of highly porous hydroxyapatite/polymer nanoscale biocomposites for bone regeneration. Acta Biomater 7:3813–3828.
13. Kitsugi T, Yamamuro T, Nagamura T, Kotani S, Kokubo T, et al. (1993) Four calcium phosphate ceramics as bone substitutes for non-weight-bearing. Biomaterials 14:216–224.
14. Phipps MC, Xu Y, Bellis SL (2012) Delivery of platelet-derived growth factor as a chemotactic factor for mesenchymal stem cells by bone-mimetic electrospun scaffolds. PLoS One 7:e40831.
15. Phipps MC, Clem WC, Grunda JM, Clines GA, Bellis SL (2012) Increasing the pore sizes of bone-mimetic electrospun scaffolds comprised of polycaprolactone, collagen I and hydroxyapatite to enhance cell infiltration. Biomaterials 33:524–534.
16. Zhang L, Tang P, Zhang W, Xu M, Wang Y (2010) Effect of chitosan as a dispersant on collagen-hydroxyapatite composite matrices. Tissue Eng Part C Methods 16:71–79.
17. Chang C, Peng N, He M, Teramoto Y, Nishio Y, et al. (2013) Fabrication and properties of chitin/hydroxyapatite hybrid hydrogels as scaffold nano-materials. Carbohydr Polym 91:7–13.
18. Rossi AL, Barreto IC, Maciel WQ, Rosa FP, Rocha-Leão MH, et al. (2012) Ultrastructure of regenerated bone mineral surrounding hydroxyapatite-alginate composite and sintered hydroxyapatite. Bone 50:301–310.
19. Kino R, Ikoma T, Yunoki S, Nagai N, Tanaka J, et al. (2007) Preparation and characterization of multilayered hydroxyapatite/silk fibroin film. J Biosci Bioeng 103:514–520.
20. Ngiam M, Liao S, Patil AJ, Cheng Z, Chan CK, et al. (2009) The fabrication of nano-hydroxyapatite on PLGA and PLGA/collagen nanofibrous composite scaffolds and their effects in osteoblastic behavior for bone tissue engineering. Bone 45:4–16.
21. Woo KM, Seo J, Zhang R, Ma PX (2007) Suppression of apoptosis by enhanced protein adsorption on polymer/hydroxyapatite composite scaffolds. Biomaterials 28:2622–2630.
22. Wang Y, Liu L, Guo S (2010) Characterization of biodegradable and cytocompatible nano-hydroxyapatite/polycaprolactone porous scaffolds in degradation in vitro. Polym Degrad Stab 95:207–213.
23. Kim HW, Lee EJ, Kim HE, Salih V, Knowles JC (2005) Effect of fluoridation of hydroxyapatite in hydroxyapatite-polycaprolactone composites on osteoblast activity. Biomaterials 26:4395–4404.
24. Xiao X, Liu R, Huang Q, Ding X (2009) Preparation and characterization of hydroxyapatite/polycaprolactone-chitosan composites. J Mater Sci Mater Med 20:2375–2383.
25. Ren J, Zhao P, Ren T, Gu S, Pan K (2008) Poly (D,L-lactide)/nano-hydroxyapatite composite scaffolds for bone tissue engineering and biocompatibility evaluation. J Mater Sci Mater Med 19:1075–1082.
26. Ali SA, Zhong SP, Doherty PJ, Williams DF (1993) Mechanisms of polymer degradation in implantable devices. I. Poly(caprolactone). Biomaterials 14:648–656.
27. Yu H, Matthew HW, Wooley PH, Yang SY (2008) Effect of porosity and pore size on microstructures and mechanical properties of poly-epsilon-caprolactone-hydroxyapatite composites. J Biomed Mater Res B Appl Biomater 86:541–547.
28. Johari N, Fathi MH, Golozar MA (2012) Fabrication, characterization and evaluation of the mechanical properties of poly (ε-caprolactone)/nano-fluoridated hydroxyapatite scaffold for bone tissue engineering. Composites Part B: Engineering 43:1671–1675.
29. Eosoly S, Vrana NE, Lohfeld S, Hindie M, Looney L (2012) Interaction of Cell Culture with Composition effects on the Mechanical Properties of Polycaprolactone-hydroxyapatite scaffolds fabricated via Selective Laser Sintering (SLS). Materials Science and Engineering: C 32:2250–2257.
30. Shor L, Guceri S, Wen X, Gandhi M, Sun W (2007) Fabrication of three-dimensional polycaprolactone/hydroxyapatite tissue scaffolds and osteoblast–scaffold interactions in vitro. Biomaterials 28:5291–5297.
31. Chim H, Hutmacher DW, Chou AM, Oliveira AL, Reis RL, et al. (2006) A comparative analysis of scaffold material modifications for load-bearing applications in bone tissue engineering. Int J Oral Maxillofac Surg 35:928–934.
32. Causa F, Netti PA, Ambrosio L, Ciapetti G, Baldini N, et al. (2006) Poly-ε-caprolactone/hydroxyapatite composites for bone regeneration: in vitro characterization and hFOB response. J Biomed Mater Res A 76:151–162.
33. Valmikinathan CM, Tian J, Wang J, Yu X (2008) Novel nanofibrous spiral scaffolds for neural tissue engineering. J Neural Eng 5:422–432.
34. Wang J, Valmikinathan CM, Liu W, Laurencin CT, Yu X (2010) Spiral-structured, nanofibrous, 3D scaffolds for bone tissue engineering. J Biomed Mater Res A 93:753–762.
35. Wang J, Yu X (2010) Preparation, characterization and in vitro analysis of novel structured nanofibrous scaffolds for bone tissue engineering. Acta Biomater 6:3004–3012.
36. Ozkan S, Kalyon DM, Yu X, McKelvey CA, Lowinger M (2009) Multifunctional proteinencapsulated polycaprolactone scaffolds: fabrication and in vitro assessment for tissue engineering. Biomaterials 30:4336–4347.
37. Kretlow JD, Mikos AG (2007) Review: mineralization of synthetic polymer scaffolds for bone tissue engineering. Tissue Eng 13:927–938.
38. Sachlos E, Czernuszka JT (2003) Making tissue engineering scaffolds work. Review: the application of solid freeform fabrication technology to the production of tissue engineering scaffolds. Eur Cell Mater 5:29–39.
39. Gross KA, Rodríguez-Lorenzo LM (2004) Biodegradable composite scaffolds with an interconnected spherical network for bone tissue engineering. Biomaterials 25:4955–4962.
40. Karageorgiou V, Kaplan D (2005) Porosity of 3D biomaterial scaffolds and osteogenesis. Biomaterials 26:5474–5491.
41. Ciapetti G, Ambrosio L, Savarino L, Granchi D, Cenni E, et al. (2003) Osteoblast growth and function in porous poly epsilon -caprolactone matrices for bone repair: a preliminary study. Biomaterials 24:3815–3824.
42. Anada T, Kumagai T, Honda Y, Masuda T, Kamijo R, et al. (2008) Dose-dependent osteogenic effect of octacalcium phosphate on mouse bone marrow stromal cells. Tissue Eng Part A 14:965–978.
43. Kawai T, Anada T, Honda Y, Kamakura S, Matsui K, et al. (2009) Synthetic octacalcium phosphate augments bone regeneration correlated with its content in collagen scaffold. Tissue Eng Part A 15:23–32.

Electrospun Poly(ester-Urethane)- and Poly(ester-Urethane-Urea) Fleeces as Promising Tissue Engineering Scaffolds for Adipose-Derived Stem Cells

Alfred Gugerell[1]*, Johanna Kober[1], Thorsten Laube[2], Torsten Walter[2], Sylvia Nürnberger[3,4],
Elke Grönniger[5], Simone Brönneke[5], Ralf Wyrwa[2], Matthias Schnabelrauch[2], Maike Keck[1]

1 Division of Plastic and Reconstructive Surgery, Department of Surgery, Medical University of Vienna, Vienna, Austria, 2 Biomaterials Department, INNOVENT e. V., Jena, Germany, 3 Ludwig Boltzmann Institute for Experimental and Clinical Traumatology, Austrian Cluster for Tissue Regeneration, Vienna, Austria, 4 Department of Traumatology, Medical University of Vienna, Vienna, Austria, 5 Research Department Applied Skin Biology, Beiersdorf AG, Hamburg, Germany

Abstract

An irreversible loss of subcutaneous adipose tissue in patients after tumor removal or deep dermal burns makes soft tissue engineering one of the most important challenges in biomedical research. The ideal scaffold for adipose tissue engineering has yet not been identified though biodegradable polymers gained an increasing interest during the last years. In the present study we synthesized two novel biodegradable polymers, poly(ε-caprolactone-co-urethane-co-urea) (PEUU) and poly[(L-lactide-co-ε-caprolactone)-co-(L-lysine ethyl ester diisocyanate)-block-oligo(ethylene glycol)-urethane] (PEU), containing different types of hydrolytically cleavable bondings. Solutions of the polymers at appropriate concentrations were used to fabricate fleeces by electrospinning. Ultrastructure, tensile properties, and degradation of the produced fleeces were evaluated. Adipose-derived stem cells (ASCs) were seeded on fleeces and morphology, viability, proliferation and differentiation were assessed. The biomaterials show fine micro- and nanostructures composed of fibers with diameters of about 0.5 to 1.3 µm. PEUU fleeces were more elastic, which might be favourable in soft tissue engineering, and degraded significantly slower compared to PEU. ASCs were able to adhere, proliferate and differentiate on both scaffolds. Morphology of the cells was slightly better on PEUU than on PEU showing a more physiological appearance. ASCs differentiated into the adipogenic lineage. Gene analysis of differentiated ASCs showed typical expression of adipogenetic markers such as PPARgamma and FABP4. Based on these results, PEUU and PEU meshes show a promising potential as scaffold materials in adipose tissue engineering.

Editor: Mário A. Barbosa, Instituto de Engenharia Biomédica, University of Porto, Portugal

Funding: The research leading to these results has received funding from the European Union Seventh Framework Programme (FP7) through the ArtiVasc 3D project under grant agreement n°263416. http://ec.europa.eu/research/fp7/index_en.cfm. http://www.artivasc.eu/. The funders had no role in study design, data collection and analysis, decision to publish, or preparation of the manuscript.

Competing Interests: The authors have declared that no competing interests exist.

* E-mail: alfred.gugerell@meduniwien.ac.at

Introduction

A high incidence of soft tissue damage due to trauma or tumor removal on the one hand and limitations in reconstructing these defects on the other hand, asks for new solutions such as tissue engineering approaches. Tissue engineering works with biomimetic methods combining material engineering with life science. It requires an engineered biodegradable and highly biocompatible scaffold which can be used as vehicle for (stem) cells, growth factors, drugs, genes or other bioactive factors. This material should serve as first artificial matrix in tissue defects supporting invading cells to produce a new extracellular matrix and stimulating them to proliferate and form the new functional tissue. Due to recent progress in the development of new biomaterials and improved scaffold processing techniques over the last decades, new promising scaffold materials can be fabricated and offered to patients [1-4]. However, the main focus in research and clinical application lies in dermal and epidermal substitutes [5], whereas the development of a subcutaneous replacement (hypodermis) is often neglected. This is in contrast

to its overall importance: the hypodermis serves not only as energy storage but also defines the shape of the body and is well known as an endocrine organ. Generation of materials with mechanical and biological properties comparable to native adipose tissue is still a challenge for researchers active in this field [6,7]. A new, appropriate biomaterial should not only be stable for several weeks to serve as a framework for invading cells, but should also be biodegradable and hold a certain thickness and elasticity to provide plasticity as filler and shock protection. Certain characteristics such as tensibility or micro- and nano-structure are important to imitate the natural extracellular matrix.

During the last years, besides hydrogel formation [8,9], electrospinning has gained much interest as processing technique providing promising scaffold structures in soft tissue engineering [10-14]. This technique not only offers a high flexibility in material selection including synthetic and also natural polymers, but also provides nano- or microstructured three-dimensional scaffolds that resemble the extracellular matrix and support the mechanical stability of tissue. Such scaffolds allow cells to detach and communicate with each other. In principle, they are able to support cell differentiation, extracellular matrix generation and vascularization [15]. In addition to biocompatibility and mechanical stability, an adequate porous structure of the matrix seems to be a crucial factor for cell differentiation and integration. Controlling the fabrication parameters of the electrospinning process to optimize fiber diameter, pore structure, as well as mesh density and thickness with regard to cell cultivation requirements is therefore an important task [16,17].

A variety of novel biocompatible copolymers have been electrospun to fabricate nanofibrous scaffolds for biomedical applications with different success. But not only the material compositions but also the fabrication process can be manipulated in order to change fiber diameter, morphology and scaffold porosity [18]. Electrospun fibrous scaffolds can be prepared with a high degree of control over their structure, creating highly porous meshes of ultrafine fibers that resemble the ECM topography [19]. The fibrillar structure can enhance cell attachment, proliferation and colony-forming capacity of stem cells in vitro in comparison with non-fibrillar tropocollagen layers [19].

Both biodegradable biopolymers and synthetic polymers can be processed by electrospinning [20]. Naturally derived materials such as collagen [21], derivatives of hyaluronic acid [22], matrigel [23], and fibrin [24] have been intensively studied as scaffolds in adipose tissue engineering. Collagen which is prevalent in the native extracellular matrix (ECM) is known to promote adipose tissue development in vivo. Unfortunately if seeded with cells collagen devices often show contraction and rapidly degrade in vivo [25]. Similar to collagen fibrin gels are able to support adipogenesis in vivo, but like collagen the material has a high degradation rate and has not been studied extensively as 3D, porous scaffold [26,27]. Hyaluronic acid which is also a component of many ECM is a readily water-soluble and degradable polymer. It therefore has to be chemically modified for use as scaffold materials. In several attempts porous sponges of hyaluronic acid esters possessing a slower degradation rate have been used in adipose tissue engineering. Human preadipocytes cultured on those scaffolds have been shown to differentiate into adipocytes in culture but their properties in vivo remain to be investigated. Several studies have demonstrated that Matrigel, a commercially available protein mixture containing ECM components like laminin is highly adipogenic in vivo when injected in mice together with growth factors like bFGF [26]. Unfortunately Matrigel is not a viable option for clinical use due to its tumor cell origin. To date no satisfying results have yet been published neither in vitro nor in vivo. We can only speculate why there is such a lack of data. Maybe because the need for a hypodermis is not as urgent as for epidermis and dermis. Besides, a final construct of generated adipose tissue will need vascularization, which makes any approach a lot more difficult.

As an advantage of synthetic polymers their mechanical properties and biodegradation behavior can be tailored over a wide range [28]. Therefore this class of materials is able to meet biological and medical requirements in tissue engineering. Especially biodegradable polymers gained an increasing interest in the last decade. Recent developments in this group of materials are focused on polyesters like polycaprolactone (PCL) [29] or poly(lactic acid) (PLA) [30–33]. Because PLA homopolymers are rather brittle materials and PCL homopolymers due to their low glass transition temperatures show only limited mechanical stability at body temperature and, in addition, due to their hydrophobic nature a low rate of degradation, often copolymers of PLA and PCL or even terpolymers containing further monomers are used as scaffold materials [34–37]. Besides that, polyurethanes

(PU) are a very promising group of polymeric materials. They often show excellent elasticity and can be designed-to-degrade at the same time, which makes them a promising alternative to poly(hydroxy acids) homopolymers or their copolyesters [38,39]. This is especially true if instead of toxicologically problematic aromatic isocyanates those, derived from natural sources, are used in polyurethane synthesis like L-lysine ethylester diisocyanate (LDI) [40–43]. In fact, the combination of polyesters with polyurethanes in related poly(ester-urethane) block co- or terpolymers should lead to (thermoplastic) polymers with excellent physical properties including elasticity and sufficient mechanical strength combined with a more variable degradation behaviour. Concerning the mechanical properties, in the present case a potentially useful scaffold materials should show viscoeleastic or viscoplastic bahaviour and a minimum tensile strength of about 5 MPa.

Adipose-derived stem cells (ASCs) are immature precursor cells located between mature adipocytes in adipose tissue [44]. These cells can serve as an ideal autologous cell source for adipose tissue engineering approaches, since they are more resistant to mechanical damage and ischemia than mature adipocytes [45]. Adipose-derived stem cells can be harvested during liposuction or resection of adipose tissue and have been shown to proliferate rapidly and differentiate into bone, adipogenic and chondrogenic lineage both in vitro and in vivo [46–48].

In the present study we synthesized two novel urethane-based polymers, a poly(ester-urethane) (PEU) and a poly(ester-urethane-urea) (PEUU). Micro-structured fleeces with adjusted morphological characteristics suitable for soft tissue engineering were fabricated from these polymers by electrospinning. We hypothesized that due to the different chemical composition and structure of the polymers the produced scaffolds strongly differ in their mechanical properties and their degradation behaviour. The main goal of this study was to elucidate the cell compatibility as well as ASCs viability, proliferation and differentiation on the two scaffolds evaluating their potential as a framework in adipose tissue engineering

The scope of this work is on the tailoring of synthetic biodegradable polymers matching the specific requirements in adipose tissue engineering. Due to the mechanical properties diverging from those of common well-known, degradable polymers, these materials have the potential to gain attraction also in other fields of soft tissue regeneration.

Materials and Methods

Ethics Statement

This study has been approved by the ethics committee of the Medical University of Vienna and the General Hospital Vienna (EK no. 957/2011). All subjects enrolled in this research have given written informed consent. Fat tissue was obtained from 10 donors undergoing abdominoplasty.

Materials and General Procedures

Dichloromethane (DCM), cyclohexane, heptane and toluene were purchased from Fisher Scientific (Schwerte, Germany) and used without further purification. Acetone, chloroform ($CHCl_3$) and methanol (MeOH) used for electrospinning and analytical purposes (HPLC grade) were obtained from VWR International (Darmstadt, Germany). Hexafluoroisopropanol (HFIP) was purchased from Apollo Scientific Ltd. (Stockport, U.K.). L-lactide was purchased by PURAC Biomaterials (Gorinchem, Netherlands). ε-Caprolactone (Sigma-Aldrich, Taufkirchen, Germany) and 2,6-diisocyanato methyl caproate (L-lysine ethyl ester diisocyanate)

(Infine Chemicals, Shanghai, China) were purified by vacuum distillation. Polyethylene glycol (M_w = 400 g mol^{-1}, PEG 400) and N-methylpyrrolidone (NMP) were purchased from Sigma Aldrich. Poly(L-lactide-co-D,L-lactide) (70%/30% (w/w), M_w = 1.35×10^6 g mol^{-1}, PLLA) was purchased from Evonik Röhm (Darmstadt, Germany).

Polymer syntheses

Representative synthesis procedures for the poly(ester-urethane) and the poly(ester-urethane-urea) materials are as follows:

Poly[(L-lactide-co-ε-caprolactone)-co-(L-lysine ethyl ester diisocyanate)-*block*-oligo(ethylene glycol)-urethane] (PEU)

Octanediol-bis(L-lactide-co-ε-caprolactone) (LLA-CL)

A mixture of 1,8-octanediol (1.0 g, 6.84 mmol), L-lactide (19.7 g, 136.78 mmol), and 16.55 µl stannous octoate (dissolved in 94 µl toluene) was stirred under nitrogen at 150°C. After 45 minuntes further 16.55 µl stannous octoate (dissolved in 94 µl toluene) was added. The mixture was then stirred for 75 minutes at this temperature followed by an addition of ε-caprolactone (58.04 ml, 547.11 mmol) and 49.9 µl stannous octoate (dissolved in 280 µl toluene). Stirring was continued at 150°C for 1 hour, followed by an addition of 49.9 µl stannous octoate. Then the solution was stirred for 3 hours, cooled to room temperature and dissolved in 75 ml dichloromethane. The solution was filtrated and the crude product was precipitated into 1000 ml cold heptane. Finally, the isolated LLA-CL was dried in vacuum at room temperature to constant weight (76.9 g).

NMR: 1.4–1.25 (m; 4.6 H); 1.65–1.40 (m; 11.8 H); 2.25 (t; 3.9 H); 4.01 (t; 4 H); 5.15–5.00 (m; 1 H)

Poly[(L-lactide-co-ε-caprolactone)-co-(L-lysine-ethyl ester-diisocyanato)urethane] prepolymer (LLA-CL-LDI)

In the next step, L-lysine ethyl ester diisocyanate (LDI) (12.77 ml, 63.24 mmol) was combined with *LLA-CL* (76.9 g) under nitrogen and reacted at 60°C with stirring for 4 hours. After cooling to room temperature the reaction mixture was dissolved in 80 ml dichloromethane, filtrated and precipitated twice into 1000 ml cold cyclohexane. Finally, the isolated isocyanate-terminated poly(L-lactide-co-ε-caprolactone) prepolymer (LLA-CL-LDI) was dried in vacuum at room temperature to constant weight (71.3 g).

Poly[(L-lactide-co-ε-caprolactone)-co-(L-lysine ethyl ester diisocyanato)-block-oligo(ethylene glycol)urethane] (PEU)

In the last step LLA-CL-LDI (71.3 g) was dissolved in N-methylpyrrolidone (300 ml) and heated to 60°C. PEG 400 (3.01 g, 4.48 mmol) was dissolved in 30 ml NMP and added dropwise. The reaction mixture was stirred overnight, cooled to room temperature and precipitated into 3000 ml of cold water. The crude product was dried in vacuum at room temperature, then dissolved in 100 ml dichloromethane, filtrated and precipitated into 1500 ml cold petrol ether. Finally, the isolated poly(ester-urethane) PEU was dried in vacuum at room temperature to constant weight (65.2 g).

Poly(ε-caprolactone-co-urethane-co-urea) (PEUU)

Octanediol-bis(ε-caprolactone) (CL)

A mixture of 1,8-octanediol (0.5 g, 3.42 mmol), ε-caprolactone (39.02 ml, 342.0 mmol), and 31.6 µl stannous octoate (dissolved in 180 µl toluene) was stirred under nitrogen at 150°C. After 30 and 90 minutes additional 15.8 µl stannous octoate (dissolved in 90 µl toluene) was added. The mixture was stirred 4 h overall, then cooled to room temperature and dissolved in 75 ml dichloromethane. The solution was filtrated and the crude product was precipitated into 1000 ml cold heptane. Finally, the isolated oligomer (CL) was dried in vacuum at room temperature to constant weight (39.9 g, 3.45 mmol according to NMR).

NMR: 1.24–1.40 (m; 50.5 H); 1.54–1.65 (m; 100.2 H); 2.25 (t; 50 H); 3.57 (t; 1 H); 4.00 (t; 49.9 H)

Poly[ε-caprolactone-co-(L-lysine ethyl ester diisocyanato)urethane] prepolymer (CL-LDI)

In the next step, LDI (6.82 ml, 33.76 mmol) was combined with CL (39.9 g, 3.45 mmol) under nitrogen and reacted at 60°C with stirring for 4 hours. After cooling to room temperature the reaction mixture was dissolved in 80 ml dichloromethane, filtrated and precipitated twice into 1000 ml cold cyclohexane. Finally, the isolated isocyanate-terminated polylactone prepolymer (CL-LDI) was dried in vacuum at room temperature to constant weight (37.0 g).

Poly(ε-caprolactone-urethane-urea) (PEUU)

In the last step CL-LDI (37.0 g) was dissolved in THF (40 ml) and LDI (495 µl, 2.45 mmol) were added and mixed thoroughly followed by an addition of DABCO-solution (2.96 ml, 2.8 M). The solvent was vaporized at 40°C. The crude product dissolved in 100 ml dichlormethane, filtrated and precipitated into 1500 ml cold petrol ether. Finally, the isolated poly(ester-urethane-urea) PEUU was dried in vacuum at room temperature to constant weight (33.2 g).

Polymer structure analytics

Molecular weights (M_n and M_w) and polydispersity indices (PDI) were determined with respect to polystyrene standards by gel permeation chromatography (GPC). The measurements were performed on a set of Shimadzu apparatuses (Shimadzu Deutschland, Duisburg, Germany) using chloroform as eluent. All samples were analysed at room temperature. Chloroform (Fisher Scientific, Germany, stabilised with 1% amylene) was used as eluent, delivered at a flow rate of 1.0 ml min^{-1}. The samples were dissolved in chloroform at a concentration of 5 mg ml^{-1}. The injection volume was 100 µl. As pre-column a PSS-SDV (100 Å, 8,0×50 mm) and as column PSS-SDV (100 Å, 8,0×300 mm), PSS-SDV (1000 Å, 8,0×300 mm) and PSS-SDV (100000 Å, 8,0×300 mm) were used. As detector a RID 10A (Shimadzu Deutschland) was used.

^{1}H- and ^{13}C- nuclear magnetic resonance (NMR) was used to characterize the chemical structures and compositions of the synthesized copolymer. The spectra were recorded on a Bruker DRX 400 spectrometer (Bruker BioSpin, Rheinstetten, Germany), using tetramethylsilane as an internal reference and CDCl$_3$ as solvent. To determine the monomer ratio or the chain length of the polymers we used the NMR. In case of polycaprolactone polymers the chain length was determined by comparison of the ^{1}H-NMR signals at 3.57 ppm being the C**H**$_2$-OH signal of the last caprolactone unit in the chain with the signal at 2.25 ppm, being the CO-C**H**$_2$-CH$_2$-signal. In case of lactide-caprolactone copolymer the monomer ratio was determined by comparison of the ^{1}H-NMR signal at 5.15–5.00 ppm, being the CH-proton of lactide, with the signal at 2.25 ppm, being the CO-C**H**$_2$-CH$_2$-signal of the caprolactone monomer.

Fabrication of foils and compact samples

Foils were fabricated by evaporation of solvent. 400 mg of the polymers were dissolved in 20 ml dichloromethane and poured into dishes (5.5 cm in diameter). The dishes were kept at room temperature for at least 3 days followed by two days at 40°C. The foils were cut into stripes of 50×5 mm^2. Compact samples (height 8 mm, diameter 8 mm) were fabricated by melting the polymers at 100–150°C in a silicone mould.

Fabrication of electrospun non-woven fleeces

A computer-aided electrospinning machine developed at Erich Huber GmbH (Gerlinden, Germany) in collaboration with INNOVENT e. V. (Jena, Germany) and recently commercialized under the trade name E-Spintronic (Erich Huber GmbH, Gerlinden, Germany) was used for the fabrication of electrospun non-woven fleeces. This machine enables the defined adjustment of major spinning parameters and a high process reproducibility. Electrospinning conditions similar to those that have been previously described [10,49] were used. A stainless-steel straight-end hollow needle of diameter of 0.4 mm was used as nozzle. A glass mirror of 2.5 mm thickness (35×35 cm^2) was used as counter electrode for collecting the electrospun non-woven fibers. The distance between the needle tip and the collector was maintained at 19–22 cm. The voltage was adjusted to 18–28 kV. The polymers dissolved in suitable solvent and appropriate concentration were fed at a constant rate of 1.5 ml h^{-1} through the syringe to the needle tip resulting in the formation of fibers with diameters of about 0.5 to 1.3 µm. The dimension of the obtained electrospun fleeces was approximately 60 cm^2.

Investigation on material properties

Mechanical Properties. Young's-modulus and tensile strength of foils were determined with a Texture Analyser TA-XT2i (Stable Micro Systems, Godalming, U.K.) with a 5 kg measuring head to characterize the mechanical properties of the polymers. Tensile tests were performed after DIN EN ISO 1798. For each material tested 5 samples were produced. Electrospun fleeces were cut into stripes (1×5 cm^2) and fixed at 5 kN wedge action grips of an Instron universal testing machine (model 4301, Norwood, MA, USA). Materials were stretched with a rate of 20 mm min^{-1} until the stripe ripped, distance and force were measured. Experiments were performed five times at minimum.

Thermal Properties. Glass transition temperatures of polymers PEUU and PEU were measured by DMA with a Perkin Elmer DMA7E instrument. Glass transition temperatures of polymer PLLA was measured by DCS with a Perkin Elmer DSC7 instrument.

Matrix Characterization. For cell experiments, fleeces were clamped into cell crowns (scaffdex, Sigma-Aldrich, St. Louis, MO, USA) to provide an even, tensed surface. Then they were incubated in 70% ethanol for 30 minutes, air-dried and washed once with PBS before cell seeding.

Scanning Electron Microscopy (SEM). Adipose-derived stem cells (ASCs) were seeded on fleeces in 24 well cell crowns (6×10^4 cells). Fleeces (with and without cells) were fixed in fixing solution containing 0.1 M sodium cacodylate and 2.5% glutaraldehyde. Subsequently, fleeces were washed (0.1 M sodium cacodylate without glutaraldehyde), dehydrated with 2,2-dimethoxypropane and dried with hexamethyldisilazane. Samples were sputter coated with palladium gold in an Emitech (Molfetta, Italy) sputter coater SC7620 and analyzed in a SEM Jeol 6510 (Jeol Ltd, Tokyo, Japan).

Polymer and fleece Degradation. Compact samples of both polymers PEU and PEUU were incubated in Sörensen phosphate buffer (pH 7.4) at 37°C for 4 weeks, rinsed, dried for 1 week in vacuum and weighted. Samples were covered with fresh buffer and procedure was repeated (total storage time 44 weeks). Furthermore, fleece samples made from PEU and PLLA were stored in simulated body fluid medium (SBF) at 37°C and released free L-lactate was monitored over 12 weeks with the ENZYTEC D-/L-lactic acid assay (R-Biopharm, Darmstadt, Germany). For comparison an elevated degradation test (refluxing of the non-

woven PEU samples for 24 hours in 1 N NaOH solution) was carried out measuring the liberated L-lactate in the same manner.

Biological characterization

ASC Isolation and Cultivation. Minced adipose tissue was washed in phosphate buffered saline (PBS, PAA Laboratories GmbH) and digested with 2 mg/ml collagenase type IV in Hanks' buffered salt solution (HBSS, both Sigma-Aldrich, St. Louis, MO, USA) for 1 h at 37°C with constant shaking. Cells were filtered through cotton gauze and centrifuged for 5 minutes at 380 g. Red blood cells in the stromal vascular fraction were lysed in 2 ml Red Blood Cell Lysing Buffer (Sigma-Aldrich) and incubated on ice for 8 minutes. Cold medium was added and suspension was filtered through a 70 µm cell filter. Cells were centrifuged for 5 minutes at 380 g and cell pellet was re-suspended in DMEM (PAA Laboratories GmbH), supplemented with 100 units ml^{-1} penicillin, 100 µg ml^{-1} streptomycin (both Life Technologies Ltd, Paisley, UK), and 10% fetal calf serum (Hyclone, Fisher Scientific GmbH, Schwerte, Germany).

For experiments, cells were counted, seeded on scaffolds and cultivated at 37°C in humidified atmosphere with 5% CO$_2$.

Staining of Live and Dead Cells. Cells were grown on top of fleeces for 48 hours in 24 well cell crowns (6×10^4 cells). For staining of live and dead cells they were incubated with fluorescein diacetate and propidium iodide fluorescent dye solution (Molecular Probes, Inc., Eugene, OR, USA) according to the manufacturers protocol. Microscopy was done on an AxioImager microscope (Zeiss, Jena, Germany).

Cell Morphology. Fleece preparation and cell seeding was conducted as described above. Cells were grown on fleeces for 48 hours. After fixation in formalin and permeabilization, cells were stained with 10 µg ml^{-1} TRITC-phalloidin/PBS (Sigma-Aldrich, St. Louis, MO, USA) and 5 µg ml^{-1} 4',6-diamidino-2-phenylindol (DAPI; SERVA Electrophoresis GmbH, Heidelberg, Germany) under light protection. Microscopic analysis was done on an AxioImager microscope (Zeiss, Jena, Germany).

MTT Assay/cell proliferation. Viability of cells was further measured using a CellTiter96 non-radioactive proliferation Assay (Promega Corporation, Madison, WI). Therefore, cells were seeded on fleeces in 24 well cell crowns (6×10^4 cells). On day one, three and seven cell viability was evaluated according to manufacturers protocol. Therefore, 15 µl dye solution was added to 100 µl medium, after two hours of incubation 100 µl stop solution was added and incubated for one hour in the dark. Fleeces were gently shaken to homogenize medium and substrate, 50 µl of the supernatant was transferred into a 96 well plate in triplicates and absorbance was measured on a Wallac 1420 VICTOR2 plate reader (PerkinElmer, Waltham, MA, USA).

Differentiation. ASCs were seeded on fleeces in 24 well cell crowns (6×10^4 cells) for 48 hours for proliferation. To induce adipocyte differentiation, cells were incubated for three days in Preadipocyte Differentiation Medium (PromoCell GmbH, Heidelberg, Germany). Afterwards cells were incubated in Adipocyte Nutrition Medium (PromoCell GmbH, Heidelberg, Germany) without IBMX, medium was changed every third day. Differentiation was evaluated on day 21. For adipored staining, viable cells were washed once with PBS and incubated with AdipoRed Assay Reagent (Lonza, Walkersville, MD, USA) according to manufacturer's protocol. After 15 minutes, cells were rinsed with PBS and fixed with 4% formalin. Cells were analyzed using an AxioImager microscope (Zeiss, Jena, Germany).

Preparation of Polymer-Coated Cell Culture Plates. 24 well plates (Becton & Dickinson, Franklin Lakes, NJ, USA) were coated with PEUU and PEU as followed: Each polymer was

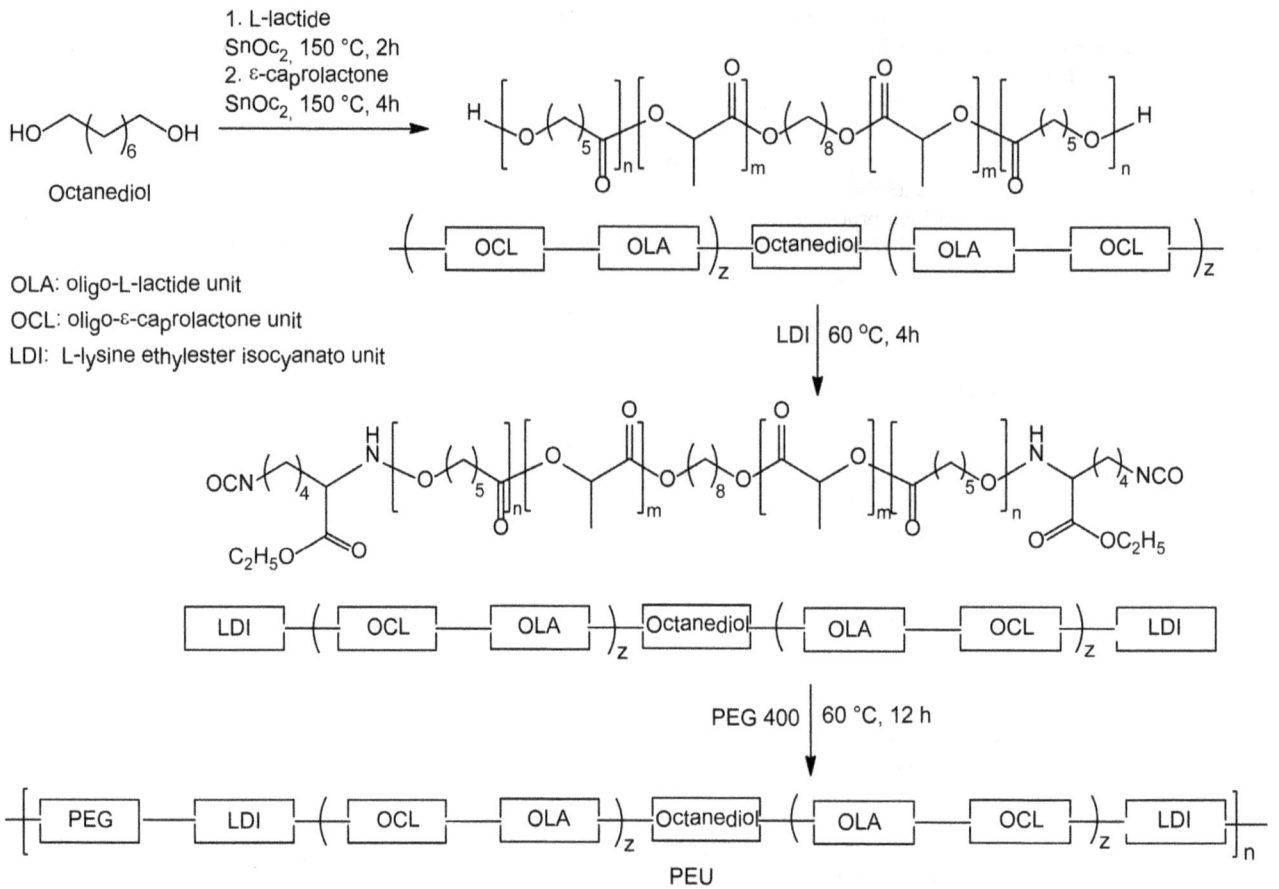

Figure 1. Schematic composition of copolymer PEU.

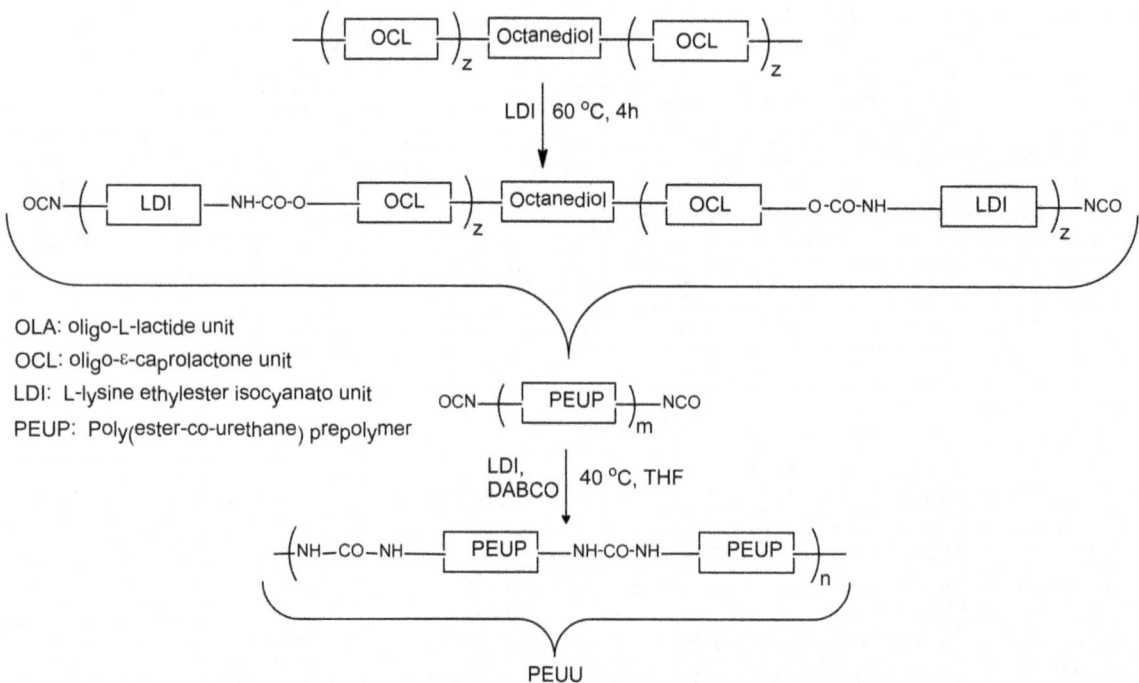

Figure 2. Schematic composition of copolymer PEUU.

Table 1. Molecular weights (M_n, M_w, M_z) and polydispersity indices of prepared oligomers and polymers determined by GPC

Sample	Molecular Weight			Dispersity Index
	M_n	M_w	M_z	PDI
CL	22,713	29,607	37,634	1.30
CL-LDI	27,306	38,936	55,006	1.43
PEUU	141,050	495,975	2,397,908	3.52
LLA-CL	22,222	32,933	45,681	1.48
LLA-CL-LDI	24,053	36,286	51,417	1.51
PEU	56,591	108,329	204,829	1.91

dissolved in hexafluoroisopropanol to gain a 1% (w/w) solution. 140 µL of this polymer solution was used to coat one well. The excessive solution was removed and the polymer-coated cell culture plates were dried under a fume hood over night. Next day, plates were stored in a cell culture incubator (37°C, 5% CO_2, 90% humidity) to use.

Gene Expression Analysis. ASCs offered by Zen-Bio were used as recommended by the supplier (Lonza, Verviers, Belgium). Cells were seeded in the polymer-coated 24 wells or uncoated control wells (6×10^4 cells) and cultivated in basal growth medium containing 10% fetal calf serum, L-glutamine and GA-1000 (PBM-2, Lonza, Verviers, Belgium) at 5% CO_2 and 37°C. To differentiate the cells into adipocytes insulin, dexamethasone, indomethacin and isobutylmethylxanthine (PBM-2, Lonza, Verviers, Belgium) were added to the medium as recommended by the supplier. After cultivating ASCs in differentiation medium, cells were harvested and RNA of ASCs, differentiated for seven days, was isolated using the RNeasy Kit whereas RNA of ASCs, differentiated for 14 days, was isolated using the RNeasy Lipid Tissue Kit (both Qiagen, Hilden, Germany) according to the manufacture's procedure. Afterwards, RNA was transcribed into cDNA using the High-Capacity cDNA Reverse Transcription Mix (Applied Biosystems, Life Technologies Ltd, Paisley, UK). To analyze gene expression Real-Time TaqMan-PCR was performed using FAM labled primers and the 7900 HT Fast and Sequence Detection System (Applied Biosystems, Darmstadt, Germany). The following PCR conditions were used: 50°C for 2 minutes, 94.5°C for 10 minutes followed by 40 cycles at 97°C for 30 sec and 59.7°C for 1 minute. Data were analyzed using the Sequence detector software supplied with the 7900 Sequence Detector and RQ Manager.

Expression levels were calculated by the $2^{-\Delta\Delta Ct}$-method, whereby GAPDH was used as endogenous reference and undifferentiated cells (donor #1) were used as calibrator and therefore were set as 1.

Statistics

Data are presented as mean ± SD of at least three independent experiments. Statistical analysis was performed with software SPSS Statistics 19 or Microsoft Excel 2010. Statistical comparisons for all experimental settings were based on two sample t-test, ANOVA or General Linear Model using Tukey's test with $p < 0.05$ considered as significant.

Results

Polymer Synthesis and Characterization

Two novel ε-caprolactone-containing polyurethane-type polymers, a poly(ester-urethane) and a poly(ester-urethane-urea) multiblock copolymer, respectively, have been synthesized in a multistep procedure adapting previously reported procedures for the synthesis of polylactones and lysine ethylester based polyurethanes [50,51]. In the first step oligolactones have been prepared by conventional ring-opening polymerization of ε-caprolactone or L-lactide/ ε-caprolactone. These oligomers have been end-capped with L-lysine ethyl ester diisocyanate (LDI). The resulting reactive prepolymers were treated with PEG 400 to afford the L-lactide- ε-caprolactone containing poly(ester-urethane), PEU, and with LDI in the presence of DABCO as catalyst to obtain the ε-caprolactone based poly(ester-urethane-urea), PEUU, respectively. Figures 1 and 2 schematically illustrate the synthesis strategy and the monomer composition of the prepared PEU and PEUU multiblock polymers.

The polymers and their intermediate prepolymers have been structurally characterized in the usual way by FT-IR spectroscopy, ^{1}H- and ^{13}C-NMR spectroscopy. Their molecular weights and polydispersity indices have be determined by GPC (Table 1).

The mechanical properties of the novel polymers were determined using polymer foils. The values for the tensile strength, Young's modulus, and maximum elongation of the synthesized polymers PEUU and PEU can be found in Table 2. For comparison the corresponding values for a commercially available PLLA are given. The stress-strain curves of the polymers are shown in Figure 3. Young's modulus was determined at the elastic deformation area at 2–10% elongation. It becomes obvious that the mechanical properties of the ε-caprolactone-containing copolymers remarkably differ from those of PLLA. Due to the presence of the ε-caprolactone soft segment, PEU and PEUU have much lower Young's modulus compared to PLLA. PEUU shows a quite different mechanical behavior in comparison to both the other polymers with regard to elongation prior break. After a steep

Table 2. Mechanical properties (tensile strength, Young's modulus, and maximum elongation) and glass transition temperatures (T_g) of synthesized polymers (PEUU, PEU) in comparison to PLLA.

Polymer	Mechanical properties			T_g
	Tensile strength [MPa]	Young's modulus [MPa]	Max. elongation [%]	Peakmaximum tan delta [°C]
PEUU	37.16 (±5,37)	184 (± 8)	1072.39 (±103.88)	–24 [a]
PEU	7.88 (±0.66)	131 (± 77)	14,64 (±0.73)	8 [a]
PLLA	50.62 (±3.40)	1393 (± 272)	5.65 (±0.60)	59 [b]

[a] peakmaximum tan delta [°C] (measured by DMA); [b] half C_p (measured by DSC).

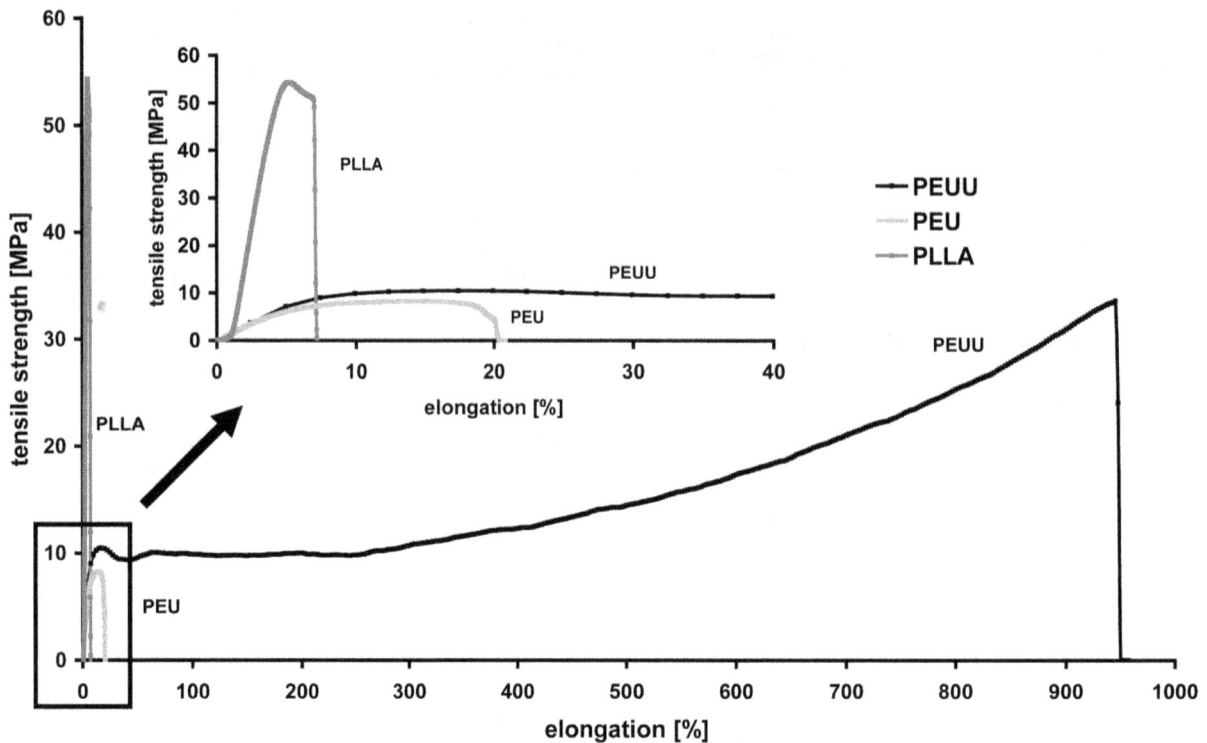

Figure 3. Stress-Strain curves of PEU and PEUU foils in comparison to conventional PLLA foils.

course of the curve at the beginning in the stress-strain diagram, similar to the course of the other polymers, at higher stress this polymer exhibits plastic deformation resulting in a high maximum elongation at break (Figure 3). A possible explanation of this behavior could be the presence of a large ε-caprolactone block within the PEUU polymer in contrast to the L-lactide/ε-caprolactone copolymer unit in PEU. A similar behaviour has been reported for copoylmers with high ε-caprolactone contents [52]. However with regard to the complex structure of PEUU further investigations are necessary to fully understand this phenomenon.

Characteristics of Electrospun Polymer Fleeces

Both polymers could be transformed into electrospun fleeces. For comparative studies, fleeces were also fabricated from commercially available PLLA using the same processing conditions. Electrospinning parameters for PLLA, PEUU and PEU are listed in Table 3.

Fiber diameters were measured on the basis of light microscopy images. It was found that fibers in PEUU are thicker and have a wider range of different microstructures compared to the fibers of biomaterials PLLA and PEU.

Surface Morphology and Characterization

The ultrastructure of the electrospun materials was observed by SEM (Figure 4, A-C; materials seeded with adipose derived stem cells for 48 hours: D-I). The thickest fibers occurred in PEUU (1.02–1.28 μm) whereas the fiber thickness of PLLA and PEU were comparable (0.65–1.14 μm) (Table 3). The latter one is woven tighter and has more branches.

To test the tensile strength the fleece materials were clamped into a tensile testing machine and stretched until break (Figure 5). For tensile strength the distance until break was measured. PLLA and PEU both expanded about 15 mm whereas PEUU could be lengthened until 45 mm and therefore showed the highest elasticity. Furthermore, the forces at break were measured whereby PEUU reached about 0.45 N, the other materials disrupted at about 0.8 N.

Table 3. Electrospinning parameters and resulting fiber diameters of the electrospun non-woven biomaterials (flow rate for all materials: 1.5 ml h^{-1})

Polymer	Solvent(s) (v/v)	Concentration in % (w/w)	Air humidity [%]	Temperature [°C]	Voltage [kV]	Electrode distance [cm]	Fiber diameter [μm]
PLLA	CHCl$_3$/MeOH (3:1)	3	25–36	23–30	20	19–22	0.56–0.89
PEUU	CHCl$_3$	4	27–40	21–25	20–22	22	1.02–1.28
PEU	acetone	23	31–55	23–29	18–24	20	0.65–1.14

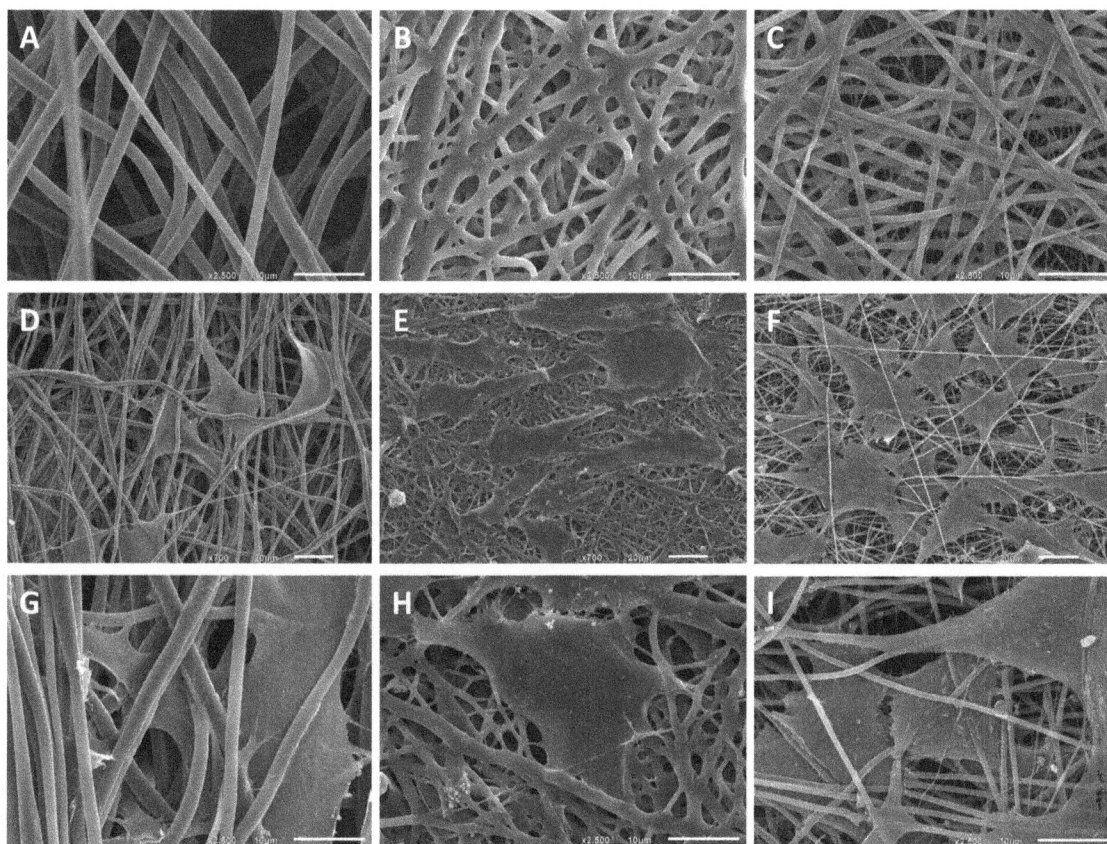

Figure 4. Ultrastructure analyses. SEM images of electrospun materials PEUU (A, D, G), PEU (B, E, H) and PLLA (C, F, I). Adipose derived stem cells were cultivated on top of the materials for 48 hours (D-I). Bars represent 10 μm (A, B, C), 20 μm (D, E, F), or 10 μm (G, H, I), respectively.

Degradation Characteristics

In a first set of experiments compact samples of polymers PEUU and PEU were incubated at 37°C in Sörensen phosphate buffer and the weight loss was determined after defined time periods. After 44 weeks polymer samples show a reduction in weight of 2.4% (PEUU) and 34.3% (PEU), respectively. The polymer with the polylactide-block (PEU) degrades significantly faster (Figure 6).

Due to the low weight of electrospun fleece materials which makes the gravimetric determination of the degradation behaviour difficult we were looking for another analytical methods better suited for electrospun polymer fleeces.

It is known that the degradation of lactide containing polymers can also be monitored by the measurement of free L-lactate formed via hydrolyis of the polylactide block in the supernatant of the degradation medium. We used this method to study the degradation of lactide-containing PEU and PLLA fleeces. The amount of free L-lactate was measured by an enzymatic assay. At first, the hydrolytic degradation of PEU under strongly elevated degradation conditions (reflux of fleeces for 24 h in 1 N NaOH solution) was investigated to determine the total (= initial) amount of releasable L-lactide in the PEU fleece to be 20.7% (w/w). As specified by the manufacturer PLLA contains 85% (w/w) L-lactide (= initial amount). Afterwards, degradation of PEU and PLLA was performed at 37°C in SBF medium measuring the amount of released L-lactate under these conditions related to the total L-lactide content of the two polymers. As shown in Figure 7, after 12 weeks, 8.8% of the initial amount of L-lactate from the poly(L-lactide) block of PEU and 13.8% of the initial amount of L-lactate from PLLA were released into the supernatant.

Based on the performed degradation studies the degradation rate of the three polymers during an initial degradation period of about 10 months is in the order PLLA > PEU > PEUU. Long-term studies on the degradation behaviour of PEU and PEUU are in progress now.

Biological Characterization

Cell Viability and Morphology. Cell viability and cell morphology are depicted in Figure 8. ASCs were cultivated on top of PEUU, PEU and PLLA for 48 hours. Viability was detected by fluorescence staining of cytoplasm of viable cells using fluorescein diacetate and the nuclei of dead cells were stained using ethidium-homodimer-1 (Figure 8). All tested electrospun materials exhibited similar ASCs viabilities. On all three materials only few dead cells could be detected and ASCs show a good viability.

To analyze cell adhesion and morphology, cytoskeleton was stained with TRITC-phalloidin. Actin fibers were well expressed and spread into cell processes. Cells aligned and spread in the direction of the polymers fibers. As already shown by scanning electron microscopy (Figure 4, D-I), ASCs grown on PEUU and PLLA scaffolds had a more spread morphology whereas cells cultivated on PEU were rather elongated.

Proliferation of ASCs. Proliferation and viability of ASCs were evaluated by MTT Assay. After 24 hours, viability of cells grown on the materials was comparable to each other. However, after seven days of proliferation, viability of cells grown on PEU and PLLA was by trend higher compared to PEUU (not significant) (Figure 9).

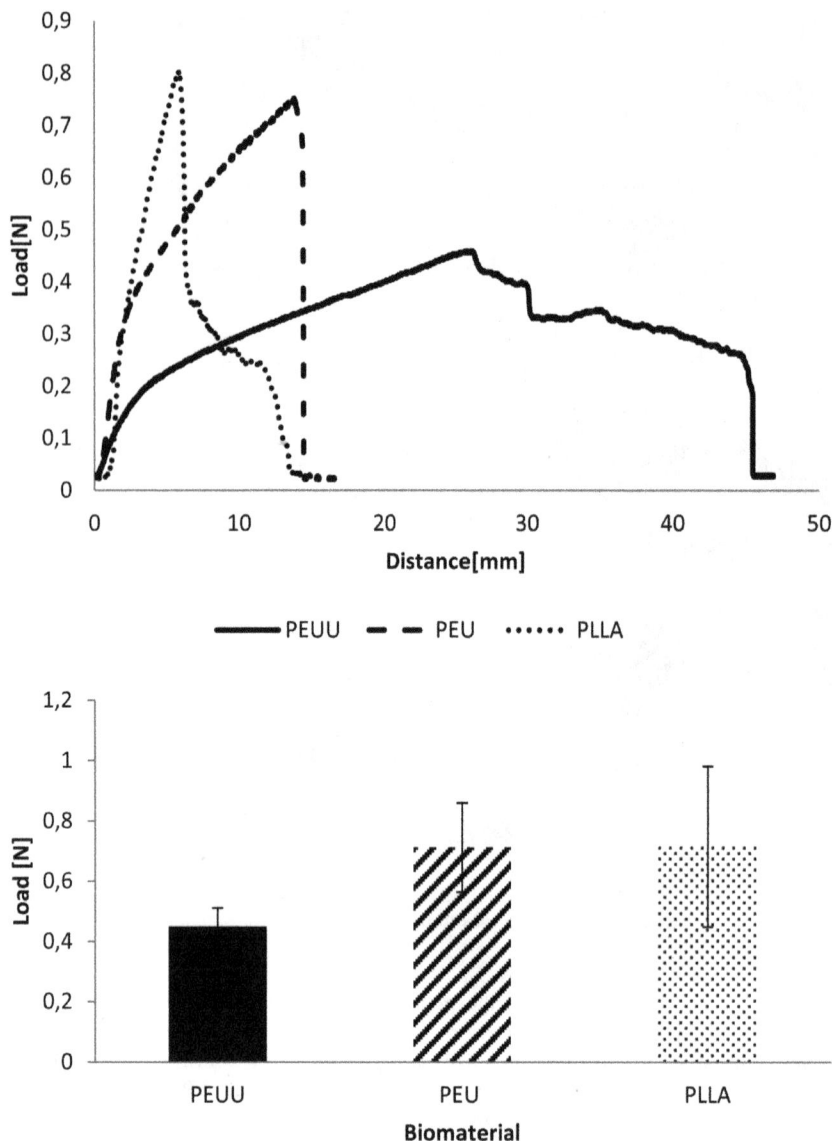

Figure 5. Tensile properties, Tensile strength (A) and tensile strength at break (B) of electrospun fleeces prepared from different polymers.

Differentiation into the adipogenic Lineage. For determination of differentiation potential ASCs were grown on biomaterials PEUU, PEU and PLLA for two days followed by induction of adipogenesis for three days and further cultivation in nutrition medium. On day 21 lipids were stained with AdipoRed. On all three materials cells were able to be differentiated into the adipogenic lineage as shown by intracellular accumulation of lipid droplets (Figure 10). No striking differences could be observed between the scaffolds.

Gene Expression Analysis of Differentiating Cells. Lastly, we sought to gain insights, whether the biomaterials have any effects on the activation or regulation of genes/receptors, which are relevant for the typical behavior of adipocytes and thus are involved in adipogenic, lipolytic or lipogenic processes. To focus on the material effects and to exclude potential effects on the cell behavior caused by 3D culturing conditions, cells were grown in polymer-coated wells (2D condition). ASCs were differentiated for 7 and 14 days into the

adipogenic lineage and the expression of 17 genes was determined by qRT-PCR. Results showed that there was no relevant difference in gene expression of the 17 analyzed genes between cells grown on polymer-coated surfaces compared to cells cultivated on the uncoated tissue culture plate or on plates coated with polymer PLLA. None of the observed genes/receptors was significantly affected by the fleece materials (Figure 11).

Discussion

In our study, poly(L-lactide-co-D,L-lactide) (PLLA) served as control, since this is an established commercially available scaffold for tissue engineering. Degradation of polylactides in the human body via a hydrolytic reaction results in L-lactic acid, a natural intermediate in metabolism [53]. PLLA has been utilized for both in vitro and in vivo studies as 3D scaffolds or grafts for adipose tissue engineering, showing potential in supporting tissue regeneration [54,55]. Above all, pure poly(L-lactide) and copolymers

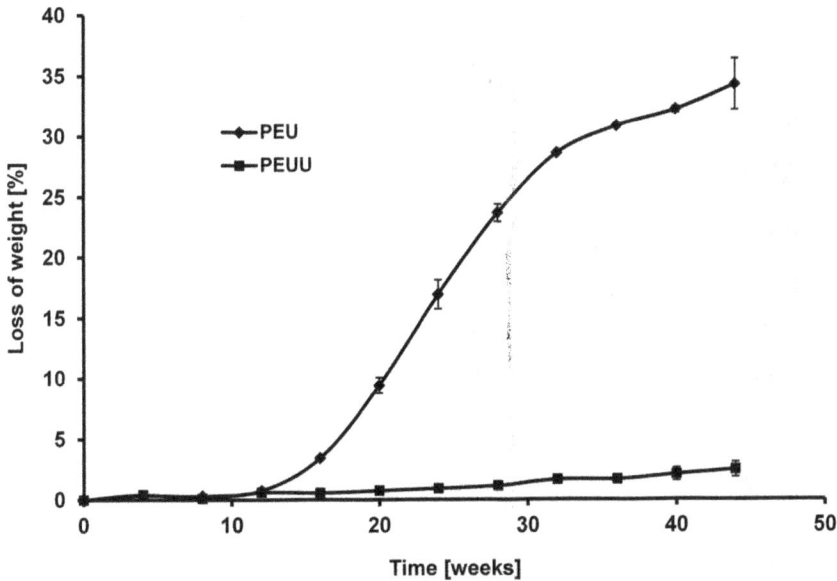

Figure 6. Degradation of compact samples of PEUU and PEU within 44 weeks in Sörensen phosphate buffer at 37°C.

such as poly(L-lactide-co-D/L-lactide) or poly(D/L-lactide-co-glycolide), are already approved by the Food and Drug Administration for clinical applications [56] and have been shown to provide favorable growth and proliferation conditions for several cell types [57]. An often discussed disadvantage of this polymer type is the formation of hydroxy acids as degradation product which may cause cytocompatibility problems by lowering the pH, especially in the late phase of degradation assuming that the degradation follows a bulk erosion mechanism [58,59].

Normally, poly-ε-caprolactone (PCL) forms softer polymer materials but shows slower degradation kinetics than PLA [60].

In this study we used ε-caprolactone- and ε-caprolactone-L-lactide containing polyester structural building blocks to construct multiblock polymers in which the polyester segments are linked together by different linking units. In the poly[(L-lactide-co-ε-caprolactone)-co-(L-lysine ethyl ester diisocyanate)-*block*-oligo(ethylene glycol)-urethane] (PEU) polyester segments are linked by urethane bonds whereas in poly(ε-caprolactone-co-urethane-co-urea) (PEUU) both urethane and urea bonds are used to elongate the polyester segments.

The chemical properties of such polymers enable degradation through hydrolytic ester cleavage similar to polylactones like PLA or its copolymers. Emerging degradation products are metabolized

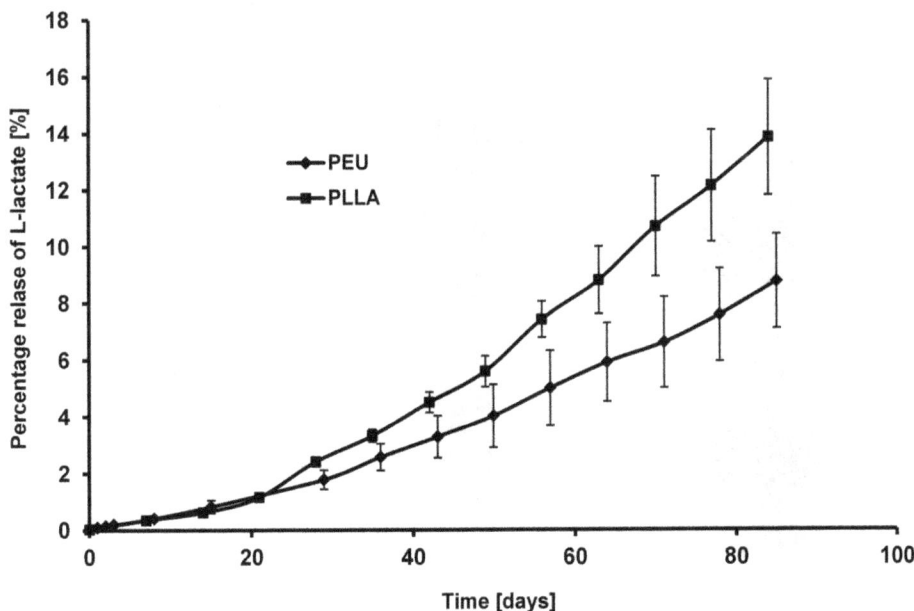

Figure 7. Percentage release of L-lactate from PEU and PLLA (referred to the total amount of L-lacide in non-degraded polymers) during degradation in SBF medium at 37°C.

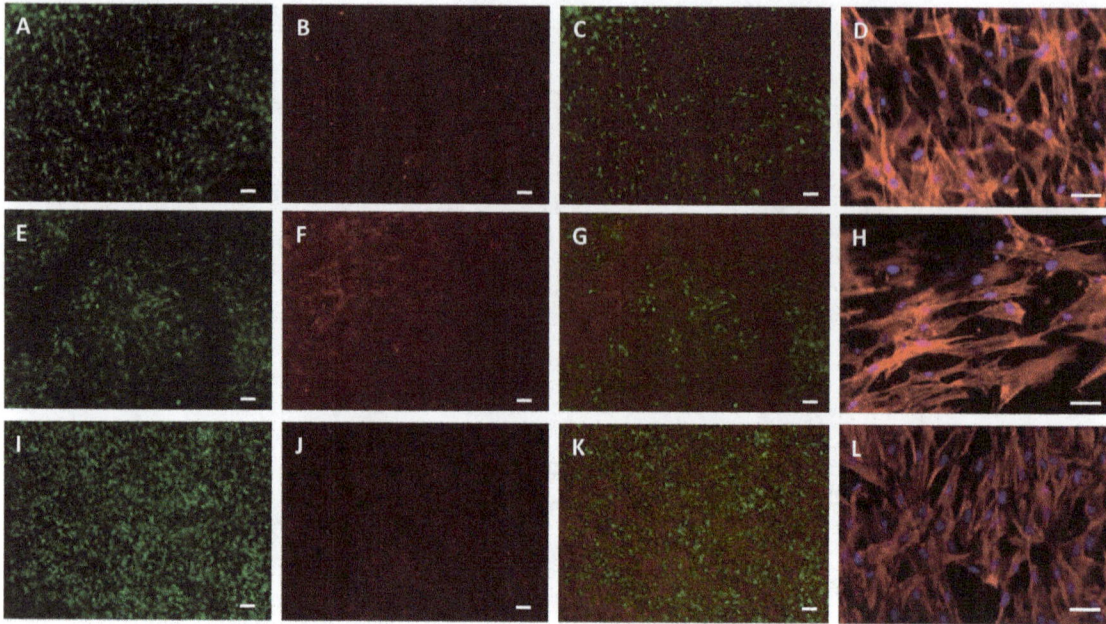

Figure 8. Cell viability and morphology. ASCs were grown on biomaterial PEUU (A-D), PEU (E-H) and PLLA (I-L) for 48 hours. Viability was detected by fluorescein diacetate (living cells; A, E, I) and ethidium-homodimer-1 (nuclei of dead cells; B, F, J) staining. C, G and K are merged pictures. Morphology was detected by TRITC-phalloidin (cytoskeleton, red) and DAPI (nuclei, blue) staining (D, H, L). Bars represent 100 µm (A-C, E-G, I-K) or 50 µm (D, H, L).

and naturally depleted by the surrounding tissue. However, due to the different linkage units in these polymers (ester (-CO-O-), urethane (-NH-CO-O-), and urea (-NH-CO-NH-) bonds) polymer degradation differs from that of PLA with respect to the different hydrolytic sensitivity of the corresponding linkages. Ester bonds are normally cleaved faster than urethane or urea ones. This can

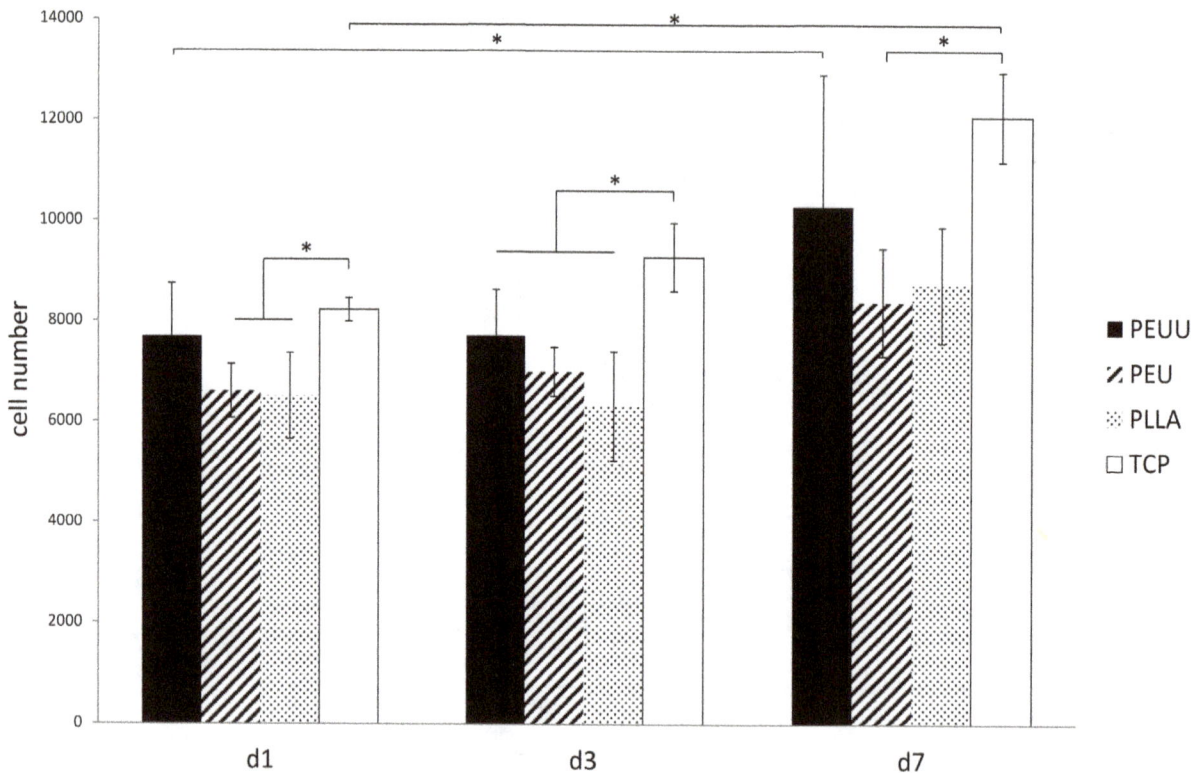

Figure 9. Cell viability of ASCs grown on biomaterials for one, three and seven days was detected by MTT Assay. One representative experiment out of four independent experiments is shown.

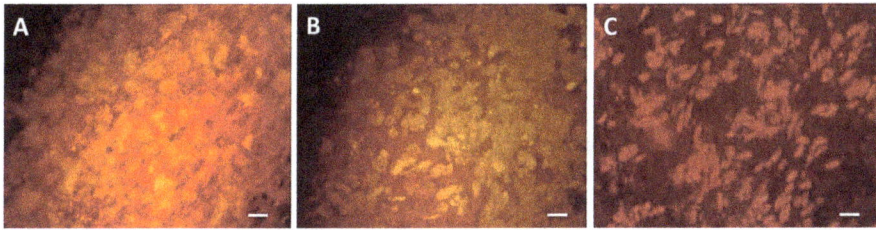

Figure 10. Adipogenic differentiation. ASCs cultivated on biomaterials PEUU (A), PEU (B) and PLLA (C) and differentiated for 21 days into an adipogenic lineage. Cells are stained with AdipoRed to show fat droplets. Bars represent 100 µm.

be also observed in the case of our polymers. Whereas during the initial degradation period of about 10 months degradation of PEU is only slightly diminished compared to PLLA which is used as a reference, PEUU degrades much slower. Varying the type of linkage between the different polymer segments may therefore represent an effective tool to tune the degradation behavior of scaffold materials. It may be an additional advantage that during the cleavage of urethanes or ureas no carboxylic acids are formed.

It is well known that not only the degradation profile but also the mechanical properties of PCL can be modified by blending or copolymerizing with other polyesters [61]. Its combination with polyurethanes or polyureas led to a higher elasticity of the material as we were able to show in the tensile strength tests. In regard to its use in soft tissue engineering a certain degree of elasticity is mandatory. Our investigations show that PEUU also led to a longer degradation time, which is eligible in soft tissue engineering, since differentiation of ASCs into mature adipose tissue takes several weeks and months. An advantage of the two new polymers is their solubility in volatile organic solvents (acetone, chloroform) well suited for electrospinning. Using this technique fibrous scaffold matrices become available possessing a structure similar to the native extracellular matrix of many tissues. The electrospun scaffolds are characterized by a high interconnective porosity and a high surface-volume ratio facilitating the transport of nutrients and waste products as well as cell adhesion and communication processes. Having this set of properties in mind, electrospun scaffolds of PEU and PEUU may also be of interest for other soft

tissue engineering applications like nerve injury repair, tendon/ligament reconstruction or vascular tissue engineering.

We could show that cell compatibility with adipose-derived stem cells was good as ASCs were able to adhere and proliferate on both new scaffolds. Morphology of the cells was slightly better on PEUU than on PEU showing a more physiological appearance. ASCs were also able to differentiate on both scaffolds which is mandatory for adipose tissue engineering. Cells were forming lipid droplets showing a physiological phenotypical appearance after 21 days. In addition gene analysis of 17 genes involved in adipocytes differentiation and lipid storage revealed no unwanted effects of the new materials on gene expression of differentiating ASCs. These include the peroxisome proliferator-activated receptor gamma (PPARgamma) known to be a master regulator of adipogenesis [62,63] and members of the C/EBP protein family also transactivating adipocytes specific genes [64]. Also perilipin A, which is supposed to prevent lipase access and therefore is important for the lipid storage of mature adipocytes [65] and the fatty acid binding protein (FABP4), which also contributes to the acquisition of the adipocytes phenotype [66] were expressed in a typical way.

Conclusion

In accordance with our initial hypothesis we found that the electrospun scaffold materials based on the two newly synthesized polymers PEU and PEUU cover a broad range of properties due

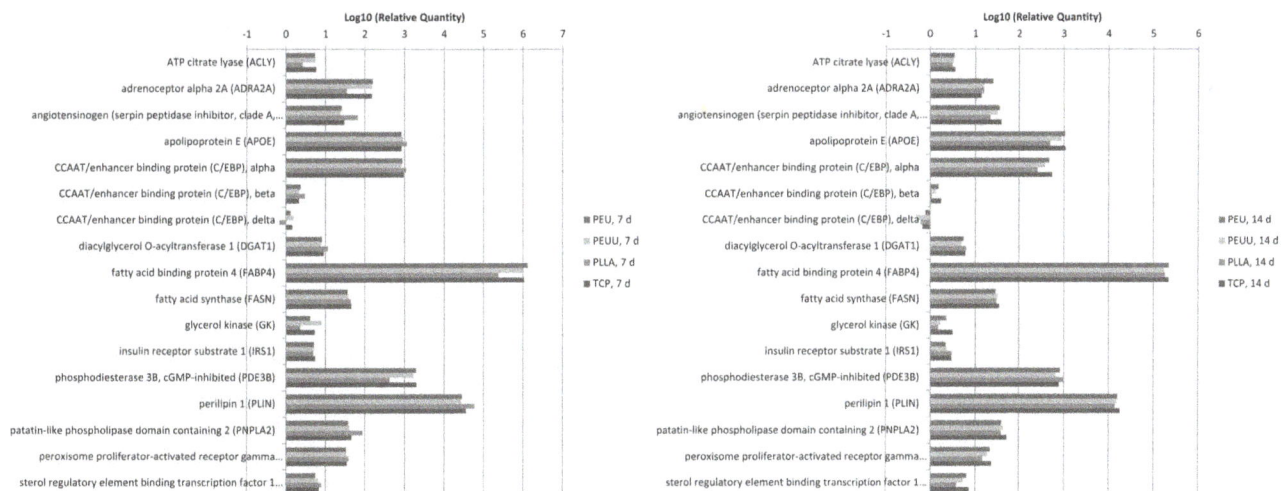

Figure 11. ASCs) were differentiated for seven and fourteen days on polymer-coated plates as well as on typical cell culture plates (TCP). Afterwards, expression of 17 genes associated with adipogenesis and lipogenesis was determined by qRT-PCR. All Ct values were normalized to GAPDH. Mean values of expression levels from 2 different donors are depicted relative to undifferentiated cells set as 1.

to their different chemical composition and structure. Especially with regard to their mechanical characteristics and their in vitro degradation behaviour the materials strongly differ from each other. PEUU electrospun scaffolds showed higher elasticity, which in our opinion is favorable in terms of soft tissue engineering. Concerning degradation the PEU scaffold disintegrated significantly faster in vitro than the PEUU material. ASCs were able to adhere, proliferate and differentiate on both polymeric scaffolds. Overall, we can conclude that both PEU and PEUU meshes can serve as a useful scaffold for adipose-derived stem cells and tissue engineering. Further investigations have to follow to verify the suitability of the tissue substitute in vivo.

Acknowledgments

Images were taken in cooperation with the Core Facility Imaging of the Medical University of Vienna.

Author Contributions

Conceived and designed the experiments: AG EG RW MS MK. Performed the experiments: AG JK TL TW SN EG SB. Analyzed the data: AG SN EG RW SB. Wrote the paper: AG SN EG MS MK SB.

References

1. Hemmrich K, von Heimburg D (2006) Biomaterials for adipose tissue engineering. Expert Review of Medical Devices 3: 635–645.
2. Ravichandran R, Sundarrajan S, Venugopal JR, Mukherjee S, Ramakrishna S (2012) Advances in Polymeric Systems for Tissue Engineering and Biomedical Applications. Macromolecular Bioscience 12: 286–311.
3. Rice JJ, Martino MM, De Laporte L, Totelli F, Briquez PS, et al. (2013) Engineering the Regenerative Microenvironment with Biomaterials. Advanced Healthcare Materials 2: 57–71.
4. Kumbar SG, James R, Nukavarapu SP, Laurencin CT (2008) Electrospun nanofiber scaffolds: engineering soft tissues. Biomedical Materials 3: 034002 (034015pp)
5. Böttcher-Haberzeth S, Biedermann T, Reichmann E (2010) Tissue engineering of skin. Burns : journal of the International Society for Burn Injuries 36: 450–460.
6. Keck M, Haluza D, Lumenta DB, Burjak S, Eisenbock B, et al. (2011) Construction of a multi-layer skin substitute: Simultaneous cultivation of keratinocytes and preadipocytes on a dermal template. Burns : journal of the International Society for Burn Injuries 37: 626–630.
7. Vermette M, Trottier V, Menard V, Saint-Pierre L, Roy A, et al. (2007) Production of a new tissue-engineered adipose substitute from human adipose-derived stromal cells. Biomaterials 28: 2850–2860.
8. Zhu J (2010) Bioactive modification of poly(ethylene glycol) hydrogels for tissue engineering. Biomaterials 31: 4639–4656.
9. Yu L, Ding J (2008) Injectable hydrogels as unique biomedical materials. Chem Soc Rev 37: 1473–1481.
10. Kluger PJ, Wyrwa R, Weisser J, Maierle J, Votteler M, et al. (2010) Electrospun poly(D/L-lactide-co-L-lactide) hybrid matrix: a novel scaffold material for soft tissue engineering. Journal of materials science Materials in medicine 21: 2665–2671.
11. Khil MS, Bhattarai SR, Kim HY, Kim SZ, Lee KH (2005) Novel fabricated matrix via electrospinning for tissue engineering. Journal of biomedical materials research Part B, Applied biomaterials 72: 117–124.
12. Ma Z, Kotaki M, Inai R, Ramakrishna S (2005) Potential of nanofiber matrix as tissue-engineering scaffolds. Tissue engineering 11: 101–109.
13. Riboldi SA, Sampaolesi M, Neuenschwander P, Cossu G, Mantero S (2005) Electrospun degradable polyesterurethane membranes: potential scaffolds for skeletal muscle tissue engineering. Biomaterials 26: 4606–4615.
14. Traurig MT, Permana PA, Nair S, Kobes S, Bogardus C, et al. (2006) Differential expression of matrix metalloproteinase 3 (MMP3) in preadipocytes/stromal vascular cells from nonobese nondiabetic versus obese nondiabetic Pima Indians. Diabetes 55: 3160–3165.
15. Fan W, Cheng K, Qin X, Narsinh KH, Wang S, et al. (2013) mTORC1 and mTORC2 Play Different Roles in the Functional Survival of Transplanted Adipose-Derived Stromal Cells in Hind Limb Ischemic Mice Via Regulating Inflammation In Vivo. Stem cells 31: 203–214.
16. Ashammakhi N, Ndreu A, Piras A, Nikkola L, Sindelar T, et al. (2006) Biodegradable nanomats produced by electrospinning: expanding multifunctionality and potential for tissue engineering. Journal of nanoscience and nanotechnology 6: 2693–2711.
17. Sill TJ, von Recum HA (2008) Electrospinning: Applications in drug delivery and tissue engineering. Biomaterials 29: 1989–2006.
18. Shenaq SM, Yuksel E (2002) New research in breast reconstruction: adipose tissue engineering. Clinics in plastic surgery 29: 111–125, vi.
19. Kral JG, Crandall DL (1999) Development of a human adipocyte synthetic polymer scaffold. Plastic and reconstructive surgery 104: 1732–1738.
20. Wendorff JH, Agarwal S, Greiner A (2012) Electrospinning. Weinheim: Wiley-VCH.
21. Gentleman E, Nauman EA, Livesay GA, Dee KC (2006) Collagen composite biomaterials resist contraction while allowing development of adipocytic soft tissue in vitro. Tissue engineering 12: 1639–1649.
22. Halbleib M, Skurk T, de Luca C, von Heimburg D, Hauner H (2003) Tissue engineering of white adipose tissue using hyaluronic acid-based scaffolds. I: in vitro differentiation of human adipocyte precursor cells on scaffolds. Biomaterials 24: 3125–3132.
23. Kawaguchi N, Toriyama K, Nicodemou-Lena E, Inou K, Torii S, et al. (1998) De novo adipogenesis in mice at the site of injection of basement membrane and basic fibroblast growth factor. Proceedings of the National Academy of Sciences of the United States of America 95: 1062–1066.
24. Schoeller T, Lille S, Wechselberger G, Otto A, Mowlavi A, et al. (2001) Histomorphologic and volumetric analysis of implanted autologous preadipocyte cultures suspended in fibrin glue: a potential new source for tissue augmentation. Aesthetic plastic surgery 25: 57–63.
25. Casadei A, Epis R, Ferroni L, Tocco I, Gardin C, et al. (2012) Adipose tissue regeneration: a state of the art. Journal of biomedicine & biotechnology 2012: 462543.
26. Young DA, Christman KL (2012) Injectable biomaterials for adipose tissue engineering. Biomedical materials 7: 024104.
27. Tanzi MC, Fare S (2009) Adipose tissue engineering: state of the art, recent advances and innovative approaches. Expert review of medical devices 6: 533–551.
28. Stride N, Larsen S, Hey-Mogensen M, Sander K, Lund JT, et al. (2013) Decreased mitochondrial oxidative phosphorylation capacity in the human heart with left ventricular systolic dysfunction. European journal of heart failure 15: 150–157.
29. Yoshimoto H, Shin YM, Terai H, Vacanti JP (2003) A biodegradable nanofiber scaffold by electrospinning and its potential for bone tissue engineering. Biomaterials 24: 2077–2082.
30. Ruwald MH, Hansen ML, Lamberts M, Hansen CM, Vinther M, et al. (2013) Prognosis among healthy individuals discharged with a primary diagnosis of syncope. Journal of the American College of Cardiology 61: 325–332.
31. Surrao DC, Waldman SD, Amsden BG (2012) Biomimetic poly(lactide) based fibrous scaffolds for ligament tissue engineering. Acta Biomater 8: 3997–4006.
32. Li WJ, Cooper JA, Jr., Mauck RL, Tuan RS (2006) Fabrication and characterization of six electrospun poly(alpha-hydroxy ester)-based fibrous scaffolds for tissue engineering applications. Acta Biomater 2: 377–385.
33. Casadei A, Epis R, Ferroni L, Tocco I, Gardin C, et al. (2012) Adipose tissue regeneration: a state of the art. J Biomed Biotechnol 2012: 462543.
34. Mun CH, Jung Y, Kim SH, Lee SH, Kim HC, et al. (2012) Three-dimensional electrospun poly(lactide-co-varepsilon-caprolactone) for small-diameter vascular grafts. Tissue Eng Part A 18: 1608–1616.
35. Chung S, Moghe AK, Montero GA, Kim SH, King MW (2009) Nanofibrous scaffolds electrospun from elastomeric biodegradable poly(L-lactide-co-epsilon-caprolactone) copolymer. Biomed Mater 4: 015019.
36. Mo XM, Xu CY, Kotaki M, Ramakrishna S (2004) Electrospun P(LLA-CL) nanofiber: a biomimetic extracellular matrix for smooth muscle cell and endothelial cell proliferation. Biomaterials 25: 1883–1890.
37. Jung Y, Lee SH, Kim SH, Lim JC, Kim SH (2013) Synthesis and characterization of the biodegradable and elastic terpolymer poly(glycolide-co-L-lactide-co--caprolactone) for mechano-active tissue engineering. J Biomater Sci Polym Ed 24: 386–397.
38. Guelcher SA (2008) Biodegradable polyurethanes: synthesis and applications in regenerative medicine. Tissue Eng Part B Rev 14: 3–17.
39. Gunatillake PA, Adhikari R (2003) Biodegradable synthetic polymers for tissue engineering. Eur Cell Mater 5: 1–16; discussion 16.
40. Bruin P, Smedinga J, Pennings AJ, Jonkman MF (1990) Biodegradable lysine diisocyanate-based poly(glycolide-co-epsilon-caprolactone)-urethane network in artificial skin. Biomaterials 11: 291–295.
41. Han J, Cao RW, Chen B, Ye L, Zhang AY, et al. (2011) Electrospinning and biocompatibility evaluation of biodegradable polyurethanes based on L-lysine diisocyanate and L-lysine chain extender. J Biomed Mater Res A 96: 705–714.
42. Tatai L, Moore TG, Adhikari R, Malherbe F, Jayasekara R, et al. (2007) Thermoplastic biodegradable polyurethanes: the effect of chain extender structure on properties and in-vitro degradation. Biomaterials 28: 5407–5417.
43. Zhang JY, Beckman EJ, Hu J, Yang GG, Agarwal S, et al. (2002) Synthesis, biodegradability, and biocompatibility of lysine diisocyanate-glucose polymers. Tissue Eng 8: 771–785.
44. Smahel J (1989) Experimental implantation of adipose tissue fragments. British journal of plastic surgery 42: 207–211.

45. von Heimburg D, Hemmrich K, Zachariah S, Staiger H, Pallua N (2005) Oxygen consumption in undifferentiated versus differentiated adipogenic mesenchymal precursor cells. Respiratory physiology & neurobiology 146: 107–116.

46. Shaodong C, Haihong Z, Manting L, Guohui L, Zhengxiao Z, et al. (2013) Research of influence and mechanism of combining exercise with diet control on a model of lipid metabolism rat induced by high fat diet. Lipids in health and disease 12: 21.

47. Kim WS, Park BS, Sung JH (2009) Protective role of adipose-derived stem cells and their soluble factors in photoaging. Archives of dermatological research 301: 329–336.

48. Kim YJ, Carvalho FC, Souza JA, Goncalves PC, Nogueira AV, et al. (2013) Topical application of the lectin Artin M accelerates wound healing in rat oral mucosa by enhancing TGF-beta and VEGF production. Wound repair and regeneration : official publication of the Wound Healing Society [and] the European Tissue Repair Society 21: 456–463.

49. Jeun JP, Kim YH, Lim YM, Choi JH, Jung CH, et al. (2007) Electrospinning of Poly(L-lactide-co-D,L-lactide). J Ind Eng Chem 13: 592–596.

50. Wu H, Li B, Wang X, Jin M, Wang G (2011) Inhibitory effect and possible mechanism of action of patchouli alcohol against influenza A (H2N2) virus. Molecules 16: 6489–6501.

51. Wacker SA, Kashyap S, Li X, Kapoor TM (2011) Examining the mechanism of action of a kinesin inhibitor using stable isotope labeled inhibitors for cross-linking (SILIC). Journal of the American Chemical Society 133: 12386–12389.

52. Park DJ, Han SK, Kim WK (2010) Is the foot elevation the optimal position for wound healing of a diabetic foot? Journal of plastic, reconstructive & aesthetic surgery : JPRAS 63: 561–564.

53. Kim K, Yu M, Zong X, Chiu J, Fang D, et al. (2003) Control of degradation rate and hydrophilicity in electrospun non-woven poly(D,L-lactide) nanofiber scaffolds for biomedical applications. Biomaterials 24: 4977–4985.

54. Mauney JR, Nguyen T, Gillen K, Kirker-Head C, Gimble JM, et al. (2007) Engineering adipose-like tissue in vitro and in vivo utilizing human bone marrow and adipose-derived mesenchymal stem cells with silk fibroin 3D scaffolds. Biomaterials 28: 5280–5290.

55. Shanti RM, Janjanin S, Li WJ, Nesti LJ, Mueller MB, et al. (2008) In vitro adipose tissue engineering using an electrospun nanofibrous scaffold. Annals of plastic surgery 61: 566–571.

56. Fu XB, Cheng B, Sheng ZY (2004) [Study in the novel function of adipose tissue on wound healing]. Zhongguo xiu fu chong jian wai ke za zhi = Zhongguo xiufu chongjian waike zazhi = Chinese journal of reparative and reconstructive surgery 18: 447–448.

57. Fu X, Fang L, Li H, Li X, Cheng B, et al. (2007) Adipose tissue extract enhances skin wound healing. Wound repair and regeneration : official publication of the Wound Healing Society [and] the European Tissue Repair Society 15: 540–548.

58. Ignatius AA, Claes LE (1996) In vitro biocompatibility of bioresorbable polymers: poly(L, DL-lactide) and poly(L-lactide-co-glycolide). Biomaterials 17: 831–839.

59. Cordewene FW, van Geffen MF, Joziasse CA, Schmitz JP, Bos RR, et al. (2000) Cytotoxicity of poly(96L/4D-lactide): the influence of degradation and sterilization. Biomaterials 21: 2433–2442.

60. Chen Q, Zhu C, Thouas GA (2012) Progress and challenges in biomaterials used for bone tissue engineering: bioactive glasses and elastomeric composites. Progress in Biomaterials 1.

61. Cohn D, Salomon AH (2005) Designing biodegradable multiblock PCL/PLA thermoplastic elastomers. Biomaterials 26: 2297–2305.

62. Chawla A, Schwarz EJ, Dimaculangan DD, Lazar MA (1994) Peroxisome proliferator-activated receptor (PPAR) gamma: adipose-predominant expression and induction early in adipocyte differentiation. Endocrinology 135: 798–800.

63. Tontonoz P, Hu E, Graves RA, Budavari AI, Spiegelman BM (1994) mPPAR gamma 2: tissue-specific regulator of an adipocyte enhancer. Genes & development 8: 1224–1234.

64. Rosen ED, Walkey CJ, Puigserver P, Spiegelman BM (2000) Transcriptional regulation of adipogenesis. Genes & development 14: 1293–1307.

65. Duncan RE, Ahmadian M, Jaworski K, Sarkadi-Nagy E, Sul HS (2007) Regulation of lipolysis in adipocytes. Annual review of nutrition 27: 79–101.

66. Gregoire FM, Smas CM, Sul HS (1998) Understanding adipocyte differentiation. Physiological reviews 78: 783–809.

The Effect of Gamma Irradiation on the Biological Properties of Intervertebral Disc Allografts: *In Vitro* and *In Vivo* Studies in a Beagle Model

Yu Ding[1], Dike Ruan[2]*, Keith D. K. Luk[3], Qing He[2], Chaofeng Wang[2]

1 Department of Rehabilitation Medicine and Pain Management Center, Navy General Hospital, Beijing, China, **2** Department of Orthopaedics, Navy General Hospital, Beijing, China, **3** Department of Orthopaedics and Traumatology, The University of Hong Kong, Pokfulam, Hong Kong, China

Abstract

Study Design: An animal experiment about intervertebral disc allograft.

Objective: To explore the feasibility to decellularize disc allografts treated by ^{60}Co Gamma Irradiation, and simultaneously, to assess the possibility to make use of the decellularized natural disc scaffold for disc degeneration biotherapy.

Summary of Background Data: Studies of both animal and human disc allograft transplantation indicated that the disc allograft may serve as a scaffold to undertake the physiological responsibility of the segment.

Methods: Experiment *in vitro*: 48 discs of beagles were harvested and divided randomly into four groups including a control group and three irradiated groups. Immediate cell viability and biomechanical properties of the discs were checked and comparisons were made among these groups. Experiment *in vivo*: 24 beagles accepted single-level allografted disc treated with different doses of gamma irradiation. Plain X-rays and MRIs were taken before and after surgery. Then, the spinal columns were harvested *en bloc* from the sacrificed beagles and were examined morphologically.

Results: There were significant differences of both the annulus fibrosus and nucleus pulposus immediate cell viabilities among the various groups. There were no obvious differences of the biomechanical properties among the four groups. The disc height and range of motion decreased significantly in all groups as time went on. The observed indexes in irradiated groups were much smaller than those in the control group, but the indexes in 18-kGy group were larger than those in 25-kGy and 50-kGy groups. Both MRI and macroscopic findings showed that the segmental degeneration in the control and 18-kGy group was less severe than that in 25-kGy and 50-kGy groups.

Conclusion: Gamma Irradiation can decellularize disc allograft successfully to provide natural scaffold for the study of degenerative disc disease therapy, and also can be used as an effective method to produce adjustable animal models.

Editor: Christof Markus Aegerter, University of Zurich, Switzerland

Funding: This work was supported by: 1. No. 30730095, the National Natural Science Foundation of China (www.nsfc.gov.cn); and 2. No. Z131107002213058, supported by the Beijing Municipal Science and Technology Commission (www.bjkw.gov.cn). The funders had no role in study design, data collection and analysis, decision to publish, or preparation of the manuscript.

Competing Interests: The authors have declared that no competing interests exist.

* Email: ruyi_2013@yeah.net

Background

Degenerative disc disease (DDD) is frequently seen in humans' during life and characterized by a multifaceted, chronic process leading to biologic and mechanical dysfunction. For the treatment, solid arthrodesis of DDD segments may bring about overload of neighboring discs causing adjacent segment degeneration (ASD) [1], [2]. Therefore, people in recent years have shown great interest in mobile prostheses to maintain stability and preserve motion of the functional spine unit (FSU) [3], [4]. Total disc allografting (TDA), as a natural mobile disc replacement, has brought about promising results in both animal studies and recent clinical trials [5–7]. However, the results showed that even though the functional spinal unit was stable and mobile, disc allograft may result in degeneration [5], [7]. This raises the research question of whether a decellularized allograft would better serve as a healthy scaffold for future biological and tissue engineered treatments.

DDD associated with the aging process is generally combined with the decrease of cell viability, loss of proteoglycan, and reduction of the ability to absorb shock between vertebrae [8]. The latest developments, though being limited in animal models, have led to promising novel approaches for the biotherapy of DDD, e. g. cell-based tissue engineering, gene therapy and the application of mesenchymal stem cells, etc [9]. Increased interest in TDA and the development in DDD biological treatments have put forward an urgent and significant need for a reliable natural scaffold for cultured cell transplantation. Compared with the degenerative autograft-disc and tissue engineering scaffold, the disc allograft has

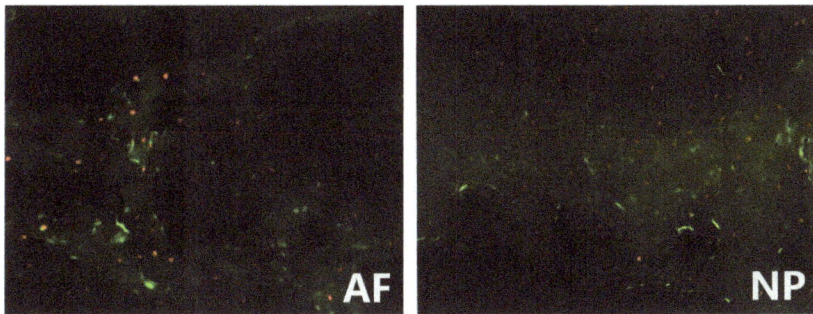

Figure 1. Disc cell viability detected by fluorescence microscope (200 times magnification). AF-annulus fibrosus, NP- nucleus pulposus. Under fluorescence microscope, the living cells were shown in green and the dead cells in orange. The cell survival rate = the number of green-stained cells/the total number of stained cells.

the theoretical advantage of providing a young, non-degenerated scaffold that could offer the best environment for the endogenous or exogenous cells to survive or regenerate [5]. Decellularized disc may provide an ideal environment for the host to culture annulus fibrosus (AF) and nucleus pulposus (NP) cells in three dimensions. Anyway, the difficulty that needs to be overcome is that the development and validation of animal model of disc degeneration continue to be a major limitation. Controllable, detectable and replicable disc degeneration will indisputably play an important role in the basic studies in clarifying and exploring the related mechanism of DDD.

Gamma Ray, which is commonly used to sterilize the allograft tissues, may be beneficial to DDD research work including disc graft decellularization and degenerative disc model preparation [10], [11]. With direct damage to the cell membrane and subsequent influence to the cellular structures, Gamma Irradiation

will similarly influence disc cell viability that may make the process of decellularization and degeneration controllable. Cell proliferation, matrix synthesis and metabolism may be altered pathologically [12], and these are probably the initiating factors of disc degeneration.

Studies of both animal and human IVD allograft transplantation indicated that the disc allograft may serve as a scaffold to undertake the physiological responsibility of the segment involved. Additionally, considering the beagle's characteristic such as super resistance to disease and moderate body size, we chose beagles as the experimental animal. In this study, we intended to explore the feasibility to decellularize disc allografts using ^{60}Co Gamma Irradiation, and meanwhile, to assess the possibility to create animal models of disc degeneration by means of in vivo disc allografting.

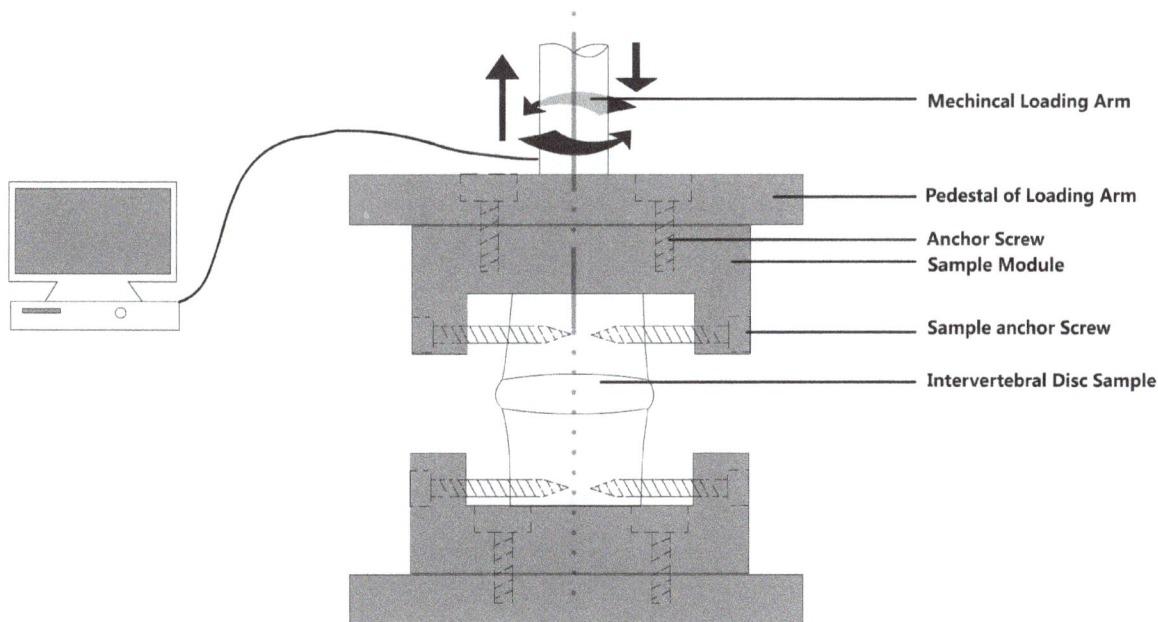

Biomechanical test of interverterbal disc

Figure 2. Biomechanical test for IVD. Discs were fixed to the fixture by special tapered screws penetrating vertebral bodies, and tested by Mechanical Testing & Simulation Machine (MTS 858 mini bionix-2, Eden Prairie).

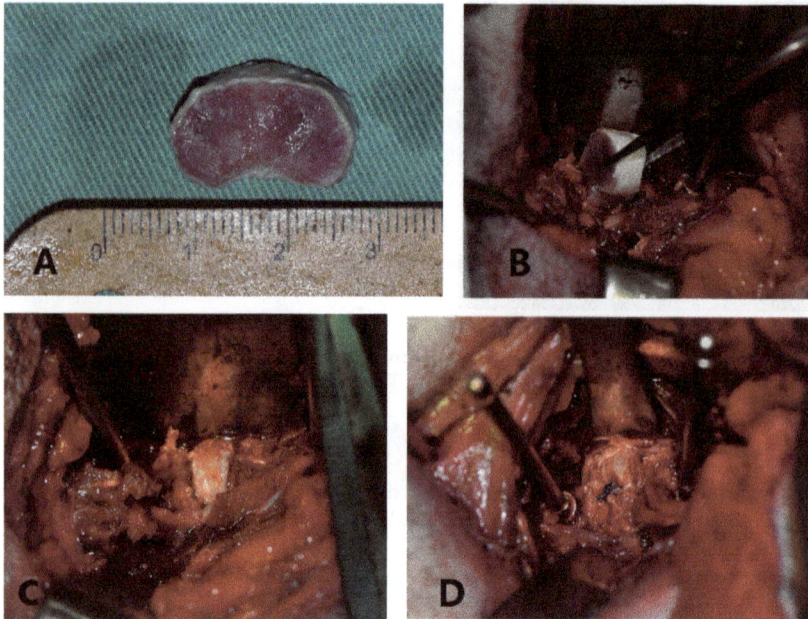

Figure 3. Surgical procedure for disc transplantation. A: disc allograft, B: transplantation, C: suitable position of the graft, D: anchoring the graft.

Materials and Methods

Thirty mature beagles (male 17, female 13) from Beijing Laboratory Animal Research Center were used in this study. The average age was 1 ± 0.17 years old, and the average weight was 10 ± 2.6 kilograms. The animal experiment protocol was approved by the Animal Ethical Committee, Navy General Hospital, PLA, China. All animal work has been conducted strictly according to the ethics guidelines. Each dog was placed in a separate cage with a relatively comfortable condition. Regular feeding and environment cleaning were done strictly in accordance with the regulations. The surgery was performed in the laboratory operating room. During surgical procedure, the beagles were generally anesthetized by intramuscular administration of Ketamin (0.1 ml/kg) and Sumianxin (compound agents consisted of xylidinothiazoline, ethylene diamine tetra-acetic acid, hydrochloric acid and haloperidol dihydroetorphine, 0.08 ml/kg). And then, Sumianxin was used continuously for 12 hours to minimize animal suffering-analgesia. During the procedure of harvesting lumbar columns, beagles were sacrificed via euthanasia with an overdose of Ketamin.

In vitro study

Disc Preparing and Freezing. Six beagles were sacrificed with an overdose of sodium pentobarbital (1 ml/kg) and the spinal columns were harvested *en bloc* from T6 to sacrum. The harvested

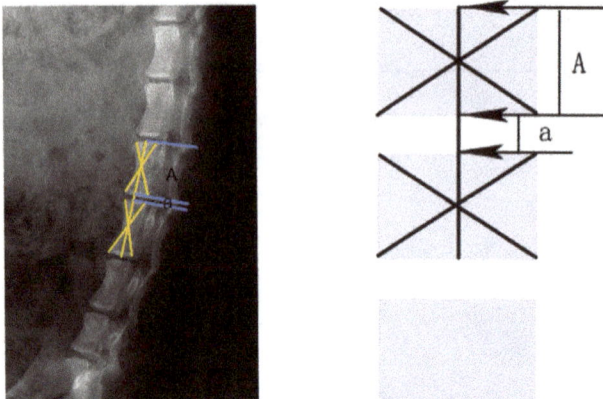

Figure 4. Radiographic measurement of DHI. Heights of the vertebral body and disc were measured using the middle vertebral line. DHI = a/A.

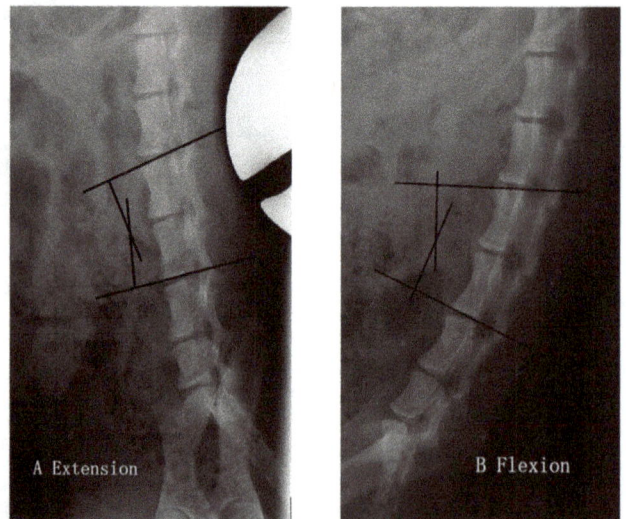

Figure 5. Method of measuring alignment using the cephalic endplate of the upper vertebrae and the caudal endplate of the lower vertebrae of the involved segment. ROM of the segment can be calculated by the difference between extension (A) and flexion (B) angles.

Table 1. Immediate cell survival rate in different groups (mean ± SD, %).

Position	control	18-kGy	25-kGy	50-kGy
AF	92.6±19.0	76.5±13.8*	50.7±12.2*	18.3±7.0**
NP	90.7±16.1	70.6±14.7*	46.9±10.5*	10.1±6.4**

Compared with the control group, all the three irradiated groups presented significantly decreased "cell viability" either in AF or NP (Wilcoxon test; *: P<0.05, **: P<0.01).

Table 2. Values of the biomechanical test (mean ± SD; axial force, N; torsional torque, N.mm).

Group	Tension	Compression	Right rotation	Left rotation
Control	52.6±25.7	72.4±39.6	734.8±409.1	676.5±332.9
18-kGy	49.2±26.0	71.6±41.4	713.1±472.6	775.3±491.2
25-kGy	53.1±35.0	67.0±30.5	791.0±557.7	666.1±374.8
50-kGy	47.7±28.6	68.2±37.6	763.4±473.9	684.5±483.3

discs from T6-L2 were used in the in vitro study, and discs from L2–L6 were preserved for the following in vivo study.

Experiment Design. The composite discs were divided randomly into four groups including a control group (discs with no irradiation) and three irradiating groups (irradiated with 18-kGy, 25-kGy or 50-kGy doses). The number of samples in each group was 12, of which six discs were chosen for cell viability checking, and another six ones were chosen for biomechanical testing.

Irradiation. Generally, 20-kGy irradiation was safe to all the kinds of allografts with satisfactoy mechanical properties [13]. To allow a safety margin, the International Atomic Energy Agency (IAEA) adopted 25-kGy as the standard irradiation dose for medical products [14], [15]. Based on the above understanding, discs in the −80°C frozen state were exposed to the ^{6}Co source in a commercial plant to accept the preset dose of irradiation, respectively (Cobalt Control Centre, Beijing University, China). Specimen bags were arranged to be rotated 180° front to back and top to bottom halfway through the irradiation period to ensure symmetrical irradiation of all grafts. Perspex dosimeters were placed adjacent to the samples to record the actual dose. Specimens were packed in dry ice all the time and stored at −80°C again after the irradiation. During the process, we took notice of two aspects in order to be sure that there was a right dose range to be given. First, the radiation dose was delivered with a

spread of 1–1.6 instead of a purely linear manner, e.g. to ensure that the dose of 25-kGy was achieved, a radiation dose range of 25~40 kGy should be proposed. Second, some evidence suggested that low temperature can influence the accuracy of the dose recording and consistently increases dosimeter readings 10~27%. So, the dosimeter was corrected with low temperature response curves accordingly.

Immediate Cell Survival Rate of the IVD after Irradiation. The discs for the immediate cell survival rate were carefully wrapped to prevent dehydration and transferred to the laboratory within half an hour. As a preliminary study, immediate cell survival rate was used to reflex indirectly the activity of the disc allograft. In fact, the living cells assessed immediately after the irradiation were not the actual disc cell viability. The gold-standard for assessing cell viability following radiation is a long-term clonogenic survival assay. The immediate cell survival rate of IVD was evaluated according to the detection of the cell's metabolic activity presented by fluorescence indicators [16]. The discs were sectioned into three to five 30-μm thick horizontal sections and placed in a solution containing 5 mg/L of Ethidium Bromide (EB) and 50 mg/L of Fluorescein Diacetate (FDA). The specimens were examined under fluorescence microscope (LEICA DMI6000 B) in order to observe the integrity of cell membranes. FDA was a kind of non-fluorescent compounds that can release fluorescence when penetrating living cells and show green stains

Table 3. DHI of the transplanted segment in various time points,

Group	Pre-op	1-month post-op	3-month post-op	6-month post-op
Control	0.2±0.09	0.16±0.05*	0.17±0.10*	0.12$^{\#}$
18-kGy Irradiation	0.19±0.07	0.15±0.06*	0.14±0.07*	0.07$^{\#}$
25-kGy Irradiation	0.19±0.06	0.12±0.09**	0.1±0.05**	fusion
50-kGy Irradiation	0.18±0.10	0.11±0.04**	0.08±0.02**	fusion

*: P<0.05;
**: P<0.01 (compared with pre-op, t test).
#: Because four beagles in each group were already sacrificed three-months postoperatively, so the average value of the two residual samples was not analyzed statistically.

Table 4. ROM of the transplanted segment in various time points (degree).

Group	Pre-op	1-month post-op	3-month post-op	6-month post-op
Control	6.3±2.4	4.6±1.9*	4.7±1.9*	2.3#
18-kGy Irradiation	5.5±2.7	3.7±1.0*	2.9±1.6**	1.6#
25-kGy Irradiation	5.8±2.8	3.5±1.3*	1.1±1.0**	fusion
50-kGy Irradiation	7.0±3.4	4.8±2.3*	0.9±0.8**	fusion

*: $P<0.05$;
**: $P<0.01$ (compared with pre-op, t test).
#: Because four beagles in each group were already sacrificed three-months postoperatively, so the average value of the two residual samples was not analyzed statistically.

under the excitation of blue light; whereas EB can only enter the defect in membrane of dead cells that showed orange fluorescent stains [17]. Figure 1 shows the signs of both living and dead disc cells under fluorescence microscope. With Image-Pro Plus Version 6.0 (network support), we can count the green and orange stains within the field of per 0.01 mm^2, and the cell survival rate was calculated as the number of green-stained cells/the total number of stained cells. Besides, the counting data were verified again by visual inspection and proofreading in order to minimize the probably error identification resulted from the "over automation". For example, the cells combined with both green and orange stains may be identified as dead cells by computer, but should be regarded as living cells because the fluorescence had entered the cells before they died. While counting, AF was divided into two parts, i.e. the lateral part and the inner annulus connecting with NP, and NP was treated as another integral part. From each part, we counted five visual fields and took the average value ultimately.

Biomechanical Testing of Disc Grafts. All the specimens were harvested, sealed and frozen in airtight bags at −80°C, and thawed at room temperature overnight before biomechanical testing. Disc allografts were fixed to the special fixture by tapered screws penetrating vertebral parts. Mechanical Testing & Simulation Machine (MTS 858 mini bionix-2, Eden Prairie) with specimen-holding apparatus and dynamic motion collecting system was used. The mechanical strength and elastic deformation of the disc allografts were measured for the evaluation of mechanical properties (Figure 2). First, the discs were imposed a vertical stretching or squeezing procedure in which the axial displacement was set to 1 mm. MTS gave the detail data of the tension or compression stress. Second, the discs were imposed left or right 5-degree rotation, and the torsional torques were measured.

In vivo study

Based on the pilot experiment, we set the desired confidential level as 95% and verified the sample size in each group. Twenty four skeletal mature beagles of 12 month years old were randomly, averagely divided into 4 groups: the control group (disc allografts with no irradiation), 18-kGy, 25-kGy, and 50-kGy irradiation group. Six L2-6 columns remaining in the first stage were disassembled into individual discs soon after the dissection and grouped according to the in vivo study design.

Surgical Technique. The beagles were generally anesthetized by intramuscular administration of Ketamin (0.1 ml/kg) and Sumianxin (0.08 ml/kg). Under aseptic conditions, the lower lumbar segments were exposed through a left retroperitoneal approach. After clearly identifying the intervertebral foramen and the posterior margin of the index IVD, osteotomy was performed at the endplates approximately 1.5 mm above and below the segment. After removal of the disc, a disc allograft of the most compatible size was selected and trimmed according to the transplanted segmental space and then positioned into the slot of the excised disc (Figure 3).

Radiologic Assessment. The beagles were examined by X-ray (AP view, lateral view and dynamic flexion-extension radiographs) and MRI before the transplantation to exclude physical abnormalities. After the surgery, the beagles were examined regularly at the time points (i. e., 0, 1, 3 and 6 months postoperatively). Disc heights and sagittal Cobb angles of the segments were measured on the digital radiograph system with the built-in software (Magic View Tools, Siemens AG, Erlangen, Germany). Intervertebral space was determined by disc height index (DHI, Figure 4) [18]. The range of motion (ROM) of the segment was calculated by the absolute value of the Cobb angles' difference displayed on the dynamic radiographs (Figure 5).

Table 5. Relative greyscales$^\triangle$ of the transplanted discs in various time points.

Group	Immediately post-op	1-month post-op	3-month post-op	6-month post-op
Control	0.96±0.59	1.00±0.52	1.11±0.86	0.45#
18-kGy Irradiation	0.96±0.78	0.88±0.69	0.91±0.65	0.52#
25-kGy Irradiation	1.02±0.67	0.87±0.51*	0.66±0.54**	fusion
50-kGy Irradiation	1.12±0.73	0.79±0.40**	0.64±0.38**	fusion

$^\triangle$Relative greyscale = transplanted disc's greyscale/mean greyscale of the adjacent discs.
*: $P<0.05$;
**: $P<0.01$ (compared with immediately post-op, t test).
#: Because four beagles in each group were already sacrificed three-months postoperatively, so the average value of the two residual samples was not analyzed statistically.

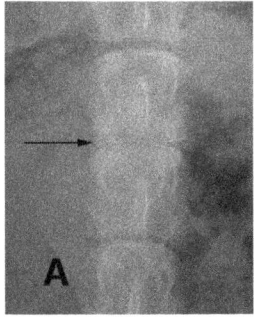

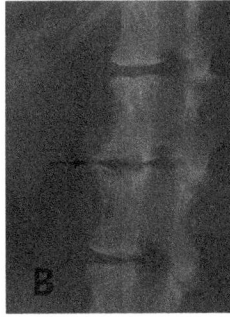

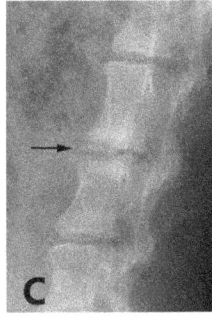

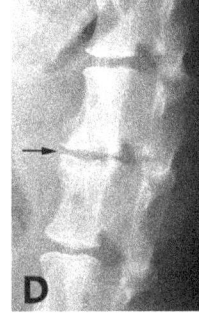

0 - KGy irradition group

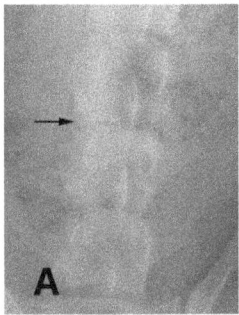

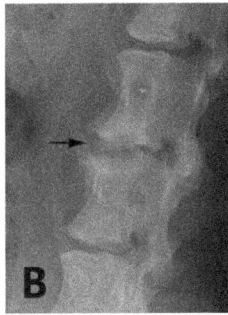

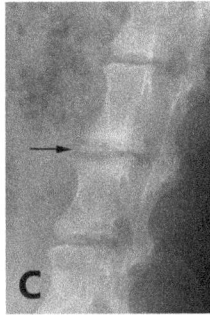

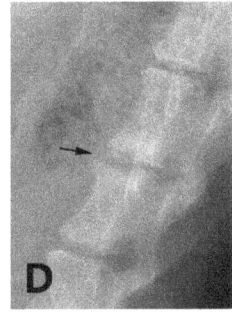

18 - KGy irradition group

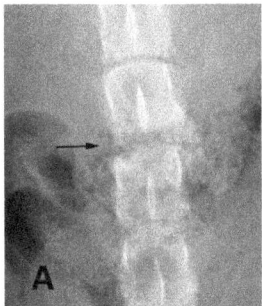

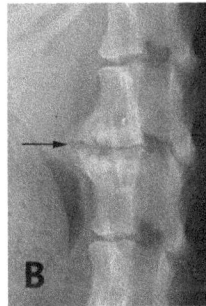

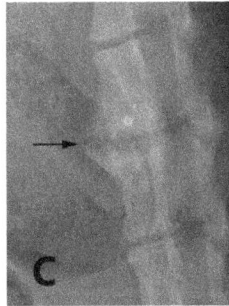

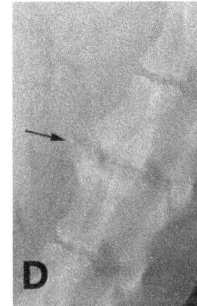

25 - KGy irradition group

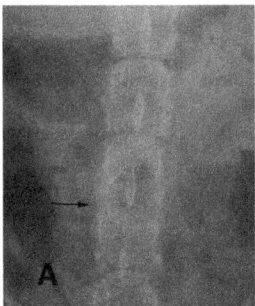

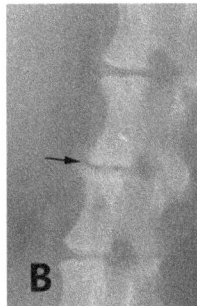

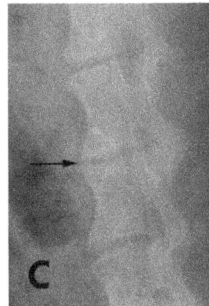

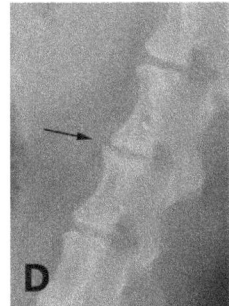

50 - KGy irradition group

Figure 6. X-rays of the disc allografts in different groups (three months postoperatively). A: AP view, B: lateral view, C: extention-lateral view, D: flexion-lateral view.

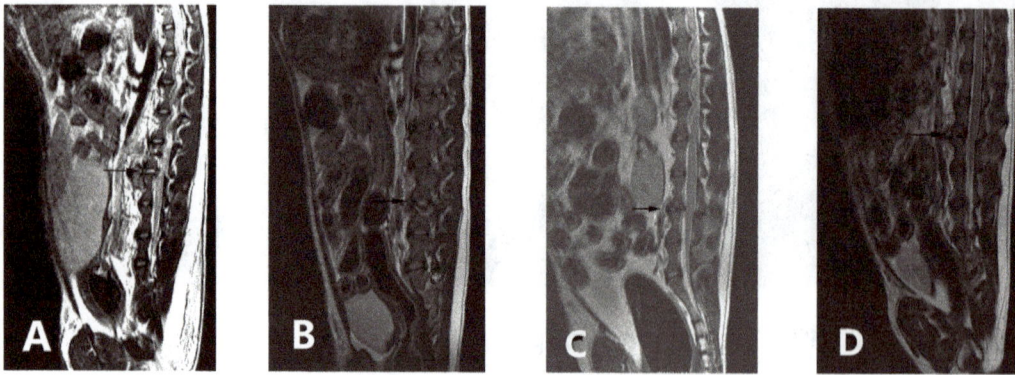

Figure 7. MRI of the disc allografts in different groups (three months postoperatively).

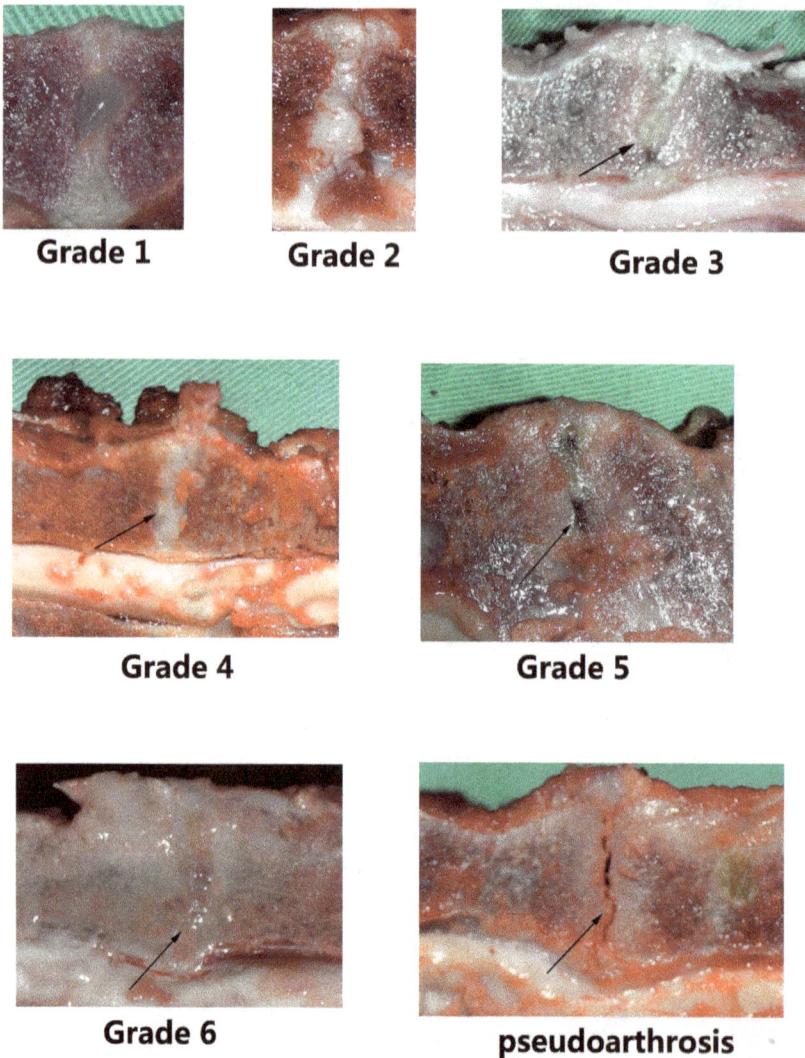

Figure 8. Representative photographs of the gross morphology of Beagle's grafted discs. Grade 1 was a normal disc with a bulging NP and clear demarcation between AF and NP. Grade 2 was the mild disc degeneration which had fibrous tissue or mucinous material peripherally. However, the transplanted disc did not display Grade 1 or Grade 2 degeneration in this study. Grade 3 presented slight narrowing of disc height and unclear demarcation between AF and NP. Grade 4 had consolidated fibrous tissue in the segment, loss of AF-NP demarcation, and early chondrophytes. Grade 5 presented horizontal clefts in NP, disruptions in AF, irregularities in the endplates, usually with small osteophytes. Grade 6 showed no visible NP, destruction of the endplate, and scar tissue, usually with larger osteophytes. Pseudoarthrosis was also regarded as Grade 6 disc degeneration.

Table 6. Classified transplanted disc degeneration at 3-month postoperatively (n).

Group	Grade III	Grade IV	Grade V	Grade VI
Control	3	3	0	0
18-kGy Irradiation	0	3	3	0
25-kGy Irradiation	0	0	2	4
50-kGy Irradiation	0	0	1	5

Correlation analysis showed that disc degeneration related significantly to the dose of irradiation (X^2 test, P<0.01).

Hydration status of the transplanted disc was evaluated on MRI scans with a modified Schneiderman's score [5]. Grayscale measurement based on MRI T2 scan was used to evaluate the extent of the disc degeneration. Higher value of disc grayscale indicated relatively less degeneration with more water content in the disc. In this study, the "relative grayscale" was calculated to normalize the graft's grayscale against that of adjacent discs for the purpose of minimizing the individual error caused by different MRI scans.

Macroscopic Findings and Morphological Evaluation. The spine columns from L2 to S1 were harvested *en bloc* from the sacrificed beagles at the final study time. Four beagles in each group were sacrificed for disc harvesting three months after the transplantation, and the residual two beagles were sacrificed six months postoperatively. The treated segment was examined morphologically to see if there's any instability or olisthesis. Then, the transplanted discs were split through the sagittal or coronal plane for morphological images to validate the radiologic findings.

Statistics

Statistical analysis was performed using SPSS 16.0 software. Analysis of variance (ANOVA) was used for hypothesis-testing of variable differences in multiple groups. Wilcoxon signed rank test was chosen for comparing two paired groups. Paired t test was used for comparisons of intra-group indexes. X^2 test was used for the correlation analysis between different variables. P-values less than 0.05 were considered significant.

Results

In vitro study

The significant decline in both AF and NP cell viability could be seen with the increase of the irradiated dose (P<0.05, Table 1). The biomechanical properties of the discs in the three irradiated groups did not change dramatically compared with those in the control group (P>0.05, Table 2).

Compared with the control group, there were no significant differences of the four kinds of biomechanical values in all the irradiated groups (Wilcoxon test; P>0.05).

In vivo study

Two weeks after the surgery, all the beagles recovered and were in good health status. Anyway, we observed the phenomenon that two grafted discs presented dissolution and absorption soon after the transplantation combining with the fracture of endplate. Another two beagles were supplemented and careful transplantation technique was emphasized to avoid the adverse event. Table 3 shows DHI values. Compared with DHI values before surgery, the significant decrease was seen in all the four groups postoperatively. There was a greater decrease both in 25-kGy and 50-kGy irradiation group. Besides, it showed that 25-kGy and 50-kGy

irradiation group had further DHI decrease at 3-month postoperatively. For the two residual beagles in 25-kGy and 50-kGy irradiation group at 6-month postoperatively, they presented the collapse of the intervertebral space with the transplanted disc disappeared and bone fusion.

Table 4 shows ROM values. The ROM values in all groups decreased significantly at 1-month postoperatively, and the irradiated groups had a further ROM decease at 3-month study. At 6-month time point, both 25-kGy and 50-kGy irradiation group had the severest degeneration with the collapse of the intervertebral space and segment bone fusion.

Table 5 shows the relative grayscales of the grafts. In the control and 18-kGy irradiation group, the disc signal could maintain at a consistent level within 1–3months postoperatively, but decreased at 6 months after transplantation. In the 25-kGy and 50-kGy irradiation group, disc signals decreased persistently at all timepoints.

Figure 6 and Figure 7 present X-ray and MRI findings of the allografted discs in the four groups. Based on Hoogendoorn's study [19], we adopted the modified disc degeneration system and classified various grades of degenerative discs in the gross morphology (Figure 8). Table 6 shows the classified transplanted disc degeneration in the four groups at 3-month postoperatively. It indicated that the extent of disc degeneration related significantly to the doses of irradiation. More irradiation being exposed, severer degeneration disc grafts presented. This confirmed the good consistency between the macroscopic and radiologic findings.

Discussion

Despite poor understanding about the relationship between IVD degeneration and clinical symptoms, many researchers worldwide are seeking biologic ways to repair degenerated discs [20]. Stem cell, NP cell and AF cell transplantations have been suggested as complementary or optional methods for the treatment of disc degeneration in recent years [21]. Differentiated cells may function more effectively in disc-mimetic conditions, as evidenced by increased cell viability, glycosaminoglycan production, and persistence of several gene expressions [22]. When evaluating the fate of cells post-implantation, the environment-dependent differences are significant and the necessity of mimicking physiological disc environment is highlighted. However, the frequently used scaffolds were degenerative autograft-disc or tissue engineered, which tended to provide a poor environment for cell seeding and proliferation. The decellularized allograft disc used as an entire, young and natural scaffold has not been noted formerly in the literature. Comparing with synthetic and polymer scaffold, the natural allograft disc has obvious advantages of structure and mechanical similarities to that of normal disc, and can be further functionalized via implanted cell proliferation, gene expression and tissue engineering reconstruction, etc. Thus, the idea of developing a natural scaffold from the allograft disc model will provide a unique and significant approach for further studies in treating DDD.

Cell transplantation is a new therapy that is based on the supplementation of matrix-producing cells [23]. Animal studies have showed promising results that imply the possibility of slowing down the process of degeneration or regenerate IVD tissues [24-26]. Although there occurred more or less degeneration of the segment receiving disc transplantation in all the four groups, it was undoubted that in vivo disc allografts can perform most of the physiological functions as expected. Treating discs with proper doses of Gamma Irradiation, we can take advantage of the IVD scaffold for the further study to explain some phenomenon regarding the mechanism of disc degeneration. As another perspective research direction, we could take advantage of the hypocellular disc for allografting experiment in order to verify the necessity of cells in preventing or postponing the progress of disc degeneration. The prevention of disc degeneration and the stimulation of the biological disc repair process will create a new category of therapy for DDD [20], [27]. A variety of therapeutic strategies for the biological treatment, aiming at the prevention of disc degeneration and/or the active stimulation of the repair process of IVD degeneration, might be applied. In all the above studies, decellularized disc grafts will be beneficial to reveal the exact changes in cell morphology specific to IVD degeneration.

The measure of the percentage of cell viability depends on the number of both the live and dead cells. With the storage method we used, both AF and NP can preserve over 90% living cells. After irradiation, the percentage decreased significantly. A significantly lower number and percentage of viable cells were found in the discs accepting higher doses of irradiation. Though the former study demonstrated that the moderate absorbed dose of irradiation was preferable for bone grafts, it is still unclear whether the disc cell viability in disc allografts is vital for sustaining the integrant bio-capability [13], [28]. In this study, three different doses of Gamma Rays, namely 18 kGy, 25 kGy and 50 kGy, were tested to irradiate beagle intervertebral disc allografts. It showed that all the fresh disc allografts survived the irradiation and transplantation and revealed a capacity of physiological activities. After implanted in vivo, transplanted segments had no signs of olisthesis and instability but showed different degrees of degeneration which was confirmed by both radiographic findings and morphological evaluation. The fresh frozen allograft group and 18-kGy irradiation group were more likely to present a slower degeneration process. Anyway, the immediate cell survival rate measurement after irradiation is not reflective of the cell death that will occur in the IVD grafts. Besides the direct damage to the cells by the free radicals, Gamma irradiation also provides the alteration of nucleic acids leading to dysfunction and destruction of the genome, which may exhibit the abnormality in the following animal experiments. We had no data about the actual cell viability of the discs for in vivo allografting, thus can't get the definite correlation between the survival rate of disc cells and the degree of disc degeneration.

Storage at the lower temperature can be regarded as a practical method for preserving disc allografts [7], [29]. Also, it is generally recognized that the biological property of allografts under irradiation is temperature-dependent. Hamer et al reported that irradiation had different effects on the plastic properties of bone and on the degree of collagen denaturation at different temperatures [30]. Gibbons et al indicated that the low temperature was prone to protect the initial mechanical properties of the bone-

patella tondon-bone unit and the tendon mid substance under 20-kGy Gamma Irradiation [29]. The composite bone-disc-bone IVD allograft is most likely destroyed by free radicals generated from water molecules, of which the mobility can be reduced by freezing and therefore the reaction of the highly-reactive oxygen free radicals are decreased.

Many tissue banks treat musculoskeletal allograft tissues with additional Gamma Irradiation in an attempt to reduce the risk of bacterial contamination and provide an additional layer of safety [31]. Radiation sterilization guarantees penetration throughout the allograft tissue and is relatively economical. To ensure a safer clinical use of the tissue allograft, Gamma Irradiation may be desirable to provide ideal sterilization targeting virus or bacteria within the allograft tissue. Establishing the appropriate irradiation dose is critical. According to the literature, both the reduction of allograft properties and inactivation of viruses and bacteria spores were dose-dependent of radiation, but the possibly proper doses reported were inconsistent [13], [32]. Some researchers suggested $15\sim20$ kGy be used because they concerned that doses of 30-kGy or greater might alter allograft mechanical properties, while the International Atomic Energy Agency adopted 25-kGy as the minimum irradiation dose for medical products [30], [33]. During the process of treating allografts, we are faced with the challenge to strike a balance during the irradiation process between achieving the designed objectives and protecting the material properties of the allograft which may otherwise result in premature transplantation failure.

This is the preliminary study focuses mainly on the feasibility of Gamma Irradiation in decellularizing and sterilizing disc allografts. Bony union between the endplate and vertebral body indicated that less than 50-kGy irradiation did no harm to the subchondral bone and endplate cartilage of the disc composite. Disc height and ROM combining with MRI disc signs can reflect intuitively the survival process of the in vivo allografts. The following macroscopic findings verified the radiographic results and this consistency implied the accuracy and rationality of disc degeneration assessment. We get the conclusion that disc allograft treated by Gamma Irradiation can be used in the animal model of disc degeneration or decellularization. This may be helpful for clarifying the mechanism of disc degeneration and delaying the aging process of human's disc. Despite the fact that appropriate doses of Gamma Irradiation may be beneficial to the disc sterilization, lots of detailed research work (e.g., pathogen inactivation assessment and long-term disc cell damage repair evaluation) should be done before the possibility of disc-banking is identified.

Acknowledgments

We wish to acknowledge Xu SX, Hou LS, Lin JN, Zhang Y, Wu JH and Shi ZY for their technical assistance in preparing the manuscript.

Author Contributions

Conceived and designed the experiments: DR KDKL. Performed the experiments: YD. Analyzed the data: CW. Contributed reagents/materials/analysis tools: QH. Wrote the paper: YD.

References

1. Xia XP, Chen HL, Cheng HB (2013) Prevalence of adjacent segment degeneration after spine surgery: a systematic review and meta-analysis. Spine 38: 597–608. doi: 10.1097/BRS.0b013e318273a2ea

2. Bertagnoli R, Yue JJ, Fenk-Mayer A, Eerulkar J, Emerson JW (2006) Treatment of symptomatic adjacent-segment degeneration after lumbar fusion with total disc arthroplasty by using the prodisc prosthesis: a prospective study with 2-year minimum follow up. Spine 4: 91–97.

3. Ghiselli J, Wang JC, Bhatia NT, Hsu WK, Dawson EG (2004) Adjacent segment degeneration in the lumbar spine. J Bone Joint Surg. Am 86: 1497–1503.

4. Röhl K, Röhrich F (2009) Artificial disc versus spinal fusion in the treatment of cervical spine degenerations in tetraplegics: a comparison of clinical results. Spinal Cord 47: 705–708. doi: 10.1038/sc.2009.31.

5. Ruan DK, Qin H, Ding Y, Hou L, Li J et al. (2007) Intervertebral disc transplantation in the treatment of degenerative spine disease: a preliminary study. The Lancet 369: 993–999.

6. Matsuzaki H, Wakabayashi K, Ishihara K, Ishikawa H, Ohkawa A (1996) Allografting intervertebral discs in dogs: a possible clinical application. Spine 21: 178–183.

7. Luk KD, Ruan DK, Chow DH, Leong JC (1997) Intervertebral disc autografting in a bipedal animal model. Clin Orthop 337: 13–26.

8. Taher F, Essig D, Lebl DR, Hughes AP, Sama AA, et al. (2012) Lumbar degenerative disc disease: current and future concepts of diagnosis and management. Adv Orthop 2: 970752. doi: 10.1155/2012/970752

9. Fassett DR, Kurd MF, Vaccaro AR (2009) Biologic solutions for degenerative disc disease. J Spinal Disord Tech 22: 297–308. doi: 10.1097/BSD.0b013e31816d5f64

10. Singh VA, Nagalingam J, Saad M, Pailoor J (2010) Which is the best method of sterilization of tumour bone for reimplantation? A biomechanical and histopathological study. Biomed Eng Online 9: 48–52. doi: 10.1186/1475-925X-9-48

11. Grieb TA, Forng RY, Bogdansky S, Ronholdt C, Parks B, et al. (2006) High-dose Gamma Irradiation for soft tissue allografts: high margin of safety with biomechanical integrity. J Orthop Res 24: 1011–1018.

12. Murrell W, Sanford E, Anderberg L, Cavanagh B, Mackay-Sim A (2009) Olfactory stem cells can be induced to express chondrogenic phenotype in a rat intervertebral disc injury model. Spine J 9: 585–594. doi: 10.1016/j.spinee.2009.02.011

13. Balsly CR, Cotter AT, Williams LA, Gaskins BD, Moore MA, et al. (2008) Effect of low dose and moderate dose gamma irradiation on the mechanical properties of bone and soft tissue allografts. Cell Tissue Bank 9: 289–298. doi: 10.1007/s10561-008-9069-0

14. Nguyen H, Morgan DA, Forwood MR (2007) Sterilization of allograft bone: is 25 kGy the gold standard for gamma irradiation? Cell Tissue Bank 8: 81–91.

15. Huq MS, Andreo P, Song H (2001) Comparison of the IAEA TRS-398 and AAPM TG-51 absorbed dose to water protocols in the dosimetry of high-energy photon and electron beams. Phys Med Biol 46: 2985–3006.

16. Ehmed SA, Gogal RM Jr, Walsh JE (1994) A new rapid and simple non-radioactive assay to monitor and determine the proliferation of lymphocytes: an alternative to [^{3}H] thymidine incorporation assay. J Immuno Methods 170: 211–214.

17. Qi AQ, Qian KX, Shao JZ (2000) A fluorometry method to determine cell viability by double staining with fluorescein diacetate and propidium iodide. Chinese Journal of Cell Biology 22: 50–53.

18. Kim KT, Park SW, Kim YB (2009) Disc height and segmental motion as risk factors for recurrent lumbar disc herniation. Spine, 34(24): 2674–2678. doi: 10.1097/BRS.0b013e3181b4aaac

19. Hoogendoorn RJ, Wuisman PL, Smit TH, Everts VE, Helder MN (2007) Experimental intervertebral disc degeneration induced by Chondroitinase ABC in the goat. Spine 32: 1816–1825.

20. Aslan H, Sheyn D, Gazit D (2009) Genetically engineered mesenchymal stem cells: applications in spine therapy. Regen Med 4: 99–108. doi: 10.2217/17460751.4.1.99

21. Henriksson HB, Svanvik T, Jonsson M, Hagman M, Horn M, et al. (2009) Transplantation of Human Mesenchymal Stems Cells into Intervertebral Discs in a Xenogeneic Porcine Model. Spine 34: 141–148. doi: 10.1097/BRS.0b013e31818f8c20

22. Kroeber MW, Unglaub F, Wang H, Schmid C, Thomsen M, et al. (2002) New in vivo animal model to create intervertebral disc degeneration and to investigate the effects of therapeutic strategies to stimulate disc regeneration. Spine 27: 2684–2690.

23. Cho H, Park SH, Lee S, Kang M, Hasty KA, et al. (2011) Snapshot of degenerative aging of porcine intervertebral disc: a model to unravel the molecular mechanisms. Exp Mol Med. 43 (6): 334–340.

24. Chiang CJ, Cheng CK, Sun JS, Liao CJ, Wang YH, et al. (2011) The effect of a new anular repair after discectomy in intervertebral disc degeneration: an experimental study using a porcine spine model. Spine 36: 761–769. doi: 10.1097/BRS.0b013e3181e08f01

25. Zeiter S, Bishop N, Ito K (2005) Significance of the mechanical environment during regeneration of the intervertebral disc. Eur Spine J 14: 874–879.

26. Ruan DK, Xin H, Zhang C, Wang C, Xu C, et al. (2010) Experimental intervertebral disc regeneration with tissue-engineered composite in a canine model. Tissue Eng Part A 16: 2381–2389. doi: 10.1089/ten.TEA.2009.0770

27. Sakai D (2008) Future perspectives of cell-based therapy for intervertebral disc disease. Eur Spine J 17: 452–458. doi: 10.1007/s00586-008-0743-5

28. Kunnev D, Rusiniak ME, Kudla A, Freeland A, Cady GK, et al. (2010) DNA damage response and tumorigenesis in Mcm2-deficient mice. Oncogene 29: 3630–3638. doi: 10.1038/onc.2010.125

29. Gibbons MJ, Butler DL, Grood ES, Bylski-Austrow DI, Levy MS, et al. (1991) Effects of Gamma Irradiation on the initial mechanical and material properties of goat bone-patellar tendon-bone allografts. J Orthopaedic Research 9: 209–218.

30. Hamer AJ, Stockley I, Elson RA (1999) Changes in allograft bone irradiated at different temperatures. J Bone and Joint Surgery 81-B: 342–344.

31. Gamero EC, Morales Pedraza J (2009) The impact of the International Atomic Energy Agency (IAEA) program on radiation and tissue banking in Peru. Cell Tissue Bank 10: 167–171. doi: 10.1007/s10561-008-9093-0

32. Akkus O, Belaney RM (2005) Sterilization by gamma radiation impairs the tensile fatigue life of cortical bone by two orders of magnitude. J Orthop Res 23: 1054–1058.

33. Yang YC, Nie L, Cheng L, Hou Y (2009) Clinical and radiographic reports following cervical arthroplasty: a 24-month study. Int Orthop 33: 1037–1042. doi: 10.1007/s00264-008-0571-6

Evaluation of Physical and Mechanical Properties of Porous Poly (Ethylene Glycol)-co-(L-Lactic Acid) Hydrogels during Degradation

Yu-Chieh Chiu[1], Sevi Kocagöz[1], Jeffery C. Larson[1,2], Eric M. Brey[1,2]*

1 Department of Biomedical Engineering, Illinois Institute of Technology, Chicago, Illinois, United States of America, 2 Research Service, Hines Veterans Administration Hospital, Hines, Illinois, United States of America

Abstract

Porous hydrogels of poly(ethylene glycol) (PEG) have been shown to facilitate vascularized tissue formation. However, PEG hydrogels exhibit limited degradation under physiological conditions which hinders their ultimate applicability for tissue engineering therapies. Introduction of poly($_L$-lactic acid) (PLLA) chains into the PEG backbone results in copolymers that exhibit degradation via hydrolysis that can be controlled, in part, by the copolymer conditions. In this study, porous, PEG-PLLA hydrogels were generated by solvent casting/particulate leaching and photopolymerization. The influence of polymer conditions on hydrogel architecture, degradation and mechanical properties was investigated. Autofluorescence exhibited by the hydrogels allowed for three-dimensional, non-destructive monitoring of hydrogel structure under fully swelled conditions. The initial pore size depended on particulate size but not polymer concentration, while degradation time was dependent on polymer concentration. Compressive modulus was a function of polymer concentration and decreased as the hydrogels degraded. Interestingly, pore size did not vary during degradation contrary to what has been observed in other polymer systems. These results provide a technique for generating porous, degradable PEG-PLLA hydrogels and insight into how the degradation, structure, and mechanical properties depend on synthesis conditions.

Editor: Mário A. Barbosa, Instituto de Engenharia Biomédica, University of Porto, Portugal

Funding: The authors would like to acknowledge support from the Veterans Administration and the National Science Foundation (IIS 1125412 and CBET 0854430). The funders had no role in study design, data collection and analysis, decision to publish, or preparation of the manuscript.

Competing Interests: The authors have declared that no competing interests exist.

* E-mail: brey@iit.edu

Introduction

Hydrogels have been investigated extensively for tissue engineering applications primarily due to mechanical properties of similar magnitude to many soft tissues [1], [2]. Poly (ethylene glycol) (PEG)-based hydrogels have received significant attention due to their biocompatibility, the relatively straightforward options for incorporation of peptide adhesion sequences [3] and growth factors, [4] and the ability to control mechanical properties based on polymerization conditions. While significant amounts of research have shown how modifications in the chemical composition of PEG hydrogels can modulate biological response, there has been little research into the role of the physical architecture.

Porous structure of biomaterials has been shown to play a role in regulating cell response and tissue integration. However, unlike ECM where cells can often migrate through pores between the solid structure, the cross-links in PEG hydrogels must be cleaved to enable cells to migrate and tissue to invade [1,8]. Introduction of pores into PEG hydrogels could be used to improve biological response and lead to improved outcomes in biomedical applications. The introduction of pores not only provides more space for cell migration and tissue invasion but also increases the surface area to volume ratio which can enhance cell seeding [5] and enable more efficient mass transport. [6] A number of strategies have been employed to generate porous materials, including gas foaming [7], polymer-polymer immiscibility [8], and particulate leaching [9]. These techniques are most commonly applied to hydrophobic polymer scaffolds, but recent studies have demonstrated that hydrogels with an interconnected porous structure can be produced using particulate leaching techniques [10,11]. PEG hydrogels generated with this technique support the formation of vascularized tissue *in vivo* in a pore size dependent manner [1]. While these materials have shown promise for tissue engineering, PEG hydrogels do not exhibit significant degradation *in vivo*.

PEG hydrogels can be made degradable under physiologic conditions through the inclusion of hydrolysable monomer units [12] or peptide sequences that are degraded by cell enzymes [12,13]. While enzymatically degradable PEG hydrogels are popular as ECM mimics, they restrict cell migration and tissue invasion to enzymatic processes and can result in significant intrasubject variability. Materials degraded by hydrolysis allow for more controlled and less variable degradation kinetics. Poly ($_L$-lactic acid) (PLLA) is a hydrophobic, biodegradable polymer [14,15] that can be introduced into PEG systems to allow for controlled degradation via hydrolysis [16]. Hydrogels formed by polymerization of poly(ethylene glycol)-co-($_L$-lactic acid) diacrylate (PEG-PLLA-DA) degrade into products that are easily processed by the body [17]. Porous PEG-PLLA-DA hydrogels could maintain many of the advantages of porous PEG systems while exhibiting controlled degradation properties.

While various biodegradable porous scaffolds have been studied extensively [18,19] these studies have largely focused on hydrophobic foams. There has been little evaluation of the structure of porous hydrogels and the influence of the degradation process on their properties. In addition, the majority of imaging techniques require destruction or modification of the samples from their native state in order to image. Autofluorescence exhibited by PEG-PLLA-DA hydrogels [20] allows the unique opportunity to monitor the 3D architecture of a porous hydrogel during the degradation process in fully swelled conditions.

Our goal is to optimize the design of porous hydrogels that coordinate the processes of vascularized tissue invasion with polymer degradation. In order to achieve this goal we must first gain an understanding into the properties of porous hydrogels and how they change during degradation. Here, we applied a particulate leaching technique to generate porous PEG-PLLA-DA hydrogels and examined the influence of polymer concentration and particulate size on the mechanical properties, pore structure, and degradation rate of the resultant hydrogels. To our knowledge, this is the first study that was able to evaluate the structure of porous hydrogels during degradation without sample processing or labeling with exogenous agents. This information could be used to help optimize hydrogel design for applications in tissue engineering.

Experimental Section

Materials

PEG (Mn ≈ 3400), stannous octoate, acryloyl chloride (98%), PKH26GL, 3,6-Dimethyl-1,4-dioxane-2,5-dione, triethylamine (99.5%), and 2-hydroxy-2-methylpropiophenone (Irgacure 1173) were obtained from Sigma (St. Louis, MO). Sodium chloride (99.5%), diethyl either, dichloromethane (DCM) anhydrous (99.9%), magnesium sulfate anhydrous (97%) and ethyl ether (anhydrous) were from Fisher Scientific (Pittsburgh, PA).

Synthesis of PEG-PLLA-DA

The method to synthesize PEG-PLLA-DA was performed as described by Chiu et al. [21] Briefly, all glassware were dried in a vacuum oven at 120°C for 24 hours and cooled under vacuum. Ten grams of PEG (MW = 3400) and 2.12 g of 3,6-Dimethyl-1,4-dioxane-2,5-dione were placed into a 50 mL centrifuge tube and lyophilized. A round bottom flask was vacuumed and filled with argon. The lyophilized PEG and 3,6-Dimethyl-1,4-dioxane-2,5-dione were placed in the flask and then 80 µL of stannous octoate added as an initiator. The flask was submerged in a constant temperature oil bath at 140°C for 4 h. The product was then dissolved in 20 mL of anhydrous DCM and filtered using a GF/F filter. The resulting polymer was precipitated in 1.5 L of ice-cold diethyl ether three times and lyophilized. Based on ^{1}HNMR analysis, these conditions result in approximately 10 lactide units per PEG macromer.

The lyophilized PEG-PLLA was acrylated as described previously. [21] Briefly, a three neck round bottom flask was vacuumed and filled with argon. Ten grams of PEG-PLLA was placed in the flask. Sixty mL of anhydrous DCM was injected into the flask, and 0.67 mL of triethylamine was added and stirred for 5 minutes. Acryloyl chloride (0.76 mL) was added dropwise. The flask was allowed to react for 24 hours at room temperature in the dark. The product was washed with 9.52 mL of 2 M K_2SO_4 and allowed to separate overnight. The organic phase was collected and precipitated in 2 L of ice-cold diethyl either. The extent of reaction, structure and purity of the products were determined by ^{1}H NMR (Advance 300 Hz; Bruker, Billerica, MA). Products were

dissolved in CDCl3 for ^{1}HNMR with 0.05% v/v tetramethylsilane (TMS) used as an internal standard. Acrylation efficiency was 93±2%.

Porous PEG-PLLA-DA Hydrogel Generation

The method for generating porous PEG-PLLA-DA hydrogels involved a salt leaching procedure with the polymer dissolved in an organic solvent. Lyophilized PEG-PLLA-DA polymer was dissolved in 1 mL of DCM and 2-hydroxy-2-methyl-propiophenone added as a photoinitiator (5% (w/v)). 250 mg of sieved salt and 250 µL of precursor were placed in a 1.5 mL centrifuge tube. The tube was vortexed for 45 seconds and placed upside down allowing the salt to settle in to the cap for 20 seconds. The concentration of PEG-PLLA-DA was varied from 12.5 to 50% (w/v) and the salt crystals used were selected by sieving in the following ranges: 150–100, 100–50 and 50–25 µm.

A microscope slide was used to cover the solution, carefully avoiding bubble formation. The solution was polymerized by irradiation under UV for 10 minutes. The sample was rotated 180° and polymerized for an additional 10 minutes. The microscope slide was removed and the DCM evaporated in a fume hood overnight. Resulting gels were placed in a 50 mL sterile centrifuge tube with 20 mL DI water with 4 mg/mL of gentamicin sulfate, and then immediately exposed to a vacuum (0.035 mBar) for 15 minutes to remove air trapped in the porous gels and to replace DCM with water. Water was changed 2 times a day until the salt was completely leached out.

Swelling Tests

The porous PEG-PLLA-DA hydrogels were placed in individual 15 mL tubes with 5 mL of PBS (2% sodium azide and 4 mg/mL of gentamicin) and incubated at 37°C. PBS was changed every day until hydrogels completely degraded. Porous PEG-PLLA-DA hydrogels were weighed at various time points.

Structural Analysis

The structure of porous PEG-PLLA-DA hydrogel could be imaged by confocal microscopy due to autofluorescence exhibited by the hydrogels [20]. A PASCAL laser scanning microscopy system from Carl Zeiss MicroImaging, Inc. (Thornwood, NY), was used for confocal imaging. The hydrogel was imaged using a 488 nm laser with a 505 nm low pass filter. Images had x and y resolution of 3.5 µm/pixel and z resolution of 1.8 µm/pixel. The samples were scanned 180 µm deep from the surface at 10 µm intervals, collecting 18 slices in total. Each stack was imported into AxioVision 4.5 (Carl Zeiss, Göttingen, Germany) in order to allow quantification of pore size. Pore size was defined as the longest axis of a given pore and was selected with the built-in caliper tool in AxioVision. Ten pores were selected at each 10 µm thick slice and pore size pooled with the values obtained from the other 19 slices. The average of the pooled values is the pore size value for that sample at that time point. This process is repeated for all samples at each time point.

Compression Testing

Compression testing was conducted at a constant strain rate of 0.5 mm/min using a RSA3 (TA Instruments) [9], [22]. Samples were formed to match the plate size (15 mm) and then compressed. The strain and normal force were recorded and used to calculate the compressive modulus for each sample. The initial diameter and area of each gel were measured and recorded before testing. Compressive moduli of the gels were found by plotting a stress-strain graph with the strain going up to 0.1 or 10%. Strain zero corresponds to the first acceptable stress value. None of the gels were fractured during compression.

RGD Conjugation

The method for conjugation of peptides to acryl-PEG was performed as described previously [21]. A solution of 50 mM NaHCO₃ (pH 8.3) was prepared as a buffer. Ten milligrams of YRGDS (American Peptide, Sunnyvale, CA) was dissolved in 5 mL of 50 mM NaHCO₃. Acryl-PEG-SVA (3400 Da; Laysan, Arab, AL) was dissolved in 7 mL of 50 mM NaHCO₃ and then added drop-wise into the stirred YRGDS solution in the dark. The molar ratio of YRGDS to acryl-PEG-SVA was 1:1.5. The solution was stirred for 2 h at 4°C in the absence of light. The final product was dialyzed (2000 Da molecular weight cut-off) in 2 L of DI water for 24 h (with replacement after 12 h). The resulting product was lyophilized and stored at −80°C until use.

Cell Culture

The cell culture and cell seeding methods have been described previously. [10] Briefly, NIH 3T3 fibroblasts (Cambrex, Walkersvile, MD) were maintained in complete media (Dulbecco's modified Eagle's medium, 10% fetal bovine serum, and 1% penicillin–streptomycin). The cells were passed when flasks reached 90% confluency. Gels were placed into 48-well plates, and incubated in complete media for 1 h. After removing media, gels were air dried in a culture hood for 1 h. Five thousand PKH26 stained 3T3 fibroblasts in 0.5 ml was added directly to the gel surface. Samples were incubated at 37°C, 5% CO_2 overnight and imagined at varying time points. Gels were imaged using confocal microscopy (488 nm laser with a 505 nm low pass filter).

Statistics

Data are presented as means ± standard deviation. Significant differences between groups of data were determined by analysis of variance with Holm-Sidak post-test. In all cases, $p < 0.05$ was considered statistically significant.

Results

The Influence of Polymer Concentration on Polymer Properties

The autofluorescence of PEG-PLLA-DA allows imaging of hydrogel structure nondestructively using confocal microscopy [20]. This is highly advantageous relative to traditional techniques for characterizing material structure, such as scanning electron microscope (SEM), because it avoids processing of samples by drying or fixation that may alter the architecture. We first investigated the effect of polymer concentration on hydrogel properties. Samples were generated with the same range of particulate size (150–100 µm) at varying polymer concentrations of 12.5, 25, and 50% PEG-PLLA-DA. Figure 1 displays confocal images of 12.5% (w/v) porous PEG-PLLA-DA hydrogels at days 1, 3, and 7. The structure of porous hydrogels at different time points can also be seen for 25% (Fig. 2) and 50% hydrogels (Fig. 3).

The confocal images show a porous structure in all hydrogels throughout the degradation process. At the first time point, the hydrogels exhibited pores with structure and size consistent with the salt crystals used as the pore agent. Interestingly, the intensity of the autofluorescence decreased as the hydrogels degraded. For example, at early time points (Fig. 2 A&D), confocal images were bright and hydrogel structures could be easily discerned. By day 14, the intensity of the images decreased and borders appeared blurry (Fig. 2 C&F). Regarding polymer concentration, the 50% group (Fig. 3 B) had higher autofluorescence compared to the 25%

group (Fig. 2 C) likely due to higher polymer concentration. However, even with the lower fluorescence signal from the later time points, quantification and analysis of scaffold architecture was still possible from the confocal images (Fig. 4). The size of the pores remained constant throughout the majority of the degradation time for all conditions. Prior to complete degradation of the hydrogels there was a slight increase in mean pore size.

The wet weight of the hydrogels was also quantified which provides information on the degradation rate (Fig. 5). The initial wet weight depended on the percentage of polymer used and increased as the materials degraded. The time to complete degradation varied with polymer concentration, with 12.5, 25, and 50% gels degraded in 7, 16, and 26 days respectively.

The Influence of Polymer Concentration on Mechanical Properties

The compressive moduli of 12.5, 25, and 50% porous PEG-PLLA-DA hydrogels, generated with salt size ranging from 150–100 µm, were quantified. Figure 6 shows typical curves generated for the hydrogels, illustrating the rapid decrease in stiffness as the hydrogels degrade. At day 1 there were significant differences in mechanical properties between 12.5, 25, and 50% porous PEG-PLLA-DA hydrogel at the same pore size (150–100 µm) (Fig. 7). The compressive modulus was higher for hydrogels with greater polymer content. The compressive moduli decreased rapidly as the hydrogels degraded (Fig. 7).

The Influence of Particulate Size on Pore Structure and Degradation

We also investigated the effect of pore size on the mechanical and degradation properties of the hydrogels. Twenty-five percent PEG-PLLA-DA hydrogels were generated with salt crystal sizes ranging from 50–25 µm, 100–50 µm, and 150–100 µm. The structure of the hydrogels imaged under swelled condition can be seen in Figure 8. In all cases, an interconnected porous structure is apparent with pore size increasing with the salt crystals used. The size and shape of the pores within the hydrogels were consistent with the salt crystals used as the pore forming agent.

Pore sizes agreed well with salt crystal sizes at day 1 (Fig. 9) and the mean pore size remained constant for most of the degradation process with a slight increase prior to complete hydrogel degradation (Fig. 9). The wet weight initially increased as the hydrogels degraded and then decreased towards the end of the process. The degradation rate of the hydrogels did not appear to depend on pore size as all conditions degraded in 15–17 days (Fig. 10).

The Influence of Particulate Size on Mechanical Properties

The stiffness of the gels decreased throughout degradation (Fig. 11). The compressive moduli were significantly different between the different pore sizes at time points up to one week ($p < 0.001$). Gels with smaller pores were stiffer then gels with larger pores, and the stiffness of the gels decreased rapidly throughout degradation.

The Influence of Copolymer Concentration on Cell Adhesion

A cell adhesion sequence (YRGDS) was incorporated in the hydrogels to determine whether the porous gels could support the adhesion of cells. Fibroblasts spread and lined the edges of pores in all polymer conditions (Fig. 12). Cells on 50% porous hydrogels made with crystal size 150–100 µm were imaged over time

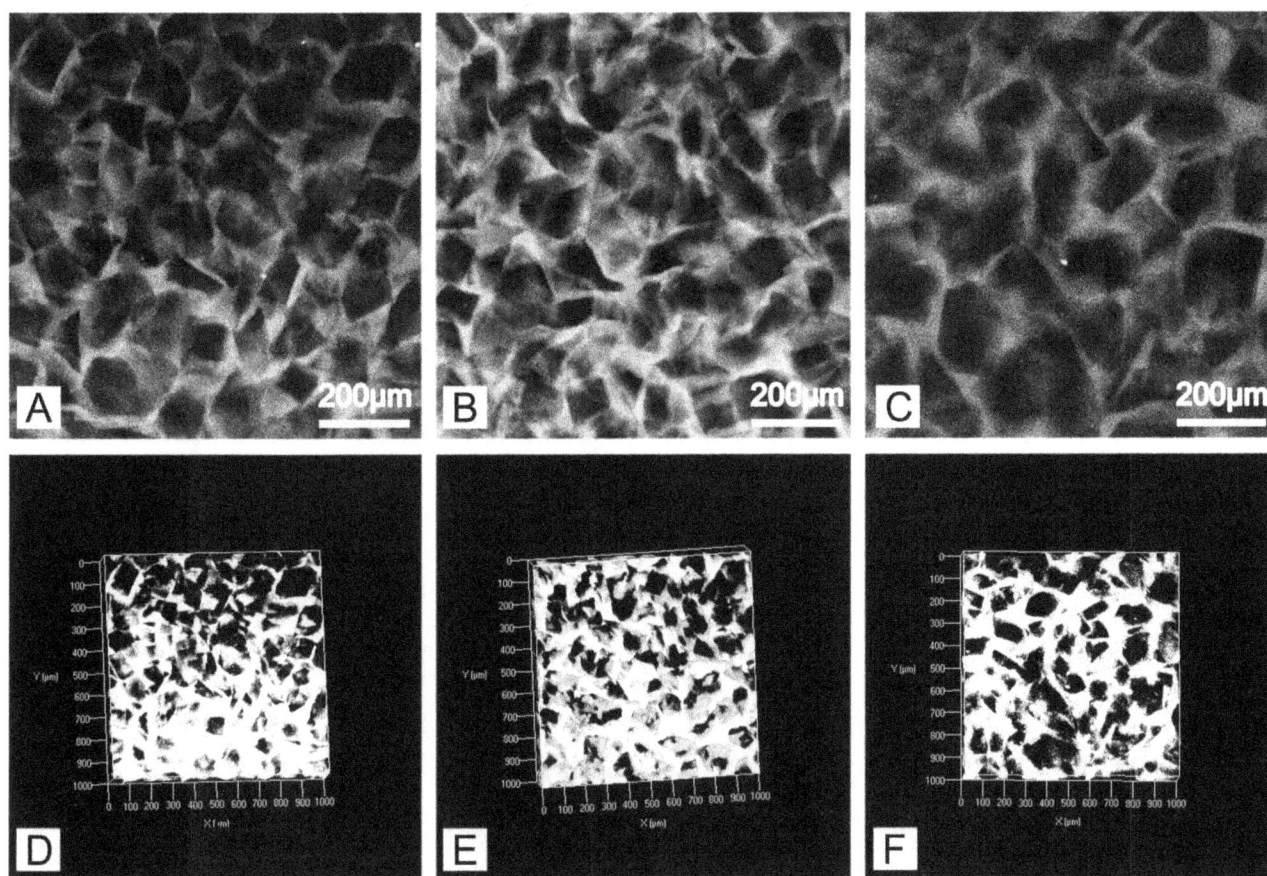

Figure 1. Two dimensional slices (A–C) and volume renderings (D–F) of 12.5% (w/v) porous PEG-PLLA-DA hydrogels generated with salt crystal size ranging from 150–100 µm at 1 (A,D), 3 (B,E), and 7 (C,F) days of incubation.

(Fig. 13). At day 1, cells appeared to line the edge of the pores (Fig. 13A). The cell organization changed as the gels degraded and, by day 22, cells had formed multicellular aggregates within the gels prior to complete degradation (Fig. 13 C).

Discussion

The ability to modulate and control tissue response to implanted biomaterials is essential to the fields of tissue engineering and regenerative medicine. We have previously investigated porous PEG hydrogels and found that these hydrogels support vascularized tissue formation [1]. Under the conditions investigated, hydrogels with pores ranging from 150–100 µm supported the most rapid vascularization *in vitro* and *in vivo*. In these studies, there were no signs of degradation exhibited by either porous or nonporous PEG hydrogels. However, the success of these materials in clinical applications requires that they degrade in a controlled fashion as new vascularized tissue develops.

Hydrogels generated from PEG-PLLA-DA copolymers have been investigated in many applications in regenerative medicine. Materials based on PEG-PLLA copolymers have been applied as biological coatings, tissue engineering scaffolds, and drug delivery systems, but there has been little investigation into the design and optimization of PEG-PLLA-DA hydrogels with porous structure [23]. The polymer conditions used here are similar to those that have been described in other studies [24]. However, the generation of pores offers a number of advantages over nonporous structures. This includes enhanced nutrient transport and higher

surface area to volume ratio. While these hydrogels do not allow invasion via protease-mediated degradation [25], the reliance on hydrolytic degradation allows the potential to decouple tissue invasion from hydrogel degradation. In this study, we generated porous PEG-PLLA-DA hydrogels by solvent casting with DCM, particulate leaching, and photopolymerization. This particulate leaching technique has been commonly used for hydrophobic polymer foams, but we show that it also serves as a simple method for generating pores in PEG-PLLA-DA hydrogels.

Studies have been performed examining the structure of hydrophobic polymer foams during degradation, but research has not been performed into porous hydrogels systems. The equal availability of water throughout the polymer volume results in differences in structural changes as the materials degrade relative to hydrophobic materials. Autofluorescence exhibited by the PEG-PLLA-DA allowed the unique ability to characterize 3D polymer structures when they are fully swelled which are the conditions used in bioreactors and cell culture [10], [20]. The origin of this autofluorescence is still not clear but it appears to result from a synergetic effect of both lactate units and diacrylate groups in the PEG-PLLA-DA backbone. However, the fluorescence not only allows imaging of the polymer structure with confocal microscopy but can be exploited to monitor degradation as the intensity is proportional to the number of PEG-PLLA chains present. [20].

The hydrogels exhibited an interconnected porous structure with initial pore size correlating well with the size of the particulates selected. The pore size and structure remained consistent throughout degradation and did not depend on polymer

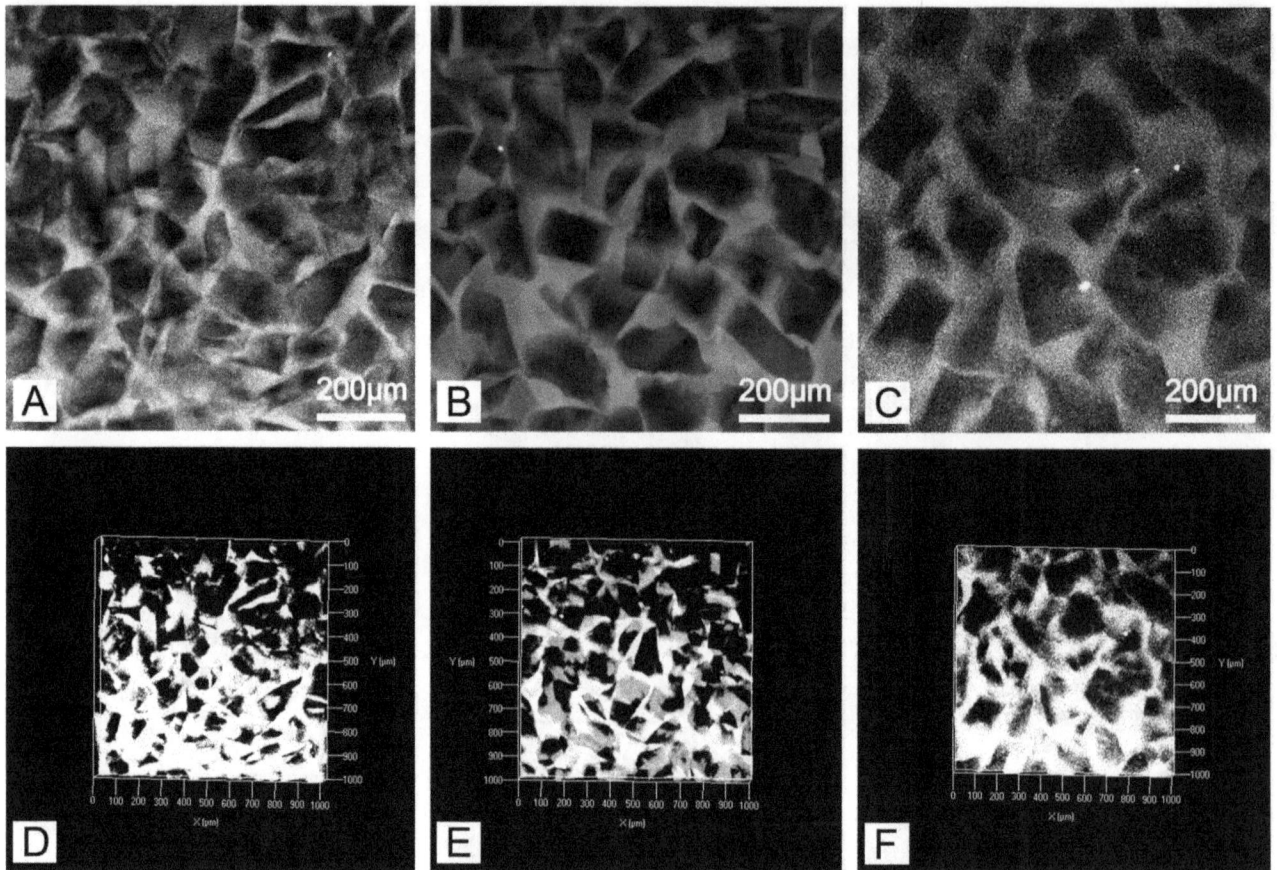

Figure 2. Two dimensional slices (A–C) and volume renderings (D–F) of 25% (w/v) porous PEG-PLLA-DA hydrogels generated with salt crystal size ranging from 150–100 μm at 1 (A,D), 7 (B,E), and 14 (C,F) days of incubation.

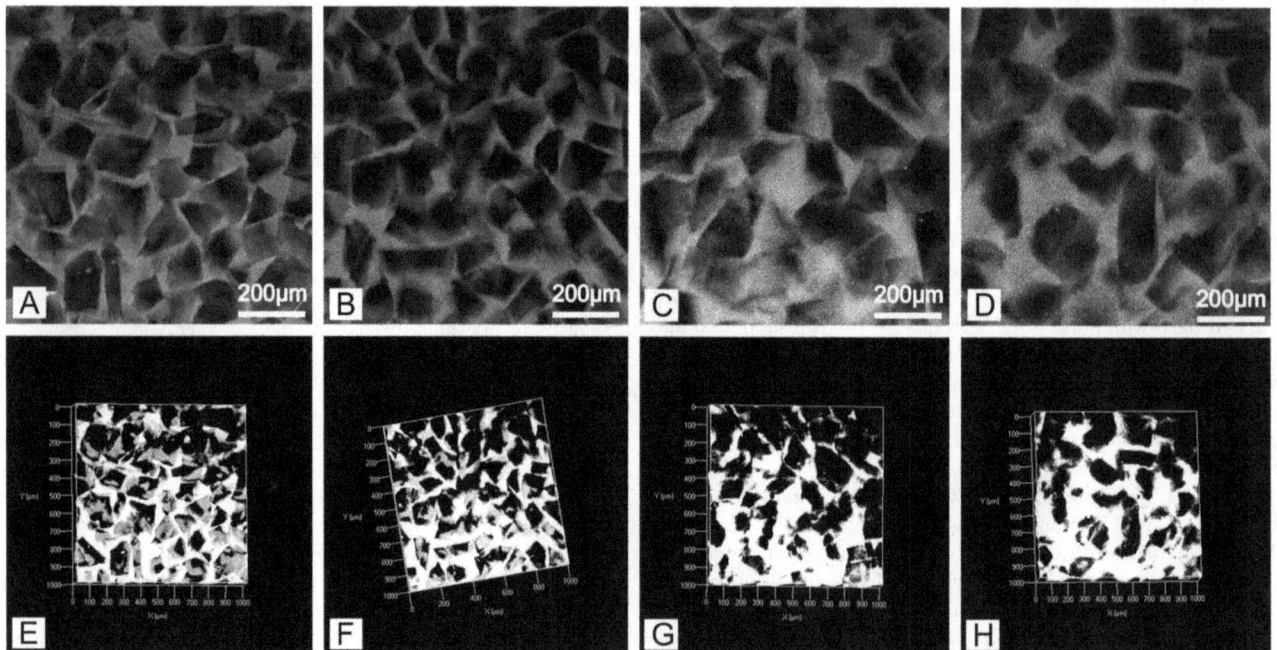

Figure 3. Two dimensional slices (A–D) and volume renderings (E–H) of 50% (w/v) porous PEG-PLLA-DA hydrogels generated with salt crystal size ranging from 150–100 μm at 1(A,E), 7(B,F), 14(C,G), and 21(D,H) days of incubation.

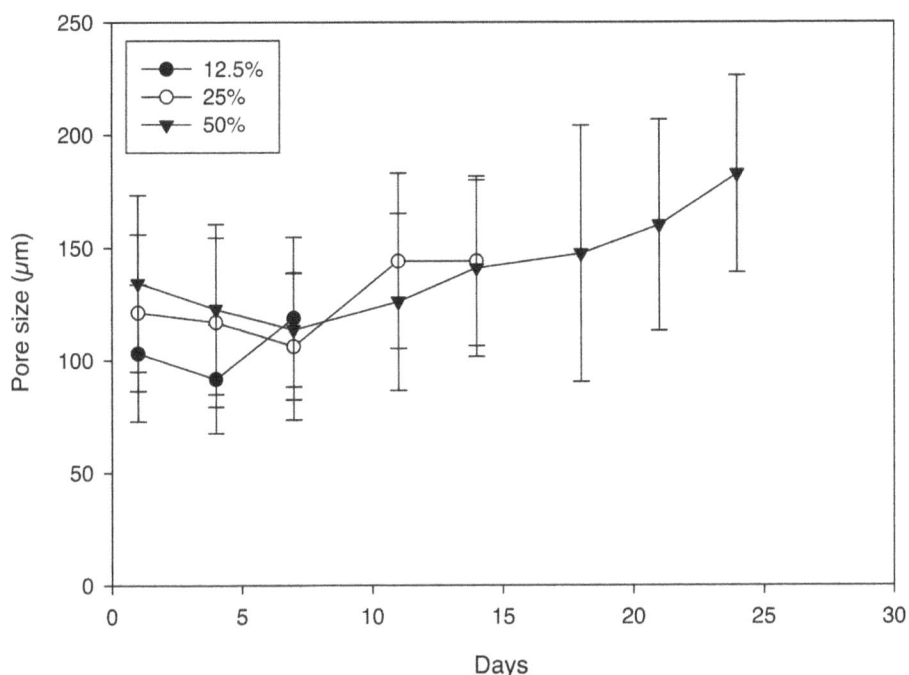

Figure 4. Mean pore size of porous PEG-PLLA-DA hydrogel generated with salt size ranging from 150–100 μm at various PEG-PLLA-DA percentages plotted versus incubation time in vitro.

concentration. In studies with hydrophobic polymer foams, pores have been shown to decrease in size and number while the scaffolds degraded [26]. In addition, the overall architecture of the pores in PLGA scaffolds changes, losing the crystal shape that results initially from the particle leaching technique. The hydrogels used in this study, however, maintained their size and structure as they degraded. While the dissolution of porous polymer foams exhibit a bulk degradation mechanism, the change in pore

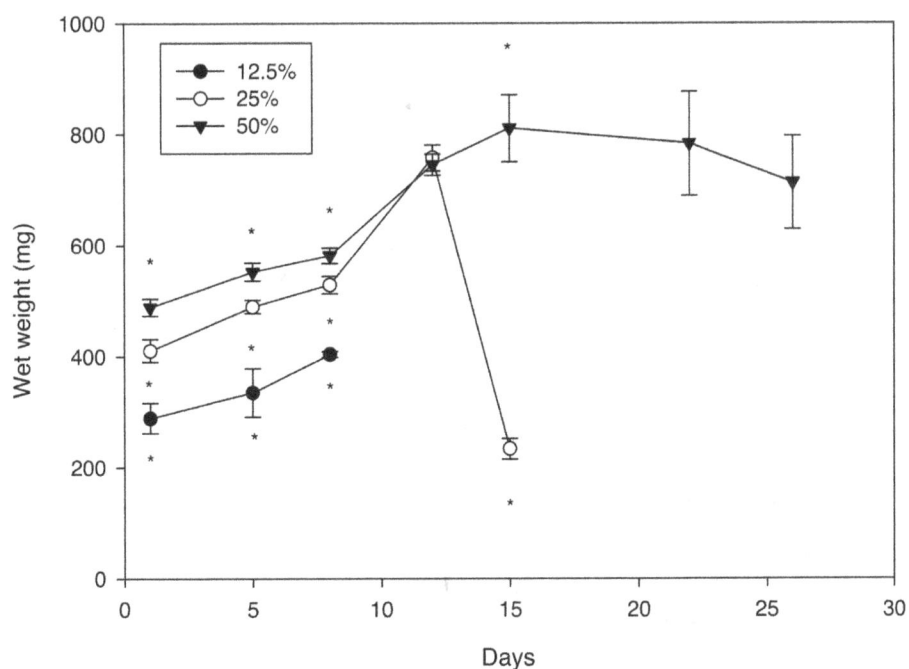

Figure 5. Wet weight of porous PEG-PLLA-DA hydrogels versus time for hydrogels generated with salt size ranging from 150–100 μm at various polymer concentrations (*indicates statistical difference between all groups at that time point, p<0.001). The significant reduction in weight seen on day 15 for 25% gels is due to the fact that these gels were highly degraded at that time point, and the gels had reduced greatly in size.

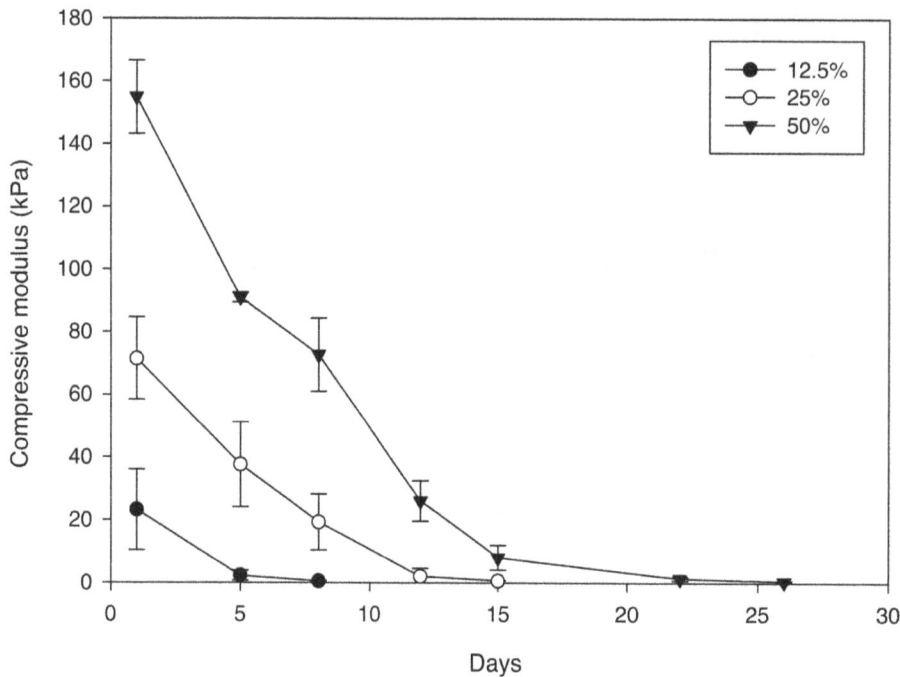

Figure 6. Sample stress-strain curves of 25% (w/v) porous PEG-PLLA-DA hydrogels generated with salt sizes ranging from 150–100 μm at days 1, 3, 5, 8,and 10.

structure occurs as the chains on the surface of the pore having greater access to water than those within the structure. This results in more rapid degradation at the surface of the pore and a change in size throughout degradation. Consistent with experimental and computational models of nonporous PEG-PLLA hydrogels [27,28], the porous hydrogels exhibit a bulk mechanism of degradation which means that the overall structure, including pore size, is maintained up to the point of a nearly instantaneous

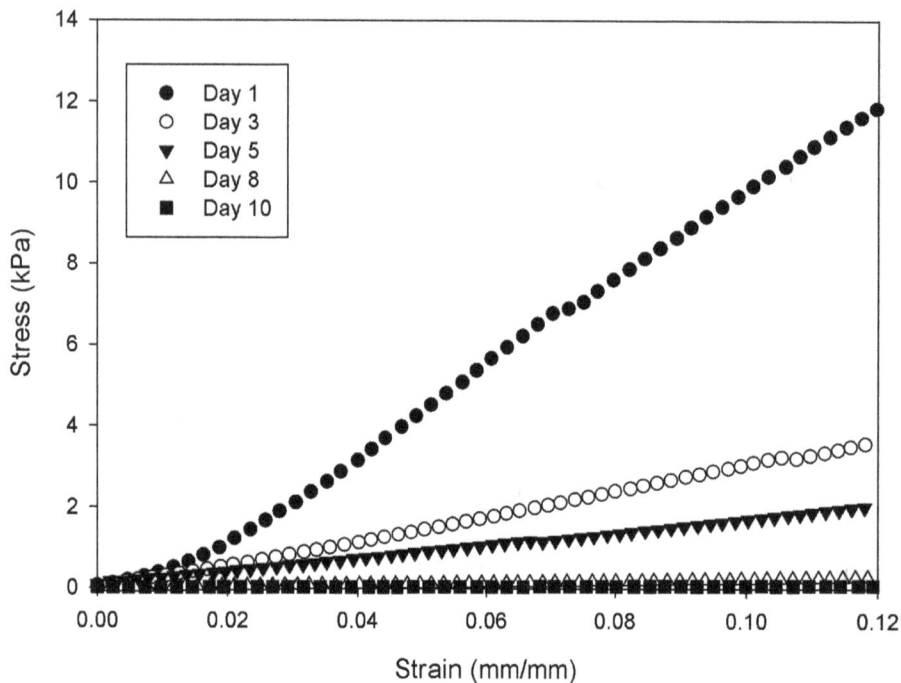

Figure 7. Compressive moduli of porous PEG-PLLA-DA hydrogels generated with salt sizes ranging from 150–100 μm with varying PEG-PLLA-DA percentages plotted versus time. Compressive moduli are statistically different points between all groups at each time point (p<0.001).

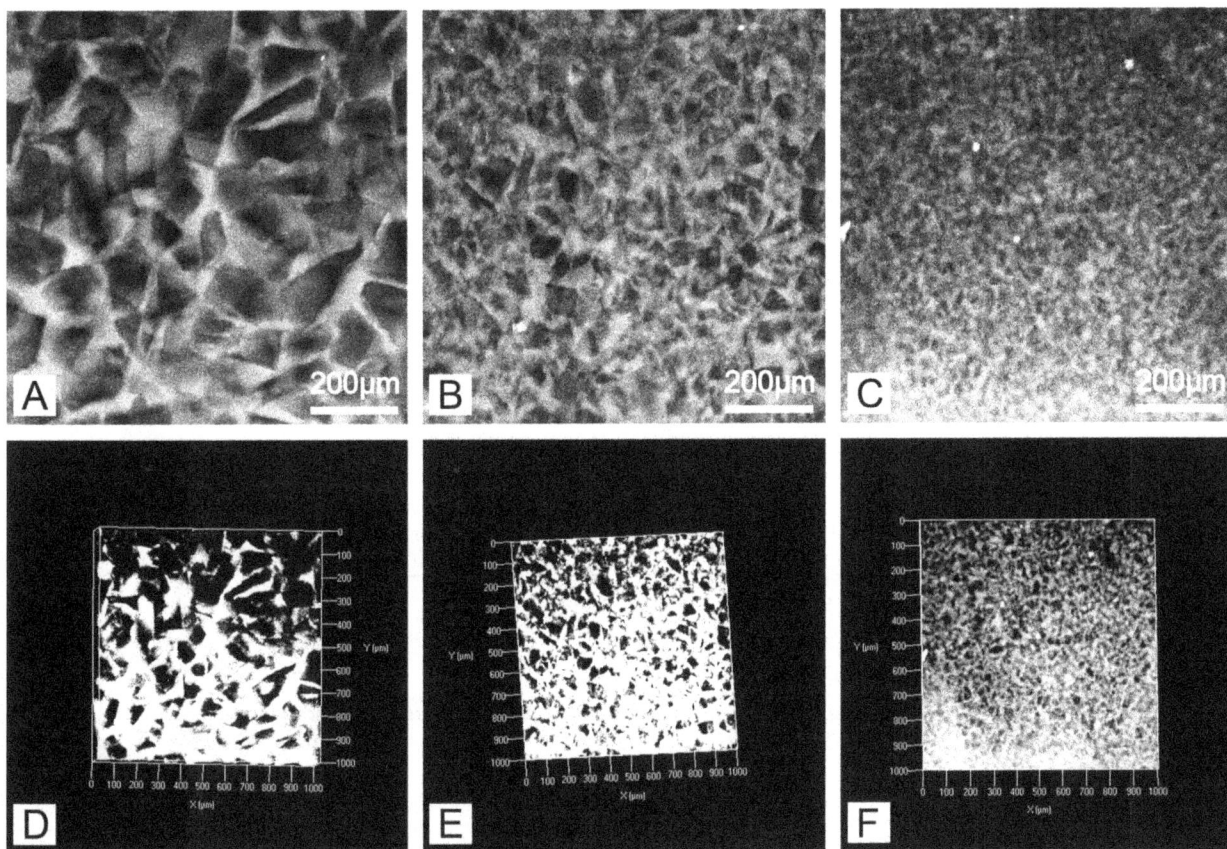

Figure 8. Two dimensional confocal sections (A–C) and three dimensional volume renderings (D–F) of porous hydrogels. The 25% (w/v) hydrogels were generated with salt crystal sizes ranging from150–100 μm (A,D), 100–50 μm (B,E), and 50–25 μm (C,F) of incubation.

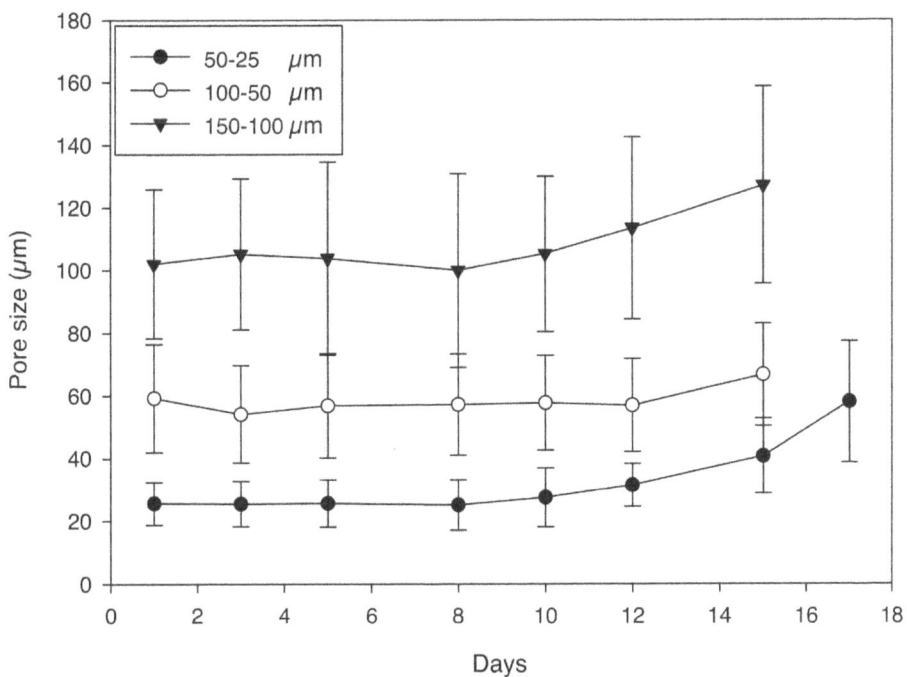

Figure 9. Mean pore size versus time for 25% (w/v) porous PEG-PLLA-DA hydrogels generated with salt crystal sizes ranging from 150–100 μm, 100–50 μm, and 50–25 μm.

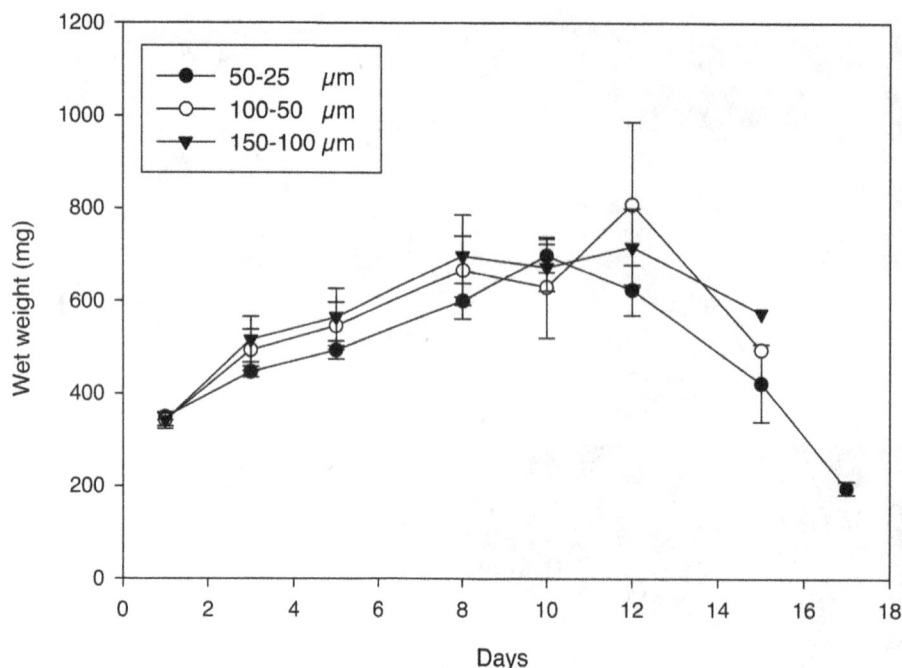

Figure 10. Wet weight versus incubation time for 25% (w/v) porous PEG-PLLA-DA hydrogel generated with varying salt sizes.

dissolution of the final volume. The rapid decrease observed in wet weight in the 25% hydrogel is an indication of this rapid dissolution. However unlike the polymer foams, hydrogels rapidly absorb water resulting in all chains having equal access to water whether they are on the surface of a pore or part of the bulk structure. Swelling ratio studies support the concept that pore size does not influence access to water. Hydrogels swell as they degrade

[29] and the wet mass was independent of pore size at all time points. The equal access of the polymer chains to water allows maintenance of pore structure as the hydrogels degrade until they reach the point of complete dissolution.

The compressive moduli of the hydrogels depended on both pore size and polymer content. As expected, the modulus increased with polymer content, which agrees with literature

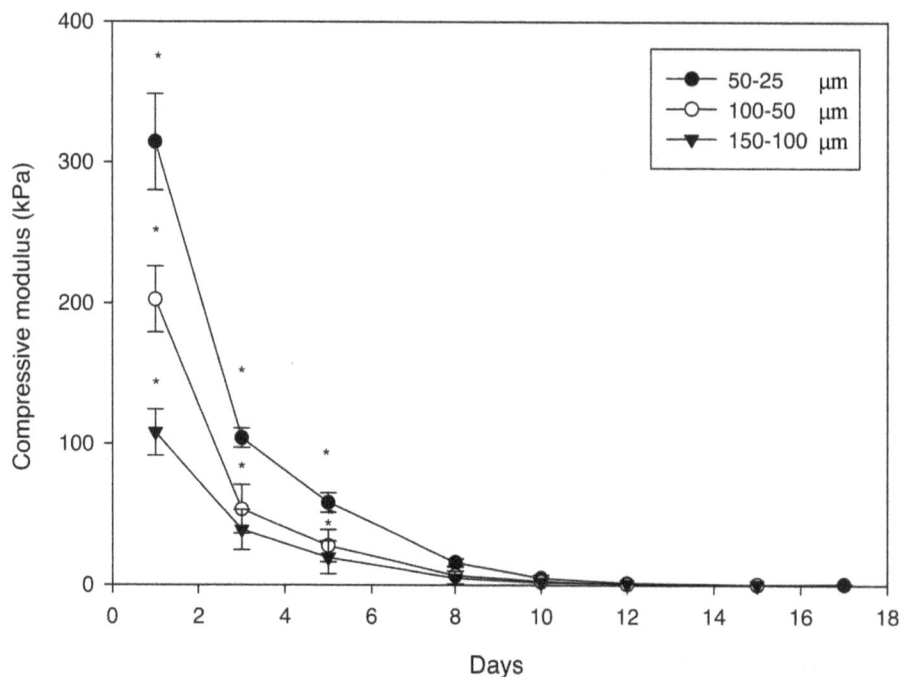

Figure 11. Compressive moduli of 25% (w/v) porous PEG-PLLA-DA hydrogels generated with varying salt sizes during in vitro degradation (*indicates statistical differences between groups) (p<0.001).

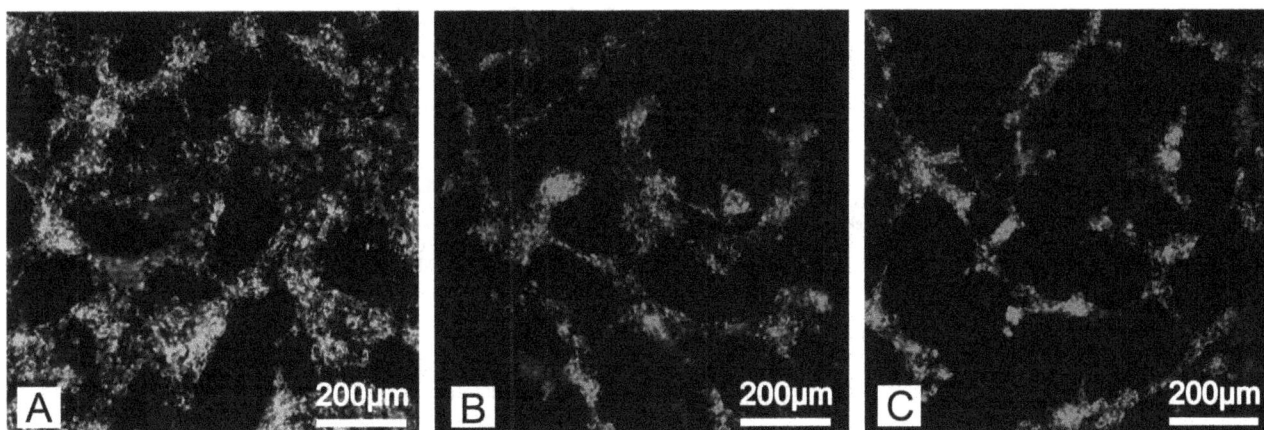

Figure 12. Confocal images of PKH 26 stained fibroblasts adhering to porous hydrogels. The hydrogels were generated using salt crystals ranging in size 150–100 μm but polymer concentrations of A) 50%, B) 25%, and C) 12.5%.

showing increasing crosslink density with polymer content [30]. In addition, the stiffness decreased with increasing pore size. This result agrees with previous studies which investigated the influence of pore size on porous poly (propylene fumarate) scaffolds [9]. Increasing polymer concentration could increase the mechanical properties of the hydrogels while keeping pore size constant. The mechanical properties also diminished rapidly during incubation suggesting a bulk mechanism of degradation, which is consistent with swelling and pore size observations. We have previously shown that pore size plays an important role in biological response to porous PEG-based hydrogels [1]. However, mechanical properties and degradation time also contribute to biological response [31], [32]. While our results suggest varying polymer concentration may allow for the design of hydrogels with mechanical properties independent of pore size, it does not appear the degradation time can be easily decoupled from mechanical properties using this approach.

Porous hydrogels supported cell attachment under all conditions following the incorporation of cell adhesion sequences. As the hydrogels degraded the structure of the cells in the gels changed. The cells eventually assembled into aggregates before the complete degradation of hydrogels. The change in cell morphology is somewhat surprising considering that the pore size and structure do not change as the gels degraded. However, the reduction in hydrogel stiffness and possible local change in ligand density with degradation could influence cell behavior. It is well-established that cell migration is influenced by hydrogel stiffness [33] and cells

exhibit a round shape and less stress-fiber formation in poly-acrylamine gels with lower stiffness [34]. Cell behavior also is dependent on ligand density [35], which may change locally as degradation reduces the dangling ligands present in the gels [36]. These results suggest how changes in hydrogel properties despite constant pore structure could change biological response to the materials. Future studies will examine cell behavior and tissue formation within porous PEG-PLLA-DA hydrogels and the role of stiffness and ligand density on the response.

Conclusion

A technique was developed for the application of salt leaching methods to generate porous PEG-PLLA-DA hydrogels. This study demonstrates the influence of polymer concentration and pore size on the mechanical, structural, and degradation properties of these porous hydrogels. The architecture of porous PEG-PLLA-DA hydrogels was monitored without drying or destruction by using the polymer's intrinsic fluorescence. Interestingly, this allowed for the determination that pore size and structure remained constant during degradation. Results from this study provide a better understanding of the mechanical property and architecture of porous PEG-PLLA-DA hydrogels during degradation, which helps to design biodegradable, porous scaffolds for tissue engineering.

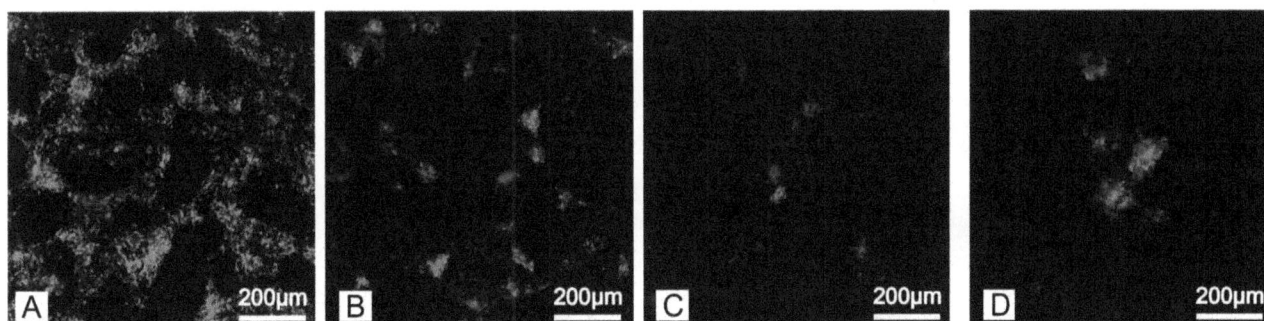

Figure 13. Confocal images of PKH 26 stained 3t3 fibroblasts on 50% porous hydrogels made with crystal size 150–100 μm at days A) 1, B) 5, C) 10, and D) 22 of incubation.

Acknowledgments

We would like to thank Dr. David Venerus for providing access to equipment for compression tests.

Author Contributions

Conceived and designed the experiments: YCC SK EMB. Performed the experiments: YCC SK JCL. Analyzed the data: YCC SK JCL EMB. Wrote the paper: YCC SK JCL EMB.

References

1. Chiu YC, Cheng MH, Engel H, Kao SW, Larson JC, et al. (2011) The role of pore size on vascularization and tissue remodeling in PEG hydrogel. Biomaterials.

2. Lin CC, Metters AT, Anseth KS (2009) Functional PEG-peptide hydrogels to modulate local inflammation induced by the pro-inflammatory cytokine TNFalpha. Biomaterials 30: 4907–4914.

3. Jun HW, West JL (2005) Endothelialization of microporous YIGSR/PEG-modified polyurethaneurea. Tissue Engineering 11: 1133–1140.

4. Yu HY, VandeVord PJ, Mao L, Matthew HW, Wooley PH, et al. (2009) Improved tissue-engineered bone regeneration by endothelial cell mediated vascularization. Biomaterials 30: 508–517.

5. Hollister SJ (2005) Porous scaffold design for tissue engineering. Nature Materials 4: 518–524.

6. Lovett M, Rockwood D, Baryshyan A, Kaplan DL (2010) Simple Modular Bioreactors for Tissue Engineering: A System for Characterization of Oxygen Gradients, Human Mesenchymal Stem Cell Differentiation, and Prevascularization. Tissue Engineering Part C-Methods 16: 1565–1573.

7. Wachiralarpphaithoon C, Iwasaki Y, Akiyoshi K (2007) Enzyme-degradable phosphorylcholine porous hydrogels cross-linked with polyphosphoesters for cell matrices. Biomaterials 28: 984–993.

8. Levesque SG, Lim RM, Shoichet MS (2005) Macroporous interconnected dextran scaffolds of controlled porosity for tissue-engineering applications. Biomaterials 26: 7436–7446.

9. Fisher JP, Holland TA, Dean D, Mikos AG (2003) Photoinitiated cross-linking of the biodegradable polyester poly(propylene fumarate). Part II. In vitro degradation. Biomacromolecules 4: 1335–1342.

10. Chiu YC, Larson JC, Isom A, Brey EM (2010) Generation of Porous Poly(Ethylene Glycol) Hydrogels by Salt Leaching. Tissue Engineering Part C-Methods 16: 905–912.

11. Betz MW, Yeatts AB, Richbourg WJ, Caccamese JF, Coletti DP, et al. (2010) Macroporous Hydrogels Upregulate Osteogenic Signal Expression and Promote Bone Regeneration. Biomacromolecules 11: 1160–1168.

12. Burdick JA, Mason MN, Hinman AD, Thorne K, Anseth KS (2002) Delivery of osteoinductive growth factors from degradable PEG hydrogels influences osteoblast differentiation and mineralization. Journal of Controlled Release 83: 53–63.

13. Moon JJ, Saik JE, Poche RA, Leslie-Barbick JE, Lee SH, et al. (2010) Biomimetic hydrogels with pro-angiogenic properties. Biomaterials 31: 3840–3847.

14. Hattori K, Tomita N, Yoshikawa T, Takakura Y (2001) Prospects for bone fixation - development of new cerclage fixation techniques. Materials Science & Engineering C-Biomimetic and Supramolecular Systems 17: 27–32.

15. Saito T, Iguchi A, Sakurai M, Tabayashi K (2004) Biomechanical study of a poly-L-lactide (PLLA) sternal pin in sternal closure after cardiothoracic surgery. Annals of Thoracic Surgery 77: 684–687.

16. Metters AT, Anseth KS, Bowman CN (2000) Fundamental studies of a novel, biodegradable PEG-b-PLA hydrogel. Polymer 41: 3993–4004.

17. Vila A, Sanchez A, Evora C, Soriano I, Jato JLV, et al. (2004) PEG-PLA nanoparticles as carriers for nasal vaccine delivery. Journal of Aerosol Medicine-Deposition Clearance and Effects in the Lung 17: 174–185.

18. Evans GRD, Brandt K, Widmer MS, Lu L, Meszlenyi RK, et al. (1999) In vivo evaluation of poly(L-lactic acid) porous conduits for peripheral nerve regeneration. Biomaterials 20: 1109–1115.

19. Patel ZS, Yamamoto M, Ueda H, Tabata Y, Mikos AG (2008) Biodegradable gelatin microparticles as delivery systems for the controlled release of bone morphogenetic protein-2. Acta Biomaterialia 4: 1126–1138.

20. Chiu YC, Brey EM, Pérez-Luna VH (2012) A Study of the Intrinsic Autofluorescence of Poly (ethylene glycol)-co-((L)-Lactic acid) Diacrylate. J Fluoresc.

21. Chiu YC, Larson JC, Perez-Luna VH, Brey EA (2009) Formation of Microchannels in Poly(ethylene glycol) Hydrogels by Selective Degradation of Patterned Microstructures. Chemistry of Materials 21: 1677–1682.

22. Bryant SJ, Chowdhury TT, Lee DA, Bader DL, Anseth KS (2004) Crosslinking density influences chondrocyte metabolism in dynamically loaded photocrosslinked poly(ethylene glycol) hydrogels. Ann Biomed Eng 32: 407–417.

23. Goraltchouk A, Freier T, Shoichet MS (2005) Synthesis of degradable poly(L-lactide-co-ethylene glycol) porous tubes by liquid-liquid centrifugal casting for use as nerve guidance channels. Biomaterials 26: 7555–7563.

24. Nuttelman CR, Rice MA, Rydholm AE, Salinas CN, Shah DN, et al. (2008) Macromolecular Monomers for the Synthesis of Hydrogel Niches and Their Application in Cell Encapsulation and Tissue Engineering. Prog Polym Sci 33: 167–179.

25. Sokic S, Papavasiliou G (2012) FGF-1 and proteolytically mediated cleavage site presentation influence three-dimensional fibroblast invasion in biomimetic PEGDA hydrogels. Acta Biomaterialia 8: 2213–2222.

26. Wu L, Ding J (2004) In vitro degradation of three-dimensional porous poly(d,l-lactide-co-glycolide) scaffolds for tissue engineering. Biomaterials 25: 5821–5830.

27. Metters AT, Bowman CN, Anseth KS (2000) A statistical kinetic model for the bulk degradation of PLA-b-PEG-b-PLA hydrogel networks. Journal of Physical Chemistry B 104: 7043–7049.

28. Metters AT, Anseth KS, Bowman CN (2001) A statistical kinetic model for the bulk degradation of PLA-b-PEG-b-PLA hydrogel networks: Incorporating network non-idealities. Journal of Physical Chemistry B 105: 8069–8076.

29. van Dijk-Wolthuis WNE, Franssen O, Talsma H, van Steenbergen MJ, Kettenes-van den Bosch JJ, et al. (1995) Synthesis, Characterization, and Polymerization of Glycidyl Methacrylate Derivatized Dextran. Macromolecules 28: 6317–6322.

30. Sokic S, Papavasiliou G (2012) Controlled Proteolytic Cleavage Site Presentation in Biomimetic PEGDA Hydrogels Enhances Neovascularization In Vitro. Tissue Eng Part A 18: 2477–2486.

31. Francis-Sedlak ME, Uriel S, Larson JC, Greisler HP, Venerus DC, et al. (2009) Characterization of type I collagen gels modified by glycation. Biomaterials 30: 1851–1856.

32. Francis-Sedlak ME, Moya ML, Huang J, Lucas SA, Chandrasekharan N, et al. (2010) Collagen glycation alters neovascularization in vitro and in vivo. Microvasc Res 80: 3–9.

33. Nemir S, Hayenga HN, West JL (2010) PEGDA hydrogels with patterned elasticity: Novel tools for the study of cell response to substrate rigidity. Biotechnol Bioeng 105: 636–644.

34. Zemel A, Rehfeldt F, Brown AEX, Discher DE, Safran SA (2010) Optimal matrix rigidity for stress-fibre polarization in stem cells. Nature Physics 6: 468–473.

35. Engler A, Bacakova L, Newman C, Hategan A, Griffin M, et al. (2004) Substrate compliance versus ligand density in cell on gel responses. Biophys J 86: 617–628.

36. Eskandari M, Brey E, Cinar A (2011) A gaussian model for substrates of entangled cross-linked poly(ethylene glycol) in biomedical applications. Biotechnol Bioeng 108: 435–445.

A Cost-Minimization Analysis of Tissue-Engineered Constructs for Corneal Endothelial Transplantation

Tien-En Tan[1,2]*, Gary S. L. Peh[3], Benjamin L. George[3], Howard Y. Cajucom-Uy[4], Di Dong[5], Eric A. Finkelstein[5,6], Jodhbir S. Mehta[2,3,7]

1 Yong Loo Lin School of Medicine, National University of Singapore, Singapore, 2 Singapore National Eye Centre, Singapore, 3 Tissue Engineering and Stem Cell Group, Singapore Eye Research Institute, Singapore, 4 Singapore Eye Bank, Singapore, 5 Health Services and Systems Research, Duke-NUS Graduate Medical School, Singapore, 6 Lien Centre for Palliative Care, Singapore, 7 Department of Clinical Sciences, Duke-NUS Graduate Medical School, Singapore

Abstract

Corneal endothelial transplantation or endothelial keratoplasty has become the preferred choice of transplantation for patients with corneal blindness due to endothelial dysfunction. Currently, there is a worldwide shortage of transplantable tissue, and demand is expected to increase further with aging populations. Tissue-engineered alternatives are being developed, and are likely to be available soon. However, the cost of these constructs may impair their widespread use. A cost-minimization analysis comparing tissue-engineered constructs to donor tissue procured from eye banks for endothelial keratoplasty was performed. Both initial investment costs and recurring costs were considered in the analysis to arrive at a final tissue cost per transplant. The clinical outcomes of endothelial keratoplasty with tissue-engineered constructs and with donor tissue procured from eye banks were assumed to be equivalent. One-way and probabilistic sensitivity analyses were performed to simulate various possible scenarios, and to determine the robustness of the results. A tissue engineering strategy was cheaper in both investment cost and recurring cost. Tissue-engineered constructs for endothelial keratoplasty could be produced at a cost of US$880 per transplant. In contrast, utilizing donor tissue procured from eye banks for endothelial keratoplasty required US$3,710 per transplant. Sensitivity analyses performed further support the results of this cost-minimization analysis across a wide range of possible scenarios. The use of tissue-engineered constructs for endothelial keratoplasty could potentially increase the supply of transplantable tissue and bring the costs of corneal endothelial transplantation down, making this intervention accessible to a larger group of patients. Tissue-engineering strategies for corneal epithelial constructs or other tissue types, such as pancreatic islet cells, should also be subject to similar pharmacoeconomic analyses.

Editor: Sanjoy Bhattacharya, Bascom Palmer Eye Institute, University of Miami School of Medicine, United States of America

Funding: The authors have no support or funding to report.

Competing Interests: The authors have declared that no competing interests exist.

* Email: jodmehta@gmail.com

Introduction

The cornea is a five-layered structure in the anterior segment of the eye. The corneal endothelium pumps fluid out of the cornea, maintaining corneal deturgescence, which is crucial for corneal transparency and clear vision. Corneal endothelial dysfunction may be inherited e.g. Fuchs' endothelial dystrophy, or acquired due to surgical trauma e.g. pseudophakic bullous keratopathy. These are the leading causes of corneal transplantation in most developed countries [1].

The field of corneal transplantation has evolved rapidly in the past 10 years. Full-thickness penetrating keratoplasty (PK) techniques have been replaced by newer partial-thickness techniques for many corneal diseases [2–6]. In particular, endothelial keratoplasty (EK) techniques like Descemet's stripping endothelial keratoplasty (DSEK) and Descemet's membrane endothelial keratoplasty (DMEK) have been very successful for treating endothelial disease [7,8]. Since 2005, the number of PK procedures performed in the USA has steadily dropped, while EK procedures have consistently risen, so much so that EK became the dominant procedure in 2012 [9]. DSEK involves

transplanting a posterior lamellar corneal graft, consisting of donor corneal endothelium, Descemet's membrane, and a layer of posterior stroma, to replace dysfunctional recipient corneal endothelium. DSEK provides faster visual recovery, greater tectonic stability, less induced astigmatism, and lower rates of immunologic rejection, with comparable 3-year graft survival and endothelial cell loss rates to PK [5,6,10]. DSEK has also been shown to be more cost-effective than PK [11]. Furthermore, DSEK grafts can be precut prior to surgery, which simplifies the procedure and reduces operating time [12,13].

However, EK, like any transplant procedure, is reliant on the availability of donor tissue [11]. Stricter tissue testing regulations and precautions against transmission of infectious disease have led to more costly processing and higher tissue discard rates, which both contribute to the rising cost of donor corneal tissue from eye banks [14,15]. With aging populations and higher incidence of age-related corneal disease, the demand for donor tissue is also likely to increase [3,16,17]. Meanwhile, supply of donor tissue is unlikely to keep up, as most eye banks do not retrieve corneal tissue from donors above 75 years of age [18]. Stricter age criteria on donor tissue imposed by surgeons could dramatically worsen this situation further [18,19]. Aging populations are therefore expected to worsen the worldwide shortage of transplantable corneal tissue [3]. Particularly in Asia, where donor retrieval rates are lower than the USA, this shortage is likely to be more acute. Such projections have engendered significant interest in alternative sources of transplantable corneal tissue [3,16].

Tissue engineering is the use of cells and materials to produce functional substitutes for damaged tissue or organs [20]. Application of tissue engineering for the corneal endothelium is attractive for two main reasons: First, corneas are the most transplanted organ worldwide [4]. Second, tissue-engineered corneal epithelial constructs are already in clinical use [21,22]. Tissue-engineered constructs comprising a monolayer of human corneal endothelial cells (HCECs), expanded within an *in vitro* environment, and seeded onto a membranous scaffold carrier should theoretically function as viable substitute graft material equivalent to precut EK donor tissues [23]. Recent advancements in reproducible culture of HCECs [3,24–26] and improvements in surgical techniques like DSEK/DMEK [7,8,10,27,28] have made the clinical application of such tissue-engineered endothelial constructs a realistic possibility [29,30]. In fact, clinical success of transplanted tissue-engineered endothelial constructs has already been demonstrated in a primate model [29,30].

Performing EK with tissue-engineered endothelial constructs is within reach. However, whether or not it is economically viable remains an open question. Cost-minimization analysis is a form of pharmacoeconomic analysis that measures and compares the costs of two competing approaches that are assumed to provide equivalent outcomes [31]. This paper uses cost-minimization analysis to assess the economic feasibility of performing EK with tissue-engineered constructs, from the perspective of an ophthalmic institution in Singapore that possesses the surgical expertise to perform EK, but is lacking established infrastructure for the acquisition of transplantable EK tissue. For such an institution, would it be more prudent to: (a) invest in a laboratory to produce tissue-engineered EK grafts, or (b) utilize donor corneal tissue procured from eye banks for EK?

Materials and Methods

A cost-minimization analysis comparing the use of tissue-engineered constructs versus procured tissue for EK was performed, based on the following assumptions:

1. EK procedures performed with either tissue-engineered constructs or precut procured tissue are identical in surgical technique, visual outcomes, complications and utility.

2. The institution possesses the surgical expertise, operating theatres and equipment for EK. However, said institution does not possess infrastructure for the acquisition of transplantable tissue. Therefore, tissue must either come from tissue engineering, or be procured from eye banks.

3. The institution does not possess an Automated Lamellar Therapeutic Keratoplasty (ALTK) system (Moria, Antony, France) for cutting EK grafts in operating theatres. Therefore, any donor corneas procured from eye banks must be precut before use in surgery.

4. Obtaining transplantable EK tissue by the tissue-engineering strategy requires the set-up of a laboratory capable of producing tissue-engineered EK grafts (henceforth referred to as the "laboratory"), while the procured-tissue strategy requires a facility capable of receiving, storing and precutting donor tissue from eye banks (henceforth called the "facility").

5. The laboratory/facility will be nested within the institution, and will have access to a surgical sterilization unit. Therefore, these costs can be excluded.

6. The amount of physical space and renovations required for either the laboratory or the facility will be similar, and these expenses can therefore be excluded.

7. Utilities costs for both the laboratory and facility will be similar, and can therefore be excluded.

8. All corneal tissue, whether used for tissue culture or for precutting, is acquired from the same source: Florida Lions Eye Bank, Miami, FL, USA.

9. All corneas arrive in matched pairs from the same donor. Therefore, corneas of a pair share delivery courier costs, and can also be handled and precut with the same set of sterile instruments, without sterilization in between.

10. There will be no tissue wastage; all corneas arrive on time, and all precutting, isolation and culture procedures will be successful. The costs of Quality Assurance (QA) testing for both tissue-engineered constructs and precut donor tissue will be included, but all will be of sufficient quality for transplant.

The authors are based in Singapore, and therefore this model was based on an institution in Singapore. All quotations were obtained in Singapore Dollars (S$), and subsequently converted to United States Dollars (US$), with S$1 equivalent to US$0.797, according to the international exchange rate on 17 June 2013.

Cost of Tissue-Engineered Constructs

Cost calculations were based on a published isolation and culture protocol that is able to reliably expand HCECs up to the third passage [25,26]. An overview of this process is illustrated in Figure 1. Briefly, HCECs were isolated from a pair of transplant-grade donor corneas by a two-step "peel-and-digest" method and were expanded using a dual-media culture system [25,26]. However, HCECs from different donors exhibit marked donor-to-donor variability [24,32]. Therefore, a more conservative estimate was used for calculations: HCECs were used at the end of the second passage (P2), ensuring that all batches maintain good morphology. Using this protocol, HCECs from one pair of corneas can be expanded to 6.0 million cells on average by the end of P2, in 6 weeks [25]. At the end of P2, 200,000 HCECs were seeded onto each 9 mm-diameter plastic compressed collagen carrier, producing 30 tissue-engineered constructs, each with a projected endothelial cell density (ECD) of 3144 cells/mm^2. From each batch of 30 constructs, 2 were selected randomly for QA testing, which involved testing for mycoplasma, endotoxin, and corneal endothelium-associated markers including Na$^+$/K$^+$-ATPase, ZO-1, GPC-4, CD200, SLC4A11, COL8A1/2, and CYYR1 [33,34]. These constructs used for QA testing were not transplanted.

It was assumed that one pair of corneas arrived every fortnight, and took 6 weeks to culture. Pairs of corneas at different stages of the protocol could be cultured simultaneously. Therefore, every fortnight, one batch of 28 transplantable constructs was ready, and one new pair of corneas arrived. At any point in time, there were 3 batches being processed concurrently, at different stages. This produced an average of 14 constructs per week. Two qualified, full-time staff should be able to manage this workload.

The monetary costs involved in setting up and running a laboratory for the implementation of this protocol were derived in consultation with senior scientists from the Tissue Engineering and Stem Cell Group of the Singapore Eye Research Institute, who have developed this protocol. Conceptually, costs were divided

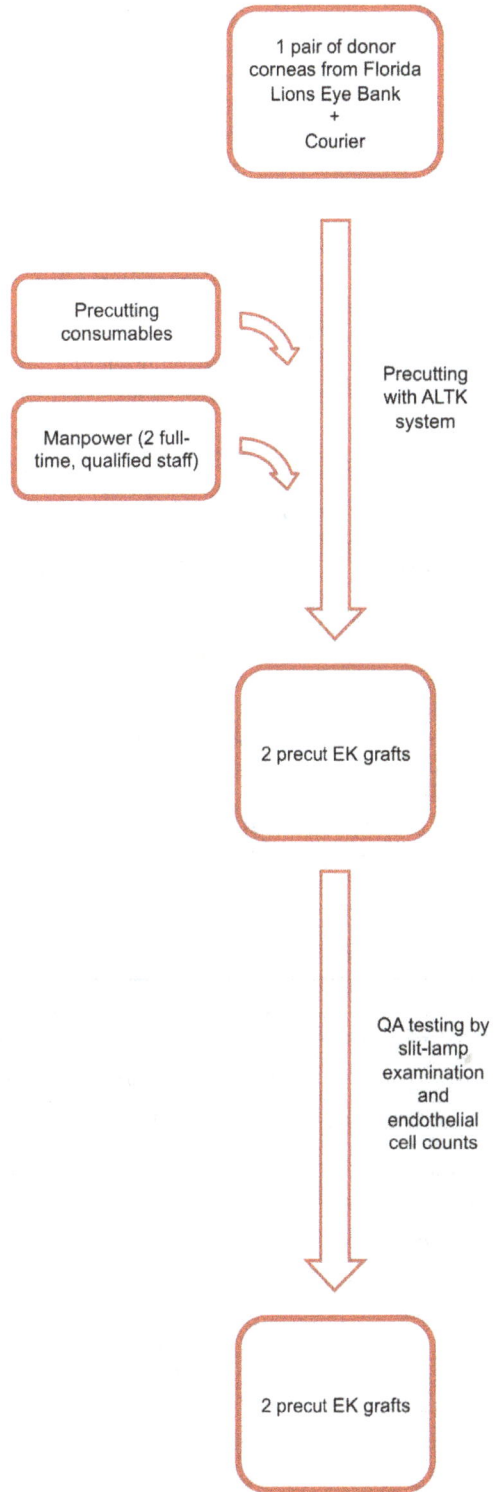

Figure 1. Overview of transplant strategies. Overview of the tissue engineering strategy (in blue) and the procured tissue strategy (in red). Abbreviations: GMP, Good Manufacturing Practice; QA, Quality Assurance; ALTK, Automated Lamellar Therapeutic Keratoplasty; EK, endothelial keratoplasty.

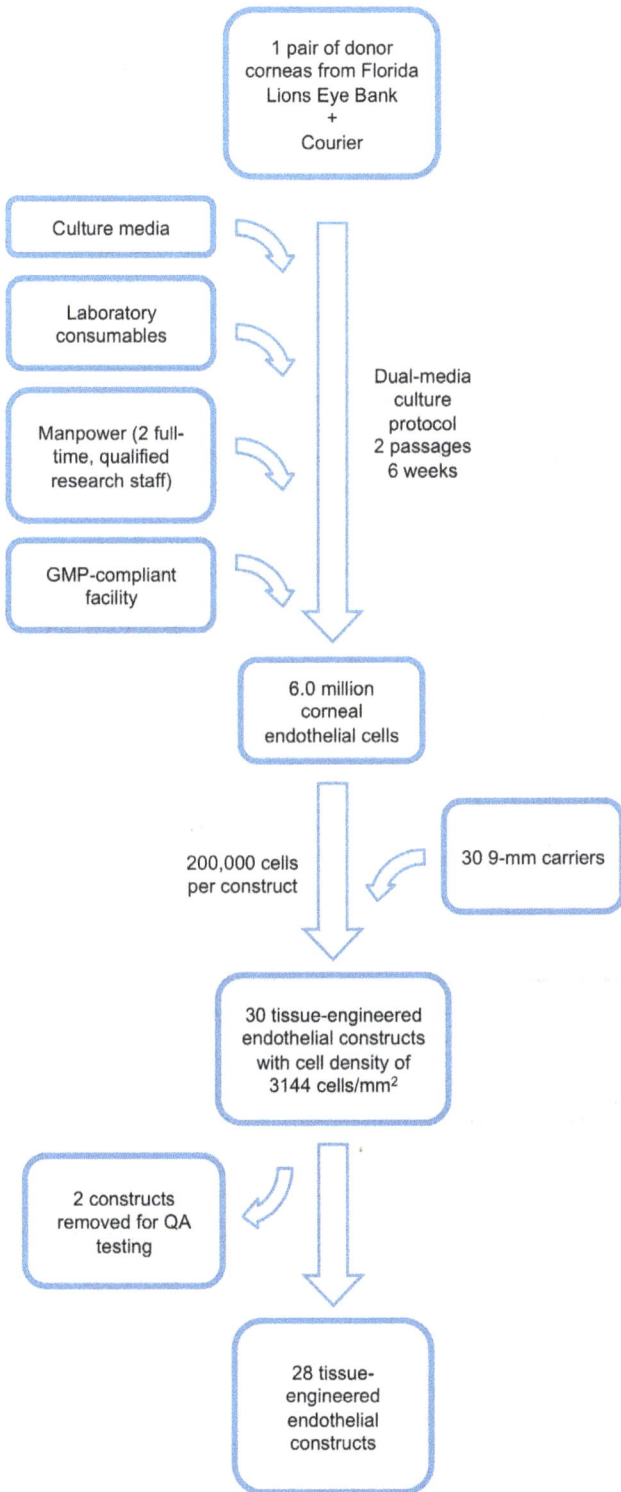

into investment costs and recurring costs. Investment costs were the capital outlay for the necessary equipment (Table 1). Recurring costs were calculated on a per-pair-of-corneas basis, and included costs of manpower, donor corneas for culture (inclusive of door-to-door courier service from the eye bank), culture media components, laboratory consumables, plastic collagen compressed carriers, QA testing, and rental of a suitable laboratory compliant with Good Manufacturing Practice (GMP) standards. Rental cost of a GMP-compliant laboratory was estimated in consultation with the director of such a laboratory in Singapore. Annual recurring costs were then derived by multiplying the recurring cost per pair of corneas by the number of cornea pairs processed in one year. Recurring costs are listed in Table 2. Applying an annual amortization rate of 20% to investment costs [35], and adding this to annual recurring costs derived total annual cost. Tissue cost per EK was calculated by dividing total annual cost by the total number of constructs produced per year, less those used for QA testing.

Cost of Utilizing Procured Corneal Tissue

Utilization of procured corneal tissue for EK requires a facility, which is able to perform precutting of donor corneal tissue, and run QA testing by slit-lamp examination and endothelial cell counts. An overview of this process can be found in Figure 1. To allow comparison with the tissue engineering strategy, it was assumed that this facility produced 14 precut EK grafts a week. Therefore, 7 pairs of corneas were processed per week. Two qualified, full-time staff, using 1 ALTK system, should be able to manage this workload.

Monetary costs involved in the setup and running of such a facility were calculated in consultation with the manager of an established local eye bank. These were similarly divided into investment costs (Table 1) and recurring costs (Table 2). Recurring costs were first calculated on a per-pair-of-corneas basis, and included costs of manpower, donor corneas (inclusive of door-to-

door courier service) and precutting consumables. Annual recurring costs were then derived by multiplying the recurring cost per pair of corneas by the number of cornea pairs processed in one year. Applying an annual amortization rate of 20% to investment costs [35], and adding this to annual recurring costs derived total annual cost. Tissue cost per transplant was calculated by dividing total annual cost by the total number of precut grafts produced per year.

Sensitivity Analysis

A number of assumptions were made in the cost-minimization analysis. In view of this, various types of sensitivity analyses were performed to test the robustness of the results. Key inputs of the model were identified (Table 3), and varied in order to examine their effect on the overall outcome of the cost-minimization analysis. Each input variable was assigned a reasonable sensitivity range (Table 3). Where possible, these sensitivity ranges were guided by available data. Otherwise, half the base case value was taken as the lower limit, and double the base case value was taken as the upper limit. One-way sensitivity analyses were performed in which each of the seven variables was varied individually, and the outcome examined (Figure 2). The outcome measured in this analysis was the cost advantage of the tissue engineering strategy over the procured-tissue strategy. Cost advantage was derived by subtracting tissue cost per transplant of the tissue engineering strategy from tissue cost per transplant of the procured-tissue strategy.

Probabilistic sensitivity analyses were also performed, in which all seven variables were varied simultaneously. Each variable was assumed to follow a triangular distribution within its designated sensitivity range (Table 3), and a specific value chosen by random sampling. These values were used to calculate the tissue cost per transplant for both strategies, which were then compared. This simulation was run 10,000 times, in order to determine which

Table 1. Investment costs.

Tissue-engineered constructs			Procured donor tissue		
Item	No.	Unit cost (US$)	Item	No.	Unit cost (US$)
4°C laboratory refrigerator	1	3,188	Eye bank refrigerator (4–8°C, with temperature-recording graph)	1	7,500
−20°C laboratory freezer	1	3,985	Specular microscope	1	40,000
Dissection microscope	1	7,970	Slit-lamp biomicroscope	1	7,000
Inverted light microscope	1	3,188	ALTK system	1	100,000
Biosafety cabinet	1	6,376	Ultrasound pachymeter	1	7,000
CO_2 incubator	1	4,782	Laminar flow hood	1	7,000
Centrifuge	1	9,564	Small box freezer	1	250
Vacuum pump	1	957			
Single channel pipettes	1	797			
Serological pipet-aid	1	399			
Forceps	1	168			
Hemocytometer	1	399			
Water bath	1	957			
Water purification system	1	7,970			
Investment cost (US$)		50,700	Investment cost (US$)		168,750

Abbreviations: ALTK system, Automated Lamellar Therapeutic Keratoplasty system (Moria, Antony, France).

Table 2. Recurring costs.

Tissue-engineered constructs		Procured donor tissue	
Item	Cost per pair of corneas (US$)	Item	Cost per pair of corneas (US$)
1 pair of donor corneas	2,900×2	1 pair of donor corneas	2,900×2
Courier	250	Courier	250
Manpower	5978	Manpower	770
Culture media components	538	Precutting consumables	250×2
Laboratory consumables	54		
Plastic compressed collagen carriers	200×30		
QA testing	66		
Rental of GMP-compliant laboratory	5,579		
Recurring cost per pair of corneas (US$)	**24,265**	**Recurring cost per pair of corneas (US$)**	**7,320**
No. of cornea pairs per year	26	No. of cornea pairs per year	364
Annual recurring cost (US$)	**630,890**	**Annual recurring cost (US$)**	**2,664,480**

Abbreviations: QA, Quality Assurance; GMP, Good Manufacturing Practice.

strategy produced transplantable tissue at a lower cost in the majority of simulations.

Additionally, in order to make the results of this study more generalizable, two alternative specific real-world scenarios were tested. First, some eye banks are able to prepare EK grafts with a DMEK technique [8,9,36]. If the facility were to prepare procured donor tissue with such a technique, they would not need to invest in an ALTK system or require precutting consumables. As an ALTK system requires a hefty investment (US$100,000), this scenario was specifically simulated and compared against the tissue engineering strategy. Second, many eye banks in the United States of America (USA) are able to precut EK grafts before distribution [9,12,13,19]. Florida Lions Eye Bank provides precut EK grafts for US$3,400 (as opposed to US$2,900 for non-precut tissue). As a result, many corneal surgeons in the USA have precut EK grafts delivered to the operating theatre just prior to surgery. In so doing,

they obviate the need for precutting equipment (including an ALTK system) and consumables, QA testing, or manpower for a dedicated facility. Their recurring costs would only include that of the precut tissue and courier service, and they would only need to invest in an eye bank refrigerator for short-term storage of procured precut EK grafts. As this is a common occurrence in the USA, specific cost calculations were also performed for this scenario and compared with the tissue engineering strategy.

Results

Investment Costs

In the tissue engineering strategy, the investment cost for the laboratory was US$50,700 (Table 1). In contrast, with the procured-tissue strategy, the investment cost needed for the facility was US$168,750 (Table 1). A large proportion of this figure came

Table 3. Sensitivity ranges.

Variable	Base case value	Sensitivity range	Remarks
Cost of each donor cornea (US$)	2,900	1,000–5,800	2,900 is the cost of a transplant-grade donor cornea from Florida Lions Eye Bank, Miami, FL, USA. 1,000 is the cost of a transplant-grade donor cornea from the National Eye Bank of Sri Lanka, Colombo, Sri Lanka.
Cost of plastic compressed collagen construct (US$)	200	100–500	
Rental of GMP-compliant laboratory per pair of corneas (US$)	5,579	2,789.5–11,158	
Cost of culture media components per pair of corneas (US$)	538	269–2,690	The GMP-compliant equivalents of some culture media components can be 5 times more expensive than their non-GMP-compliant counterparts. Therefore, the upper limit of the sensitivity range was set at 5 times the base case value.
Cost of QA testing per pair of corneas (US$)	66	33–132	
Culture yield (no. of constructs produced per pair of corneas, before QA testing)	30	15–60	
No. of transplantable EK grafts produced per week (for both strategies)	14	7–28	

Abbreviations: GMP, Good Manufacturing Practice; QA, quality assurance; EK, endothelial keratoplasty.

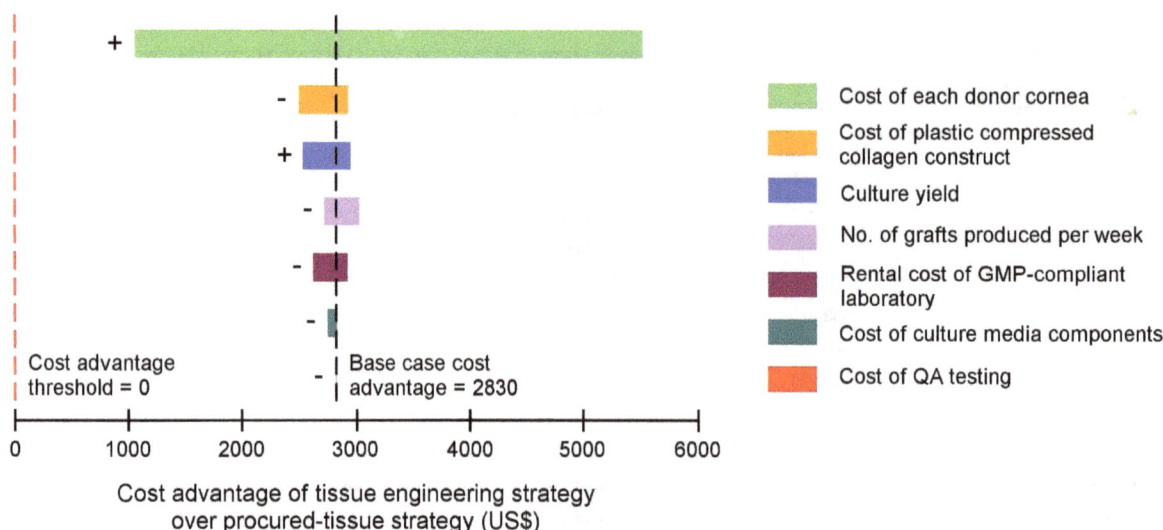

Figure 2. One-way sensitivity analysis. One-way sensitivity analysis of selected variables on the cost advantage of the tissue engineering strategy over the procured-tissue strategy. + indicates that the cost advantage increases as that variable increases, while - indicates that the cost advantage decreases as that variable increases. Abbreviations: GMP, Good Manufacturing Practice; QA, Quality Assurance.

from the US$100,000 cost of the ALTK system used for precutting grafts.

Recurring Costs

In order to produce 14 tissue-engineered constructs per week, the laboratory processed 1 pair of donor corneas every 2 weeks. The recurring cost of processing each pair was US$24,270. In 1 year, the laboratory operating at this rate processed 26 pairs of corneas, producing 728 transplantable constructs. Therefore, the annual recurring cost for the laboratory was US$630,890 (Table 2).

In contrast, the facility processed 7 pairs of donor corneas every week to produce 14 precut grafts per week. The recurring cost of processing each pair was US$7,320. At this rate, the facility processed 364 pairs of corneas in 1 year to produce 728 precut grafts. The annual recurring cost for the facility was US$2,664,480 (Table 2).

Tissue Cost per Transplant

Producing 728 transplantable constructs per year, the laboratory required a total annual cost of US$641,030. Therefore, the

laboratory produced transplantable tissue for US$880 per transplant on average (Table 4). At the same rate of production, the facility required a total annual cost of US$2,698,230. Therefore, the facility produced transplantable tissue for US$3,710 per transplant (Table 4).

Sensitivity Analysis

Sensitivity analyses performed support the outcome of the cost-minimization analysis. In one-way sensitivity analyses, varying individual variables (Table 3) in the model still resulted in the same conclusion. Within the assigned sensitivity ranges, the tissue engineering strategy was always the preferred strategy; cost advantage never crossed the threshold of 0 (Figure 2). As an illustration, if one were to ignore the assigned sensitivity ranges, the cost of the plastic compressed collagen carrier would have had to be multiplied 14 times in order for both strategies to be equal in cost. The equivalent multipliers for rental cost of a GMP-compliant laboratory, cost of culture media components and QA testing cost were 15, 150 and 1200 respectively. With the remaining three variables, there was no positive value at which cost advantage reached 0. Two of the variables (cost of donor cornea and culture yield) were positively related to the cost advantage (labeled "+" in Figure 2); as these variables increased, the cost advantage of the tissue engineering strategy increased. The other five variables were inversely related to the cost advantage (labeled "-"). In terms of the magnitude of effect, it was found that the largest determinant of the cost advantage was the cost of each donor cornea. This was followed by the cost of the plastic compressed collagen carrier, and by the culture yield from each pair of corneas.

In probabilistic sensitivity analyses, all seven variables were varied simultaneously across 10,000 simulations. The tissue engineering strategy produced transplantable tissue at a lower cost than the procured-tissue strategy in 100% of simulations.

As an additional form of sensitivity analysis, two specific alternative scenarios were tested. First, if the facility in the procured-tissue strategy were to prepare EK grafts with a DMEK technique (without the need for an ALTK system or precutting consumables), the tissue cost per transplant would be reduced to

Table 4. Tissue cost per transplant.

	Tissue-engineered constructs	Procured donor tissue
Investment cost (US$)	50,700	168,750
Amortization of investment cost (20%) (US$)	10,140	33,750
Annual recurring cost (US$)	630,890	2,664,480
Total annual cost (US$)	641,030	2,698,230
No. of transplants per year	728	728
Tissue cost per transplant (US$)	**880**	**3,710**

Abbreviations: None.

US$3,430. Second, if eye surgeons were to accept delivery of precut EK grafts direct from the eye bank, as is the case for many corneal surgeons in the USA, the tissue cost per transplant would be US$3,660. Both of these figures are still significantly more than the US$880 per transplant produced by the tissue engineering strategy.

Discussion

The technique of cultivating HCECs from one donor *ex vivo* and transplanting them on a carrier to treat endothelial disease in the recipient was conceptualized over 30 years ago [23]. Since then, however, efforts to translate this into clinical practice have been hampered by difficulty in culturing HCECs and the lack of effective surgical techniques to transplant them. Recently though, greater understanding in the cell biology of HCECs [37–40] has led to the establishment of reliable HCEC culture protocols [3,24–26]. HCECs can now be consistently expanded up to the third passage, while retaining their unique cellular morphology and the expression of characteristic markers indicative of the corneal endothelium [24,25]. Suitable carriers for these cultured HCECs have also been successfully tested [41–45]. Finally, the advent and success of EK techniques like DSEK/DMEK enables the effective surgical delivery of these constructs [7,8,10,27,28]. Such recent advances may have made tissue-engineered endothelial constructs a realistic prospect, but can they be produced at a competitive cost?

The results of this cost-minimization analysis indicate that tissue engineering can produce transplantable corneal endothelial tissue at a fraction of current costs (Table 4). Both investment costs and recurring costs were lower for tissue-engineered constructs compared to tissue procured from eye banks. Although certain assumptions were necessary to allow comparison between the two strategies, we performed multiple sensitivity analyses to attempt to account for uncertainty in our estimates, and variation in costs across different settings or countries. These sensitivity analyses performed demonstrate that the results of this study are robust and may even be generalizable across various settings. In particular, even if the substantial cost of the ALTK system was removed from the procured-tissue strategy, or if one were to procure already-precut tissue from eye banks (as is common practice in the USA), the tissue engineering strategy still provides a significant cost advantage. This reduction in the cost of transplantable tissue could potentially make corneal endothelial transplants more accessible for a large number of patients. Furthermore, one-way sensitivity analyses have identified two key variables that are positively related to the cost advantage of the tissue engineering strategy over procured corneal tissue: the cost of donor corneas, and the culture yield (indicated "+" in Figure 2). If the costs of donor corneal tissue from eye banks continue to rise (as they have done historically) [14], and as isolation and culture protocols for HCECs are refined and culture yield improves, the cost of corneal endothelial transplantation could fall even further.

A key advantage of the tissue engineering strategy is that *in vitro* expansion of HCECs allows tissue from one donor to treat multiple recipients [3,41]. Widespread adoption of this strategy should increase the overall supply of transplantable corneal endothelial tissue, and help to meet the current and expected future shortfalls in tissue for corneal endothelial transplants [3]. In our center, and in many others in Asia, demand for corneal endothelial transplants exceeds supply. There are always patients on the waiting list, and donor tissue is not always readily available. However, if the supply of donor tissue was more predictable, like in the tissue engineering strategy, patients on the waiting list could

be scheduled for surgery based on when the grafts will be ready, reducing waiting time for patients, and minimizing week-to-week fluctuation in transplant numbers. Also, because multiple transplants can be performed from one donor cornea, serological and other Quality Assurance (QA) tests need only be done on the one source cornea, resulting in significant cost savings, and further reducing delays in transplantation.

The implications of this analysis are not limited to corneal endothelial transplantation alone. This analytic model could potentially be applied to other fields, including: other types of tissue-engineered constructs e.g. epithelial constructs for ocular surface reconstruction, as well as transplantation of other tissues like pancreatic islets for the treatment of Type 1 Diabetes Mellitus [46,47]. Stem cell-based alternatives to donor pancreatic islet tissue are currently an area of active research [48–53]. As cell culture protocols and techniques become more clearly defined, it would be important to conduct similar pharmacoeconomic analyses of these stem cell-based strategies for pancreatic islet transplantation as well. To the best of our knowledge, no such studies have been attempted yet.

The results of this study should be interpreted with a measure of caution. Cost-minimization analysis is grounded in the assumption that the competing therapies produce equivalent outcomes [31]. While tissue-engineered endothelial constructs should, in theory, perform as well as precut donor tissue in terms of surgical ease, complication rates and outcomes, this has yet to be proven. We do not know if the two approaches will have equivalent success rates, long-term graft survival or quality of life after transplantation. There is currently no data from human studies that proves their therapeutic equivalence, and future clinical trials in this area are needed. Our pharmacoeconomic analyses are complementary to clinical trials in this area, and should be interpreted alongside such work. If data arises showing that the two therapies do indeed differ in outcome, then a full cost-effectiveness analysis will be necessary. Nevertheless, the authors feel that for the purposes of this analysis, this assumption is reasonable. From a surgical perspective at least, corneal endothelial transplantation with tissue-engineered grafts is technically feasible, and should not be significantly different from precut EK grafts in terms of surgical ease or complication rates. Levis et al found that their plastic compressed collagen carrier could be successfully manipulated and deployed using current graft delivery devices without complication [41]. Also, cultured HCECs express cellular markers characteristic of *in vivo* HCECs, such as Na^+/K^+-ATPase, that are in part responsible for the function of the corneal endothelium in maintenance of corneal clarity [24]. Therefore, although more clinical research is needed in this field, the authors feel that the above assumptions are not unreasonable.

Some corneal transplant surgeons have reported using single donor corneas for both deep anterior lamellar keratoplasty (DALK) and EK procedures in two separate recipients [54–58]. This technique of split cornea transplantation, if widely adopted by surgeons or eye banks, could potentially help to alleviate the shortage of donor corneal tissue. Also, the use of a single donor cornea for two separate transplants would significantly reduce the cost of procured donor tissue for each transplant procedure. However, the impact of this concept is limited by the fact that one of the two procedures would have to be a DALK, which is less commonly performed than EK procedures like DSEK. Even though DALK is performed quite often in Singapore [59], it only accounts for 1.89% of corneal transplant procedures performed in the United States [9]. Furthermore, DALK is often performed with tissue that has endothelium that would be unsuitable for EK surgery [60] and hence, this is unlikely to have significant impact on the results of this cost-minimization analysis.

If tissue-engineered endothelial constructs are to be produced on a large scale, the validity of the QA tests used must be well established. Presently, there are a few markers specific to HCECs that can be tested for, such as Na^+/K^+-ATPase, ZO-1, GPC-4, CD200, SLC4A11, COL8A1/2, and CYYR1 [33,34]. However, none of these have been directly linked to clinical success of the transplant. Corneal epithelial cultures, in contrast, have a validated QA test using expression of $\Delta Np63\alpha$ that has been directly linked to successful clinical outcomes [22]. Ideally, a similarly validated QA test should be developed for tissue-engineered endothelial constructs before use becomes widespread.

Finally, there are limits to the generalization of these results. This analysis was performed in the authors' setting of Singapore. While we have performed different sensitivity analyses in an attempt to make these results generalizable across various settings, there are certain scenarios that are beyond the scope of this paper. First, this analysis was designed from the perspective of an institution that currently has no means of acquiring EK tissue. Therefore, the results are not immediately applicable to an institution that has already made investments in an eye bank and/ or an ALTK system. Also, there are many considerations unique to healthcare institutions in the developing world, and various cost factors may be quite different. Therefore, these results may not be directly applicable to institutions in the developing world. In addition, we acknowledge that institutions in different countries operate within vastly different healthcare settings. Our analysis

simply shows that a tissue engineering strategy can potentially increase the supply of transplantable tissue, which will in turn lower the cost of tissue per corneal endothelial transplant. Whether these cost savings will be passed on to patients or enjoyed as additional revenue for the institution will vary based on market characteristics. There are many economic factors besides the production cost of a therapy for a particular institution to consider in its investment decisions, such as government reimbursement or subsidy rates. More detailed market and budget impact analyses would have to be performed by each institution, taking into account the unique healthcare environment they operate in. Such issues are beyond the scope of this paper.

In conclusion, tissue-engineered constructs for corneal endothelial transplantation can be produced at a lower cost than donor tissue procured from eye banks. Future clinical trials are needed to establish the therapeutic equivalence of both these tissue sources for EK. Tissue-engineered endothelial constructs have the potential to increase the supply of transplantable corneal endothelial tissue and bring the overall costs of corneal endothelial transplantation down.

Author Contributions

Conceived and designed the experiments: TET GSLP BLG EAF JSM. Performed the experiments: TET GSLP BLG HYC DD. Analyzed the data: TET DD EAF JSM. Contributed reagents/materials/analysis tools: GSLP BLG HYC DD. Wrote the paper: TET GSLP BLG HYC DD EAF JSM.

References

1. Mannis MJ, Holland EJ, Beck RW, Belin MW, Goldberg MA, et al. (2006) Clinical profile and early surgical complications in the Cornea Donor Study. Cornea 25: 164–170.
2. Tan DT, Anshu A, Mehta JS (2009) Paradigm shifts in corneal transplantation. Ann Acad Med Singapore 38: 332–338.
3. Peh GS, Beuerman RW, Colman A, Tan DT, Mehta JS (2011) Human corneal endothelial cell expansion for corneal endothelium transplantation: an overview. Transplantation 91: 811–819.
4. Tan DT, Dart JK, Holland EJ, Kinoshita S (2012) Corneal transplantation. Lancet 379: 1749–1761.
5. Anshu A, Price MO, Tan DT, Price FW Jr (2012) Endothelial keratoplasty: a revolution in evolution. Surv Ophthalmol 57: 236–252.
6. Price MO, Gorovoy M, Price FW Jr, Benetz BA, Menegay HJ, et al. (2013) Descemet's stripping automated endothelial keratoplasty: three-year graft and endothelial cell survival compared with penetrating keratoplasty. Ophthalmology 120: 246–251.
7. Price FW Jr, Price MO (2005) Descemet's stripping with endothelial keratoplasty in 50 eyes: a refractive neutral corneal transplant. J Refract Surg 21: 339–345.
8. Melles GR, Ong TS, Ververs B, van der Wees J (2006) Descemet membrane endothelial keratoplasty (DMEK). Cornea 25: 987–990.
9. Eye Bank Association of America (2013) 2012 Eye Banking Statistical Report.
10. Gorovoy MS (2006) Descemet-stripping automated endothelial keratoplasty. Cornea 25: 886–889.
11. Bose S, Ang M, Mehta JS, Tan DT, Finkelstein E (2012) Cost-Effectiveness of Descemet's Stripping Endothelial Keratoplasty versus Penetrating Keratoplasty. Ophthalmology.
12. Woodward MA, Titus M, Mavin K, Shtein RM (2012) Corneal donor tissue preparation for endothelial keratoplasty. J Vis Exp: e3847.
13. Kitzmann AS, Goins KM, Reed C, Padnick-Silver L, Macsai MS, et al. (2008) Eye bank survey of surgeons using precut donor tissue for descemet stripping automated endothelial keratoplasty. Cornea 27: 634–639.
14. Cole G (1994) Managing the future of cornea supply and demand: the costs affecting eye banking. Cornea 13: 87–89.
15. Badenoch PR, Coster DJ (2009) Is viral nucleic acid testing of eye donors cost-effective? Med J Aust 191: 408.
16. Sabater AL, Guarnieri A, Espana EM, Li W, Prosper F, et al. (2013) Strategies of human corneal endothelial tissue regeneration. Regen Med 8: 183–195.
17. Keenan TD, Carley F, Yeates D, Jones MN, Rushton S, et al. (2011) Trends in corneal graft surgery in the UK. Br J Ophthalmol 95: 468–472.
18. Woodward MA, Ross KW, Requard JJ, Sugar A, Shtein RM (2013) Impact of surgeon acceptance parameters on cost and availability of corneal donor tissue for transplantation. Cornea 32: 737–740.
19. Li JY, Mannis MJ (2010) Eye banking and the changing trends in contemporary corneal surgery. Int Ophthalmol Clin 50: 101–112.

20. Langer R, Vacanti JP (1993) Tissue engineering. Science 260: 920–926.
21. Ezhkova E, Fuchs E (2010) Regenerative medicine: An eye to treating blindness. Nature 466: 567–568.
22. Rama P, Matuska S, Paganoni G, Spinelli A, De Luca M, et al. (2010) Limbal stem-cell therapy and long-term corneal regeneration. N Engl J Med 363: 147–155.
23. Jumblatt MM, Maurice DM, McCulley JP (1978) Transplantation of tissue-cultured corneal endothelium. Invest Ophthalmol Vis Sci 17: 1135–1141.
24. Peh GS, Toh KP, Wu FY, Tan DT, Mehta JS (2011) Cultivation of human corneal endothelial cells isolated from paired donor corneas. PLoS One 6: e28310.
25. Peh GS, Toh KP, Ang HP, Seah XY, George BL, et al. (2013) Optimization of human corneal endothelial cell culture: density dependency of successful cultures in vitro. BMC Res Notes 6: 176.
26. Peh GS, Chng Z, Ang HP, Cheng TY, Adnan K, et al. (2013) Propagation of Human Corneal Endothelial Cells? A Novel Dual Media Approach. Cell Transplant.
27. Melles GR, Wijdh RH, Nieuwendaal CP (2004) A technique to excise the descemet membrane from a recipient cornea (descemetorhexis). Cornea 23: 286–288.
28. Khor WB, Mehta JS, Tan DT (2011) Descemet stripping automated endothelial keratoplasty with a graft insertion device: surgical technique and early clinical results. Am J Ophthalmol 151: 223–232 e222.
29. Koizumi N, Sakamoto Y, Okumura N, Okahara N, Tsuchiya H, et al. (2007) Cultivated corneal endothelial cell sheet transplantation in a primate model. Invest Ophthalmol Vis Sci 48: 4519–4526.
30. Koizumi N, Sakamoto Y, Okumura N, Tsuchiya H, Torii R, et al. (2008) Cultivated corneal endothelial transplantation in a primate: possible future clinical application in corneal endothelial regenerative medicine. Cornea 27 Suppl 1: S48–55.
31. Jolicoeur LM, Jones-Grizzle AJ, Boyer JG (1992) Guidelines for performing a pharmacoeconomic analysis. Am J Hosp Pharm 49: 1741–1747.
32. Zhu C, Joyce NC (2004) Proliferative response of corneal endothelial cells from young and older donors. Invest Ophthalmol Vis Sci 45: 1743–1751.
33. Cheong YK, Ngoh ZX, Peh GS, Ang HP, Seah XY, et al. (2013) Identification of cell surface markers glypican-4 and CD200 that differentiate human corneal endothelium from stromal fibroblasts. Invest Ophthalmol Vis Sci 54: 4538–4547.
34. Chng Z, Peh GS, Herath WB, Cheng TY, Ang HP, et al. (2013) High throughput gene expression analysis identifies reliable expression markers of human corneal endothelial cells. PLoS One 8: e67546.
35. Bohringer D, Maier P, Sundmacher R, Reinhard T (2009) Costs and financing. A cost calculation of an up-to-date eye bank in Germany. Dev Ophthalmol 43: 120–124.

36. Schlotzer-Schrehardt U, Bachmann BO, Tourtas T, Cursiefen C, Zenkel M, et al. (2013) Reproducibility of Graft Preparations in Descemet's Membrane Endothelial Keratoplasty. Ophthalmology 120: 1769–1777.

37. Okumura N, Koizumi N, Ueno M, Sakamoto Y, Takahashi H, et al. (2012) ROCK inhibitor converts corneal endothelial cells into a phenotype capable of regenerating in vivo endothelial tissue. Am J Pathol 181: 268–277.

38. Zhu YT, Chen HC, Chen SY, Tseng SC (2012) Nuclear p120 catenin unlocks mitotic block of contact-inhibited human corneal endothelial monolayers without disrupting adherent junctions. J Cell Sci 125: 3636–3648.

39. Hirata-Tominaga K, Nakamura T, Okumura N, Kawasaki S, Kay EP, et al. (2013) Corneal endothelial cell fate is maintained by LGR5 through the regulation of hedgehog and Wnt pathway. Stem Cells 31: 1396–1407.

40. Shima N, Kimoto M, Yamaguchi M, Yamagami S (2011) Increased proliferation and replicative lifespan of isolated human corneal endothelial cells with L-ascorbic acid 2-phosphate. Invest Ophthalmol Vis Sci 52: 8711–8717.

41. Levis HJ, Peh GS, Toh KP, Poh R, Shortt AJ, et al. (2012) Plastic compressed collagen as a novel carrier for expanded human corneal endothelial cells for transplantation. PLoS One 7: e50993.

42. Watanabe R, Hayashi R, Kimura Y, Tanaka Y, Kageyama T, et al. (2011) A novel gelatin hydrogel carrier sheet for corneal endothelial transplantation. Tissue Eng Part A 17: 2213–2219.

43. Bayyoud T, Thaler S, Hofmann J, Maurus C, Spitzer MS, et al. (2012) Decellularized bovine corneal posterior lamellae as carrier matrix for cultivated human corneal endothelial cells. Curr Eye Res 37: 179–186.

44. Wang TJ, Wang IJ, Lu JN, Young TH (2012) Novel chitosan-polycaprolactone blends as potential scaffold and carrier for corneal endothelial transplantation. Mol Vis 18: 255–264.

45. Liang Y, Liu W, Han B, Yang C, Ma Q, et al. (2011) Fabrication and characters of a corneal endothelial cells scaffold based on chitosan. J Mater Sci Mater Med 22: 175–183.

46. Shapiro AM (2011) State of the art of clinical islet transplantation and novel protocols of immunosuppression. Curr Diab Rep 11: 345–354.

47. Zinger A, Leibowitz G (2013) Islet transplantation in type 1 diabetes: hype, hope and reality - a clinician's perspective. Diabetes Metab Res Rev.

48. Aguayo-Mazzucato C, Bonner-Weir S (2010) Stem cell therapy for type 1 diabetes mellitus. Nat Rev Endocrinol 6: 139–148.

49. Dominguez-Bendala J, Inverardi L, Ricordi C (2011) Stem cell-derived islet cells for transplantation. Curr Opin Organ Transplant 16: 76–82.

50. Wen Y, Chen B, Ildstad ST (2011) Stem cell-based strategies for the treatment of type 1 diabetes mellitus. Expert Opin Biol Ther 11: 41–53.

51. Leung PS, Ng KY (2013) Current progress in stem cell research and its potential for islet cell transplantation. Curr Mol Med 13: 109–125.

52. Viswanathan C, Sarang S (2013) Status of stem cell based clinical trials in the treatment for diabetes. Curr Diabetes Rev 9: 429–436.

53. Zhang WJ, Xu SQ, Cai HQ, Men XL, Wang Z, et al. (2013) Evaluation of islets derived from human fetal pancreatic progenitor cells in diabetes treatment. Stem Cell Res Ther 4: 141.

54. Sharma N, Agarwal P, Titiyal JS, Kumar C, Sinha R, et al. (2011) Optimal use of donor corneal tissue: one cornea for two recipients. Cornea 30: 1140–1144.

55. Heindl LM, Riss S, Bachmann BO, Laaser K, Kruse FE, et al. (2011) Split cornea transplantation for 2 recipients: a new strategy to reduce corneal tissue cost and shortage. Ophthalmology 118: 294–301.

56. Heindl LM, Riss S, Laaser K, Bachmann BO, Kruse FE, et al. (2011) Split cornea transplantation for 2 recipients - review of the first 100 consecutive patients. Am J Ophthalmol 152: 523–532 e522.

57. Heindl LM, Riss S, Adler W, Bucher F, Hos D, et al. (2013) Split cornea transplantation: relationship between storage time of split donor tissue and outcome. Ophthalmology 120: 899–907.

58. Vajpayee RB, Sharma N, Jhanji V, Titiyal JS, Tandon R (2007) One donor cornea for 3 recipients: a new concept for corneal transplantation surgery. Arch Ophthalmol 125: 552–554.

59. Mohamed-Noriega K, Angunawela RI, Tan D, Mehta JS (2011) Corneal transplantation: changing techniques. Transplantation 92: e31–32; author reply e32–33.

60. Koo TS, Finkelstein E, Tan D, Mehta JS (2011) Incremental cost-utility analysis of deep anterior lamellar keratoplasty compared with penetrating keratoplasty for the treatment of keratoconus. Am J Ophthalmol 152: 40–47 e42.

Dermal Substitutes Support the Growth of Human Skin-Derived Mesenchymal Stromal Cells: Potential Tool for Skin Regeneration

Talita da Silva Jeremias[1], Rafaela Grecco Machado[1], Silvia Beatriz Coutinho Visoni[1], Maurício José Pereima[2,3], Dilmar Francisco Leonardi[4,5], Andrea Gonçalves Trentin[1]*

1 Departamento de Biologia Celular, Embriologia e Genética, Centro de Ciências Biológicas, Universidade Federal de Santa Catarina, Florianópolis, Santa Catarina, Brasil, 2 Departamento de Pediatria, Centro de Ciências da Saúde, Universidade Federal de Santa Catarina, Florianópolis, Santa Catarina, Brasil, 3 Hospital Infantil Joana de Gusmão, Florianópolis, Santa Catarina, Brasil, 4 Hospital Governador Celso Ramos, Florianópolis, Santa Catarina, Brasil, 5 Departamento de Cirurgia, Universidade do Sul de Santa Catarina, Florianópolis, Santa Catarina, Brasil

Abstract

New strategies for skin regeneration are needed in order to provide effective treatment for cutaneous wounds and disease. Mesenchymal stem cells (MSCs) are an attractive source of cells for tissue engineering because of their prolonged self-renewal capacity, multipotentiality, and ability to release active molecules important for tissue repair. In this paper, we show that human skin-derived mesenchymal stromal cells (SD-MSCs) display similar characteristics to the multipotent MSCs. We also evaluate their growth in a three-dimensional (3D) culture system with dermal substitutes (Integra and Pelnac). When cultured in monolayers, SD-MSCs expressed mesenchymal markers, such as CD105, Fibronectin, and α-SMA; and neural markers, such as Nestin and βIII-Tubulin; at transcriptional and/or protein level. Integra and Pelnac equally supported the adhesion, spread and growth of human SD-MSCs in 3D culture, maintaining the MSC characteristics and the expression of multilineage markers. Therefore, dermal substitutes support the growth of mesenchymal stromal cells from human skin, promising an effective tool for tissue engineering and regenerative technology.

Editor: Graca Almeida-Porada, Wake Forest Institute for Regenerative Medicine, United States of America

Funding: This work was supported by the Ministério da Saúde (MS-SCTIE-DECIT), Ministério da Ciência, Tecnologia e Inovação/Conselho Nacional de Desenvolvimento Científico e Tecnológico (MCTI/CNPq/Brazil), CNPq/PIBIC/PIBIT (Brazil), Coordenação de Aperfeiçoamento de Pessoal de Nível Superior (CAPES, Brazil), and Fundação de Amparo à Pesquisa do Estado de Santa Catarina (FAPESC-PPSUS, Brazil). The funders had no role in study design, data collection and analysis, decision to publish, or preparation of the manuscript.

Competing Interests: The authors have declared that no competing interest exists.

* E-mail: andrea.trentin@ufsc.br

Introduction

Full-thickness skin injuries, such as extensive burns, chronic ulcers and deep wounds result in numerous physiological and functional problems. Their treatment requires a coverage that supports repair and restoration of skin functionality. Traditional procedures use autologous skin grafts for that purpose. However, removal of patients' healthy skin is a highly invasive procedure and impossible to perform in some cases, such as extensive burns [1]. Several efforts in the field of tissue engineering have been made in order to develop a more effective and feasible treatment [2]. Such studies aim to identify molecules involved in tissue repair and to develop biomaterials that resemble the skin tissue architecture for cell therapy [3]. Notably, dermal substitutes have been used to cover the lesion to facilitate cell colonization, thereby promoting dermal regeneration [4]. Commercially available dermal substitutes have been developed with different matrices, including the dermal regeneration template Integra (Integra Lifescience, Plainsboro, NJ, USA) and Pelnac (Gunze Limited, Kyoto, Japan). Integra is a bilaminar membrane with an external silicone layer, which simulates epidermal function, and an inner layer consisting of bovine collagen fibers attached to chondroitin-6-sulfate glycosaminoglycan (shark cartilage-derived) with a mean pore diameter of 80 μm [4]. Similarly, Pelnac matrix is a bilaminar membrane with an external silicone layer and a porcine collagen sponge matrix with pore diameter in the range of 60–110 μm [5]. In both matrices, the inner layer (dermal layer) serves as a scaffold for vascularization and colonization by dermal fibroblasts [4]; [5].

Recent cell-based therapies for cutaneous lesions have combined bioartificial scaffolds and stem cells in order to improve skin regeneration [6]. In this scenario, mesenchymal stem cells (MSCs) provide several advantages, such as multipotentiality and the ability to expand *in vitro* for long periods [7]. MSCs constitute a population of adherent cells with fibroblast-like morphology and self-renewal capacity and the potential to differentiate into osteocytes, chrondrocytes and adipocytes [8]; [9]. These cells are also characterized by the expression of a specific pattern of cell-surface markers, both positive, including CD105, CD73 and CD90, and negative, including CD45, CD34, and CD14 or CD11b [9]. Moreover, MSCs modulate immune and inflammatory responses, and they also release active molecules that affect cell migration, proliferation and survival at the site of lesion. Therefore, MSCs play an active role in inflammatory, proliferative and remodeling phases of skin regeneration, thus improving tissue repair [10]; [11]. Importantly, stromal cells with functional and phenotypic proprieties similar to MSCs have been identified in skin [12].

It has been suggested that the combination of stem or progenitors cells with synthetic or natural scaffolds can provide an improved microenvironment for cell survival and functions compared with the inoculation of isolated cells directly at the site of lesion [13]. In this paper, we assessed the multipotent characteristics of human skin-derived mesenchymal stromal cells (SD-MSCs) cultured in monolayers and evaluated their integration with the dermal substitutes Integra and Pelnac through a three-dimensional (3D) culture system. Our results suggest that the association of human SD-MSC with dermal templates provides a cell-based therapeutic potential for skin regeneration.

Materials and Methods

Isolation and Culture of Human Skin-derived Mesenchymal Stromal Cells (SD-MSCs)

Human tissue fragments were obtained by written informed consent from healthy patients undergoing facial lifting. The procedure was approved by the Ethics Committee of the Federal University of Santa Catarina, Brazil. Tissue samples were digested with dispase (12.5 U/mL, 15 h, 4°C; BD), and the dermis was mechanically removed and digested again with 0.25% trypsin–EDTA (30 min, 37°C, Invitrogen). The obtained cell suspension was filtered through a 70-μm mesh (BD), centrifuged (500 g; 10 min; 22°C) and plated in 25-cm^2 flasks (Corning) in standard medium consisting of Dulbecco's Modified Eagle's Medium-F12 (DMEM-F12; Invitrogen) supplemented with 15% fetal bovine serum (FBS; Vitrocell), penicillin (200 U/mL; Invitrogen) and streptomycin (10 mg/mL; Invitrogen). Cells were maintained until confluence at 37°C in a humidified 5% CO_2 atmosphere with medium changed every 3 days and expanded up to 20 multiple passages.

Culture of Human SD-MSCs on Dermal Substitutes

The dermal substitutes Integra and Pelnac were cut into fragments of 3–4 mm^2 with a surgical punch, washed with PBS (10 min) and then with standard medium (10 min), followed by placement in 96-well culture plates with the dermal layer up. Cells ($1 \times 10^{4/}$well) were seeded and maintained as described above.

Osteogenic Differentiation

Osteogenic differentiation was performed as previously described with some modifications [14]. Briefly, cells (3×10^4/well in 24-well plates) were cultured in DMEM supplemented with dexamethasone (10^{-9} M; Sigma-Aldrich), ascorbate-2-phosphate (50 ug/mL; Sigma-Aldrich), b-glycerolphosphate (3.15 mg/mL; Sigma-Aldrich), 10% FBS and antibiotics. The medium was changed every 3 days. Control cells were cultured in standard medium. After 30 days, cells were fixed in 4% paraformaldehyde (Sigma-Aldrich) and stained with 2% Alizarin Red (Sigma-Aldrich) solution for 5 min.

Adipogenic Differentiation

Adipogenic differentiation was performed as previously described (Coura et al. 2008). Briefly, cells (3×10^4/well in 24-well plates) were cultured in DMEM supplemented with dexamethasone (10^{-8} M; Sigma-Aldrich), indomethacin (100 uM; Sigma-Aldrich), insulin (2.5 ug/mL; Sigma-Aldrich), 10% FBS and antibiotics. The medium was changed every 3 days. Control cells were cultured in standard medium. After 30 days, cells were fixed in 4% paraformaldehyde and stained with 2% Oil Red O (Sigma-Aldrich) solution for 5 min.

Flow Cytometry

Isolated skin cells (10^5 cells) were harvested by trypsinization and incubated (60 min; 4°C) with the following antibodies (all from BD Bioscience): fluorescein isothiocyanate (FITC)-conjugated anti-CD73 or -CD45; phycoerythrin (PE)-conjugated anti-CD90 or peridin chlorophyll protein (PerCP)-conjugated anti-CD-105. Negative control staining was performed by using FITC-, PE- or PerCP-conjugated mouse IgG isotype antibodies. Cells were analyzed in a FACSCalibur flow cytometer (BD Bioscience), and data were examined by FLOWJO software.

Immunofluorescence Staining

Cells were fixed in 4% formaldehyde, washed in PBS, and permeabilized for the analysis of intracellular markers (20 min, 0.25% Triton X-100; Sigma). The monolayers were then incubated with a blocking solution (PBS with 5% FBS) (45 min, room temperature), followed by incubation (overnight at 4°C) with the primary antibodies: anti-CD105 (Southern Biotech), anti-Nestin (Abcam), anti-α-SMA (α-smooth muscle actin, Sigma-Aldrich), anti-βIII-Tubulin (Promega) and anti-Fibronectin (Dako). After extensive washing in PBS, a second incubation (1 h; 37°C) with Alexa Fluor-488- or Alexa Fluor-547-specific anti-mouse or anti-rabbit secondary antibodies (all from Invitrogen) was performed. Cell nuclei were stained with 40, 6-diamino-2-phenylindole (DAPI; Sigma-Aldrich). Florescence labeling was observed using an epifluorescent microscope (Olympus IX71).

Proliferation and Viability Assay (MTS Assay)

Cell proliferation and viability were analyzed by the CellTiter 96 AQueous Non-Radioactive Cell Proliferation Assay (Promega) according to the manufacturer's instructions. Briefly, cells (1×10^4/ well in a 96-well plate) were seeded on dermal substitutes or on the plastic surface, and after 1, 4 and 7 days, the culture was incubated with MTS/PMS solution diluted in standard medium (4 h, 37°C, humidified 5% CO_2 atmosphere). The formazan product was quantified by absorbance at 490 nm in a microplate reader (Tecan Infinite M200).

Confocal Microscopy

Three days after cell seeding on dermal substitutes, cultures were fixed in 4% paraformaldehyde for 30 min, stained with DAPI and visualized under confocal microscopy (Leica DMI6000). Three-dimensional reconstruction of the scaffold was performed using ImageJ software.

Scanning Electron Microscopy

Forty-eight hours after cell seeding on dermal substitutes, cultures were fixed (2.5% glutaraldehyde in 0.1 M sodium cacodylate buffer, Sigma-Aldrich) (12 h, 4°C), washed (sodium cacodylate buffer) and post-fixed (1% osmium tetroxide solution, 2 h). After dehydration (30%, 50%, 70%, 90% and 100% ethanol), cultures were dried in a critical point of CO_2 (Leica MS CPD 030), metalized with 30 nm gold overlay (Leica EM SCD 500), and analyzed in a scanning electron microscope (Jeol JSM-6390LV) under capturing electrons at 15 kV by side illumination.

Reverse Transcription-Polymerase Chain Reaction

Total RNA was isolated using the TRIZOL reagent following the manufacturer's instructions (Invitrogen). Samples were then treated with DNase RQ1 RNase Free (Promega), according to the manufacturer's instructions, to avoid any contaminating

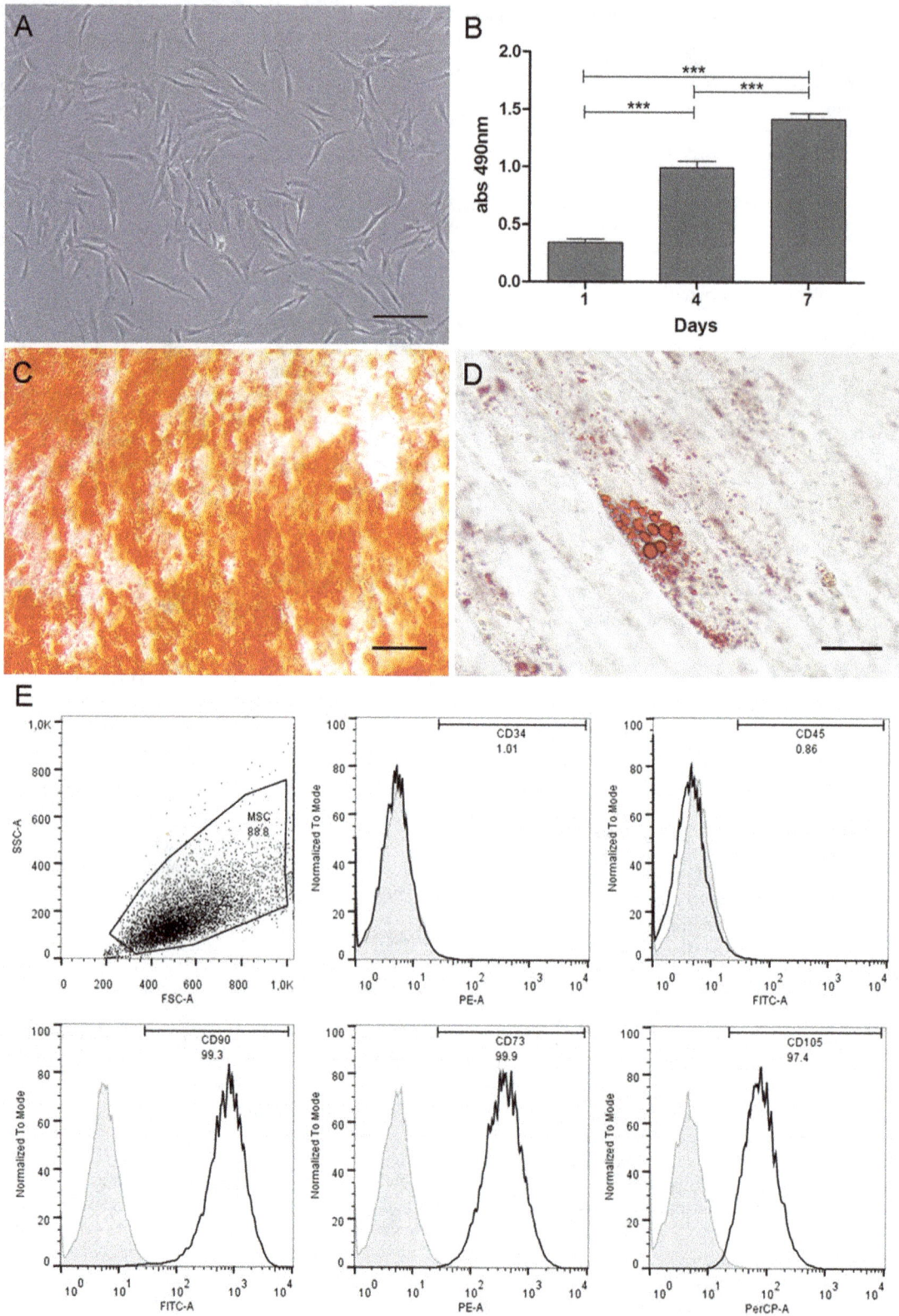

Figure 1. MSC phenotypic characterization of human skin-derived cells. (A) Morphological analysis of skin-derived cells by phase contrast microscopy. (B) MTS cell proliferation/viability assay. (C) Osteogenic and (D) adipogenic differentiation. (C) Cells cultured in inductive medium formed Alizarin Red S-stained mineralized nodules and (D) Oil red O-stained lipid clusters. (E) Flow cytometry analysis of hematopoietic (CD34, CD45) and MSC (CD90, CD73, CD105) markers. Specific markers are shown by black curves and controls by gray curves. ***$p < 0.001$. Scale bar: (C–F): 50 μm. Other pictures: 200 μm.

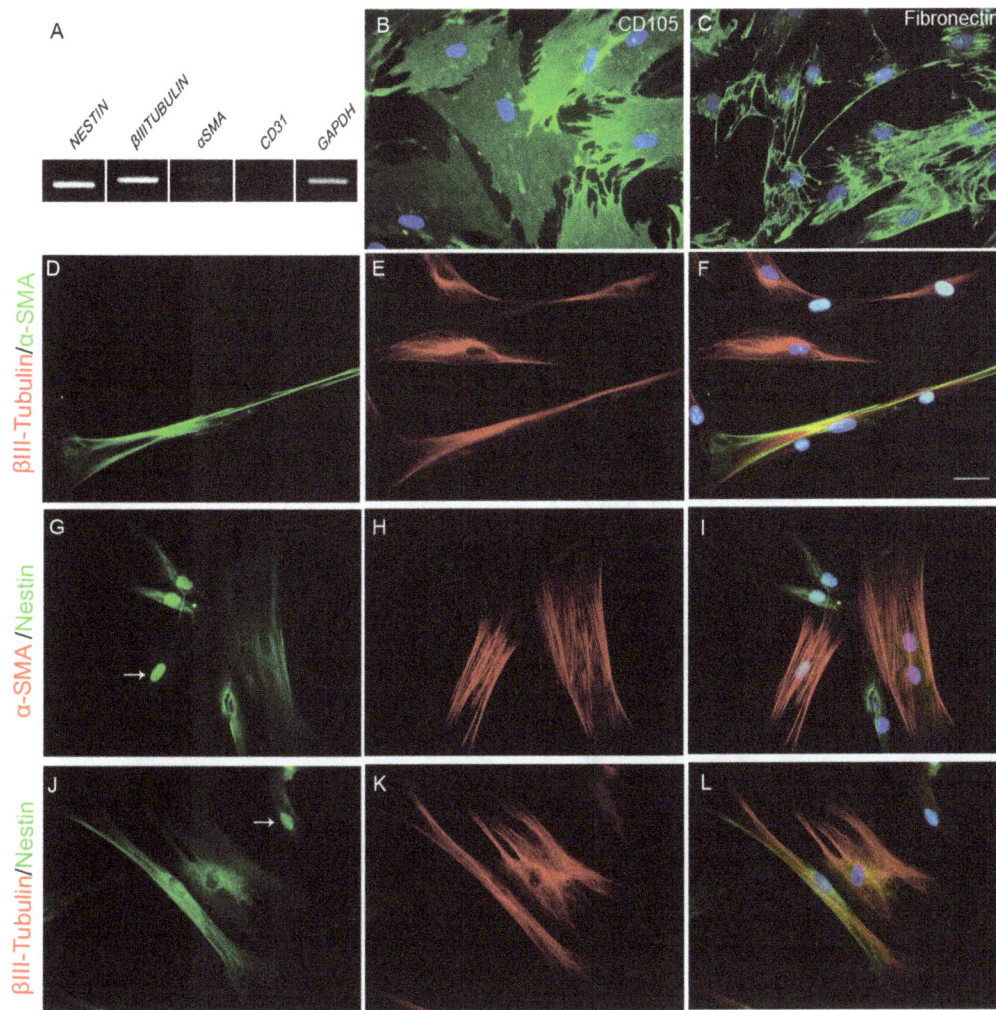

Figure 2. Multilineage potential of SD-MSCs. (A) Gene expression profile by RT-PCR of SD-MSCs cultivated in standard medium. (B–E) Immunofluorescence staining of (B) CD105 and (C) Fibronectin. (D–F) Double-staining of α-SMA and βIII-Tubulin, (G–I) α-SMA and Nestin and (J–L) βIII-Tubulin and Nestin co-expression. (F, I and L): Merged pictures of D–E, G–H, J–K, respectively. Cell nuclei were stained with DAPI (blue). Arrows: Nestin nuclear staining. Scale bar: 50 μm.

DNA. The RNA was quantified by spectrophotometry, and 2 μg of RNA were used for reverse transcription using the Thermoscript RT-PCR system for first-strand cDNA synthesis (Invitrogen). The PCR reactions were done using 1 μL of the RT reaction mixture and specific oligonucleotide primers for, GAPDH (glyceraldehyde-3-phosphate dehydrogenase), Nestin, βIII-Tubulin, α-SMA, CD31 and CD90. Oligonucleotide sequences and PCR conditions are shown in Table S1. GAPDH expression was used as an internal control of RNA integrity and efficiency of the reverse transcription process. Seven microliters of the PCR mixture were separated by electrophoresis on 2% agarose gel, and the reaction products were visualized under ultraviolet-induced fluorescence. All experiments were performed in triplicate.

Results

Phenotypic Characterization of Human SD-MSCs

Skin-derived cells with a fibroblast-like morphology adhered to the plastic culture dish (Figure 1A). Cells could be maintained in this condition for at least 20 passages. The high expansion capacity

in vitro was confirmed by the progressive increase in the values of MTS assay during the 7 days of culture (Figure 1B).

The mesenchymal stem cell (MSC) characteristics were evaluated by the ability to originate osteogenic (Figure 1C), adipogenic (Figure 1D) and chondrogenic (data not shown) phenotypes. Skin-derived cells upon induction, were capable of differentiating into osteoblasts and adipocytes, characterized by dense extracellular matrix with calcium deposition (Figure 1C), and intracellular lipids (Figure 1D), respectively; nevertheless strong chondrogenic differentiation (data not shown) could not be observed, as previously reported [15].

Following MSC characterization, the expression of hematopoietic (CD34 and CD45) and mesenchymal (CD90, CD73 and CD105) stem cell markers was evaluated by flow cytometry (Figure 1E). Skin-derived cells were negative for CD34 and CD45 and positive for CD90 (99.3%), CD73 (99.9%) and CD105 (97.4%). The results were similar at both low (P1–P4) and high passages (P10–P20) (data not shown). Therefore, human skin-derived cells display such MSC characteristics as plastic adhesion, fibroblast morphology, mesodermal differentiation capacity, expression of mesenchymal markers, and absence of hematopoi-

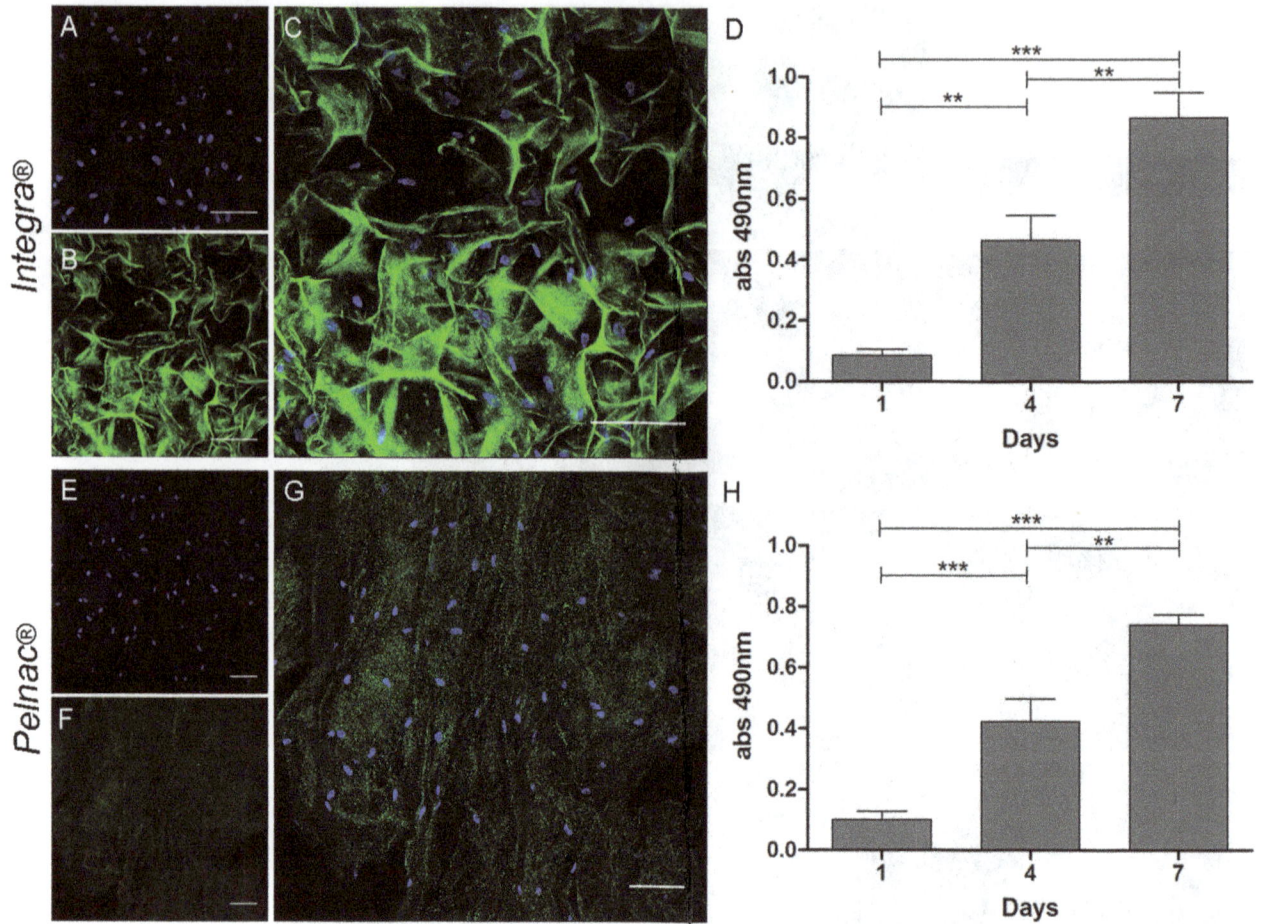

Figure 3. 3D cultures of human SD-MSCs in (A–D) Integra and (E–H) Pelnac. Confocal microscopy of SD-MSCs cultured in (A–C) Integra and (D–F) Pelnac. (A and E) DAPI nuclear staining of SD-MSCs. (B and F) Integra and (F) Pelnac dermal substitutes, respectively (green autofluorescence). (C and G) merged images of (A and B) and (E and F), respectively. (D and H) MTS cell viability assay of SD-MSCs cultivated in Integra and Pelnac, respectively. ***p<0.001, **p<0.01. Scale bar: 100 um.

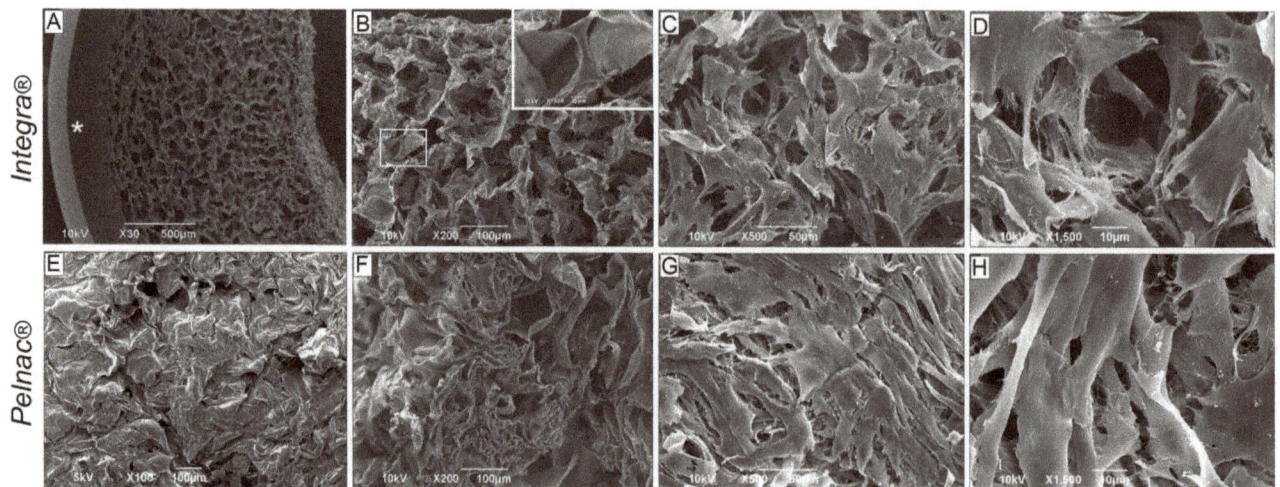

Figure 4. Scanning electron microscopy (SEM) images of SD-MSCs cultured in (A–D) Integra and (E–H) Pelnac. (A) Cross-sectional view of Integra dermal substitute alone showing the silicone (*) and the inner layer. (B–D) 3D culture of SD-MSCs in Integra 48 hours after seeding. (E) Surface and (F) cross-sectional views of Pelnac showing the collagen layer. (G–H) 3D culture of SD-MSCs in Pelnac 48 hours after seeding. Insets in B: different magnifications of a SD-MSC.

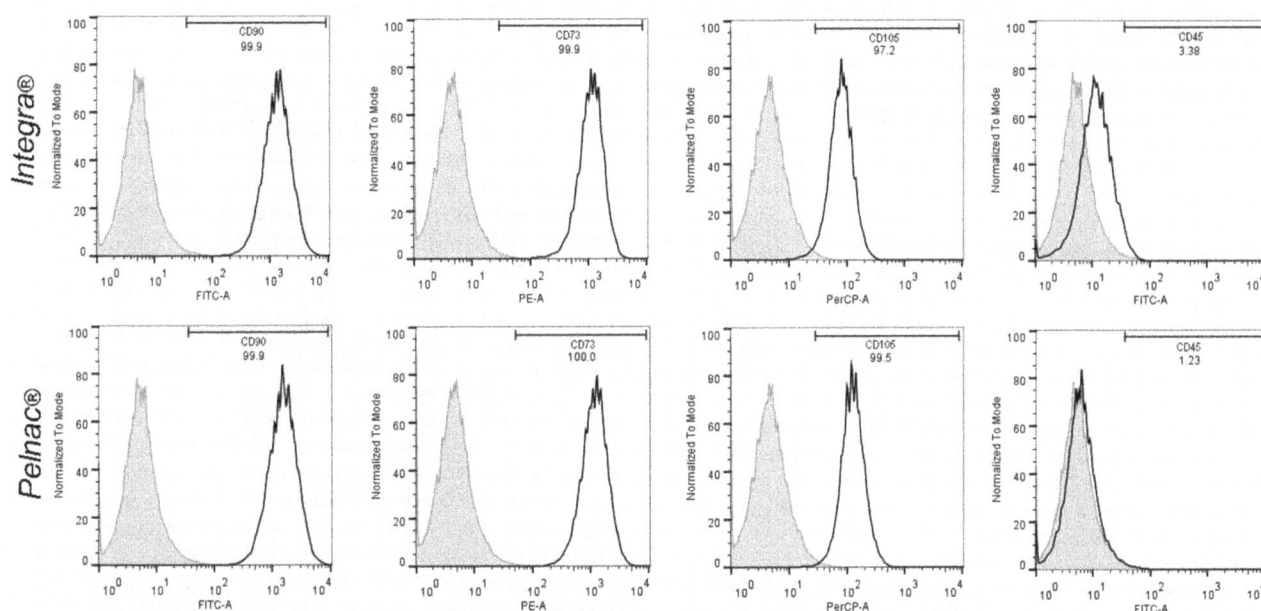

Figure 5. Immunophenotypic profile of human SD-MSCs cultured in Integra and Pelnac. Flow cytometry analysis of SD-MSCs (CD90, CD73, CD105) and hematopoietic (CD45) markers in SD-MSCs after 3 days of culture in Integra (upper panel) and Pelnac (lower panel). Curves in black show the specific markers, and gray curves correspond to controls.

etic markers. Thereafter, these cells were termed skin-derived mesenchymal stromal cells (SD-MSCs).

Expression of Multilineage Markers

Next, the expression of markers of mesenchymal (α-SMA, CD105 and Fibronectin), neural (Nestin and βIII-Tubulin), and endothelial cells (CD31) by SD-MSCs was investigated (Figure 2). RT-PCR assay demonstrated the mRNA expression of Nestin, βIII-Tubulin and α-SMA, but not CD31 (Figure 2A). At the protein level, as revealed by immunofluorescence staining, most SD-MSCs expressed CD105 (Figure 2B), Fibronectin (Figure 2C), βIII-Tubulin (Figure 2E-F and K–L) and Nestin (Figure 2 G-I and J–L). Surprisingly, a nuclear distribution of Nestin was visualized in some cells (arrows in Figure 2G and J). In addition, about 40%

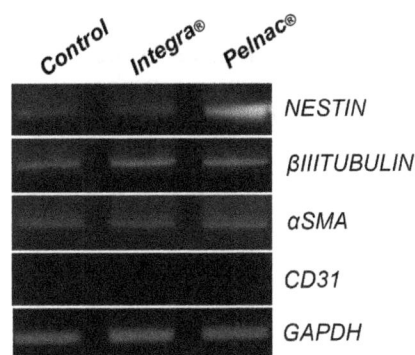

Figure 6. Gene expression profile of human SD-MSCs cultured in Integra and Pelnac. RT-PCR analysis showing mRNA expression of Nestin, βIII-Tubulin, α-SMA, CD31 and GAPDH (as internal control) in SD-MSCs cultured in dermal substitutes. Control: SD-MSCs cultured on plastic surfaces.

of cells were positive to α-SMA (Figure 2D–H). Co-expression of Nestin, βIII-Tubulin and α-SMA is shown in Figure 2D–L.

Viability/proliferation of Human SD-MSCs in 3D Culture with Dermal Substitutes

The ability of the dermal substitutes Integra and Pelnac to support the survival and growth of SD-MSCs was assessed by MTS assay. Significant and progressive increase in viability/proliferation of SD-MSCs with both dermal substitutes was detected during the 7-day period of culture (Figure 3D and H for Integra and Pelnac, respectively). Differences (P>0.05) in the MTS values between Integra and Pelnac were not observed at any time point of the analysis.

Adhesion and Morphology of Human SD-MSCs in 3D Culture with Dermal Substitutes

Confocal microscopy analysis revealed that SD-MSCs adhere and migrate within both Integra (Figure 3A–D) and Pelnac (Figure 3 E–H) matrices, although mostly distributed in the upper portion (Figure S1). The structure and porosity of Integra (Figure 4A) and Pelnac (Figure 4E–F) inner layers were shown by scanning electron microscopy (SEM). SEM analysis also revealed that SD-MSCs similarly attached and spread at both dermal matrices with the fibroblast-like fusiform morphology (Figure 4B-D and G-H for Integra and Pelnac, respectively). Together with MTS assay, these findings indicate that both dermal substitutes equally support the adhesion, spread and growth of SD-MSCs *in vitro*.

Phenotypic Profile of Human SD-MSCs in 3D Cultures with Dermal Substitutes

Most SD-MSCs cultured within Integra (Figure 5, upper panel) and Pelnac (Figure 5, lower panel) matrices were positive to the CD90 (99.9% and 99.9%, respectively), CD73 (99.9%, and 100%, respectively) and CD105 (97.2% and 99.5%, respectively) markers, but negative to CD45 (less than 3.38%) by flow cytometry. Consequently, SD-MSCs expressed the mRNA of Nestin, βIII-Tubulin, and α-SMA, but not the CD31 mRNA (Figure 6). Therefore, SD-MSCs in 3D culture with both Integra and Pelnac matrices are able to maintain phenotypic characteristics similar to MSCs.

Discussion

In this study, mesenchymal stromal cells were isolated from human skin. These skin-derived mesenchymal stromal cells share the criteria established by the International Society of Cellular Therapy for Human MSC [9] including: adhesion to plastic culture dish, fibroblast morphology, adipogenic and osteogenic differentiation, expression of mesenchymal markers and absence of hematopoietic markers. Indeed, stromal cells have been demonstrated to exhibit MSC characteristics such as cell phenotype, transdifferentiation potential and immunosuppressive properties and thus are functionally comparable to MSCs, being difficult to discriminate between them [12].

In addition, despite the absence of a neuron-like morphology, most SD-MSCs expressed the neural proteins Nestin [16]; [17] and βIII-Tubulin [18]. Both were also reported in MSCs [19] [20]; [21]. Interestingly, similar to some brain tumor cell lines [16], we detected in some SD-MSCs the presence of Nestin by nuclear staining, which suggests the possible involvement of gene expression. Therefore, a possible neural potential of human SD-MSCs could be implicated. Future studies are needed to determine the role of these cytoskeleton proteins in SD-MSCs [22].

In our study, about 40% of SD-MSCs were positive to α-SMA, a cytoskeleton marker of smooth muscle cells, also related to vasculogenesis [23]. In fact, smooth muscle markers have been reported in MSCs with multilineage potential [24] which could be related to the multipotentiality of these cells. Efforts to track the identity of tissue-resident MSCs have suggested their perivascular origin [25]; [7], evidenced by the pericyte-like features [26]. Moreover, the three cytoskeletal proteins, α-SMA, Nestin and βIII-Tubulin, which we found in human SD-MSCs were also observed in pericytes [27]; [28] [29], suggesting a possible perivascular origin.

Dermal Substitutes Integra and Pelnac Allow Survival, Adhesion and Spreading of Human SD-MSCs in a 3D Culture System

The combination of stem cells with scaffolds has been used with some success in several injuries [30]. In this paper, we investigated in vitro the feasibility and efficacy of combining human SD-MSCs with dermal substitutes for clinical use. The biocompatibility of Integra and Pelnac with SD-MSCs was demonstrated by 3D imaging and MTS assay that showed the infiltration, distribution and proliferation of these cells into both matrices. Although the cells were mostly concentrated in the upper portion of the matrices, possibly as a result of the top-down cell-seeding procedure, we believe that a uniform distribution could be achieved in the scaffolds by improving the technique. In addition, SD-MSCs could adhere to both matrices, establishing strong connections and cytoplasmic extensions and maintaining the typical MSC-fibroblast morphology, as revealed by SEM. Similar results were reported in human MSCs from bone marrow and adipose tissue, which were able to grow and proliferate in Integra [31]; [32]. Importantly, although Pelnac is largely used in Japan, its combination with MSCs was first demonstrated in the present study. Together, these observations confirmed that both dermal

substitutes are suitable substrates for the colonization of SD-MSCs.

Human SD-MSCs Maintain MSC Characteristics in 3D Culture with Integra and Pelnac Dermal Substitutes

It is well established that the composition and mechanical properties of the extracellular environment regulate intracellular signaling and thus influence different aspects of cell behavior, including cell growth, migration, survival and phenotypic fate [33]; [34]. Therefore, the effective attachment and maintenance of stem cells onto scaffolds is a critical step to be considered in tissue engineering [35]. Hence, to further investigate the efficiency of Integra and Pelnac in supporting the growth and maintenance of SD-MSCs, their phenotypic profile in 3D culture was studied. In fact, in this culture condition, most SD-MSCs were positive for the MSC markers CD105, CD73 and CD90, but negative for the hematopoietic stem cell marker CD45. In addition, they also expressed neural (Nestin and βIII-Tubulin) and mesenchymal (α-SMA) markers, but did not express the endothelial marker CD31. These results suggest that the phenotypic profile of SD-MSCs observed in conventional 2D plastic culture (monolayer) was maintained in a 3D culture system with dermal substitutes.

Despite differences in the composition of Integra and Pelnac, as explained above, our findings, when taken together, show that the growth and phenotypic characteristics of SD-MSCs were similar in both templates. Therefore, both dermal substitutes are biocompatible scaffolds for SD-MSCs growth, and the 3D culture system reported here represents a potential tool for tissue engineering.

Conclusions

In this study, we isolated and characterized a population of cells from human skin termed herein as skin-derived mesenchymal stromal cells (SD-MSCs) with similar characteristics to MSCs. These cells express in vitro multilineage markers (mesenchymal and neural, but not hematopoietic or endothelial). We developed a 3D culture system of human SD-MSCs with the dermal substitutes Integra and Pelnac, in which these cells were found to adhere and proliferate, while maintaining the MSC properties and gene expression of multilineage markers. In conclusion, 3D culture systems using dermal substitutes described in this study provide an efficient in vitro environment for human SD-MSCs and could therefore be useful for tissue engineering and cell therapy.

Supporting Information

Figure S1 3D reconstruction of confocal images of human SD-MSC cultures in Integra and Pelnac. (A) SD-MSCs cultured in Integra and (B) Pelnac. Blue: DAPI nuclear staining of SD-MSCs. Green: autofluorescence of dermal substitutes.

Acknowledgments

We wish to thank Hospital Ilha e Maternidade (Florianópolis) and Dr. Rogério Gomes for providing the skin fragments. The authors are also grateful to the *Laboratórios Multiusuários de Estudos em Biologia (LAMEB/UFSC)* and *Laboratório Central de Microscopia Eletrônica (LCME/UFSC)* for the technical support.

Author Contributions

Conceived and designed the experiments: TdSJ AGT DFL MJP. Performed the experiments: TdSJ RGM SBV. Analyzed the data: TdSJ SBV AGT. Contributed reagents/materials/analysis tools: AGT MJP DFL. Wrote the paper: TdSJ AGT.

References

1. Wong DJ, Chang HY, Biology E (2009) Skin tissue engineering. StemBook: 1–9.
2. Serpooshan V, Julien M, Nguyen O, Wang H, Li A, et al. (2010) Reduced hydraulic permeability of three-dimensional collagen scaffolds attenuates gel contraction and promotes the growth and differentiation of mesenchymal stem cells. Acta Biomater 6: 3978–3987.
3. Gurtner GC, Werner S, Barrandon Y, Longaker MT (2008) Wound repair and regeneration. Nature 453: 314–321.
4. Burke JF, Yannas IV, Quinby WC, Bondoc CC, Jung WK (1981) Successful use of a physiologically acceptable artificial skin in the treatment of extensive burn injury. Ann Surg 194: 413–428.
5. Suzuki S, Matsuda K, Maruguchi T, Nishimura Y, Ikada Y (1995) Further applications of "bilayer artificial skin". Br J Plast Surg 48: 222–229.
6. Ko SH, Nauta A, Wong V, Glotzbach J, Gurtner GC, et al. (2011) The role of stem cells in cutaneous wound healing: what do we really know? Plast Reconstr Surg 127 Suppl: 10S–20S.
7. Da Silva Meirelles L, Caplan AI, Nardi NB (2008) In search of the in vivo identity of mesenchymal stem cells. Stem Cells 26: 2287–2299.
8. Pittenger MF, Mackay AM, Beck SC, Jaiswal RK, Douglas R, et al. (1999) Multilineage potential of adult human mesenchymal stem cells. Science 284: 143–147.
9. Dominici M, Le Blanc K, Mueller I, Slaper-Cortenbach I, Marini F, et al. (2006) Minimal criteria for defining multipotent mesenchymal stromal cells. The International Society for Cellular Therapy position statement. Cytotherapy 8: 315–317.
10. Maxson S, Lopez EA, Yoo D, Danilkovitch-Miagkova A, Leroux MA (2012) Concise review: role of mesenchymal stem cells in wound repair. Stem Cells Transl Med 1: 142–149.
11. Sharma RI, Snedeker JG (2012) Paracrine interactions between mesenchymal stem cells affect substrate driven differentiation toward tendon and bone phenotypes. PLoS One 7: e31504.
12. Vishnubalaji R, Al-Nbaheen M, Kadalmani B, Aldahmash A, Ramesh T (2012) Skin-derived multipotent stromal cells–an archrival for mesenchymal stem cells. Cell Tissue Res 350: 1–12.
13. Hamdi H, Furuta A, Bellamy V, Bel A, Puymirat E, et al. (2009) Cell delivery: intramyocardial injections or epicardial deposition? A head-to-head comparison. Ann Thorac Surg 87: 1196–1203.
14. Coura GS, Garcez RC, de Aguiar CBNM, Alvarez-Silva M, Magini RS, et al. (2008) Human periodontal ligament: a niche of neural crest stem cells. J Periodontal Res 43: 531–536.
15. Vaculik C, Schuster C, Bauer W, Iram N, Pfisterer K, et al. (2012) Human dermis harbors distinct mesenchymal stromal cell subsets. J Invest Dermatol 132: 563–574.
16. Krupkova O, Loja T, Redova M, Neradil J, Zitterbart K, et al. (2011) Analysis of nuclear nestin localization in cell lines derived from neurogenic tumors. Tumour Biol 32: 631–639.
17. Wiese C, Rolletschek A, Kania G, Blyszczuk P, Tarasov K V, et al. (2004) Nestin expression–a property of multi-lineage progenitor cells? Cell Mol Life Sci 61: 2510–2522.
18. Katsetos CD, Legido A, Perentes E, Mörk SJ (2003) Class III beta-tubulin isotype: a key cytoskeletal protein at the crossroads of developmental neurobiology and tumor neuropathology. J Child Neurol 18: 851–66; discussion 867.
19. Bossolasco P, Montemurro T, Cova L, Zangrossi S, Calzarossa C, et al. (2006) Molecular and phenotypic characterization of human amniotic fluid cells and their differentiation potential. Cell Res 16: 329–336.
20. Foudah D, Redondo J, Caldara C, Carini F, Tredici G, et al. (2013) Human mesenchymal stem cells express neuronal markers after osteogenic and adipogenic differentiation. Cell Mol Biol Lett 18: 163–186.
21. Barnabé GF, Schwindt TT, Calcagnotto ME, Motta FL, Martinez G, et al. (2009) Chemically-induced RAT mesenchymal stem cells adopt molecular properties of neuronal-like cells but do not have basic neuronal functional properties. PLoS One 4: e5222.
22. Birbrair A, Zhang T, Wang Z–M, Messi ML, Enikolopov GN, et al. (2013) Skeletal muscle pericyte subtypes differ in their differentiation potential. Stem Cell Res 10: 67–84.
23. Owens GK, Kumar MS, Wamhoff BR (2004) Molecular regulation of vascular smooth muscle cell differentiation in development and disease. Physiol Rev 84: 767–801.
24. Liu Y, Deng B, Zhao Y, Xie S, Nie R (2013) Differentiated markers in undifferentiated cells: Expression of smooth muscle contractile proteins in multipotent bone marrow mesenchymal stem cells. Dev Growth Differ. doi:10.1111/dgd.12052.
25. Crisan M, Corselli M, Chen WCW, Péault B (2012) Perivascular cells for regenerative medicine. J Cell Mol Med 16: 2851–2860.
26. Nombela-Arrieta C, Ritz J, Silberstein LE (2011) The elusive nature and function of mesenchymal stem cells. Nat Rev Mol Cell Biol 12: 126–131.
27. Crisan M, Yap S, Casteilla L, Chen C–W, Corselli M, et al. (2008) A perivascular origin for mesenchymal stem cells in multiple human organs. Cell Stem Cell 3: 301–313.
28. Alliot F, Rutin J, Leenen PJ, Pessac B (1999) Pericytes and periendothelial cells of brain parenchyma vessels co-express aminopeptidase N, aminopeptidase A, and nestin. J Neurosci Res 58: 367–378.
29. Stapor PC, Murfee WL (2012) Identification of class III β-tubulin as a marker of angiogenic perivascular cells. Microvasc Res 83: 257–262.
30. Nie C, Yang D, Morris SF (2009) Local delivery of adipose-derived stem cells via acellular dermal matrix as a scaffold: a new promising strategy to accelerate wound healing. Med Hypotheses 72: 679–682.
31. Egaña JT, Fierro FA, Krüger S, Bornhäuser M, Huss R, et al. (2009) Use of human mesenchymal cells to improve vascularization in a mouse model for scaffold-based dermal regeneration. Tissue Eng Part A 15: 1191–1200.
32. Formigli L, Benvenuti S, Mercatelli R, Quercioli F, Tani A, et al. (2011) Dermal matrix scaffold engineered with adult mesenchymal stem cells and platelet-rich plasma as a potential tool for tissue repair and regeneration. J Tissue Eng Regen Med 6: 125–134.
33. Birgersdotter A, Sandberg R, Ernberg I (2005) Gene expression perturbation in vitro–a growing case for three-dimensional (3D) culture systems. Semin Cancer Biol 15: 405–412.
34. Godier AFG, Marolt D, Gerecht S, Tajnsek U, Martens TP, et al. (2008) Engineered microenvironments for human stem cells. Birth Defects Res C Embryo Today 84: 335–347.
35. Masaeli E, Morshed M, Nasr-Esfahani MH, Sadri S, Hilderink J, et al. (2013) Fabrication, characterization and cellular compatibility of poly(hydroxy alkanoate) composite nanofibrous scaffolds for nerve tissue engineering. PLoS One 8: e57157.

Fabrication, Characterization and Cellular Compatibility of Poly(Hydroxy Alkanoate) Composite Nanofibrous Scaffolds for Nerve Tissue Engineering

Elahe Masaeli[1,2,3], Mohammad Morshed[1], Mohammad Hossein Nasr-Esfahani[2], Saeid Sadri[4], Janneke Hilderink[3], Aart van Apeldoorn[3], Clemens A. van Blitterswijk[3], Lorenzo Moroni[3]*

1 Department of Textile Engineering, Isfahan University of Technology, Isfahan, Iran, **2** Department of Cell and Molecular Biology, Cell Science Research Center, Royan Institute for Biotechnology, ACECR, Isfahan, Iran, **3** Department of Tissue Regeneration, University of Twente, Enschede, The Netherlands, **4** Department of Electrical and Computer Engineering, Isfahan University of Technology, Isfahan, Iran

Abstract

Tissue engineering techniques using a combination of polymeric scaffolds and cells represent a promising approach for nerve regeneration. We fabricated electrospun scaffolds by blending of Poly (3-hydroxybutyrate) (PHB) and Poly (3-hydroxy butyrate-co-3- hydroxyvalerate) (PHBV) in different compositions in order to investigate their potential for the regeneration of the myelinic membrane. The thermal properties of the nanofibrous blends was analyzed by differential scanning calorimetry (DSC), which indicated that the melting and glass temperatures, and crystallization degree of the blends decreased as the PHBV weight ratio increased. Raman spectroscopy also revealed that the full width at half height of the band centered at 1725 cm^{-1} can be used to estimate the crystalline degree of the electrospun meshes. Random and aligned nanofibrous scaffolds were also fabricated by electrospinning of PHB and PHBV with or without type I collagen. The influence of blend composition, fiber alignment and collagen incorporation on Schwann cell (SCs) organization and function was investigated. SCs attached and proliferated over all scaffolds formulations up to 14 days. SCs grown on aligned PHB/PHBV/collagen fibers exhibited a bipolar morphology that oriented along the fiber direction, while SCs grown on the randomly oriented fibers had a multipolar morphology. Incorporation of collagen within nanofibers increased SCs proliferation on day 14, GDNF gene expression on day 7 and NGF secretion on day 6. The results of this study demonstrate that aligned PHB/PHBV electrospun nanofibers could find potential use as scaffolds for nerve tissue engineering applications and that the presence of type I collagen in the nanofibers improves cell differentiation.

Editor: Thomas Claudepierre, University of Leipzig, Germany

Funding: EM wishes to acknowledge Isfahan University of Technology and Ministry of Science, Research and Technology (Iran) for the scholarship and financial supports. The funders had no role in study design, data collection and analysis, decision to publish, or preparation of the manuscript.

Competing Interests: The authors have declared that no competing interests exist.

* E-mail: l.moroni@utwente.nl

Introduction

Neural injuries are very common in clinical practice and may lead to permanent disabilities in patients. Existing tissue engineering approaches focus on finding alternative procedures for nerve regeneration using polymeric scaffolds 1]. Various strategies have been employed to create a biodegradable nerve guidance scaffold to assist regenerating axons by serving as a growth substrate for neural cells, 2]. Highly porous electrospun nanofiber matrices are a logical choice because of the physical and structural similarities to the extracellular matrix (ECM) components such as collagen fibers and their high surface area 3]. Recent reports on neural regeneration also highlight the promise of using electrospun nanofibrous scaffolds in combination with mesenchymal stem cells (MSCs) 4], human adipose tissue-derived stem cells (hASCs) 5], nerve precursor cells (NPCs) 6], neural stem cells 7] or Schwann cells (SCs) 8]. Among the physical and chemical cues that can be imparted to improve neural regeneration, nanofiber orientation has been shown to increase ECM production. Fiber alignment greatly influenced cell growth and related functions in different cell

sources such as neurons and human coronary artery smooth muscle cell (SMCs) 9,10]. As a result of contact guidance, a cell has the maximum probability of migration in preferred directions associated with chemical, structural and mechanical properties of the substrate 11]. It has been reported in different studies that, unidirectional aligned nanofibers can provide better contact guidance effects towards neurite outgrowth and help in providing cues to enhance SCs extension and axon regeneration 9,12–14].

SCs are the main glial cells of the peripheral nervous system which can promote neuronal regeneration by at least three routes: (i) an increase in cell surface adhesion molecular synthesis, (ii) production of a basement membrane which consists of ECM proteins, and (iii) production of neurotrophic factors and their corresponding receptors 15,16]. Therefore, an ideal scaffold onto which SCs attach, proliferate, and migrate plays a key role in neural tissue engineering.

Several biomaterials have been investigated for neural tissue engineering 1,16–18]. Among these, poly (hydroxyalkanoates) (PHAs) are a family of biological polyesters produced by microorganisms as intracellular carbon and energy sources. The physical and chemical properties of PHAs can be controlled by changing the monomer composition 19]. Poly (3-hydroxybutyrate) (PHB) and its copolymer with 3-hydroxyvalerate (PHBV) are two among the most common members of PHAs that proved to possess favorable physicochemical and biological properties and have, thus, found increasing applications in the fabrication of tissue engineering scaffolds 17,20]. PHB is known as a rigid and highly crystalline polymer with slow degradation rate that results in a poor processing window and higher cost, while PHBV is more flexible and easier to process than PHB as a result of its lower glass transition and melting temperatures 21,22].

Blending of PHB with PHBV results in a decrease of the melting temperature, leading to the possibility to process the materials at lower temperature in order to form specific anatomical shapes while avoiding or limiting the degradation23], Thus, PHBV can be blended with PHB to obtain a scaffold with improved physical and mechanical characteristics, as well as cell adhesion, proliferation and degradation properties.

Electrospinning of PHB(50)/PHBV(50) blends as scaffolds for bone tissue engineering were reported by Sombatmankhong et al. 24,25]. The 50/50 wt PHBV/PHB mats showed no cytotoxicity to human osteoblasts and the highest alkaline phosphatase activity in comparison to those of neat PHB and PHBV scaffolds. Zonari et al. 5] also reported that PHB(30)/PHBV(70) nanofibrous scaffolds can be used as a substrate for endothelial differentiation of hASCs and vascularization of bone tissue.

PHB and PHBV scaffolds have also shown a progressive potential for use in neural tissue engineering as a substrate for SCs culture. For instance, Sangsanoh et al. 18] and Swantong et al. 26] investigated the in vitro response of SCs on various types of electrospun nanofibers including neat PHB and PHBV and confirmed that these materials are non-toxic to cells. In vitro studies have also shown increased proliferation of SCs on PHB microfibrous conduits coated with ECM molecules 27,28]. In vivo studies proved that PHB conduits seeded with SCs could bridge nerve gaps in both spinal cord and peripheral nerves 27,29]. Despite both PHB and PHBV having hydrophobic surfaces and lacking functional groups, their support of limited cell adhesion and differentiation still shows promise for the field of neural tissue engineering 30]. Blending synthetic and natural polymers is one of the most effective methods for providing scaffolds with improved cell adhesion, bioactivity, and degradation rate for tissue engineering applications 9]. So far, many attempts have been made to design and develop appropriate electrospun nanofibrous scaffolds for neural tissue engineering by solution blending of synthetic biopolymers with natural polymers such as hyaluronic acid 31,32], gelatine 9], collagen 2,31,32] and chitosan 13]. It is well known that collagen is the major element of natural ECM, induces low immune response, is a good substrate for cell adhesion, and can be remodelled 32,33]. Therefore, collagen is an optimal candidate among ECM proteins to be blended with PHAs to obtain a scaffold that can enhance cell adhesion and differentiation.

Despite several studies which have aimed at finding an optimal biodegradable three-dimensional scaffold for re-myelination and aid in nerve regeneration, PHB/PHBV blended nanofibrous scaffolds displaying bioactive cues have not been investigated in depth. Hence, electrospun PHB/PHBV random and aligned nanofibers were fabricated in the present study in order to optimize the effect of polymer composition and nanofiber alignment on SCs adhesion, proliferation, and differentiation. The further effect of collagen motifs on cell activity was also assessed by fabricating blended PHB/PHBV/collagen type I aligned nanofibrous scaffolds.

Results

Scaffolds Characterization

SEM images of the surface morphology, microstructure of the fabricated scaffolds and diameter distribution of nanofibers are shown in Figure 1. Briefly, all five PHB/PHBV blend compositions were successfully electrospun at concentration of 6, 8 and 10% wt. The variation of fiber diameter with blend ratio and polymer concentration is shown in Figure S1 (supplementary information). For a fixed voltage, flow rate and collecting conditions, a decrease in polymer concentration is expected to decrease the fiber diameter, which was statistically significant for PHB(100)/PHBV(0), PHB(50)/PHBV(50) and PHB(0)/PHB(100) blends. For the PHB(75)/PHBV(25) blend, diameter differences were statistically significant by decreasing concentration from 8% to 6% and for PHB(25)/PHBV(75) from 10% to 8% concentration. With increasing the ratio of PHBV in the blends, a decrease of the fiber diameter was observed at all polymer concentrations. For a fixed concentration and composition, maximum ranges of fiber diameter were seen at a concentration of 6% in case of PHB(0)/PHBV(100). When the PHBV ratio increased, the fiber diameter changed from 1208 ± 149 nm to 451 ± 65 nm, while keeping the polymer concentration at 6% wt. Similarly, when the polymer concentration increased, the fiber diameter changed from 451 ± 65 nm to 853 ± 126 nm while keeping the polymer composition constant.

The miscibility of blend components was analyzed by evaluating the changes in the T_m and in the crystallization percentage as a function of composition ratio using DSC. The temperature range chosen for the heating step was from $-50°C$ to $200°C$ to cover the melting and crystallization processes of the PHB/PHBV blends. The trends obtained from first heating and cooling scans are presented in Figure 2. By increasing the amount of PHBV, T_m decreased from $178.96°C$ to $166°C$, T_g from $15.513°C$ to $-0.565°C$ and T_c from $83.51°C$ to $63.86°C$. Similarly, ΔH_m decreased from 81.65 J/g to 57.35 J/g and χ from 74.91% to 39.28%.

Raman spectra of pure PHB, PHBV and their blends together with main Raman shifts and assignments of the Raman bands of PHB/PHBV samples are shown in Figure 3 and Table S2. The Raman spectra of different blends are similar to each other, because of the structural resemblance of PHB and PHBV. As a result of their different crystallinity degrees, different scaffold compositions presented distinct full width at half maximum at 1725 cm^{-1}.

To analyse the effect of speed of mandrel rotation on fiber alignment, the PHB(50)/PHBV(50) blend was selected, after also considering the analysed mechanical properties and initial cell response. The morphology of aligned and randomly oriented PHB(50)/PHBV(50) nanofibers was examined by SEM. As shown in Figure 4, the degree of fiber alignment was evaluated as the number of fibers oriented in a determined direction with respect to a horizontal base line. In the case of random nanofibers, the broad distribution of fiber angles indicates a random orientation. Conversely, aligned nanofibers displayed a narrow distribution of fiber orientation. Briefly, the distribution of fiber angles was narrower and narrower with increasing the speed of the collector drum from 1000 to 5000 rpm. Average fiber diameters of aligned and randomly oriented nanofibers collected with speeds of 1000, 3000 and 5000 rpm were 963 ± 117 nm, 986 ± 112 nm, and 925 ± 156 nm, respectively.

As shown in Figure S2 (supplementary information) and Table 1, the mechanical properties of the randomly oriented and aligned PHB/PHBV nanofibers differed considerably. While the percent

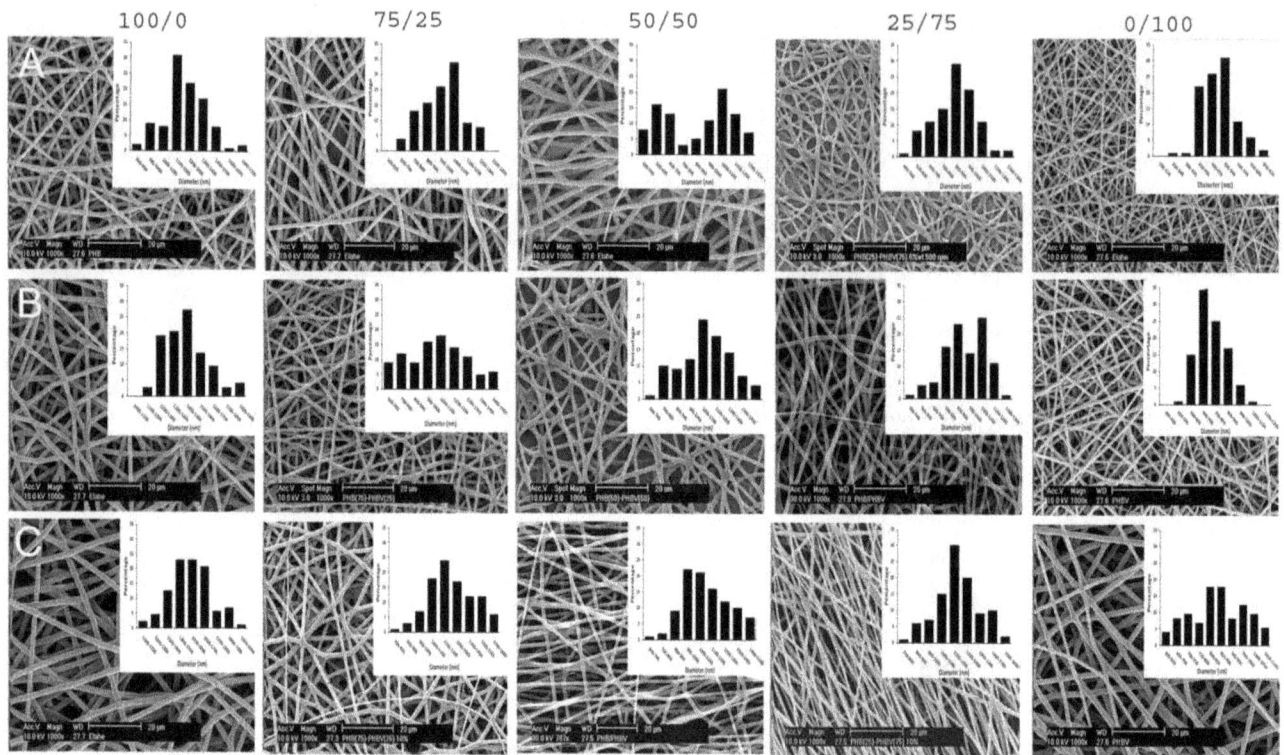

Figure 1. SEM images and histograms illustrating the diameter distribution of PHB/PHBV nanofibers. Polymer concentrations of (A) 6, (B) 8 and (C) 10% wt/wt were used. The voltage, flow rate and collecting conditions were fixed at 16 kV, 1.5 ml/h and 15 cm, respectively. Scale bars represent 2 0 μm for SEM images. Y axis range is 0–35% for all histograms. X axis ranges are 800–1700 nm (A 100/0), 500–1400 nm (A 75/25), 400–1300 nm (A 50/50), 400–1300 nm (A 25/75), 200–650 nm (A 0/100), 1000–1900 nm (B 100/0), 600–1500 nm (B 75/25), 600–1500 nm (B 50/50), 400–1300 nm (B 25/75), 300–1200 nm (B 0/100), 1100–2000 nm (C 100/0), 700–1600 nm (C 75/25), 600–1500 nm (C 50/50), 700–1600 nm (C 25/75) and 600–1100 nm (C 0/100).

of elongation of aligned nanofibers is approximately ten times more than randomly oriented nanofibers, the young modulus and tensile strength are lower than that of random mats. As expected, the mechanical properties of PHB(50)/PHBV(50) random nano-

fibers were between the mechanical properties of PHB(100)/ PHBV(0) and PHB(0)/PHBV(100) random nanofibers.

The surface wettability of different PHB(50)/PHBV(50) scaffolds was investigated through a water contact angle method. The

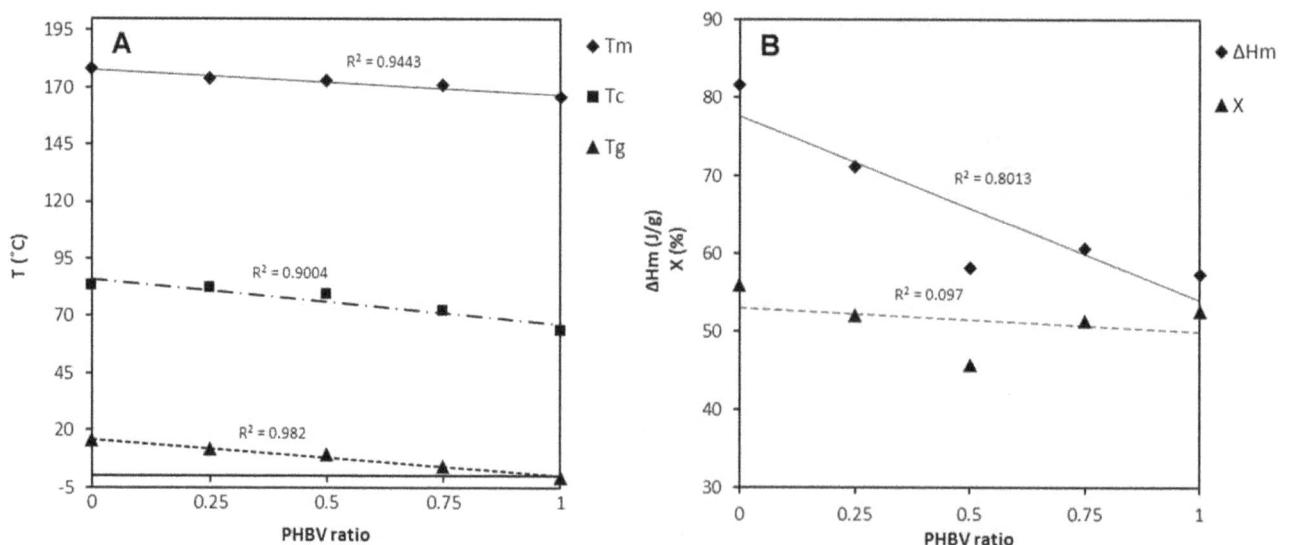

Figure 2. Thermal properties of PHB/PHBV nanofibers versus blend compositions. (A) melting temperatures (T_m), glass transition temperature (T_g) and crystallization temperatures (T_c); (B) melting enthalpy (ΔH_m) and crystallization percentage (χ).

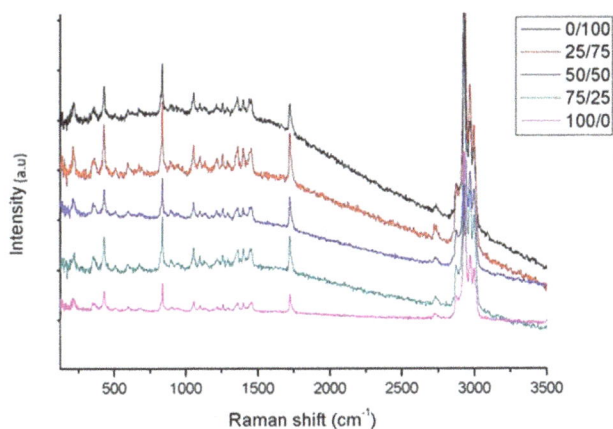

Figure 3. Raman spectra of PHB/PHBV nanofibers in different blend compositions.

results showed that the water contact angle values were $123.5\pm6.3°$ for PHB(50)/PHBV(50) random nanofibers and $107.4\pm5.0°$ for PHB(50)/PHBV(50) aligned nanofibrous scaffolds, which further decreased to $94.7\pm9.4°$ for PHB(45)/PHBV(45)/Col(10) blend nanofibers.

Schwann cell activity on PHB/PHBV solution blended random nanofibers

Figure S3 (supplementary information) shows SEM analysis of SCs spreading and morphology on the surface of different PHB/PHBV nanofibrous scaffolds after cell seeding for 1 and 7 days. A normal bipolar extensions and spindle-shaped morphology of SCs on PHB/PHBV nanofibrous scaffolds was observed. At day 1, SCs located on the surface of the scaffolds were still rounded, while after 7 days cells have expanded and proliferated considerably.

Alamar blue and DNA quantification assays were performed during the 14 days of cell culture to investigate the metabolic activity and proliferation of SCs, respectively. As shown in Figure S4A (supplementary information) the metabolic activity of cells in all PHB/PHBV blended nanofibrous scaffolds increased with time, except at day 14 where a plateau was reached. On day 1, the metabolic activity of PHB(25)/PHBV(75) was significantly higher than that of the other samples. With increasing culturing time, all samples were rather similar except PHB (75)/PHBV (25) which showed less metabolic activity in comparison to the other blends. Between 1 and 3 days after cell seeding, the largest increase in the metabolic activity of cells was observed on the PHB(100)/PHBV(0) and PHB(50)/PHBV(50) nanofibrous scaffolds, while between day 3 and day 7 the increase of metabolic activity measured on PHBV(75)/PHB(25) composition was the largest observed.

DNA amount increased for each scaffold composition from day 1 to day 14 (Figure S4B, supplementary information), indicating SCs proliferation during interval time. The largest increase in DNA amount was observed on the PHB (100)/PHBV (0) scaffolds as well as PHB(50)/PHBV(50) composition. On day 14 the amount of DNA for both PHB (100)/PHBV (0) and PHB(50)/PHBV(50) samples was significantly more than that of other compositions.

Schwann cell activity on PHB/PHBV and PHB/PHBV/Col aligned or random nanofibers

Figure 5 shows SEM analysis of SCs spreading and morphology on the surface of randomly oriented and aligned PHB(50)/PHBV(50) nanofibrous scaffolds after cell seeding for 1, 3, and 7 days. These SEM images also showed normal bipolar and tripolar extensions and spindle shaped morphology of SCs on both nanofibrous scaffolds. Aligned nanofibers showed cells oriented along the direction of fibers and clustered around the aligned fibers in a longitudinal fashion, while the random fibers showed cells oriented in different directions. Figure 6 shows outputs of digital image processing analysis to investigate alignment of

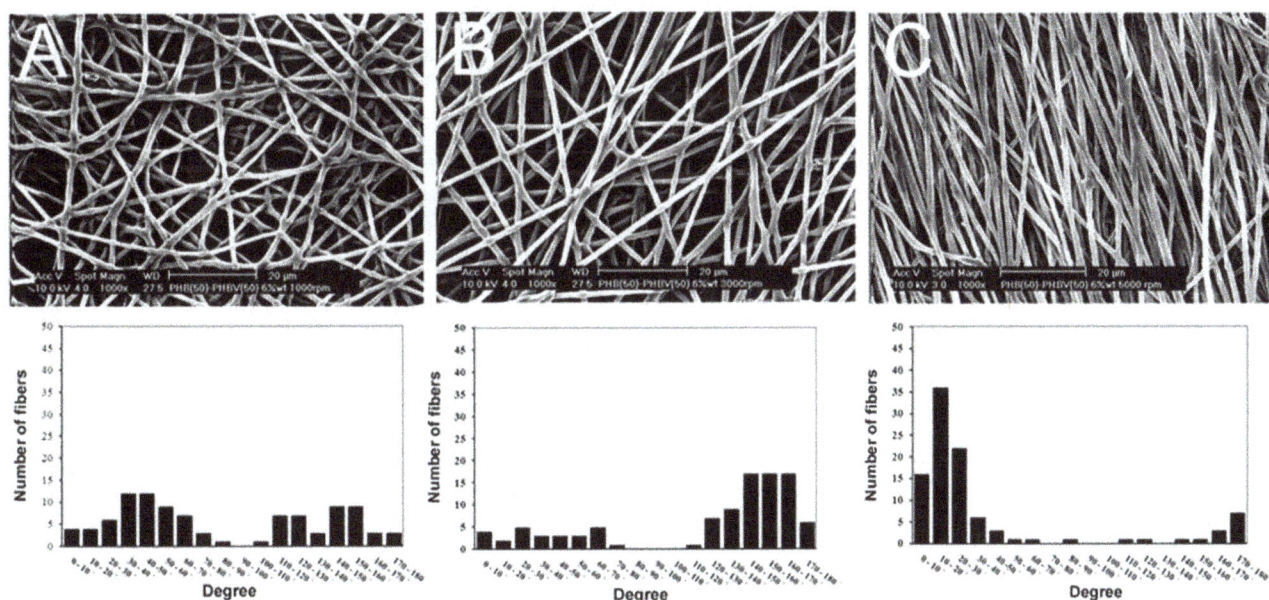

Figure 4. SEM images and histograms illustrating the orientation of PHB(50)/PHBV(50) nanofibrous mats. Nanofibers collected with speed of (A) 1000 rpm, (B) 3000 rpm, (C) 5000 rpm. The voltage, flow rate and collecting conditions were fixed at 16 kV, 1.5 ml/h and 15 cm, respectively. Scale bars represent 20 μm for SEM images.

Table 1. Mechanical properties of PHB, PHBV and PHB(50)/PHBV(50) random and uniaxially oriented nanofibers.

	Tensile stress (MPa)	Tensile strain (%)	Tensile modulus (MPa)	Thickness (mm)
PHB random nanofibers	1.7 ± 0.3	5.3 ± 0.7	67.8 ± 23.5	0.31 ± 0.02
PHB/PHBV random nanofibers	3.4 ± 0.3	8.6 ± 2.0	59.3 ± 11.3	0.26 ± 0.02
PHBV random nanofibers	4.9 ± 0.5	46.1 ± 20.0	254.0 ± 25.4	0.19 ± 0.01
PHB/PHBV aligned nanofibers	1.3 ± 0.1	91.7 ± 23.6	45.4 ± 4.7	0.25 ± 0.01

cultured SCs on uniaxially and randomly oriented nanofibrous substrates. A dominant peak around 40° with cumulative intensity of 1 for Radon transform output of aligned nanofibers was observed, while in the case of random substrates this dominant peak was not seen and its cumulative intensity was zero.

SCs functionality was also identified with p75LNGFR staining, which is one of the most common cell markers for SCs 15]. Immunocytochemistry indicated that cultured SCs were also positive for p75LNGFR, thus confirming normal cell functionality (Figure 7). Furthermore, fluorescent miscroscopy also confirmed spread cell morphology and tight attachment to the electrospun mats in addition to orientation of SCs in the direction of the fiber alignment, similar to what was observed by SEM and image processing algorithms.

Metabolic Activity and Proliferation of SCs on Scaffolds

SCs were cultured for 14 days on PHB/PHBV random and aligned nanofibers to evaluate the influence of scaffold topography on cell metabolic activity and proliferation. SCs metabolic activity increased with time in both scaffolds. While metabolic activity of SCs on random nanofibers was significantly higher than on aligned fibers at day 3, this trend was reversed by day 14 (Figure 8A). DNA amounts (Figure 8C) increased for each scaffold from day 1 to day 14. After 2 weeks a significant difference between cell proliferation on aligned and randomly oriented nanofibers was measured.

Similarly, Figures 8B and 8D present the effect of collagen incorporation on metabolic activity and proliferation of SCs. No significant difference between metabolic activity of SCs on PHB/PHBV/Col and PHB/PHBV aligned nanofibers was observed, while proliferation of SCs on PHB/PHBV/Col scaffold was significantly higher than PHB/PHBV aligned nanofibrous scaffolds at day 14.

The gene expression of different SCs markers was analysed using quantitative RT-PCR (Figure 9A). No statistically significant difference was observed on SCs cultured on these groups for GDNF, BDNF, CNTF, NGF-F and PMP22, while the presence of collagen in the scaffolds resulted in a significant up-regulation of GDNF expression (Figure 9B) in comparison to aligned nanofibers in the absence of collagen, indicating that collagen might have accelerated SCs differentiation.

To further analyse cell differentiation on these scaffolds, we measured the amount of NGF secreted by the cells. There was no significant difference between NGF released from SCs on different groups at days 3, 6 and 12, whereas aligned nanofibers supported a statistically higher production of NGF at days 9 and 14 in comparison to random nanofibers (Figure 9C) ($p \leq 0.05$). Collagen containing scaffolds exhibited the highest concentration of secreted NGF at day 6 compared to unmodified aligned PHB(50)/PHBV(50) nanofibers, while after 14 days of culture the concentration of NGF on collagen containing nanofibers was significantly lower than unmodified PHB/PHBV scaffolds (Figure 9D).

Discussion

Characterization of Scaffolds

As previously mentioned, the use of artificial polymeric scaffolds to prepare a suitable environment for nerve regeneration has become a promising technique for the restoration of major nerve injuries 1]. To achieve this goal, mimicking the architecture of ECM is one of the critical properties of tissue engineering scaffold 35]. 36]. It is generally accepted that SCs play a crucial role during nerve regeneration through the production of neurotrophic factors, formation of myelin sheath and extraction of extracellular proteins, as well as acting as a physical guide for the newly regenerating axons 18,28]. Recently, there has been interest in the use of polymeric substrates that are pre-seeded with SCs to act as a persistent source of neurotrophic factors and an effective scaffold for enhancing nerve regeneration [14,23]. PHB and PHBV

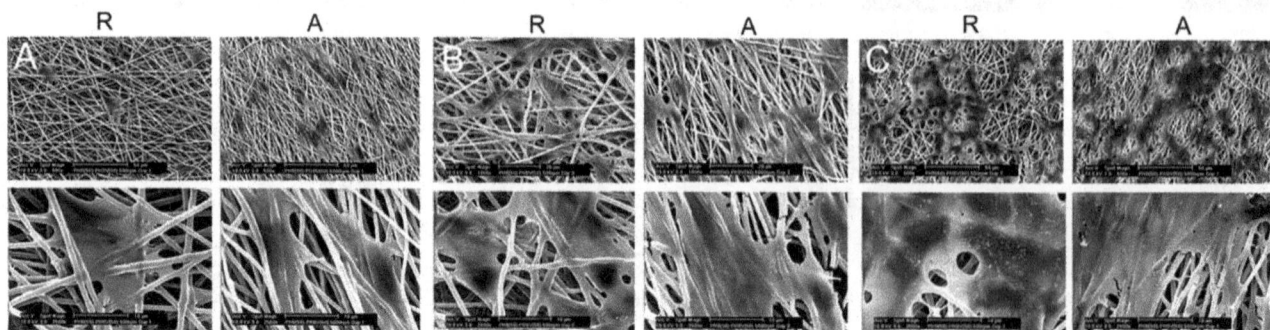

Figure 5. SEM images of SCs on PHB(50)/PHBV(50) random (R) and aligned (A) nanofibers. (A) 1 day, (B) 3 days, and (C) 7 days after cell seeding. Scale bars represent 50 μm for top and 10 μm for button pictures, respectively.

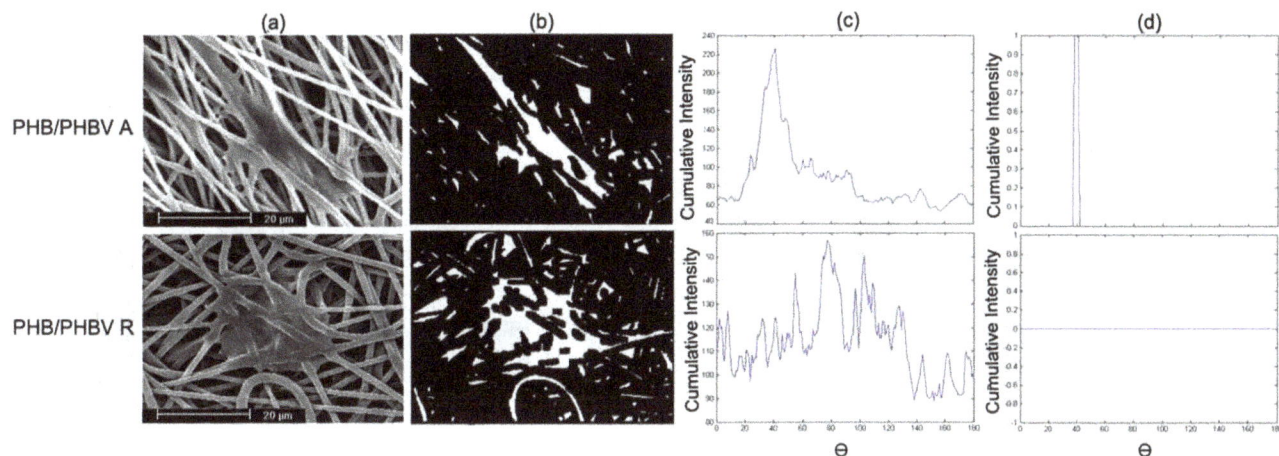

Figure 6. Image processing algorithms on SEM images of cultured SCs on PHB(50)/PHBV(50) random (PHB/PHBV R group) and PHB(50)/PHBV(50) aligned (PHB/PHBV A group) nanofibrous scaffolds. (A) Original SEM image of SCs on scaffold after 1 day of cell culture; (B) binary image showing the processing output of nonlinear diffusion filtering; (C) Radon transform results of binary image for a range of [0,180°] and with a resolution of $\Delta\theta = 1°$; (D) dominant peak of the maximum of Radon transform matrix.

polymers have been experimentally used as an alternative to direct epineural repair and bridge short and long-gap nerve injury models. In this study, the potential of PHB/PHBV and PHB/PHBV/collagen electrospun nanofiber blends pre-seeded with SCs has been investigated. Normally, the effects of different processing parameters in electrospinning should be assessed for individual polymers and solution blends. Hereby, electrospinning of various solution blend compositions of PHB/PHBV (100/0, 75/25, 50/50, 25/75 and 0/100) was successfully used in three concentrations of 6%, 8% and 10 % wt/wt for the generation of nanofibrous scaffolds (Figure 1). Control of fiber diameter was possible by changing the polymer concentration and blend composition (Figure S1, supplementary information). The fiber diameter is expected to increase with increasing polymer concentration due to

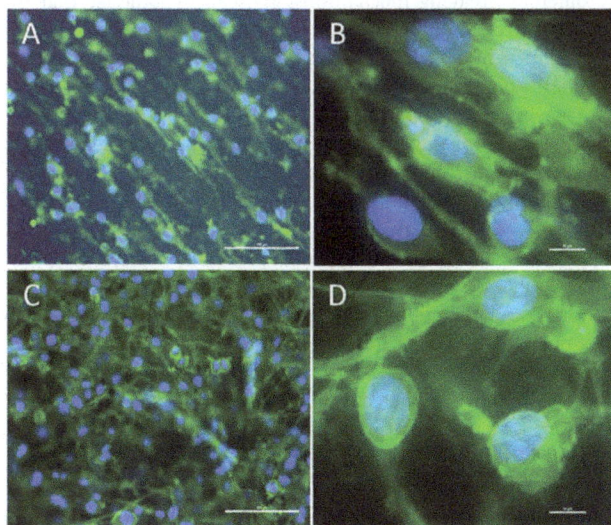

Figure 7. Immunostaining identification of adult rat SCs by p75LNGFR and counterstained by DAPI. Cells on aligned (A and B) and random nanofibers (C and D) showed p75LNGFR positive staining after 7 days of culture. Scale bars show 20×(100 µm) for A and C, and 100×(10 µm) for B and D pictures.

an increase in the viscosity and surface tension of polymers solution as also demonstrated by other researchers 37–40]. Additionally, fiber diameter decreased by increasing the amount of PHBV in the solution blend, probably due to a consequent decrease in the viscosity of polymer solution as a result of the lower molecular weight of PHBV.

The homogeneity of the polymeric blends is important in terms of their internal structural and mechanical integrity 41]. Therefore, DSC and Raman spectroscopy were used to investigate the thermal and physical properties of blended polymeric nanofibers. The curves shown in Figure 2 revealed that the addition of PHBV to PHB, even in quantities as small as 25% in mass, decreased glass transition, melting temperature and crystallinity of the resulting nanofibers as well as broadened the processing window, since there was improved melt stability at lower processing temperatures, similar to the results obtained by other researchers 34,42–44]. Evidently, good flexibility is an important factor for a nerve tissue-engineered conduit to resist tearing and stretching forces, and retain stable shape during the regeneration process. The influence of PHBV content on thermal behavior of PHB/PHBV blends proves that the mixture of hydroxybutyrate with hydroxyvalerate is less crystalline, not as fragile, more flexible and more readily processable than PHB itself 45].

The Raman spectra of PHB/PHBV electrospun scaffolds indicated that the band centered at 1725 cm^{-1} for different samples depicted distinct full width at half maximum: PHB(100)/PHBV(0) (30 cm^{-1}), PHB(75)/PHBV(25) (40 cm^{-1}), PHB(50)/PHBV(50) (44 cm^{-1}), PHB(25)/PHBV(75) (46 cm^{-1}) and PHB(0)/PHBV(100) (52 cm^{-1}). The difference between band widths can be assigned to the different crystallinity of these two polymers. These results are in agreement with those obtained from DSC, and the study of Izumi *et al.* 46] showing that crystallinity degree of PHBV decreased as the hydroxyvalerate (HV) content increased. Despite similarities among PHB and PHBV Raman spectra, the relative intensity of the bands at some wave lengths is different. However, it is not possible to correlate the changes of these bands with the respective blend composition, since the samples have different crystallinity that may influence the Raman intensity of these bands.

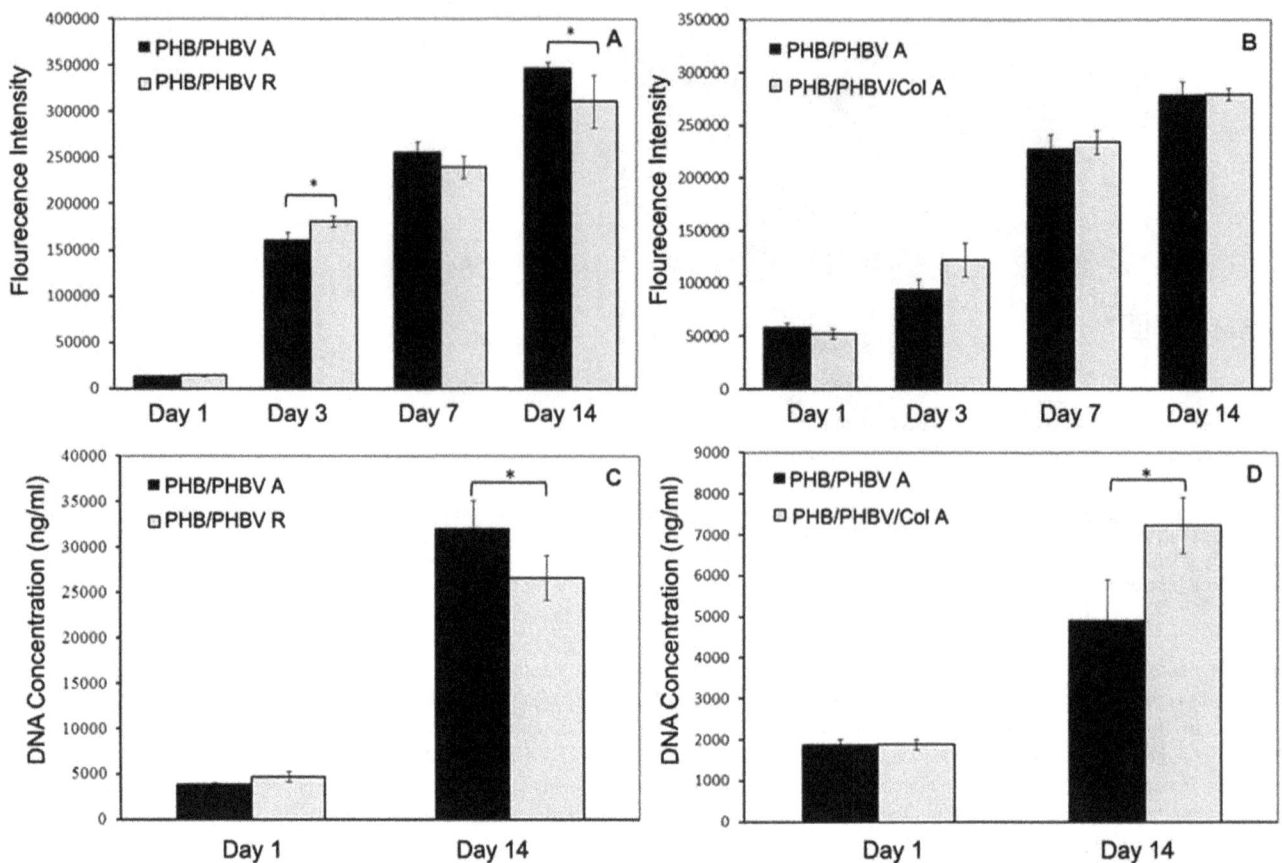

Figure 8. Alamar blue and DNA quantification assay results to compare metabolic activity and proliferation of SCs. (A, C) PHB(50)/PHBV(50) random nanofibers (PHB/PHBV R group) and PHB(50)/PHBV(50) aligned nanofibers (PHB/PHBV A group); (B, D) PHB(50)/PHBV(50) aligned nanofibers (PHB/PHBV A group) and PHB(45)/PHBV(45)/collagen(10) aligned nanofibers (PHB/PHBV/Col group) during 14 days of culture. Asterisks represent significant difference at $p \leq 0.05$.

It has well been established that fiber orientation influences cell adhesion and growth, and guides cellular elongation 13]. In this study, the influence of the drum rotating speed on orientation of nanofibrous webs was examined to design aligned PHB/PHBV nanofibrous scaffolds. The distribution of fiber angles was narrower in proportion to the increase of rotating speed, thus resulting in improvement of fiber alignment (Figure 4). These results indicate that the higher speed of rotation of the drum collector exerts a larger tangential force on the polymer solution jet, inducing alignment of fibers without considerable influence on the fiber diameter. Similar effect was also observed by Wang *et al.* and Cooper *et al.* during the fabrication of chitosan and chitosan-PCL nanofibers, respectively 13,47].

Also, an ideal nerve conduit should have a Young's modulus approaching that of nerve tissues to withstand manual manipulation, *in vivo* physiological loading during nerve regeneration, and retain its structure after implantation 48]. It is generally accepted that Young's modulus of a rabbit tibial nerve and a rat sciatic nerve in the longitudinal direction, for example, is in the range of 0.50 and 2.72 Mpa, respectively 13,48].

As shown in Table 1, the tensile strength of PHB(50)/PHBV(50) random and aligned nanofibrous scaffolds were 337 ± 0.3 and 1.30 ± 0.1 MPa, respectively. Moreover, the percent elongation of aligned fibrous scaffolds was considerably higher in comparison to the randomly oriented scaffolds which indicated their good elasticity and flexibility. The difference between mechanical

properties of random and aligned nanofibers is probably caused by different breaking mechanisms. We can hypothesize that the increased strength of random fibers results from more compactness of nanofibers in the web, which can assist them to withstand higher loadings. Another interesting result was observed when comparing the mechanical properties of unmodified PHB and PHBV nanofibers. Surprisingly, mechanical properties (tensile strength, tensile modulus and elongation percentage) of PHBV and PHB(50)/PHBV(50) nanofibrous mats were considerably higher than that of PHB fibers. Many studies have confirmed that PHB copolymers with 3-hydroxyvalerate (PHBV) are more flexible and more easily processed than PHB as a result of its lower glass transition and melting temperatures 21,22], which is in agreement with our DSC results (Figure 2).

In this study, we fabricated five different composite of PHB/PHBV solution blending electrospun nanofibers for nerve tissue engineering and investigated physico-thermal properties of all solution blends. In conclusion, PHBV copolymer partly fills the gap of toughness and flexibility of PHB; however, PHB/PHBV composite nanofibrous mats exhibit lower melting points with respect to PHB, narrowing the utilization temperature range.

Interaction between SCs and Scaffolds

Although the *in vitro* assays illustrated the cellular biocompatibility of the fabricated scaffolds, direct cellular interactions with the material surface also plays an important role in tissue

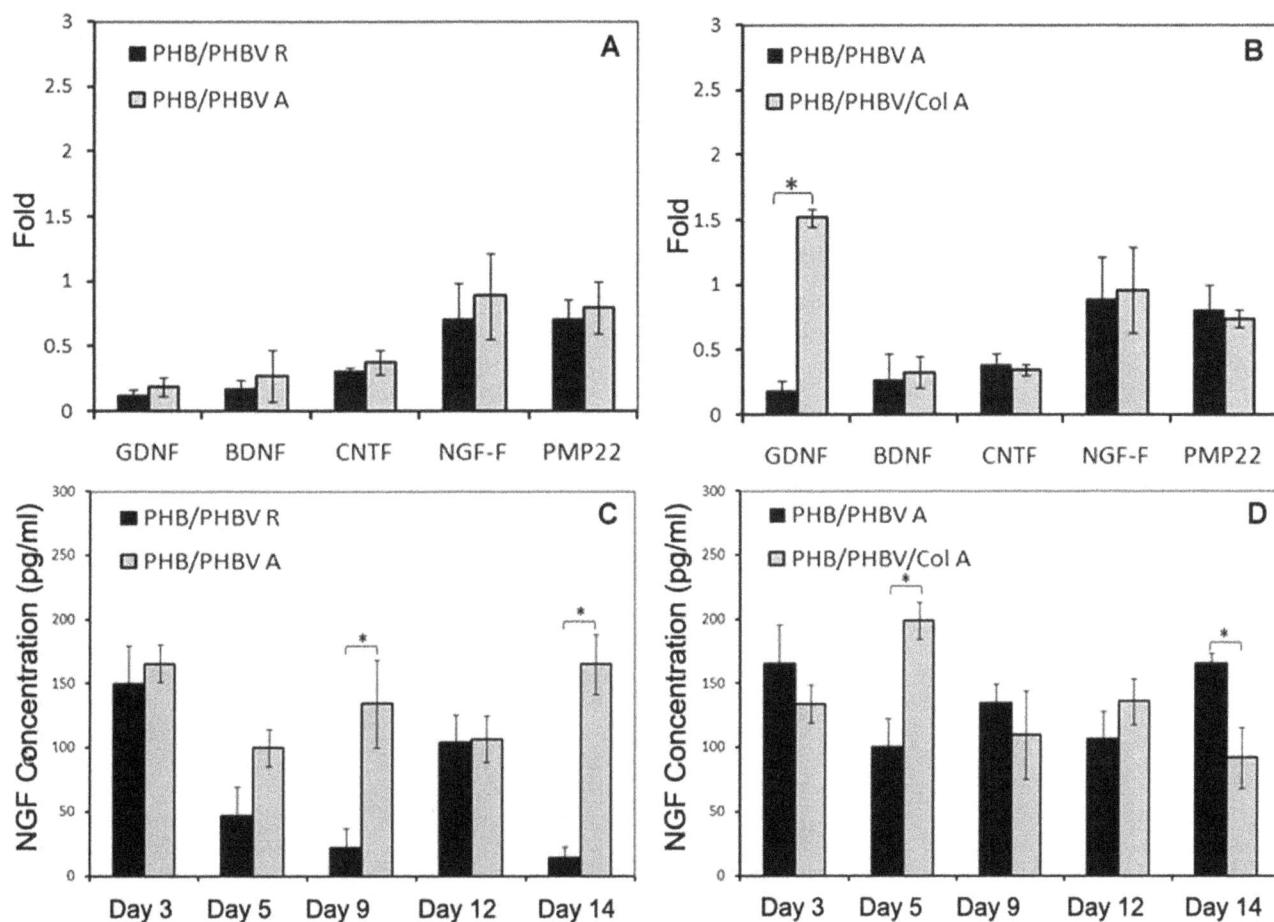

Figure 9. Real-time PCR analysis and NGF assay results to compare gene expression and secretions of NGF released from SCs. (A, C) PHB(50)/PHBV(50) random nanofibers (PHB/PHBV R group) and PHB(50)/PHBV(50) aligned nanofibers (PHB/PHBV A group); (B, D) PHB(50)/PHBV(50) aligned nanofibers (PHB/PHBV A group) and PHB(45)/PHBV(45)/collagen(10) aligned nanofibers (PHB/PHBV/Col group) during 14 days of culture. Asterisks represent significant difference at $p \leq 0.05$.

regeneration. Effective attachment of SCs onto scaffolds is a critical step for the fabrication of tissue engineered nerve conduits. For example, SCs morphology and directionality are two key factors contributing to neurogenesis 13]. SEM analysis showed physiological cell morphology in PHB/PHBV blended samples in addition to increased proliferation between day 1 and day 7 (Figure S3, supplementary information). Bi-polar and tri-polar extensions with spindle shaped morphology are characteristics of normal SCs 14], also observed in SEM images of day 1. The high surface area and extracellular-like physical environment provided by nanofibers compared to that of other non-fibrillar surfaces may have led to an increase in cellular attachment and the observed cell polarity. For instance, Leong et al. 49] indicated that the cell-matrix adhesion in nanofibrous mats is stronger due to an increase in the surface area of the nanofibrous mats and their three-dimensional features. Similar results were also observed by Sangsanoh et al. 18] on different polymeric nanofibers composed of neat PHB, PHBV, polycaprolactone (PCL), poly(L-lactic acid) (PLLA), and chitosan seeded with SCs.

Aligned fibrous scaffolds could exhibit SCs columns (Figure 5), also known as bands of Büngner. Bands of Büngner are formed when SCs proliferate and the remaining connective tissue basement membrane lines endoneurial tubes 50,51].

Indeed, bands of Büngner comprise longitudinally aligned SCs strands that guide selectively regrowing axons.13,50]. Wang et al. have suggested that a presumed mechanism for the formation of band of Bunger is the polarized expression of adhesion proteins such as dystronlycan and α6β4 integrin on the SCs axis 47]. It can also be expected that SCs alignment obtained on the oriented nanofibers might induce oriented axonal growth as a result of classical contact guidance 14]. Similar results have been observed by different researchers, who have also reported oriented arrangement of SCs on aligned electrospun nanofibrous mats 13,14,47,52].

In addition to ordinary SEM imaging, we made use of digital image processing algorithms to evaluate the orientation of cultured SCs on aligned and random fibrous scaffolds. After binarization and calculation of Radon transform, it was possible to determine the main orientation angle of cultured SCs. Results confirmed cell orientation on aligned fibers with a high resolution and an accuracy of 95%.

Immunostaining with p75 low-affinity NGF receptor (p75LNGFR) marker also showed normal topography and activity of SCs on aligned nanofibrous scaffolds (Figure 7). p75LNGFR is one of the two receptor types for neurotrophins, a family of protein growth factors that stimulate neuronal cells to survive and differentiate. It is well known that SCs in the distal stump of injured peripheral nerves synthesize the p75LNGFR receptor. You et al, 53] showed that the presence of p75 protein in SCs is necessary for reinnervation and the expression of p75 by SCs provides a suitable environment for the regenerating axons, especially in the early stages of regeneration. Zorick and Lemke showed the effective role of LNGFR in all stages of SCs differentiation and myelination. Here, we can hypothesize that the positive staining of cultured SCs on the designed blended scaffolds indicates SCs maintain their potential to support axonal regeneration 54].

The next step to consider PHB/PHBV nanofibers in nerve tissue engineering was to ensure that attached SCs could proliferate and perform their normal functions. To evaluate the influence of blend composition on SCs metabolic activity and cell proliferation, SCs were cultured for 14 days on PHB/PHBV random nanofibers. As shown in Figure S4B (supplementary information), proliferation of SCs increased within 14 days of culture and the amount of proliferation for PHB(100)/PHBV(0) and PHB(50)/PHBV(50) compositions was higher than that of other samples. Likewise, Suwantong et al. [15] showed that SCs can proliferate on PHB or PHBV nanofibers within 5 days. As previously mentioned, experimental evidence indicates that grafts seeded with SCs prior to implantation have shown enhanced nerve regeneration in vitro and in vivo 55]. For example, Yu et al. 56] successfully used collagen/PCL nanofibers seeded with SCs to regrow axons. Schnell et al. 57] also indicated that electrospun collagen/PCL substrates pre-seeded with SCs can support oriented neurite outgrowth and glial migration from dorsal root ganglia explants.

To further evaluate the synergistic effect of alignment with bioactive motifs from the basement membrane, we cultured SCs aligned electrospun PHB(50)/PHBV(50) nanofibers with or without blending with collagen type I. In general, aligned nanofibers enhanced SCs proliferation on day 14, and metabolic activity on days 3 and 14, in comparison to random fibers. Gupta et al. 14] believed that, rough surface topography and more interconnected pores make random nanofibers better scaffolds for SCs proliferation. Our morphological studies showed that the direction of SCs is parallel to the direction of fiber alignment. Therefore, it can be concluded that the arrangement of cells in controlled architecture has beneficial effects on SCs metabolic activity.

Incorporation of collagen onto PHB(50)/PHBV(50) electrospinning solution did not result in any significant increase of SCs metabolic activity compared to unmodified nanofibrous scaffolds (Figure 8B), while a significant increase in SCs proliferation was observed (Figure 8D). The improved proliferation observed on collagen blended nanofibrous scaffolds could be firstly explained as a result of increased hydrophilicity, which increases serum proteins adsorption. Secondly, the presence of collagen molecules on the surface of nanofibers could be associated to a preferential adhesion of SCs to these protein motifs, better mimicking their native environment. Previous studies also confirmed our results indicating increased cells proliferation and attachment on collagen functionalized scaffolds. For example, Meng et al. reported that the PHBV/collagen nanofibrous scaffolds accelerated the adhesion and growth of NIH-3T3 cells more effectively than PHBV nanofibrous scaffolds 45]. Enhanced attachment and viability of neural stem cell on nanofibrous collagen immobilized scaffolds was also reported by Li et al. 58].

Previous studies have shown that the terminal chemical groups present on the surface of scaffolds could control cell growth and differentiation 6]. Wang et al. 59] believed that a fraction of N-containing groups such as amine introduced to the surface may be positively charged at physiological pH because of protonation in the culture medium, which helps the adhesion of cells that carry negative charges on the membrane surface. Ren et al. 60] also showed that amino groups are more effective than hydroxyl and carboxyl groups towards nerve stem cell proliferation and migration. Hence, it can be concluded that the presence of amino groups on the surface of PHB/PHBV/Col nanofibrous scaffolds and the higher efficiency of amino groups with respect to hydroxyl and carboxyl groups may account for the observed enhancement in SCs proliferation.

Once axonal growth is initiated, cell survival and continued axonal regeneration depend on the supply of trophic physiological growth factors, which are synthesized by SCs 61,62]. The expression of these cytokines changes during SCs differentiation. PMP22 and CNTF expression increase during myelination, while low levels of BDNF are present in SCs of either developing or mature nerves 63]. Therefore, real-time PCR analysis of the genetic expression of these neural cytokines gives more information about SCs maturity and differentiation on the designed scaffolds. In this study, there was no significant difference between all gene expressions of random and aligned PHB/PHBV nanofibrous scaffolds, suggesting substrate orientation has not increased SCs differentiation and myelination (Figure 9A). With incorporation of collagen in the scaffolds, GDNF expression of SCs on PHB/PHBV/Col aligned scaffolds significantly increased compared to neat nanofibers (Figure 9B). GDNF is an important growth promoting factors for dopaminergic and motor neurons, especially in the case of chronic differentiation 62]. Li et al. 64] showed that increased expression of GDNF by genetically modified SCs improved peripheral nerve regeneration. Piirso et al. suggested that GDNF promotes myelination of small caliber axons that normally do not myelinate and enhances myelination in neuron-SCs co-cultures 65]. In this study, high expression of GDNF factor on collagen containing scaffolds suggests a possible positive role of collagen in SCs differentiation.

Another important neurocytokin for SCs differentiation and myelination is NGF. It has been proved that NGF influences myelination of axons by both SCs and oligodendrocytes 65,66]. Indeed SCs are commonly recognized as the major source of NGF synthesis in the axotomized nerve 67]. In this study, ELISA assay results of NGF secretion showed significant increase of NGF released by SCs cultured on aligned nanofibers on days 9 and 14 in comparison to that of random mats (Figure 9C). Collagen containing scaffolds also showed significantly more NGF secretion of SCs in comparison to that of PHB(50)/PHBV(50) scaffolds on day 6. At day 14, NGF secretion was, however, significantly lower on collagen containing scaffolds than on PHB/PHBV ones (Figure 9D). This might be explained by a faster differentiation of SCs when cultured in presence of collagen, thereby reducing the production of NGF at later culture times.

Therefore, PHB/PHBV and PHB/PHBV/Col aligned nanofibrous scaffolds fulfil the requirement of being a stable mechanical and physical support for axonal growth, whereas the seeded SCs produce the required neurotrophic factors, which are helpful for stimulating the outgrowth of axons. Further studies should aim at deepening our understanding whether this composition could also be optimal for SCs differentiation and myelinic membrane formation during extended culturing periods and in preclinical animal models.

Materials and Methods

Materials

Poly(R-3-hydroxybutyrate) (PHB), $M_w = 437000$, Poly(3-hy-droxybutyrate)-co-(R-3-hydroxyvalerate) (PHBV) with 5% wt poly (3-hydroxyvalerate), $M_w = 150000$, and solvents (chloroform and N, N-dimethyl formamide (DMF) and Hexaflouroisopropanol (HFIP)) were purchased from Sigma Aldrich (St. Louis, MO, USA) and Merck Co. (Germany), respectively. Acid soluble collagen type I powder of bovine origin was a generous gift from Kensey Nash Corporation (USA, Catalogue number 20003–04). Purchased reagents for the cell culture were as follows: fetal bovine serum (FBS) from Hyclone (Logan, UT, USA), Dulbecco Modified Eagle Medium (DMEM), phosphate buffered saline (PBS), penicillin, streptomycin and trypsin-EDTA from Gibco BRL (Gaithersburg, MD, USA).

Electrospinning

All polymeric solutions of PHB/PHBV blends were prepared and stirred overnight at 60°C before use by dissolving two components in a mixture of chloroform (90) and DMF (10) solvents at concentrations of 6, 8 and 10% wt. The desired solution was loaded into a syringe and the flow rate was controlled using a syringe pump (KDS 100, KD Scientific). The other end of the syringe was connected to a needle, on which a positive high voltage (16 kV) was applied using a high voltage generator (Gamma High Voltage Research Inc., USA). The polymeric solution was fed at a rate of 1.5 ml*min^{-1} through a 10 cc syringe with a 23 G needle placed 15 cm from a rotating mandrel collector with speed of 1000 and 5000 rpm to collect PHB/PHBV random and aligned nanofibers, respectively. Temperature and humidity were monitored during the process and ranged between 24–26 °C and 37–42%, respectively.

PHB/PHBV/Col nanofibers were also fabricated by solution blending of PHB, PHBV and collagen at a ratio of 45:45:10 (wt/wt), and at a total concentration of 4% (wt/wt). In short, collagen was dissolved in HFIP at room temperature, and then PHB and PHBV were added to the collagen solution. The final solution was stirred for 24 h at room temperature. The solution was electrospun with mentioned flow rate and voltage and aligned PHB/PHBV/Col nanofibers were collected with a speed of 5000 rpm.

Scanning electron microscopy (SEM)

The morphology of electrospun fibers spun under the above mentioned conditions was observed using scanning electron microscopy (SEM) (XL 30 ESEM-FEG, Philips). Fiber diameters were calculated from SEM micrographs by measuring 100 fibers per condition using Manual Microstructure Distance Measurement software (Nahamin Pardazan Asia Co., Iran).

Differential scanning calorimetry (DSC)

The DSC experiments were conducted in a Q100 calorimeter with refrigerated cooling system (TA Instruments). Prior to the scans, temperature and energy calibrations were performed with an indium standard. All the samples were (a) held at −50°C for 5 min, (b) heated from −50°C to 200°C, (c) held at 200°C for 5 min to erase the thermal history, (d) cooled down from 200°C to −50°C, (e) held again at −50°C for 5 min, and (f) re-heated to 200°C. The heating and cooling rate was of 10°Cmin^{-1}. The glass transition temperature (T_g), melting temperature (T_m), crystallization temperature (T_c) and melting enthalpy (ΔH_m) of PHB/PHBV blends were determined from the heating and cooling scans. The crystallinity degree (χ) can be calculated by applying the following

equations (1 and 2):

$$(1) \quad x = \frac{\Delta H_m}{\Delta H_{ref}}$$

$$(2) \quad \Delta H_{ref} = \Delta H_{ref,PHB} \times x + \Delta H_{ref,PHBV} \times (1-x)$$

Where ΔH_{ref} is the melting enthalpy of 100% crystalline polymer (i.e. $146 \, Jg^{-1}$ and $109 \, Jg^{-1}$ for PHB and PHBV, respectively 34]) and x is the weight fraction of PHB in the blend.

Raman Spectroscopy

Confocal non-resonant Raman microspectroscopy was performed using a home-built laser-scanning Raman microspectrometer. In brief, the 647.1 nm excitation light from a Krypton ion laser was focused onto the sample through a 40x air objective (Olympus). Raman spectra were acquired in 2 s at a laser power of 35 mW under the objective. For each sample, spectra were taken at 5 randomly chosen spots and averaged. Data preprocessing was performed using routines written in MATLAB 6.5 (The Math-Works Inc., Natick, MA).

Mechanical properties of nanofibrous scaffolds

For mechanical strength measurements, scaffolds were cut into 20×10 mm rectangular strips. Tensile test on the scaffolds was performed with a tensile testing machine (Zwick Z050, Germany) at a crosshead speed of 10 mm/min. Tensile properties were calculated from the stress-strain curves as the means of five measurements and the average value was reported with standard deviation (± SD).

Surface wettability measurements

Water contact angle (θ) of the nanofibrous scaffolds was determined on a contact angle goniometer (Dataphysics OCA-20). Deionised water was used as a probe liquid. Water contact angle was measured by at least five independent measurements and was presented as mean ± standard deviation.

Cell Culture

The in vitro experiments were performed using a rat Schwann cell line RT4-D6P2T (ATCC, USA), cultured in high glucose DMEM, and supplemented with 10% fetal bovine serum (FBS) and 1% Penicillin/Streptomycin. The cells were incubated at 37°C in a humidified atmosphere containing 5% CO_2 and the cultured medium was changed once every 3 days. Electrospun discs with a diameter of 15 mm were soaked in 70% ethanol for 2 hr, washed twice with PBS, transferred to a non-treated 24 well plate (NUNC) and incubated overnight in basic cell culture medium. Rubber O-rings (Eriks BV, The Netherlands) were used to secure the scaffolds in place and prevent them from floating. After removing the medium, each scaffold was seeded with 50'000 cells in 50 µl basic medium and incubated for 30 min to allow cell attachment and topped up to 1 ml with culture medium.

Cell Morphology Study

On days 1 and 7, one sample from each group was used for SEM analysis. The media were removed and the scaffolds were washed twice with PBS and fixed in 4% formalin for 30 min. After rinsing with PBS, the scaffolds were dehydrated in a series of increasing ethanol concentrations (70, 80, 90, 96, and 100%), 30 min in each concentration, before being dried using a critical point dryer (Balzers CPD-030). The samples were then sputter coated with gold (Cressington) for SEM observation (XL 30 ESEM-FEG, Philips).

Assessment of Cell Alignment Using Digital Image Processing

To investigate cells alignment, we applied digital image processing algorithms to the SEM images of cultured SCs on scaffolds. Processing was performed on a dataset of 50 SEM images of cultured SCs on both randomly and uniaxially oriented nanofibrous scaffolds, which have been captured 1 day after cell culture and cropped to the same size of 295×235 pixels. All of the images had the same magnification of 1000x and were captured with the same brightness and contrast to ease further image analysis.

The fallowing algorithm was used to verify the probable existence of SCs orientation on fibrous scaffolds. Firstly, by Nonlinear Diffusion Filtering (NDF) the original images were smoothed. NDF preserved the high gradient objects of the image and smoothed low gradient parts of it. After that, by using binary algorithms of image processing, a binary image of the NDF output was made and Radon transform of the binary image was calculated for a range of $[0,180°]$ and with a resolution of $\Delta = 1°$. After calculation of the maximum of Radon transform matrix columns and verification of a detectable dominant peak, it was possible to specify the localization of angle.

Cell Metabolism

Cell metabolism was assessed using Alamar blue assay according to the manufacturer's protocol. Briefly, culture medium was replaced with medium containing 10% (v/v) Alamar blue solution (Biosource, Camarillo, CA, USA) and the cells were incubated at 37°C for 4 hr. Absorbance was measured at 590 nm using a Perkin Elmer Victor3 1420 Multilabel plate reader. Cell metabolism was analyzed on day 1, 3, 7 and 14 (n = 3), and medium containing Alamar blue solution was replaced with fresh medium after each measurement.

Cell Proliferation

After 1 and 14 days of cell culture, the scaffolds were taken from the culture medium, washed in PBS and frozen at −80°C until further processing. Subsequently, they were digested at 56°C(>16 h) in a Tris-EDTA buffered solution containing Proteinase-K (1 mgml^{-1}), 18.5 pepstatin A (18.5 µgml^{-1}), and iodoacetamide (1 µgml^{-1}) (Sigma-Aldrich).

DNA quantification assay was performed with CyQuant dye kit according to the manufacturers description (Molecular Probes, Eugene, Oregon, U.S.A.), using a spectrofluorometer (Victor3, Perkin Elmer, U.S.A.), at an excitation wavelength of 480 nm and an emission wavelength of 520 nm.

ELISA Assay of NGF Secretion

To quantify the concentration of nerve growth factor (NGF) in cell cultured supernatant, commercially available ELISA kits were used according to the manufacturer's instruction (Promega). The plates were read at 450 nm and analysed using Elisa machine (Lightcycler II, Roche Diagnostics GmbH, Germany). Secretions of NGF were measured following 1, 3, 6, 9, 12 and 14 days of culture.

RNA extraction and quantitative real-time RT-PCR

To analyse the expression of neural markers by SCs, total RNA was isolated using a combination of the TRIzol® method with the NucleoSpin®RNA II isolation kit (Bioké). Briefly, at day 7 of cell culture, scaffolds (n = 3) were washed with PBS once and 1 ml of TRIzol reagent (Invitrogen) was added to the samples. After five minutes, the samples were stored at −80 °C for RNA isolation. After chloroform addition and phase separation by centrifugation, the aqueous phase containing the RNA was collected, mixed with equal volume of 75% ethanol and loaded onto the RNA binding column of the kit. Subsequent steps were in accordance with the manufacturer's protocol. RNA was collected in RNAse-free water. The quality and quantity of RNA was analysed by gel electrophoresis and spectrophotometry. Seven hundred fifty nanograms of RNA were used for first strand cDNA synthesis using iScript (Bio-Rad) according to the manufacturer's protocol. One µl of undiluted cDNA was used for subsequent analysis. Time reverse transcription polymerase chain reaction (real-time PCR) was performed on an iQ5 real time PCR machine (Bio-Rad) using SYBR Green supermix (Bio-Rad). Expression of neural marker genes was normalised to β-Actin levels and fold inductions were calculated using the comparative ΔCT method. The expressions of brain derived neurotrophic factor (BDNF), glia derived neurotrophic factor (GDNF), nerve growth factor (NGF), ciliaryneurotrophic factor (CNTF), and peripheral myelin protein 22 (PMP22) were detected by the real-time PCR (Table S1).

Immunofluorescent staining

The cells were washed in PBS and fixed with paraformaldehyde (4% v/v)(Sigma-Aldrich) in PBS for 30 min. Fixed cells were washed with PBS. Blocking was carried out in a blocking buffer that consisted of 10% w/v BSA solution in PBS (1X) for 1 h. Primary antibodies against p75 low affinity NGF receptor (p75LNGFR, 1:500, Abcam; ab6172) were applied in dilute buffer consisting of 10% BSA in PBS at room temperature for 2 h or overnight at 4°C. Cells were then washed and the secondary antibody, goat anti-mouse IgG conjugated-fluorescein isothiocyanate (FITC, 1:50), was applied for 45 min at 37°C. Cells were counterstained with DAPI for 10 min and observed under fluorescence microscope. For negative controls, the primary antibody was excluded.

Statistical analysis

All data presented are expressed as mean ± standard deviation (SD). Statistical analysis was carried out using one-way analysis of variance (ANOVA) followed by a Tukey's post hoc test. A value of $p < 0.05$ was considered statistically significant.

Conclusion

In this study, various PHB/PHBV solution blended compositions were electrospun and the effect of blend composition on fiber diameter was studied. SEM investigations showed that the diameter of nanofibers decreased with increasing the PHBV ratio. Characterization of samples by DSC and Raman spectroscopy showed that these electrospun nanofibers are homogenous. Crystallinity and melting temperature of nanofibers decreased with increasing the PHBV ratio. PHB/PHBV nanofibers also showed better mechanical properties and flexibility in comparison to that of neat PHB nanofibers. SEM images showed that degree of orientation increased by using a higher speed rotating rotator.

When SCs were seeded on these scaffolds, bipolar cell morphology was observed. In the case of aligned fibers, SCs oriented themselves in the direction of fiber alignment probably due to contact guidance phenomenon.

The potential of these fibrous mats as nerve scaffolds was further assessed by observing the cellular activity. Aligned PHB/PHBV nanofibrous scaffolds showed higher SCs proliferation after 14 days in comparison to random nanofibers. Incorporation of collagen to PHB/PHBV also increased SCs proliferation as well as neurotrophin secretion and GDNF expression. In conclusion, aligned PHB/PHBV and PHB/PHBV/Col nanofibrous mats provide a favorable environment for Schwann cell growth and myelin sheath regeneration.

Supporting Information

Figure S1 Variation of fiber diameters of PHB/PHBV nanofibers with polymer concentration and blend composition. The voltage, flow rate and collecting conditions were fixed at 16 kV, 1.5 ml/h and 15 cm, respectively. Scaffolds prepared at 6% concentration were chosen for *in vitro* cell culture experiments.

Figure S2 Tensile stress-strain curves of PHB, PHBV and PHB(50)/PHBV(50) random and uniaxially oriented nanofibers.

Figure S3 SEM images of SCs on different PHB/PHBV electrospun solution blending nanofibrous scaffolds. (A) 1 day, and (B) 7 days after cell seeding. Scale bars represent 50 μm for top and 10 μm for bottom pictures, respectively.

Figure S4 Metabolic activity and proliferation of SCs on different PHB/PHBV electrospun solution blending nanofibrous scaffolds during 14 days of culture. (A) alamar blue assay; (B) DNA quantification assay. Asterisks represent significant difference at $p \leq 0.05$.

Table S1 The designed primers of genes for real-time PCR.

Table S2 Raman shift (cm^{-1}) and assignment of the Raman bands of PHB/PHBV nanofibers.

Acknowledgments

Authors are also grateful to Paul Wieringa and Maqsood Ahmed for critical discussion.

Author Contributions

Conceived and designed the experiments: EM MM CvB LM. Performed the experiments: EM JH. Analyzed the data: EM MM MHN-E SS JH AvA CvB LM. Contributed reagents/materials/analysis tools: JH AvA. Wrote the paper: EM MM MHN-E SS JH AvA CvB LM.

References

1. Johnson EO, Soucacos PN (2008) Nerve repair: Experimental and clinical evaluation of biodegradable artificial nerve guides. Injury 395: 530–536.
2. Prabhakaran MP, Venugopal JR, Chyan TT, Hai LB, Chan CK, et al. (2008) Electrospun Biocomposite Nanofibrous Scaffolds for Neural Tissue Engineering. Tissue Engineering A 14: 1787–1797.
3. Schiffman JD, Schauer CL (2008) A Review: Electrospinning of Biopolymer Nanofibers and their Applications. Polymer Reviews 48: 317–345.
4. Prabhakaran MP, Venugopal JR, Ramakrishna S (2009) Mesenchymal stem cell differentiation to neuronal cells on electrospun nanofibrous substrates for nerve tissue engineering. Biomaterials 30: 4996–5003.
5. Zonari A, Novikoff S, Electo NRP, Breyner NM, Gomes DA, et al. (2012) Endothelial Differentiation of Human Stem Cells Seeded onto Electrospun Polyhydroxybutyrate/Polyhydroxybutyrate-Co-Hydroxyvalerate Fiber Mesh. Plos One 7: 1–9.
6. Ghasemi-Mobarakeh L, Prabhakaran MP, Morshed M, Nasr-Esfahani MH, Ramakrishna S (2010) Bio-functionalized PCL nanofibrous scaffolds for nerve tissue engineering. Materials Science and Engineering C 30: 1129–1136.
7. Xu X-Y, Li X-T, Peng S-W, Xiao J-F, Liu C, et al. (2010) The behaviour of neural stem cells on polyhydroxyalkanoate nanofiber scaffolds. Biomaterials 31: 3967–3975.
8. Chewa SY, Mi R, Hokec A, Leong KW (2008) The effect of the alignment of electrospun fibrous scaffolds on Schwann cell maturation. Biomaterials 29: 653–661.
9. Ghasemi-Mobarakeh L, Prabhakaran MP, Morshed M, Nasr-Esfahani MH, Ramakrishna S (2008) Electrospun poly(3-caprolactone)/gelatin nanofibrous scaffolds for nerve tissue engineering. Biomaterials 29: 4532–4539.
10. Xu CY, Inai R, Kotaki M, Ramakrishna S (2004) Aligned biodegradable nanofibrous structure: a potential scaffold for blood vessel engineering. Biomaterials 25: 877–886.
11. Xu CY, Inai R, Kotaki M, Ramakrishna S (2004) Aligned biodegradable nanofibrous structure: a potential scaffold for blood vessel engineering. Biomaterials 25: 877–886.
12. Koh HS, Yong T, Chan CK, Ramakrishna S (2008) Enhancement of neurite outgrowth using nano-structured scaffolds coupled with laminin. Biomaterials 29: 3574–3582.
13. Cooper A, Bhattarai N, Zhang M (2011) Fabrication and cellular compatibility of aligned chitosan–PCL fibers for nerve tissue regeneration. Carbohydrate Polymers 85: 149–156.
14. Gupta D, Venugopal JR, Prabhakaran MP, Dev VRG, Low S, et al. (2009) Aligned and random nanofibrous substrate for the in vitro culture of Schwann cell for neural tissue engineering. ActaBiomaterialia 5: 2560–2569.
15. Niapour A, Karamali F, Karbalaie K, Kiani A, Mardani M, et al. (2010) Novel method to obtain highly enriched cultures of adult rat Schwann cells. Biotechnology Letter 32: 781–786.
16. Griffin J, Delgado-Rivera R, Meiners S, Uhrich KE (2011) Salicylic acid-derived poly(anhydride-ester) electrospun fibers designed for regenerating the peripheral nervous system. Journal of Biomedical Materials Research A 97: 230–242.
17. Masaeli E, Morshed M, Rasekhian P, Karbasi S, Karbalaie K, et al. (2012) Does the tissue engineering architecture of poly(3-hydroxybutyrate) scaffold affects cell–material interactions? Journal of Biomedical Materials Research A 100: 1907–1918.
18. Sangsanoh P, Waleetorncheepsawat S, Suwantong O, Wutticharoenmongkol P, Weeranantanapan O, et al. (2007) In Vitro Biocompatibility of Schwann Cells on Surfaces of Biocompatible Polymeric Electrospun Fibrous and Solution-Cast Film Scaffolds. Biomacromolecules 8: 1587.
19. Kim DY, Kim HW, Chung MG, Rhee YH (2007) Biosynthesis, Modification, and Biodegradation of Bacterial Medium-Chain-Length Polyhydroxyalkanoates. Journal of Microbiology 45: 87–97.
20. Chen G-Q, Wu Q (2005) The application of polyhydroxyalkanoates as tissue engineering materials. Biomaterials 26: 6565–6578.
21. Rentsch C, Rentsch B, Breier A, Hofmann A, Manthey S, et al. (2010) Evaluation of the osteogenic potential and vascularization of 3D poly(3)hydroxybutyrate scaffolds subcutaneously implanted in nude rats. Journal of Biomedical Materials Research A 92: 185–195.
22. Yu W, Lan CH, Wang SJ, Fang PF, Sun YM (2010) Influence of zinc oxide nanoparticles on the crystallization behavior of electrospun poly(3-hydroxybutyrate-co-3-hydroxyvalerate) nanofibers. Polymer 51: 2403–2409.
23. Avella M, Martuscelli E, Raimo M (2000) Properties of blends and composites based on poly(3-hydroxy)butyrate (PHB) and poly(3-hydroxybutyrate-hydroxyvalerate) (PHBV) copolymers. Journal of Materials Science 35: 523–545.
24. Sombatmankhong K, Sanchavanakit N, Pavasant P, Supaphol P (2007) Bone scaffolds from electrospun fiber mats of poly(3-hydroxybutyrate), poly(3-hydroxybutyrate-co-3-hydroxyvalerate) and their blend. Polymer 48: 1419–1427.
25. Sombatmankhong K, Suwantong O, Waleetorncheepsawat S, Supaphol P (2006) Electrospun fiber mats of poly(3-hydroxybutyrate), poly(3-hydroxybutyrate-co-3-hydroxyvalerate), and their blends. Journal of Polymer Science B 44: 2923–2933.
26. Suwantong O, Waleetorncheepsawat S, Sanchavanakit N, Pavasant P, Cheepsunthorn P, et al. (2007) In vitro biocompatibility of electrospun poly(3-hydroxybutyrate) and poly(3-hydroxybutyrate-co-3-hydroxyvalerate) fiber mats. Biological Macromolecules 40: 217–223.
27. Novikova LN, Pettersson J, Brohlin M, Wiberg M, Novikov LN (2008) Biodegradable poly-b-hydroxybutyrate scaffold seeded with Schwann cells to promote spinal cord repair. Biomaterials 29: 1198–1206.
28. Armsterong SJ, Wiberg M, Terenghi G, Kingham P (2007) ECM Molecules Mediate Both Schwann Cell Proliferation and Activation to Enhance Neurite Outgrowth. Tissue Engineering 13: 2863–2870.
29. Kalbermaten DF, Erba P, Mahay D, Wiberg M, Pierer G, et al. (2008) Schwann cell strip for peripheral nerve repair. Journal of Hand Surgery 33E: 587–594.
30. Pompe T, Keler K, Mothes G, Tesse M, Zimmemman R, et al. (2007) Surface modification of polyhydroxyalkanoates to control cell-matrix adhesion. Biomaterials 28: 28–37.

31. Nandakumar A, Fernandes H, de Boer J, Moroni L, Habibovic P, et al. (2010) Fabrication of Bioactive Composite Scaffolds by Electrospinning for Bone Regeneration. Macromolecular Bioscience 10: 1365–1373.

32. Cao D, Wu Y-P, Fu Z-F, Tian Y, Li C-J, et al. (2011) Cell adhesive and growth behavior on electrospun nanofibrous scaffolds by designed multifunctional composites. Colloids and Surfaces B 84: 26–34.

33. Lee H, Yeo M, Ahn SH, Kang D-O, Jang CH, et al. (2011) Designed hybrid scaffolds consisting of polycaprolactone microstrands and electrospun collagen-nanofibers for bone tissue regeneration. Journal of Biomedical Materials Research B 97B: 263–270.

34. Gunaratne LMWK, Shanks RA (2005) Melting and thermal history of poly(hydroxybutyrate-cohydroxyvalerate) using step-scan DSC. Thermochimica Acta 430: 183–190.

35. Ayres C, Jha B, Sell S, Bowlin G, Simpson D (2010) Nanotechnology in the design of soft tissue scaffolds: innovations in structure and function. WIREs Nanomedicine and Nanobiotechnology 2: 20–34.

36. Mohanna PN, Young RC, Wiberg M, Terenghi G (2003) A composite poly-hydroxybutyrate-glial growth factor conduit for long nerve gap repairs. Journal of Anatomy 203: 553–565.

37. Sill TJ, von Recum HA (2008) Electrospinning: Applications in drug delivery and tissue engineering. Biomaterials 29: 1989–2006.

38. Ndreu A, Nikkola L, Yikauppila H, Ashmmakhi N, Hasirci V (2008) Electrospun biodegradable nanofibrous mats for tissue engineering. Nanomedicine 3: 45–60.

39. Lee IS, Kwon OH (2004) Nanofabrication of Microbial Polyester by Electrospinning Promotes Cell Attachment. Macromolecular Research 12: 374–378.

40. Tong H-W, Wang M (2011) Electrospinning of Poly(Hydroxybutyrate-co-hydroxyvalerate) Fibrous Scaffolds for Tissue Engineering Applications: Effects of Electrospinning Parameters and Solution Properties. Journal of Macromolecular Science, Part B 50: 1535–1558.

41. Neves SC, Moreira Texeira LS, Moroni L, Reis RL, Van Blitterswijk CA, et al. (2011) Chitosan/Poly(3-caprolactone) blend scaffolds for cartilage repair. Biomaterials 32: 1068–1079.

42. Tan SM, Ismail J, Kummerlowe C, Kammer HW (2006) Crystallization and Melting Behavior of Blends Comprising Poly(3-hydroxy butyrate-co-3-hydroxy valerate) and Poly(ethylene oxide). Journal of Applied Polymer Science 101: 2776–2783.

43. Chan CH, Leummerlowe C, Kammer H-W (2004) Crystallization and melting behavior of Poly(30hydroxybutyrate)-based blends. Macromolecular Chemistry and Physics 205: 664–675.

44. Saito M, Inoue Y, Yoshie N (2001) Cocrystallization and phase segregation of blends of poly (3-hydroxybutyrate) and poly(3-hydroxybutyrate-co-3-hydroxy-valerate). Polymer 42: 5573–5580.

45. Meng W, Kim S-Y, Yuan J, Kim JC, Kwon OH, et al. (2007) Electrospun PHBV/collagen composite nanofibrous scaffolds for tissue engineering. Journal of Biomaterials Science 18: 81–94.

46. Izumi CMS, Temperini MLA (2010) FT-Raman investigation of biodegradable polymers: Poly(3-hydroxybutyrate) and poly(3-hydroxybutyrate-co-3-hydroxy-valerate). Vibrational Spectroscopy 54: 127–132.

47. Wang W, Itoh S, Kanno K, Kikkawa T, Ichinose S, et al. (2009) Effect of Schwann cell alignment along the oriented electrospun chitosan nanofibers on nerve regeneration. Journal of Biomedical Materials Research A 91: 994–1005.

48. Prabhakaran MP, Venugopal JR, Ramakrishna S (2009) Mesenchymal stem cell differentiation to neuronal cells on electrospun nanofibrous substrates for nerve tissue engineering. Biomaterials 30: 4996–5003.

49. Leong MF, Chian KS, Mhaisalkar PS, Ong WF, Ratner BD (2009) Effect of electrospun poly (D, L-lactide) fibrous scaffold with nanoporous surface on attachment of porcine esophageal epithelial cells and protein adsorption. Journal of Biomedical Materials Research A 89: 1040–1048.

50. Campbell WW (2008) Evaluation and management of peripheral nerve injury. Clinical Neurophysiology 119: 1951–1965.

51. Ribeiro-Resende VT, Koenig B, Nichterwitz S, Oberhoffner S, Schlosshauer B (2009) Strategies for inducing the formation of bands of Bungner in peripheral nerve regenerationq. Biomaterials 30 (2009) 5251–5259 30: 5251–5259.

52. Huanga C, Chena R, Kea Q, Morsic Y, Zhanga K, et al. (2011) Electrospun collagen–chitosan–TPU nanofibrous scaffolds for tissue engineered tubular grafts. Colloids and Surfaces B 82: 307–315.

53. You S, Petrov T, Chung PH, Gordon T (1997) The expression of the low affinity nerve growth factor receptor in long-term denervated Schwann cells. Glia 20: 87–100.

54. Zorick TS, Lemket G (1996) Schwann cell differentiation. Current Opinion in Cell Biology 8: 870–876.

55. Valmikinathan CM, Hoffman J, Yu X (2011) Impact of scaffold micro and macro architecture on Schwann cell proliferation under dynamic conditions in a rotating wall vessel bioreactor. Materials Science and Engineering, Part C 31: 22–29.

56. Yu W, Zhao W, Zhu C, Zhang X, Ye D, et al. (2011) Sciatic nerve regeneration in rats by a promising electrospun collagen/poly(ε-caprolactone) nerve conduit with tailored degradation rate. BMC Neuroscience 12: 68–82.

57. Schnell E, Klinkhammer K, Balzer S, Brook G, Klee D, et al. (2007) Guidance of glial cell migration and axonal growth on electrospun nanofibers of poly-e-caprolactone and a collagen/poly-e-caprolactone blend. Biomaterials 28: 3012–3025.

58. Li W, Guo Y, Wang H, Shi D, Liang C, et al. (2008) Electrospun nanofibers immobilized with collagen for neural stem cells culture. Journal of Materials Science 19: 847–854.

59. Wang Y-Y, Lu L-X, Shi J-C, Wang H-F, Xiao Z-D, et al. (2011) Introducing RGD Peptides on PHBV Films through PEG-Containing Cross-Linkers to improve the Biocompatibility. BioMacromolecules 12: 551–559.

60. Ren YJ, Zhang H, Huang H, Wang XM, Zhou ZY, et al. (2009) In vitro behavior of neural stem cells in response to different chemical functional groups. Biomaterials 30: 1036–1044.

61. Feng S, Shen X, Fu Z, Shao M (2011) Preparation and Characterization of Gelatin–Poly(L-lactic) Acid/Poly(hydroxybutyrate-co-hydroxyvalerate) Composite Nanofibrous Scaffolds. journal of Macromolecular Science B 50: 1705–1713.

62. Mey J, Brook G, Hodde D, Kriebel A (2012) Electrospun Fibers as Substrates for Peripheral Nerve Regeneration. Advanced Polymer Science 246: 131–170.

63. Friedman HC, Jelsma TN, Bray GM, Aguayo AJ (1996) A Distinct Pattern of Trophic Factor Expression in Myelin-Deficient Nerves of Trembler Mice: Implications for Trophic Support by Schwann Cells. Journal of Neuroscience 16: 5344–5350.

64. Li Q, Ping P, Jiang H, Liu K (2006) Nerve conduit filled with GDNF gene-modified schwann cells enhances regeneration of the peripheral nerve. Microsurgery 26: 116–121.

65. Piirsoo M, Kaljas A, Tamm K, Timmusk T (2010) Expression of NGF and GDNF family members and their receptors during peripheral nerve development and differentiation of Schwann cells in vitro. Neuroscience Letters 469: 135–140.

66. Chan JR, Watkins TA, Cosgaya JM, Zhang CZ, Chen L, et al. (2004) NGF Controls Axonal Receptivity to Myelination by Schwann Cells or Oligodendrocytes. Neuron 43: 183–191.

67. Shakhbazau A, Kawasoe J, Hoyng SA, Kumar R, van Minnen J, et al. (2012) Early regenerative effects of NGF-transduced Schwann cells in peripheral nerve repair. Molecular and Cellular Neuroscience 50: 103–112.

9

Regional Variations in the Cellular, Biochemical, and Biomechanical Characteristics of Rabbit Annulus Fibrosus

Jun Li[1], Chen Liu[1], Qianping Guo[2], Huilin Yang[1,2], Bin Li[1,2]*

1 Department of Orthopaedics, The First Affiliated Hospital of Soochow University, Suzhou, Jiangsu, China, **2** Orthopedic Institute, Soochow University, Suzhou, Jiangsu, China

Abstract

Tissue engineering of annulus fibrosus (AF), the essential load-bearing disc component, remains challenging due to the intrinsic heterogeneity of AF tissue. In order to provide a set of characterization data of AF tissue, which serve as the benchmark for constructing tissue engineered AF, we analyzed tissues and cells from various radial zones of AF, i.e., inner AF (iAF), middle AF (mAF), and outer AF (oAF), using a rabbit model. We found that a radial gradient in the cellular, biochemical, and biomechanical characteristics of rabbit AF existed. Specifically, the iAF cells (iAFCs) had the highest expression of collagen-II and aggrecan genes, while oAF cells (oAFCs) had the highest collagen-I gene expression. The contents of DNA, total collagen and collagen-I sequentially increased from iAF, mAF to oAF, while glycosaminoglycan (GAG) and collagen-II levels decreased. The cell traction forces of primary AFCs gradually decreased from iAFCs, mAFCs to oAFCs, being 336.6 ± 155.3, 199.0 ± 158.8, and 123.8 ± 76.1 Pa, respectively. The storage moduli of iAF, mAF, and oAF were 0.032 ± 0.002, 2.121 ± 0.656, and 4.130 ± 0.159 MPa, respectively. These measurements have established a set of reference data for functional evaluation of the efficacy of AF tissue engineering strategies using a convenient and cost-effective rabbit model, the findings of which may be further translated to human research.

Editor: Bart O. Williams, Van Andel Institute, United States of America

Funding: This study was supported by grants from the National Natural Science Foundation of China (81171479), Natural Science Foundation of Jiangsu Province (BK2011291), the Graduate Research and Innovation Program of Jiangsu Province (CXZZ12_0843), and the Jiangsu Provincial Special Program of Medical Science(BL2012004). The funders had no role in study design, data collection and analysis, decision to publish, or preparation of the manuscript.

Competing Interests: The authors have declared that no competing interests exist.

* E-mail: binli@suda.edu.cn

Introduction

Disc degeneration disease (DDD), the major cause of low back pain, has become a serious health problem and significantly contributes to healthcare expenditures [1]. Current conservative or surgical treatments for DDD can hardly reverse the biological function of degenerated intervertebral disc (IVD) cells and tissue, and may even lead to degenerative changes in adjacent vertebrae, let alone the high post-surgery recurrence rate [2,3]. Instead, tissue engineering has emerged as a promising approach for DDD therapy by using engineered disc replacements [4,5]. However, despite the considerable progress in engineering the nucleus pulposus (NP) of IVD, none has led to translation to clinical implementation. One of the reasons is lack of effective strategies to repair damaged annulus fibrosus (AF) [4,6]. As a component which plays a critical role in the biomechanical properties of IVD, the structural integrity of AF is essential to confining NP and maintaining physiological intradiscal pressure upon loading [4]. Injuries of AF tissue, small or large, can lead to substantial deterioration of whole IVD which characterizes DDD [7]. Therefore, repairing/regenerating AF is essential in order to achieve effective disc repair/regeneration [8].

Nonetheless, AF tissue engineering has remained challenging because of the remarkable complexity of AF tissue [9–11]. Unlike NP and cartilage end plate (CEP), AF is an intrinsically heterogeneous tissue which consists of a series of oriented concentric layers surrounding NP. The biological, biochemical, and biomechanical characteristics significantly vary along its radial direction. An ideal tissue engineered AF, therefore, should recapitulate the biochemical, microstructural, and cellular characteristics of native AF tissue. This requires systematic comprehension of the regional variations of AF on both qualitative and quantitative basis, which provides well-defined native cellular and tissue benchmarks for evaluating the functional equivalence of engineered tissues [12]. Unfortunately, except for human, there are limited characterization data for AFs of other mammals. For example, rabbit is a commonly used model for IVD research taking advantage of its moderate size, ease of surgery and post-surgery analyses [13,14]. However, lacking information of the regional difference of rabbit AF tissue has been an obstacle for appropriate construction of engineered AF.

To this end, we characterized the cellular, biochemical, and biomechanical specifics of different regions of rabbit AF tissue in this study. We isolated cells from various AF regions along its radial direction, i.e., inner, middle, and outer AF, to perform measurements at cellular level. We then performed histological and biochemical analyses of each region. Importantly, we, for the first time, employed a novel cell traction force microscopy (CTFM) technique to measure the cell traction forces (CTFs) of individual AF cells, which may serve as useful biophysical markers for characterizing AF cells from various regions. In addition to tensile test through traditional macro-scale approach, we also determined the region-wise biomechanical properties of AF tissues using nanoindentation technique. Results from this study may provide a foundation for further comprehension of AF biology as well as fundamental benchmarks for functional evaluation of the strategies of AF tissue engineering.

Materials and Methods

AF Tissue Harvesting

The spinal column of a six-month-old New Zealand white rabbits was harvested by dissecting from the surrounding muscles under a sterile environment. The column was sectioned transversally in the middle of each disc after the muscles and ligaments were removed, and IVDs from T10 to L5 were isolated (**Fig. S1**). With the NP removed, the inner (iAF), middle (mAF), and outer AF (oAF) were carefully separated into three equal-thickness sections along its radial direction under a binocular dissection microscope. In addition, discrimination of iAF, mAF, and oAF was also possible based on the gloss appearance and the amount of hydrated matrix between the lamellae (**Fig. 1**). The animal surgery protocol was approved by the Institutional Animal Care and Use Committee (IACUC) of Soochow University.

Isolation and Culture of AF Cells

AF tissues were minced and cells were isolated by digesting the tissue with 2 mg/ml collagenase type I (Sigma, C0130) at 37°C. After being filtered through a 200-μm filter and centrifuged, cells were re-suspended in Alpha's modified Eagle's medium (Hyclone, SH30265.01B) supplemented with 10% fetal bovine serum (Hyclone, SV30087.02), 100 U/ml penicillin, and 100 μg/ml streptomycin, and seeded at a density of 5×10^3 cells/cm². The cells were maintained in a humidified incubator at 37°C and 5% CO_2. The medium was changed every 3 days.

Cell Proliferation Analysis

The isolated primary cells from each region of AF were seeded in a 96-well plate at a density of 2×10^3 cells per well. After predetermined periods of time, the cells were washed twice with PBS. Then 100 μl medium and 10 μl Cell Counting Kit-8 (CCK-8) (KeyGEN Biotech, KGA317) solution was added to each well.

After 2 h of incubation, the absorbance at 450 nm was measured using a microplate reader (BioTek Instruments, USA).

Gene Expression Analysis

The primary cells from each AF region were lysed and total RNA was extracted using the TRIZOL isolation system (Invitrogen). Reverse transcription was then performed using the RevertAid™ First-Strand cDNA Synthesis Kit (Fermentas, K1622) and oligo(dT) primers for 60 minutes at 42°C on a reverse transcription PCR system (Eastwin Life Science, Beijing). Primers for GAPDH, collagen-I, collagen-II and aggrecan were designed using the mRNA sequences deposited in Gene Bank (**Table 1**). Real-time quantitative PCR (RT-qPCR) was performed with a Bio-Rad CFX96™ Real-Time System using the SsoFast™ EvaGreen Supermix Kit (Bio-Rad). The copy numbers were normalized to GAPDH, and the fold differences were calculated using △△Ct method by referencing to the gene expression of oAFCs.

Histological Analysis

Freshly harvested specimens were rinsed in PBS and placed in a formalin solution for 48 h. After being rinsed with distilled water for 24 h, the specimens were decalcified for 5 days, followed by sequential dehydration in 70%, 95% and 100% alcohol and xylene, respectively. After being embedded in paraffin, the specimens were cut into 5 μm slices and mounted on slides. They were then dipped in xylene to remove the paraffin, and rinsed with alcohol of gradual concentrations. For *hematoxylin & eosin (H&E) staining*, the specimens were rinsed in water for 4 min, dipped in hematoxylin for 6 min, and rinsed with water for 5 min. Then they were placed in eosin for 1 min. *Safranin Orange - Fast Green staining* followed a similar pattern. Slides were cleaned with xylene and alcohol and then rinsed with water, after which they were dipped in 0.02% Fast Green for 3 min, 1% acetic acid for 15 sec, and finally 0.1% aqueous Safranin Orange for 3 min. Stained slides were viewed with a Zeiss Axiovert 200 inverted phase contrast microscope (Carl Zeiss Inc., Thornwood, NY) and images were recorded using an AxioVision software.

Immunohistochemistry

After paraffin was removed, the tissue sections were rinsed with water for 2 min and then incubated in 1% hydrogen peroxide in

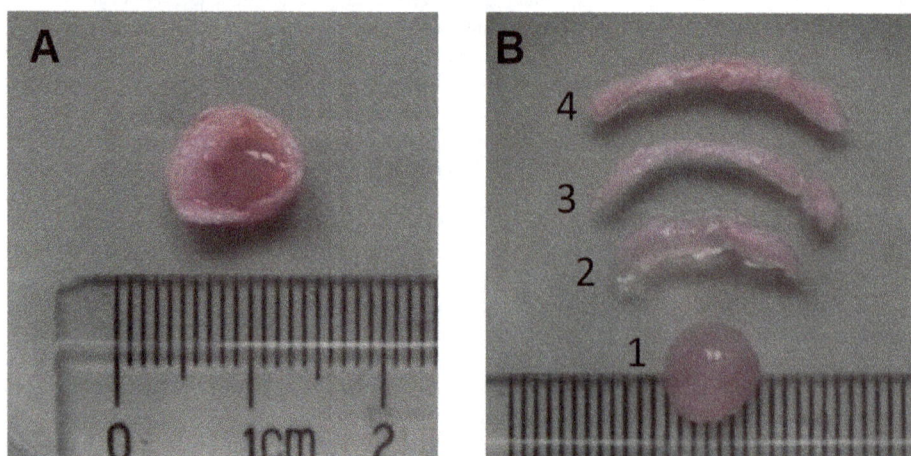

Figure 1. Pictures of an IVD and its sections. (A) A whole rabbit IVD. (B) With the NP (1) being removed, the AF was separated into three equal parts, i.e., iAF (2), mAF (3) and oAF (4), respectively.

Table 1. Sequences of primers for RT-PCR.

Gene	Sequence	Accession number
Collagen-I	Forward: 5′-CTGACTGGAAGAGCGGAGAGTAC-3′	AY633663
	Reverse: 5′-CCATGTCGCAGAAGACCTTGA-3′	
Collagen-II	Forward: 5′-AGCCACCCTCGGACTCT-3′	NM_001195671
	Reverse: 5′-TTTCCTGCCTCTGCCTG-3′	
Aggrecan	Forward: 5′-ATGGCTTCCACCAGTGCG-3′	XM_002723376
	Reverse: 5′-CGGATGCCGTAGGTTCTCA-3′	
GAPDH	Forward: 5′-ACTTTGTGAAGCTCATTTCCTGGTA-3′	NM_001082253
	Reverse: 5′-GTGGTTTGAGGGCTCTTACTCCTT-3′	

methanol for 30 min, followed by washing with Tris buffer. Then they were incubated in 10 mM citrate buffer at 60°C for 10 min and then 2% goat serum (Gibco) for 10 min. Monoclonal antibodies to collagen-I (1:200) (Abcam, 90395) and collagen-II (1:200) (Millipore, II-4C11) were used. After washing with Tris, they were incubated in streptavidin peroxidase followed by 3,3′-diaminobenzidine (DAB) using an EnVision Detection Kit. Finally they were washed with water and counterstained with hematoxylin. Mouse mAb IgG1 isotype control (Cell Signaling Technology, 5415S) was used. Rabbit tendon and articular cartilage tissues were used as positive controls for collagen-I and collagen-II, respectively (**Fig. S2**).

Biochemical Assays

Tissues from iAF, mAF, and oAF regions were weighed, homogenized, and then subjected to biochemical analyses to determine the contents of DNA, proteoglycan (PG), hydroxyproline (HYP), collagen-I and collagen-II, respectively. In brief, the DNA content was determined using quantitative fluorescence measurement of the homogenate hydrolysate mixed with bisbenzimidazole (Hoechst 33258, Sigma) [15]. The total glycosaminoglycan (GAG) content was quantified through the 1,9-dimethylmethylene blue (DMMB) dye-binding assay using a commercially available kit (Genmed Scientifics Inc, USA, GMS 19239.2) [16]. The HYP content was determined using a kit (Genmed Scientifics Inc., GMS50133.1) as described previously [17]. The contents of collagen-I and collagen-II were quantified using enzyme-linked immunosorbant assay (ELISA) kits (R&D Systems, USA). All assays were performed according to the instructions provided by the manufacturers and the absorbance was measured using a microplate reader. All measurements were normalized by the wet weight of tissues.

Cell Traction Force Microscopy (CTFM)

The glass surface of a glass-bottomed petri dish was treated with NaOH solution (0.1 M) for 1 day and then coated with 3-aminopropyltrimethoxysilane for 5 min. The dish was then washed with deionized (DI) water, followed by incubation with 0.5% glutaraldehyde for 30 min. Next, the dish was thoroughly washed with DI water and air-dried. After that, a mixture (acrylamide, 5%; bis-acrylamide, 0.1%) was vacuumed for 20 min and thoroughly mixed with 0.2 μm fluorescent microbeads (volume ratio: 80/1), 40 μl ammonium persulfate, and 4 μl N,N,N′,N′-tetramethylethylenediamine. Eleven microliter of the mixture was then added to the center of pretreated dish and allowed to cure for 30 min at room temperature, which led to formation of a polyacrylamide gel (PAG) on the glass. Before conjugating collagen-I to PAG, its surface was first activated using N-sulfosuccinimidyl-6-[4′-azido-2′-nitrophenylamino] hexanoate (Sulfo-SANPAH, Pierce, Rockland, IL) under ultraviolet (UV) exposure. In brief, 100 μl Sulfo-SANPAH in 30 mM HEPES solution was added to the surface of PAG and exposed to UV for 5 min and then PAG was washed with PBS twice. Next, 130 μl collagen-I solution (100 μg/ml) was added to PAG surface, followed by overnight incubation at 4°C.

The collagen-coated PAG was thoroughly washed with PBS before a cell suspension containing about 3,000 cells was added to it. The primary AFCs were allowed to attach and spread on the gel for 6 h before cell images were taken. In order to acquire the images of cell and microbeads for CTF measurement, a region where individual cells resided on PAG was selected and then a phase contrast image of cells was taken. This was followed by taking a fluorescence image of microbeads, referred to as "force-loaded" image. After the culture medium was carefully extracted and the cells were removed using 2 ml 0.5% trypsin, an image of fluorescent microbeads under the same view, i.e., "null-force" image, was taken. Based on the three images, a custom-made MATLAB program was used to determine the displacement fields and compute CTFs [18].

Nanoindentation Test

First, a motion segment including the L4 and L5 vertebrae and the IVD between them (L4–L5) was harvested from a six-month-old New Zealand white rabbit using an orthopedic saw. The L4–L5 IVD was isolated by carefully removing vertebral bone until the remaining L4 bone was approximately 1 mm thick (**Fig. 2**). Under dissection microscope and with the L5 rigidly fixed in a vice, a sharp scalpel was used to cut the IVD at a level that was 1 mm inferior to the endplate of L4 vertebrae. After being embedded with paraffin, the L5 vertebrae and L4–L5 IVD were mounted on a metal disc and then finely polished by graded sandpapers before being mounted on the nanoindentation system (**Fig. S3**). The test points were selected from inner, middle, and outer regions of AF using an optical microscope attached to the system. In order to measure the complex modulus of AF tissue, a dynamic nanoindentation protocol for viscoelastic solids was applied as previously described [19,20]. Briefly, dynamic indentations with five running frequencies (10, 5.62, 3.16, 1.78 and 1 Hz) and 50 nm oscillation amplitude were performed using the "G-Series XP CSM flat punch complex modulus" module of the NanoSuite method on a nanoindentation facility (Agilent, Nano Indenter G200) at the axial direction using a flat indenter (diameter, 215 μm) at approximately 22°C. The samples were

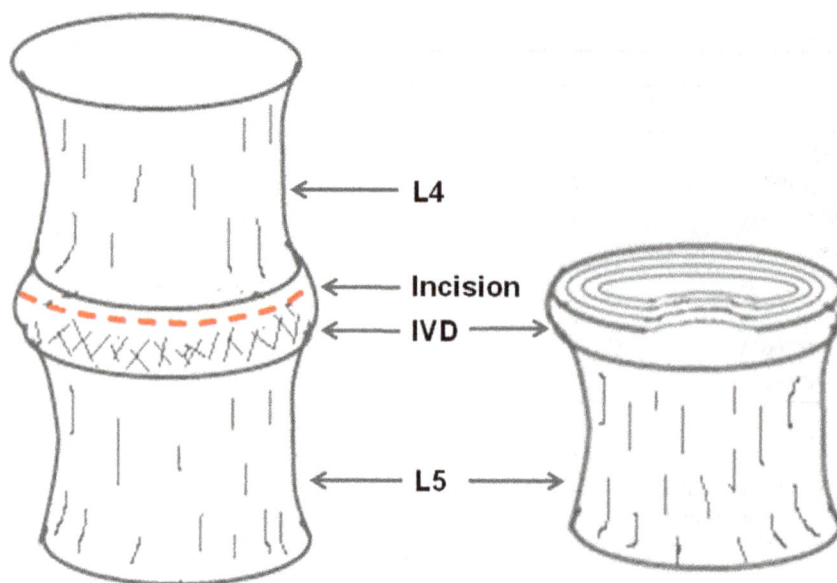

Figure 2. Schematic illustration of the sample preparation for nanoindentation test.

irrigated with PBS during test to avoid dehydration. For each AF region, at least ten points were measured.

Tensile Test

After a whole AF was separated into three sections, i.e., iAF, mAF, and oAF, they were carefully tailored to achieve the same size for testing (**Fig. S4A**). With both ends being fixed, a sample was clamped by two fixtures (**Fig. S4B**). Tensile test was performed on an Instron 3365 testing system using a load cell of 10 N at a displacement rate of 1.5 mm/min. A validated and constant load was applied on the sample until its breakage (**Fig. S4C**). At least ten samples from each AF region were measured.

Statistical Analysis

All data are represented as mean±SD. Statistical analyses were performed using SPSS software. Analyses of gene expression, biochemical assay, tensile test, and CTFM results were performed using Kruskal-Wallis test. Nanoindentation results were analyzed using multivariate analysis of variance (MANOVA). Difference between groups is considered statistically significant if p is less than 0.05.

Results

AF Cells

The primary cells in various AF regions showed distinctively different morphology (**Fig. 3A–C**). Cells in iAF (iAFCs) had rounded, chondrocyte-like morphology, while cells in oAF (oAFCs) predominantly displayed spindle-shaped, fibroblastic morphology. The cells in mAF region (mAFCs), on the other hand, represented a mixture of iAFCs and oAFCs. The density of cells gradually increased from iAF, mAF to oAF, being 2×10^{4}, 3×10^{4}, and 5×10^{4} cells per gram of tissue, respectively. However, these region-specific AFCs appeared to proliferate at similar rate during culture (**Fig. 3D**). According to RT-qPCR analysis, iAFCs exhibited the least expression of collagen-I gene, while oAFCs exhibited the greatest (**Fig. 3E**). In contrast, iAFCs exhibited the greatest expression of collagen-II and aggrecan genes, while oAFCs had the least expression of them (**Fig. 3F–G**). Not

surprisingly, expression of the above genes in mAFCs was between those in iAFCs and oAFCs.

Histological Analysis

Morphologically, the rabbit AF appeared distinctive at different zones along its radial direction. The iAF was loose and fully hydrated, while oAF was denser and less hydrated. The mAF appeared as a transition zone between iAF and oAF (**Fig. 1**). According to H&E staining, the AF tissue was mainly composed of collagen and iAF was less dense compared to mAF and oAF (**Fig. 4A–D**). The cells in iAF were round, while cells in oAF region were elongated and peripherally oriented. In addition, upon Safranin Orange and Fast Green staining iAF was intensively stained orange, indicating the existence of high PG content (**Fig. 4E**). In contrast, oAF was markedly stained with Fast Green, evidencing the presence of massive collagen fibers. It should be noted that the staining of oAF was not typical green, but bluish instead, meaning that this region was simultaneously stained with Safranin-O and Fast Green, indicative of the co-existence of PG and collagen in oAF. However, compared to PG, collagen dominated in oAF, and its content continuously decreased inward along the radial direction. Similarly, in mAF, there was gradual increase of collagen content but decrease of PG, with PG being the dominating matrix in this region. From the immunohistochemical staining of AF it was clear that there was less collagen-I in iAF compared to oAF (**Fig. 4F**). On the contrary, distribution of collagen-II showed an exactly opposite trend, i.e., more collagen-II in iAF but less in oAF (**Fig. 4G**). Again, mAF represented a transition zone between iAF and oAF, with its matrix being a mixture of collagen-I and collagen-II.

Biochemical Analysis

Further, the biochemical compositions of various AF regions were quantified. Clearly, the DNA content gradually increased from iAF, mAF to oAF, being 0.49±0.12, 0.90±0.18, and 1.57±0.17 µg/mg tissue, respectively (**Fig. 5A**). These measurements echoed the findings from AF cell counting and histological evaluation and confirmed that there are more cells residing in oAF

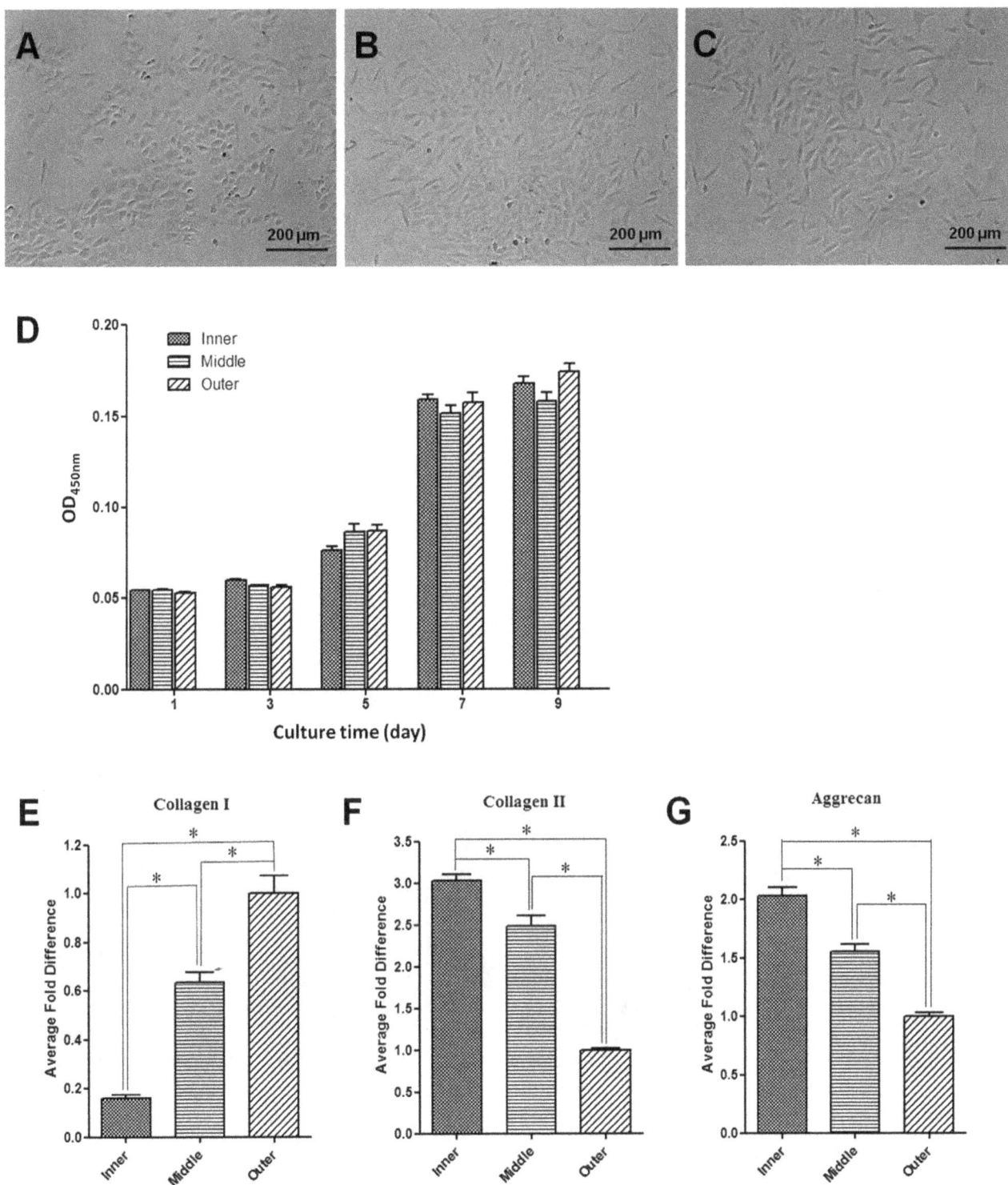

Figure 3. Morphology, proliferation, and gene expression of AF cells. (A–C) Phase contrast images of primary iAFCs, mAFCs, and oAFCs. (D) Proliferation of primary AFCs. (E–G) Real time quantitative PCR analysis of collagen-I, collagen-II, and aggrecan genes, respectively, for primary AFCs of various regions. Gene expression was normalized to GAPDH expression, and fold differences were calculated using the $\triangle\triangle Ct$ method by comparing to gene expression of oAFCs. All data are presented as mean±SD. Asterisk (*) indicates significant difference between groups ($p<0.01$, n = 3).

than those in iAF. Meanwhile, the GAG content decreased from iAF, mAF to oAF, being 12.97±1.38, 7.19±1.39, and 1.46±0.44 μg/mg tissue, respectively (**Fig. 5B**). On the other hand, the content of total collagen protein, indicated by HYP measurement, gradually increased, being 10.37±1.72, 15.33±1.54, and 19.66±0.99 μg/mg tissue, for iAF, mAF, and oAF, respectively (**Fig. 5C**). Moreover, the content of collagen-I, as determined by ELISA, increased from 1.56±0.41, 2.56±0.42,

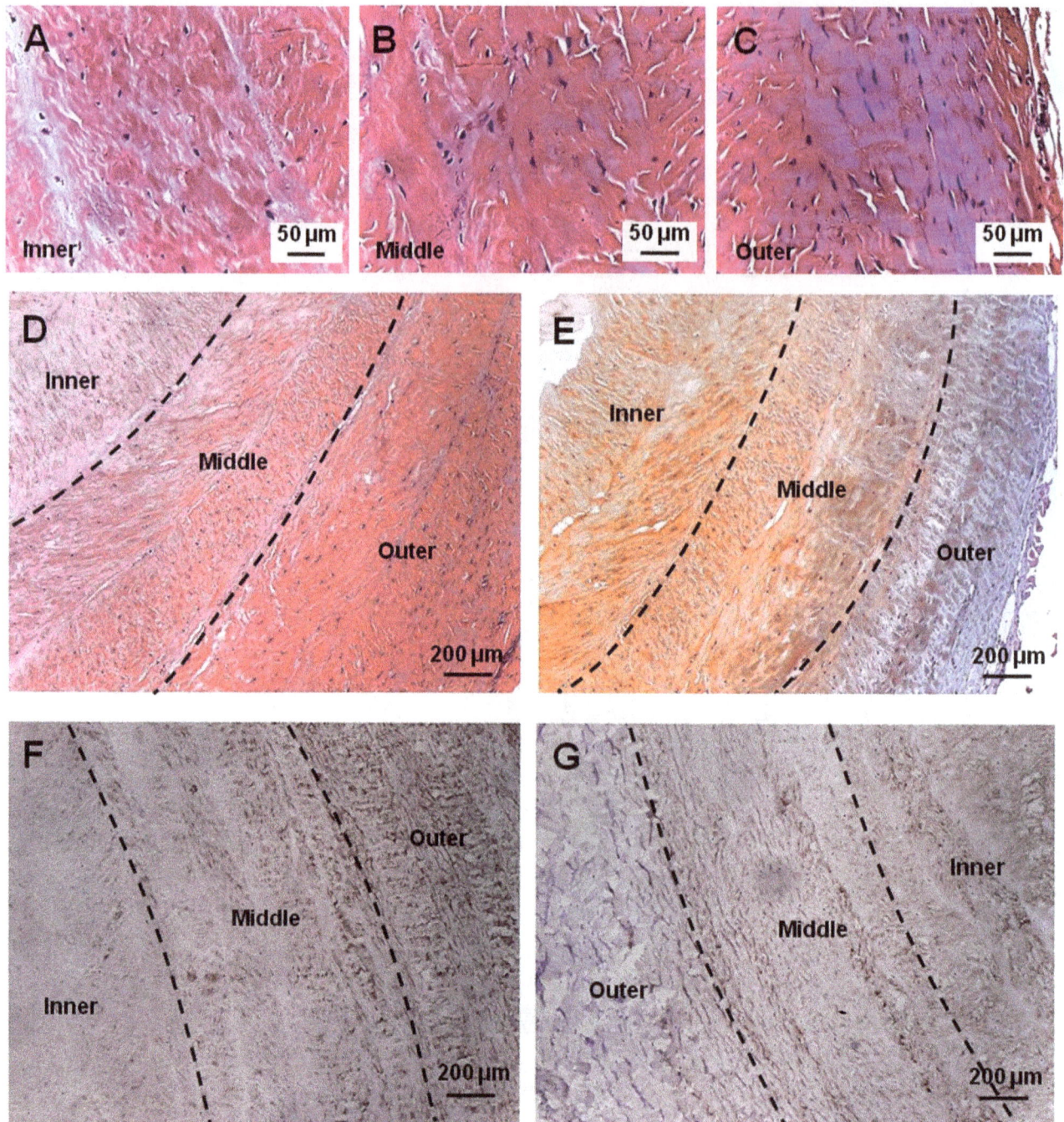

Figure 4. Histological and immunohistochemical analysis of AF tissue. (A–C) H&E stain of iAF, mAF and oAF, respectively. (D) H&E stain of a whole AF tissue section. (E) Safranin Orange–Fast Green stain of a whole AF tissue section. (F–G) Immunohistochemical stain for collagen-I and collagen-II expression of whole AF tissue sections.

to 6.39 ± 0.84 µg/mg tissue, for iAF, mAF, and oAF, respectively (**Fig. 5D**). In contrast, collagen-II content decreased from 3.94 ± 0.30, 1.43 ± 0.36, to 0.69 ± 0.21 µg/mg tissue from iAF, mAF to oAF (**Fig. 5E**).

Cell Traction Force Measurement

The cell traction forces (CTFs) of primary AFCs from different regions were measured using cell traction force microscopy (CTFM) technology (**Fig. 6A**). Apparently, the CTF gradually decreased from iAFCs, mAFCs to oAFCs, being 336.6 ± 155.3,

199.0 ± 158.8, and 123.8 ± 76.1 Pa, respectively (**Fig. 6B**). However, the spread area of cells kept increasing from iAFCs, mAFCs to oAFCs (**Fig. 6C**).

Mechanical Tests

According to the nanoindentation results measured at different indentation frequencies, oAF and iAF consistently showed the highest and lowest storage moduli (equivalent to Young's moduli), respectively, while mAF had moderate storage modulus in between (**Fig. 7A–D**). For instance, the storage moduli of iAF,

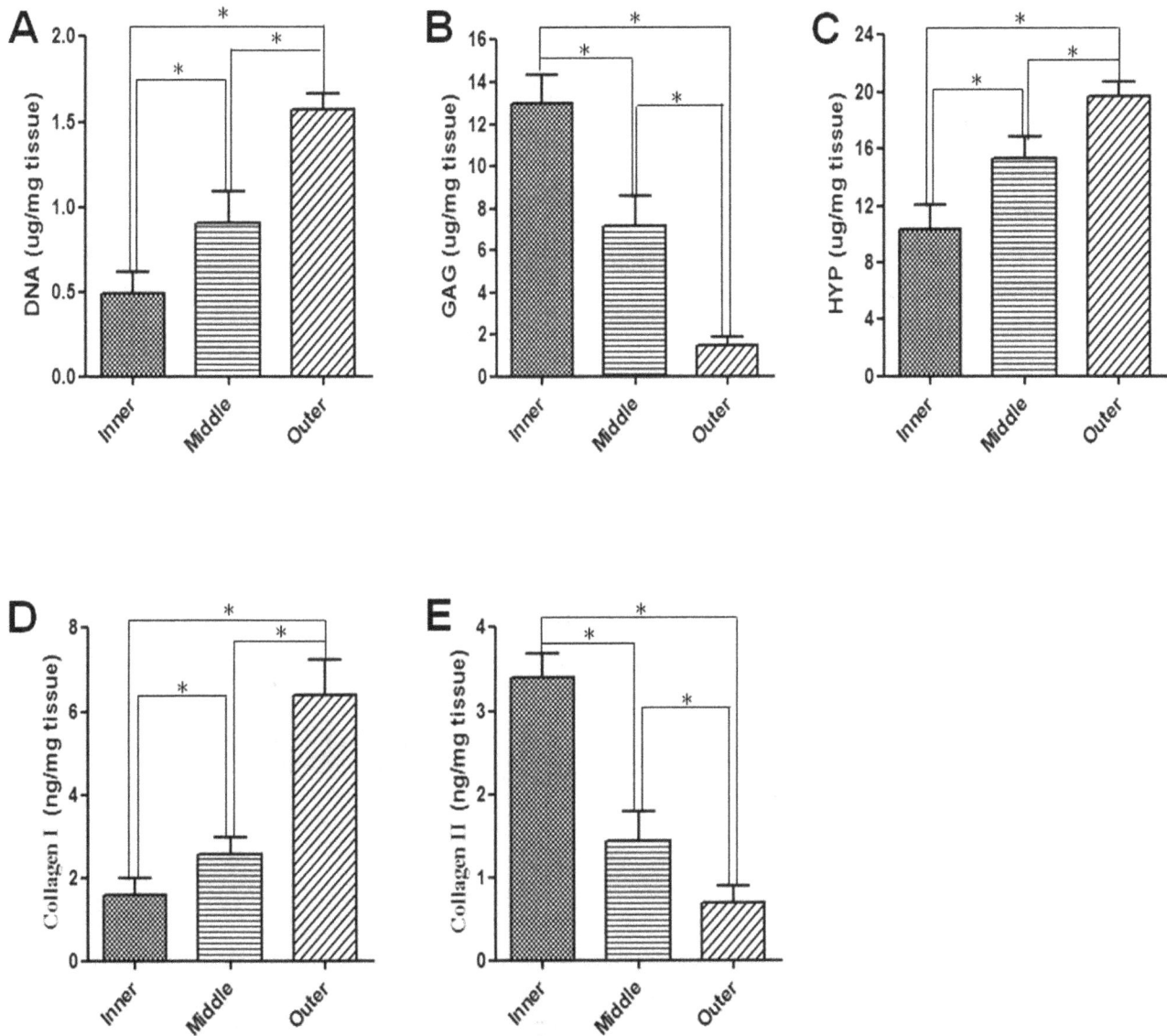

Figure 5. Biochemical analysis of AF tissues. (A–C) The contents of DNA, GAG, and HYP, respectively, in iAF, mAF and oAF tissues. (D–E) The contents of collagen-I and collagen-II as measured by ELISA. All measurements were normalized by wet tissue weight. All data are presented as mean±SD. Asterisk (*) indicates significant difference between groups ($p<0.05$, $n\geq3$).

mAF, and oAF were 0.032±0.002, 2.121±0.656, and 4.130±0.159 MPa, respectively, when measured at a frequency of 1 Hz. Interestingly, the fluctuation of storage modulus appeared to be largest for mAF compared to iAF and oAF. Such regional variations in the modulus of AF tissue were further confirmed by the results of tensile tests, according to which the Young's moduli were 0.509±0.199, 1.790±0.328, and 2.984±0.406 MPa for iAF, mAF, and oAF, respectively (**Fig. 7E and Fig. S4**).

Discussion

Despite recent exciting advancement in AF tissue engineering [21–25], major challenge remains toward fabricating AF replacements that are both biologically and functionally equivalent to native AF tissue. The tremendous complexity of AF at cellular, biochemical, microstructural, and biomechanical levels constitutes formidable technical hurdles [26]. In order to appropriately construct an AF alternative, such regional variations must be fully

acknowledged and used as well-defined guiding benchmarks for AF tissue engineering. While there have been overwhelming studies characterizing human AF, little is known about the AF of rabbit, a very useful model for IVD studies [13,14]. Therefore, this study set out to acquire the region-wise characteristics of rabbit AF from cellular, biochemical, structural, and mechanical aspects.

Among the various regions of AF, we found that iAF was relatively loose and highly hydrated, presumably for assisting NP to absorb axial compressive stress (**Fig. 1**) [27]. On the other hand, oAF was less hydrated and significantly more compact, which helped resist circumferential and torsion stresses. Morphologically, mAF represented a smooth transition from iAF to oAF. Histological studies revealed that iAF contained high level of PG and collagen-II, whereas oAF was fibrous and rich in collagen-I, and mAF was a mixture of its two neighbor zones (**Fig. 4**). Quantitative biochemical studies further indicated that oAF contained more DNA, total collagen, and collagen-I, while iAF contained more PG and collagen-II (**Fig. 5**). Such distinctions in

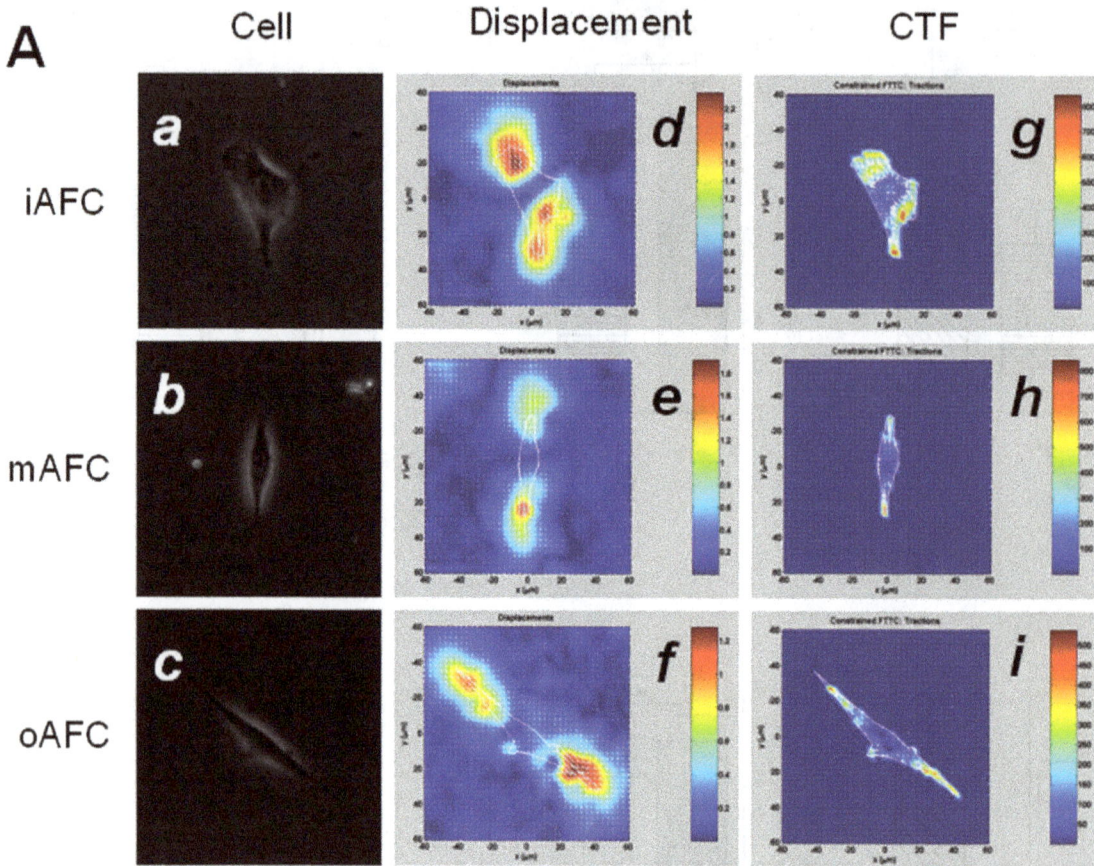

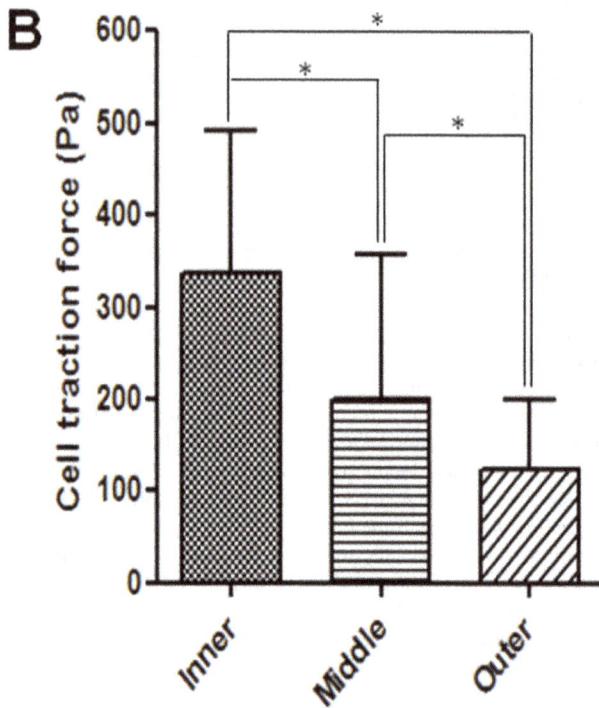

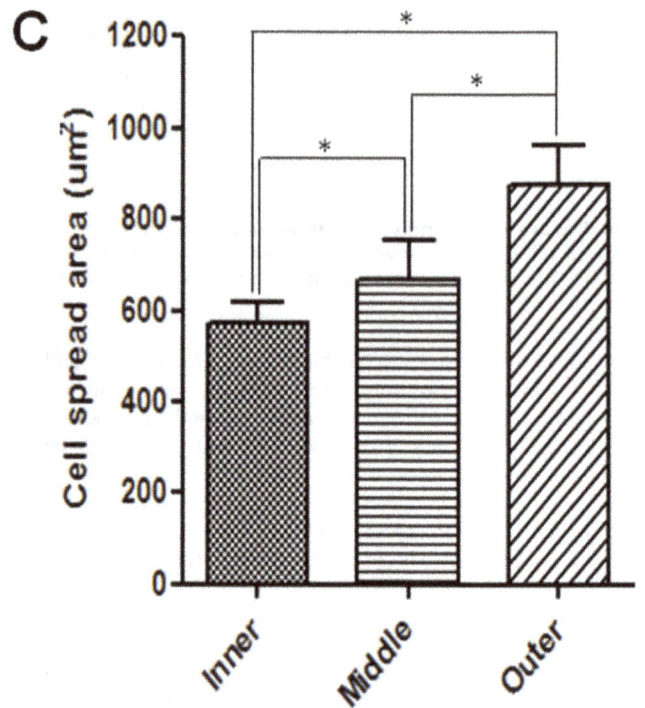

Figure 6. CTFM measurement of AF cells. (A) CTFM for measuring CTFs of iAFC, mAFC and oAFC. *a–c*, primary cells from each region; *d–f*, the substrate displacement fields; *g–i*, the CTF maps. (B–C) CTFs and spread areas of iAFC, mAFC and oAFC, respectively. All data are presented as mean±SD. Asterisk (*) indicates significant difference between groups (*p*<0.05, n≥30).

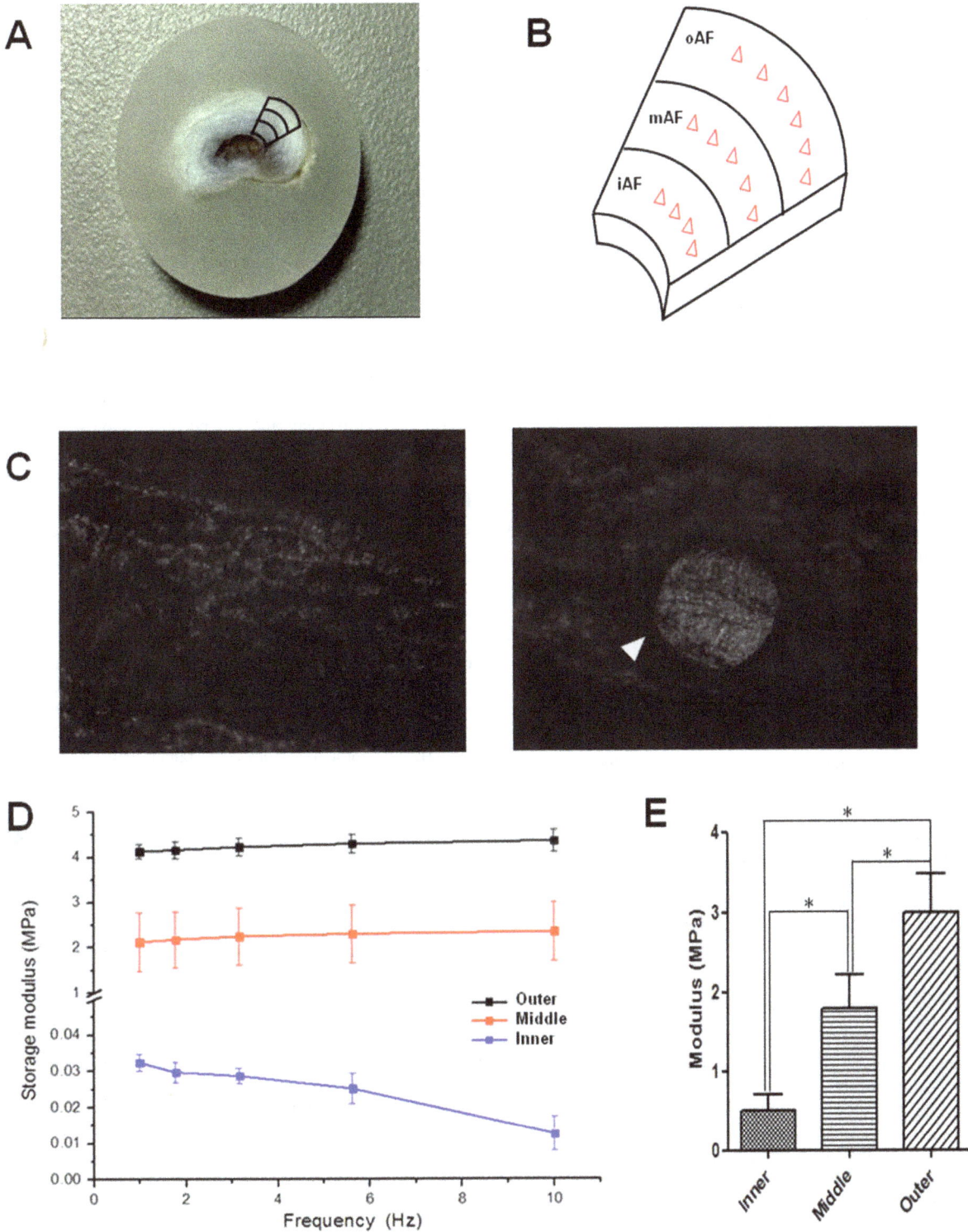

Figure 7. Mechanical tests of AF tissues. (A–B) The picture of a paraffin-embedded AF sample and a schematic showing how the testing regions and points were selected. Note NP was removed from the IVD. (C) The pre-test and post-test images of a sample under nanoindentation test, from which an imprint is clearly seen (shown by the arrowhead). (D) The storage moduli of oAF, mAF and iAF measured using nanoindentation at different frequencies. (E) The Young' moduli of oAF, mAF and iAF measured using tensile test. All data are presented as mean±SD. Asterisk (*) indicates significant difference between groups ($p<0.01$, $n \geq 10$).

the matrix composition of various AF regions were a result of the different phenotypes of cells, which produced different types of extracellular matrix (ECM) corresponding to the zone where they resided (**Fig. 3**). These observations resemble the findings of human AF studies [11,28,29].

While distinctions exist in the morphology and gene profiles of iAFC, mAFC, and oAFC, it is commonly agreed that there still lack specific biological markers for discriminating them. Therefore, we, for the first time, tried to characterize these cells using a novel biophysical approach, i.e., cell traction force microscopy (CTFM) (**Fig. 6**). Being the forces produced by cells and exerted on ECM, cell traction forces (CTFs) function to maintain cell shape, enable cell migration, and initiate mechanotransduction signals, and therefore play a vital role in many biological processes [30]. A close examination of CTFs of various AFCs may help better understand the cellular and molecular mechanisms of their biological roles. We founded that CTF gradually changed among AFCs of different regions (**Fig. 6B**). Interestingly, the rounded iAFCs exhibited much higher CTF than spindle-shaped oAFCs although they had much smaller size (**Fig. 6C**). This is consistent with our earlier finding and is likely related to the difference in their actin cytoskeletal structure [31]. Moreover, difference in the mechanical environment of cells in different AF regions may also count. The iAFCs were mainly subjected to high axial compressive stress and hydrostatic pressure [32], whereas oAFCs underwent peripheral tensile stresses [33]. As a result, cells from various AF regions responded with different cellular mechanical activity and in turn, different matrix synthesis and turnover of cells [34]. Nonetheless, being able to discriminate different types of AFCs, CTFM appears to be an effective technology for characterizing AFC phenotype through a simple and convenient biophysical approach.

Recently, the elasticity of ECM has proven to effectively regulate the differentiation and lineage commitment of stem cells [35,36]. Given the fact that different types of cells from the same origin reside at various regions of AF tissue, we performed region-wise biomechanical test of AF to check whether elasticity difference existed. We employed nanoindentation technique to measure the elastic modulus of AF tissue in a belief that measurements at micro−/nano- level, a scale comparable to cell size, more accurately reflect the mechanical environment that cells indeed sense [37]. As predicted, the stiffness of iAF, mAF, and oAF sequentially increased (**Fig. 7D**), echoing the observations from previous studies using macro-scale mechanical tests [10,38]. In addition, we also performed tensile tests for AF tissues and found the results validated the efficacy and accuracy of nanoindentation tests (**Fig. 7E**). Such a stiffness gradient of AF is correlated to the spatial distribution of its matrix components (**Figs. 3–5**) and directly associated with the mechanical loads at specific anatomical sites [10,39]. For instance, the relatively soft iAF mainly functions to assist NP in absorbing compressive stress in axial direction. In contrast, the much tougher oAF functions to resist circumferential shearing and torsion stresses and to enable uniform hoop stress upon disc compression [10]. Therefore, instead of a simplified homogenous model, anisotropic models which take into account the biomechanical anisotropy of AF can better reflect the real situation of this tissue and should be used for AF tissue engineering [40]. In addition, the regional variation of elasticity in AF also implies that AF tissue-specific stem/progenitor cells may differentiate into various AFCs according to the stiffness of their residential region to maintain physiological turnover of matrix and repair/

regenerate AF upon injury [35,41]. It is worth noting that when the measuring frequency was increased from 1 to 10 Hz, the stiffness of both mAF and oAF slightly increased, while the stiffness of iAF remarkably decreased. Such a pattern of stiffness change of iAF resembles the pseudoplastic behavior of hydrogel, i.e., shear thinning at high frequency, again implying the NP-like nature of iAF.

In summary, the rabbit AF tissue is anisotropic and shows a typical gradient behavior of cellular, biochemical, and biomechanical characteristics along the radial direction. As a result, structurally and mechanically homogenous AF replacements are likely not able to provide the biological function and mechanical stability necessary for preventing further degeneration of other spinal components [37]. On the other hand, engineered AF tissues that recapitulate the regional variations represent a more appropriate and promising solution toward this problem. Findings from this study, therefore, have established a set of reference data for functional evaluation of the efficacy of AF tissue engineering strategies using a convenient and cost-effective rabbit model, the findings of which may be further translated to human research.

Supporting Information

Figure S1   Harvest of rabbit IVD tissues. (A) A portion of spinal column from T10 through L5. (B) IVD harvesting. (C) A whole rabbit IVD.

Figure S2   Positive and isotype controls for immunohistochemistry. (A–B) Rabbit tendon tissue was stained with anti-collagen-I antibody. Positive expression of collagen-I was seen in the tissue. (C–D) Rabbit articular cartilage was stained with anti-collagen-II antibody. Positive expression of collagen-II was seen in the tissue. (E–F) Rabbit AF tissues were stained with IgG1 isotype control antibody. Negative stain was seen in the tissues.

Figure S3   The setup for nanoindentation test of AF tissue. (A) Overview of the nanoindentation test system. (B) Mounting of a paraffin-embedded AF sample on the system for nanoindentation. After the sample was mounted, its surface was checked and specific locations for indentation were identified using a microscope attached to the system. Then the optical lens was switched away and the indenter was placed for indentation test. Note the indenter was out of focus in the picture.

Figure S4   Tensile test of AF tissue. (A) A whole AF was separated into three layers, being iAF, mAF and oAF, respectively. (B) A piece of AF tissue sample was fixed for testing. (C) Elongation of AF tissue during tensile test. (D) The setup for tensile test.

Author Contributions

Conceived and designed the experiments: BL JL HY. Performed the experiments: JL CL QG. Analyzed the data: JL BL. Contributed reagents/materials/analysis tools: JL CL QG BL. Wrote the paper: JL BL.

References

1. Hegewald AA, Knecht S, Baumgartner D, Gerber H, Endres M, et al. (2008) Regenerative treatment strategies in spinal surgery. Front Biosci 13: 1507–1525.
2. Hakkinen A, Kiviranta I, Neva MH, Kautiainen H, Ylinen J (2007) Reoperations after first lumbar disc herniation surgery; a special interest on residives during a 5-year follow-up. BMC Musculoskelet Disord 8: 2.
3. Hughes SP, Freemont AJ, Hukins DW, McGregor AH, Roberts S (2012) The pathogenesis of degeneration of the intervertebral disc and emerging therapies in the management of back pain. J Bone Joint Surg Br 94: 1298–1304.
4. Hudson KD, Alimi M, Grunert P, Hartl R, Bonassar LJ (2013) Recent advances in biological therapies for disc degeneration: tissue engineering of the annulus fibrosus, nucleus pulposus and whole intervertebral discs. Curr Opin Biotechnol 24: 872–879.

5. O'Halloran DM, Pandit AS (2007) Tissue-engineering approach to regenerating the intervertebral disc. Tissue Eng 13: 1927–1954.

6. Adams MA, Dolan P (2012) Intervertebral disc degeneration: evidence for two distinct phenotypes. J Anat 221: 497–506.

7. Iatridis JC (2009) Tissue engineering: Function follows form. Nat Mater 8: 923–924.

8. Bron JL, Helder MN, Meisel HJ, Van Royen BJ, Smit TH (2009) Repair, regenerative and supportive therapies of the annulus fibrosus: achievements and challenges. Eur Spine J 18: 301–313.

9. Cortes DH, Han WM, Smith LJ, Elliott DM (2013) Mechanical properties of the extra-fibrillar matrix of human annulus fibrosus are location and age dependent. J Orthop Res 31: 1725–1732.

10. Skaggs DL, Weidenbaum M, Latridis JC, Ratcliffe A, Mow VC (1994) Regional variation in tensile properties and biochemical composition of the human lumbar annulus fibrosus. Spine 19: 1310–1319.

11. Bruehlmann SB, Rattner JB, Matyas JR, Duncan NA (2002) Regional variations in the cellular matrix of the annulus fibrosus of the intervertebral disc. J Anat 201: 159–171.

12. Nerurkar NL, Elliott DM, Mauck RL (2010) Mechanical design criteria for intervertebral disc tissue engineering. J Biomech 43: 1017–1030.

13. Masuda K, Aota Y, Muehleman C, Imai Y, Okuma M, et al. (2005) A novel rabbit model of mild, reproducible disc degeneration by an anulus needle puncture: correlation between the degree of disc injury and radiological and histological appearances of disc degeneration. Spine 30: 5–14.

14. Kroeber MW, Unglaub F, Wang H, Schmid C, Thomsen M, et al. (2002) New in vivo animal model to create intervertebral disc degeneration and to investigate the effects of therapeutic strategies to stimulate disc regeneration. Spine 27: 2684–2690.

15. Kim YJ, Sah RLY, Doong JYH, Grodzinsky AJ (1988) Fluorometric Assay of DNA in Cartilage Explants Using Hoechst-33258. Anal Biochem 174: 168–176.

16. Farndale RW, Sayers CA, Barrett AJ (1982) A direct spectrophotometric microassay for sulfated glycosaminoglycans in cartilage cultures. Connect Tissue Res 9: 247–248.

17. Stegemann H, Stalder K (1967) Determination of hydroxyproline. Clin Chim Acta 18: 267–273.

18. Li B, Lin M, Tang Y, Wang B, Wang JH (2008) A novel functional assessment of the differentiation of micropatterned muscle cells. J Biomech 41: 3349–3353.

19. Chou AI, Bansal A, Miller GJ, Nicoll SB (2006) The effect of serial monolayer passaging on the collagen expression profile of outer and inner anulus fibrosus cells. Spine 31: 1875–1881.

20. Herbert EG, Oliver WC, Pharr GM (2008) Nanoindentation and the dynamic characterization of viscoelastic solids. Journal of Physics D-Applied Physics 41.

21. Driscoll TP, Nakasone RH, Szczesny SE, Elliott DM, Mauck RL (2013) Biaxial mechanics and inter-lamellar shearing of stem-cell seeded electrospun angle-ply laminates for annulus fibrosus tissue engineering. J Orthop Res 31: 864–870.

22. Koepsell L, Remund T, Bao J, Neufeld D, Fong H, et al. (2011) Tissue engineering of annulus fibrosus using electrospun fibrous scaffolds with aligned polycaprolactone fibers. J Biomed Mater Res A 99: 564–575.

23. Bowles RD, Williams RM, Zipfel WR, Bonassar LJ (2010) Self-assembly of aligned tissue-engineered annulus fibrosus and intervertebral disc composite via collagen gel contraction. Tissue Eng Part A 16: 1339–1348.

24. Wan Y, Feng G, Shen FH, Laurencin CT, Li X (2008) Biphasic scaffold for annulus fibrosus tissue regeneration. Biomaterials 29: 643–652.

25. Nerurkar NL, Baker BM, Sen S, Wible EE, Elliott DM, et al. (2009) Nanofibrous biologic laminates replicate the form and function of the annulus fibrosus. Nat Mater 8: 986–992.

26. Nerurkar NL, Sen S, Huang AH, Elliott DM, Mauck RL (2010) Engineered disc-like angle-ply structures for intervertebral disc replacement. Spine 35: 867–873.

27. Yu J, Tirlapur U, Fairbank J, Handford P, Roberts S, et al. (2007) Microfibrils, elastin fibres and collagen fibres in the human intervertebral disc and bovine tail disc. J Anat 210: 460–471.

28. Horner HA, Roberts S, Bielby RC, Menage J, Evans H, et al. (2002) Cells from different regions of the intervertebral disc: effect of culture system on matrix expression and cell phenotype. Spine 27: 1018–1028.

29. Smith LJ, Fazzalari NL (2006) Regional variations in the density and arrangement of elastic fibres in the anulus fibrosus of the human lumbar disc. J Anat 209: 359–367.

30. Li B, Wang JH (2010) Application of sensing techniques to cellular force measurement. Sensors 10: 9948–9962.

31. Li F, Li B, Wang QM, Wang JH (2008) Cell shape regulates collagen type I expression in human tendon fibroblasts. Cell Motil Cytoskeleton 65: 332–341.

32. McNally DS, Adams MA (1992) Internal intervertebral disc mechanics as revealed by stress profilometry. Spine 17: 66–73.

33. Stokes IA (1987) Surface strain on human intervertebral discs. J Orthop Res 5: 348–355.

34. Hutton WC, Elmer WA, Boden SD (1999) The effect of hydrostatic pressure on intervertebral disc metabolism. Spine 24: 1507–1015.

35. Engler AJ, Sen S, Sweeney HL, Discher DE (2006) Matrix elasticity directs stem cell lineage specification. Cell 126: 677–689.

36. Discher DE, Janmey P, Wang YL (2005) Tissue cells feel and respond to the stiffness of their substrate. Science 310: 1139–1143.

37. Lewis NT, Hussain MA, Mao JJ (2008) Investigation of nano-mechanical properties of annulus fibrosus using atomic force microscopy. Micron 39: 1008–1019.

38. Adams MA, Dolan P (2005) Spine biomechanics. J Biomech 38: 1972–1983.

39. Buckwalter JA, Woo SL, Goldberg VM, Hadley EC, Booth F, et al. (1993) Soft-tissue aging and musculoskeletal function. J Bone Joint Surg Am 75: 1533–1548.

40. Guerin HL, Elliott DM (2007) Quantifying the contributions of structure to annulus fibrosus mechanical function using a nonlinear, anisotropic, hyperelastic model. J Orthop Res 25: 508–516.

41. Feng G, Yang X, Shang H, Marks IW, Shen FH, et al. (2010) Multipotential differentiation of human anulus fibrosus cells: an in vitro study. J Bone Joint Surg Am 92: 675–685.

10

Comparison of Decellularization Protocols for Preparing a Decellularized Porcine Annulus Fibrosus Scaffold

Haiwei Xu[1,2◑], **Baoshan Xu**[1*◑], **Qiang Yang**[1,2*◑], **Xiulan Li**[3], **Xinlong Ma**[1], **Qun Xia**[1], **Yang Zhang**[3], **Chunqiu Zhang**[4], **Yaohong Wu**[1,2], **Yuanyuan Zhang**[5]

1 Department of Spine Surgery, Tianjin Hospital, Tianjin, China, **2** Graduate School, Tianjin Medical University, Tianjin, China, **3** Cell Engineering Laboratory of Orthopaedic Institute, Tianjin Hospital, Tianjin, China, **4** School of Mechanical Engineering, Tianjin University of Technology, Tianjin, China, **5** Wake Forest Institute for Regenerative Medicine, Wake Forest University School of Medicine, Winston-Salem, North Carolina, United States of America

Abstract

Tissue-specific extracellular matrix plays an important role in promoting tissue regeneration and repair. We hypothesized that decellularized annular fibrosus matrix may be an appropriate scaffold for annular fibrosus tissue engineering. We aimed to determine the optimal decellularization method suitable for annular fibrosus. Annular fibrosus tissue was treated with 3 different protocols with Triton X-100, sodium dodecyl sulfate (SDS) and trypsin. After the decellularization process, we examined cell removal and preservation of the matrix components, microstructure and mechanical function with the treatments to determine which method is more efficient. All 3 protocols achieved decellularization; however, SDS or trypsin disturbed the structure of the annular fibrosus. All protocols maintained collagen content, but glycosaminoglycan content was lost to different degrees, with the highest content with TritonX-100 treatment. Furthermore, SDS decreased the tensile mechanical property of annular fibrosus as compared with the other 2 protocols. MTT assay revealed that the decellularized annular fibrosus was not cytotoxic. Annular fibrosus cells seeded into the scaffold showed good viability. The Triton X-100–treated annular fibrosus retained major extracellular matrix components after thorough cell removal and preserved the concentric lamellar structure and tensile mechanical properties. As well, it possessed favorable biocompatibility, so it may be a suitable candidate as a scaffold for annular fibrosus tissue engineering.

Editor: Abhay Pandit, National University of Ireland, Galway, Ireland

Funding: This work was supported by the National Natural Science Foundation of China (grant numbers 31000432 and 81272046), China Postdoctoral Science Foundation funded project (grant numbers 2011M500530, 2012T50235), and the Research Foundation of the Tianjin Health Bureau (grant number 2010KR08). The funders had no role in study design, data collection and analysis, decision to publish, or preparation of the manuscript.

Competing Interests: The authors have declared that no competing interests exist.

* E-mail: xubaoshan99@126.com (BX); yangqiang1980@126.com (QY)

◑ These authors contributed equally to this work.

Introduction

Disc degenerative disease is generally thought to be the main cause of chronic low back pain, which has a lifetime prevalence of 80% in the general population and causes a huge public health burden in industrialized countries [1]. Current treatments ranging from conservative management to invasive procedures are primarily palliative and seek to eliminate the pain generated by ruptured or herniated disks but do not attempt to restore disc structure and function [2]. Tissue-engineering techniques have emerged as a promising therapeutic approach to treat degenerative discs by replacing the damaged tissue with a biomaterial and appropriate cells [3].

The scaffold is a major component in tissue engineering. Cells live and proliferate in the scaffold, which can perform a variety of functions lacking in damaged tissue *in vivo*. An ideal scaffold is necessary in annulus fibrosus (AF) tissue engineering. It should have good biocompatibility, moderate porosity and proper degradation rate and be similar to natural AF in composition, shape, structure and mechanical properties [4].

The AF is a multi-lamellar fibrocartilagenous ring, comprised primarily of collagen and proteoglycans. It consists of 15–25 concentric layers within which the collagen fibers lie parallel to each other at approximately a 30° angle to the transverse plane of the disc but in alternate directions in successive layers [5]. The widths of lamellae in AF differ from outer to inner layers, being thicker in the inner than the outer layers. Meanwhile, the numbers of lamellae vary circumferentially, with the greatest number in the lateral region of the disc and the smallest in the posterior region [6]. The AF contains mainly types I and II collagen. The outer AF contains mostly type I and the inner AF contains mainly type II, for a decrease in ratio of types I to II collagen from the outer to inner AF [7]. However, water and proteoglycan content increase from the outer to inner AF [8].

The structure of AF is complicated and the components are distributed unevenly, so fabricating an artificial scaffold identical to AF in components and structure is difficult. To date, none of the scaffold designs used for AF tissue engineering, including polyamide nanofibers, alginate/chitosan hybrid fiber, demineralized bone matrix gelatin/polycaprolactone triol malate, and demineralized and decellular bone, have been able to replicate the composition and lamellar structure of AF. An ideal AF scaffold is the goal.

With the development of decellularization technology, tissue-specific extracellular matrix (ECM) as a complete novel biomaterial has attracted the attention of many researchers. ECM scaffolds and substrates are ideal candidates for tissue engineering because in our body, cells are surrounded by ECM. The ECM functions as a support material and also regulates cellular functions such as cell survival, proliferation, morphogenesis and differentiation. Moreover, the ECM can modulate signal transduction activated by various bioactive molecules such as growth factors and cytokines. Ideally, scaffolds and substrates used for tissue engineering and cell culture should provide the same or similar microenvironment for seeded cells as existing ECM *in vivo*. Decellularized matrices have been widely used for engineering functional tissues and organs such as cartilage, skin, bone, bladder, blood vessels, heart, liver, and lung [9–14] and have achieved impressive results.

Because acellular matrixes have been used for tissue engineering and clinical purposes, we wondered whether acellular AF could preserve the ECM, microstructure and biomechanical properties of native AF as ideal scaffold material for tissue-engineered AF. We found no evidence of decellularized AF in the literature, so we investigated a decellularization method suitable for AF. We compared 3 decellularization methods that are widely used and are effective in tissue or organ decellularization. We aimed to determine which method was advantageous in cell removal and preserving the ECM components, structure and mechanical properties of natural AF for an ideal scaffold for AF tissue engineering.

Materials and Methods

AF Preparation

We obtained animal material from the Animal Experimental Room of Tianjin Hospital. All animal experiments were approved by the Animal Experimental Ethics Committee of Tianjin Hospital and the animals were treated according to the experimental protocols under its regulations. Fresh pig tails were transported to the laboratory within 2 h after slaughter. AF were dissected from the intervertebral discs in pig tails. All surrounding tissues were carefully removed by use of scissors, and then AF samples were washed in phosphate-buffered saline (PBS) to remove excess blood. Specimens (external diameter 9~11 mm, thickness 4.5~5.5 mm) were randomly divided into 4 groups and treated as follows.

Decellularization Methods

Triton X-100. Pig AF was placed in hypotonic Tris-HCl buffer (10 mM, pH 8.0) with 0.1% ethylenediamine tetraacetic acid (EDTA; Sigma) and 10 KIU/ml aprotinin (Sigma) at 4°C for 48 h. Then AF samples were agitated in Tris-HCl buffer with 3% Triton X-100 (Sigma), 0.1% EDTA and 10 KIU/ml aprotinin at 4°C for 72 h. The solution was changed every 24 h. Then AF samples were incubated with 0.2 μg/mL ribonuclease A (RNase A; Sigma) and 0.2 mg/mL desoxyribonclease I (DNase I; Sigma) at 37°C for 24 h. Finally, decellularized AF was washed with PBS for 24 h to remove residual reagents. All steps were conducted under continuous shaking [11,15–17].

SDS. Pig AF was frozen at −80°C for 3 h and thawed at room temperature for 4 h. After 3 cycles of freezing-dissolving, AF samples were decellularized with 10 mM Tris-HCl buffer containing 0.5% SDS (Sigma), 0.1% EDTA and 10 KIU/ml aprotinin at room temperature for 72 h. The decellularization solution was refreshed every 24 h. Decellularized AF was incubated with 0.2 μg/mL RNase A and 0.2 mg/mL DNase I at 37°C for 24 h, then washed with PBS for 24 h to remove residual reagents. All steps were conducted under continuous shaking [12,14,18].

Trypsin. Pig AF were incubated under continuous shaking in trypsin/EDTA (0.5% trypsin and 0.2% EDTA; both Sigma) in hypotonic Tris-HCl buffer, together with RNase A (20 μg/ml) and DNase I (0.2 mg/ml) at 37°C for 72 h. The trypsin/EDTA solution was changed every 24 h. Then decellularized AF was washed with PBS for 24 h under shaking for removal of residual substances [19–21].

Control Group. Fresh pig AF was stored at −20°C.

Histology

After decellularization, tissue specimens (n = 10) were fixed in 10% (v/v) neutral buffered formalin, dehydrated with a graded ethanol and embedded in paraffin wax, cut into sections of 5.0 μm by use of a microtome and mounted on glass slides. Haematoxylin and eosin (H&E) staining was used to evaluate the cellular content and general structure of the AF. Nucleic acids were stained with Hoechst 33258 dye (Sigma). Proteoglycan was visualized by Toluidine blue staining and Safranin O staining. Sirius red stain was used to visualize collagen distribution and orientation.

Immunofluorescence Examination

Specimens for immunofluorescence stain were mounted with OCT compound and cryosectioned at 10 μm thick. After rehydration by immersion in PBS for 10 min, sections were incubated with a monoclonal antibody against collagen I (Shiankexing, Beijing) at 4°C overnight, followed by extensive washes with PBS, then incubated with FITC-conjugated IgG antibody (Sigma) for 1 h at room temperature. After 3 washes in PBS, sections were observed by fluorescence microscopy.

Scanning Electron Microscopy (SEM)

Decellularized or control AF samples were freeze-dried, cut along the transverse plane by use of a sharp blade, then loaded onto aluminum studs, coated with gold and examined under a field emission scanning electron microscope (1530VP, LEO, Germany). Morphological changes were compared before and after treatment.

Rehydration Analysis

Water imbibition was quantified to compare potential changes in imbibition properties of decellularized and natural AF. Fresh and decellularized AF (n = 15) was immersed in PBS containing 10 KIU/ml aprotinin at 4°C for 24 h to achieve fully swollen and hydrated states. Samples were then freeze-dried, and the weight before and after freeze-drying was measured. The swelling ratio (%) of samples was calculated as (Ws-Wd)/Wd, where Ws is the sample weight after immersion in PBS and Wd is the sample weight after freeze-drying [13].

Collagen Content

Collagen content was measured as described [22]. Samples (n = 10) were first lyophilized to a constant weight, then samples (30 mg dry weight) were acid-hydrolyzed with hydrochloric acid (HCl) at 100°C for 20 min and neutralized with sodium hydroxide (NaOH). Oxidation of standard and test solution was achieved by adding N-chloro-p-toluenesulfonamide sodium salt (Chloramine T; Sigma) followed by p-dimethylamino-benzaldehyde (Sigma), and the absorbance was read at 570 nm. The amount of hydroxyproline present in the test samples was determined against a standard curve.

Glycosaminoglycan (GAG) Content

GAG content was quantified by the DMMB assay as described [23]. Briefly, samples (n = 10) were freeze-dried to a constant weight, and samples (10 mg) were digested in papain buffer (125 mg/ml papain, 5 mM cysteine–HCl, 5 mM disodium EDTA in PBS) at 60°C for 24 h. Then, 50 µl of each sample was mixed with 250 µl 1, 9-dimethyl-methylene blue (Sigma) in a 96-well microtiter plate and the absorbance was measured at 530 nm. The amount of GAG content was calculated by reference to a standard curve prepared using different concentrations of chondroitin sulfate sodium salt from shark cartilage (Sigma).

Biomechanical Testing

Mechanical test samples $15 \times 4 \times 1$ mm were dissected from the outer anterior section of AF along circumferential direction (Fig. 1A). Before testing, samples were immersed in PBS (pH 7.4) for 4 h, then strips were mounted under zero strain onto frozen fixtures in a mechanical apparatus (Bose, Boston, USA) and the initial specimen length was recorded. The samples were then stretched to tensile failure at a rate of 1 mm/min. Samples were kept moist during testing by dropping normal saline solution on the specimens. All testing was conducted at room temperature.

For each specimen, ultimate load, stress, and strain; toughness; elastic modulus; and mechanical work to fracture were determined by computer and compared with the curve of load-displacement. A schematic diagram of the load-displacement curve is shown in Fig. 1B.

Ultimate load refers to the largest load value in the tensile process that can be read at the highest point of the load-displacement curve. It is a straightforward reflection of tissue strength but affected by the cross-sectional area of specimens. Under the same condition, ultimate load is positively related to the cross-sectional area. So, the ultimate load can be compared only in the same cross-sectional area.

Ultimate stress is a tensile parameter that excludes the influence of cross-sectional area. It refers to the amount of force per unit of initial cross-sectional area at tensile failure. Ultimate stress was calculated by dividing the maximum load by the original cross-sectional area of the specimen.

Ultimate strain was calculated by dividing the change in length by the initial length of the specimen.

Toughness is the slope of the ascending linear portion of the load-displacement curve. The greater the toughness, the harder the specimens is pulled off.

Elastic modulus refers to the stress needed to produce per unit of elastic deformation. It is one of the most commonly used indicators reflecting the tensile properties. Elastic modulus was calculated from the slope of the ascending linear region of the stress-strain curve.

Mechanical work to fracture is the work performed when the AF is stretched to fracture. Mechanical work to fracture was calculated by numerical integration of the area under the load-displacement curve in the left of breaking point.

Cytotoxicity Assay

Depending on the above results, cytotoxicity study and subsequent experiments were conducted with samples of the Triton X-100 Group.

3(4,5-dimethylthiazole-2-yl)-2,5-diphenyltetrazolium-bromide (MTT; Sigma) assay was performed to determine the cytotoxicity of decellularized AF. Briefly, rabbit AF cells were seeded onto wells of flat-bottomed 96-well plates at 5×10^3 cells/mL (200 µl per well). The plates were incubated for 24 h before the medium was replaced with control medium (positive control) and different concentrations (25%, 50%, 100%) of extracts prepared as described [24]. At days 1–6, the proliferation activity of the cells was determined by MTT assay. The optical density (OD) absorbance at 570 nm was determined with use of a microplate reader (RT-6000, Rayto, USA). Five replicates were considered per sample.

Isolation and Culture of AF Cells

Lumbar spines were dissected aseptically from New Zealand white rabbits (female, 6 weeks old) killed under the guidelines specified by the Animal Experimental Ethics Committee of Tianjin Hospital. AF was separated from intervertebral discs with use of a blade, and all surrounding tissues (including muscles, tendons and nucleus pulposus) were carefully removed. The

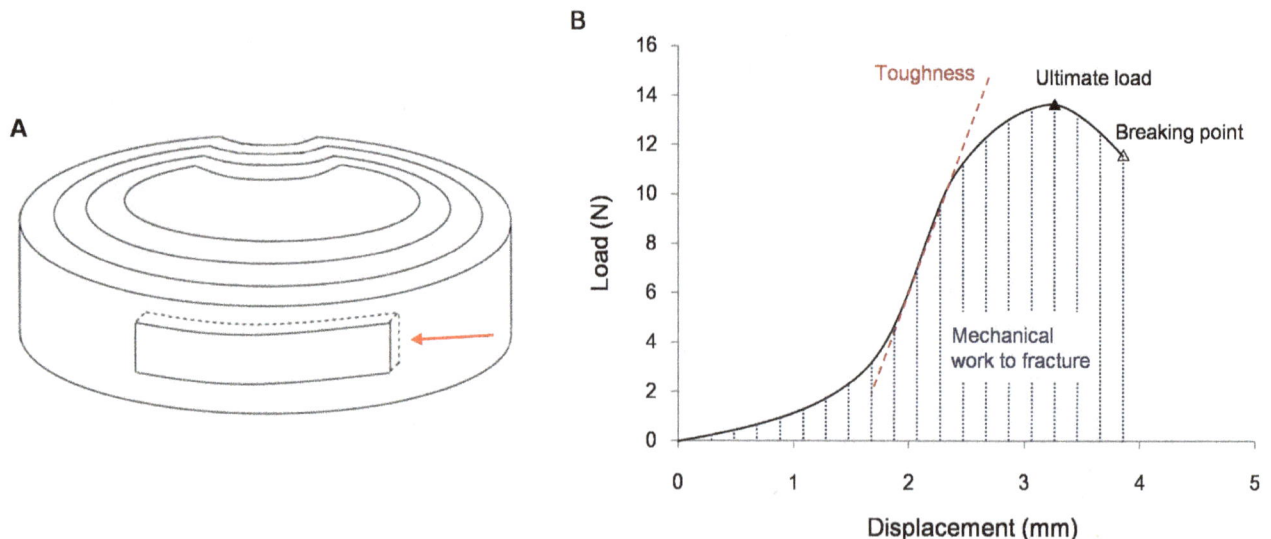

Figure 1. Schematic diagrams of specimens for tensile testing and load-displacement curve. (A) Schematic diagram of the intervertebral disc and locations of annulus fibrosus (AF) specimens for tensile testing. AF samples (arrow) were dissected from the outer zones of anterior regions, with the longest dimension in the circumferential direction. (B) Schematic diagram of load-displacement curve.

collected AF was cut into small pieces and digested with 0.25% collagenase (Sigma) for 6 h at 37°C. Cell suspensions were filtered through a nylon mesh and cultured in Dulbecco's modified Eagle's medium (DMEM; Gibco) containing 10% fetal bovine serum (FBS; Gibco) and 1% antibiotics at 37°C in a humidified atmosphere of 5% CO_2. The medium was changed every 3 days. Cells at passage 2 were used in this study.

Cell Seeding

Decellularized AF (Triton X-100 Group) was disinfected with 70% ethanol, thoroughly rinsed in sterile PBS for 24 h, and immersed in DMEM containing 10% FBS and 1% antibiotics for 24 h. The liquid on the surface of decellularized AF was dried by use of sterile filter paper, then 100 μl cell suspension containing 1×10^6 AF cells was seeded into each decellularized AF by drop-wise addition onto the surface of the decellularized AF. At 1 h later, the decellularized AF was turned over and another 100 μl cell suspension was seeded onto the surface. The cell-containing constructs were incubated for 2 h before the culture medium was supplemented slowly for further culture. Culture medium was changed every 2 days.

Cell Distribution and Viability Assessment

After 7 days of culture, the cell-seeded constructs were fixed in 10% (v/v) neutral buffered formalin, dehydrated with ethanol and embedded in paraffin wax. They were cut into sections of 5.0 μm by use of a microtome and stained with H&E to observe cell distribution in decellularized AF. The viability of cells seeded into scaffolds was detected by a live/dead assay kit (Invitrogen): live cells were stained with calcein AM (green) and dead cells with ethidium homodimer (EthD-1) (red). The constructs were incubated with live/dead dye at 37°C, 5% CO_2, with saturated humidity for 30 min, then constructs were observed under a confocal microscope (TCS SP5 II, Leica, Germany) for cell viability.

Statistical Analysis

Data analysis involved SPSS 16.0 (SPSS, Chicago, IL, USA). Results were expressed as mean ± SD. Differences between groups were assessed by one-way ANOVA, followed by Sceffe or Tamhane's T2 tests for multiple comparisons. $P < 0.05$ was considered statistically significant.

Results

Morphology and History

Macroscopically, after decellularization, AF swelled and the central voids became smaller as compared with natural AF (Fig. 2A–D). The 3 decellularization groups did not differ macroscopically.

On H&E staining, control AF showed many cells scattered among collagen fibers, which were compact with an ordered arrangement (Fig. 3). Decellularized Triton X-100, SDS or trypsin samples showed no cells, and the mesh of collagen fibers was looser than in control samples. Triton X-100 and trypsin samples retained the concentric lamellar arrangements of collagen, similar to natural AF, but some fractured collagen fibers could be seen in trypsin samples. In SDS samples, lamellar arrangements of collagen were disturbed, with gaps between the collagen fibers. Results were similar with Hoechst 33258 staining (Fig. 4). Many blue fluorescent dots representing DNA were evenly distributed in natural AF, with none in Triton X-100, SDS or trypsin samples.

Toluidine blue and Safranin O staining showed that both natural AF and decellularized AF were rich in proteoglycans, but staining was less dense in decellularized than natural AF (Fig. 5,6). Proteoglycan content may have decreased during the decellularization process. Sirius red staining showed enriched collagen content in both natural and decellularized AF (Fig. 7).

Immunohistochemistry

All samples were positive for collagen type I (Fig. 8), with no differences in staining density.

SEM

In control samples, collagen fibers were arranged orderly, with a concentric lamellar structure (Fig. 9). Triton X-100 samples showed a concentric lamellar structure, with no difference from natural AF. However, the arrangement of collagen fibers was severely disturbed in SDS samples, with no lamellar structure. Trypsin samples retained the concentric lamellar structure, but the arrangement of collagen fibers was somewhat disorganized as compared with control and Triton X-100 samples.

Hydration Results

The decellularized AF showed a high capacity to absorb water (Fig. 10A). The swelling ratios for decellularized AF in Triton X-100, SDS, and trypsin samples did not differ from each other (11.65±2.56, 9.97±1.68, 9.71±1.04 mg water/mg sample dry weight respectively), but swelling was greater than for control samples (7.81±1.13) ($p < 0.05$), so decellularized AF contained significantly more water than natural AF. This water uptake was likely responsible for "pushing apart" areas of the collagen matrix throughout decellularized AF, leading to the appearance shown on H&E, Toluidine blue and Safranin O staining.

Quantification of Collagen

The content of hydroxyproline was detected in samples for calculating collagen content. Control and decellularized AF samples did not differ in mean collagen content per mg of tissue (Fig. 10B).

Quantification of GAG

GAG content was lower in decellularized than control AF samples ($p < 0.05$; Fig. 10C). The GAG content in Triton X-100 samples was closest to that in natural AF, and higher than that in SDS or trypsin samples ($p < 0.05$). GAG content was lower in SDS and trypsin than control samples.

Biomechanical Testing

The ultimate load and stress values decreased as follows: Triton X-100> control>trypsin>SDS samples, with no significant difference between control and Triton X-100 or trypsin samples but a difference between control and SDS samples ($P = 0.004$, $P = 0.012$, Table 1). The ultimate strain values decreased as follows: Triton X-100> SDS>control>trypsin samples, with no significant difference among the 4 groups ($P = 0.078$). The toughness and elastic modulus values decreased as follows: trypsin>control>Triton X-100> SDS samples, with no significant difference between control and Triton X-100 or trypsin samples but a difference between control and SDS samples ($P = 0.003$, $P = 0.008$). The mechanical work to fracture values decreased as follows: trypsin>Triton X-100> control>SDS samples, with no difference between control and Triton X-100 or trypsin samples but a difference between control and SDS samples ($P = 0.027$).

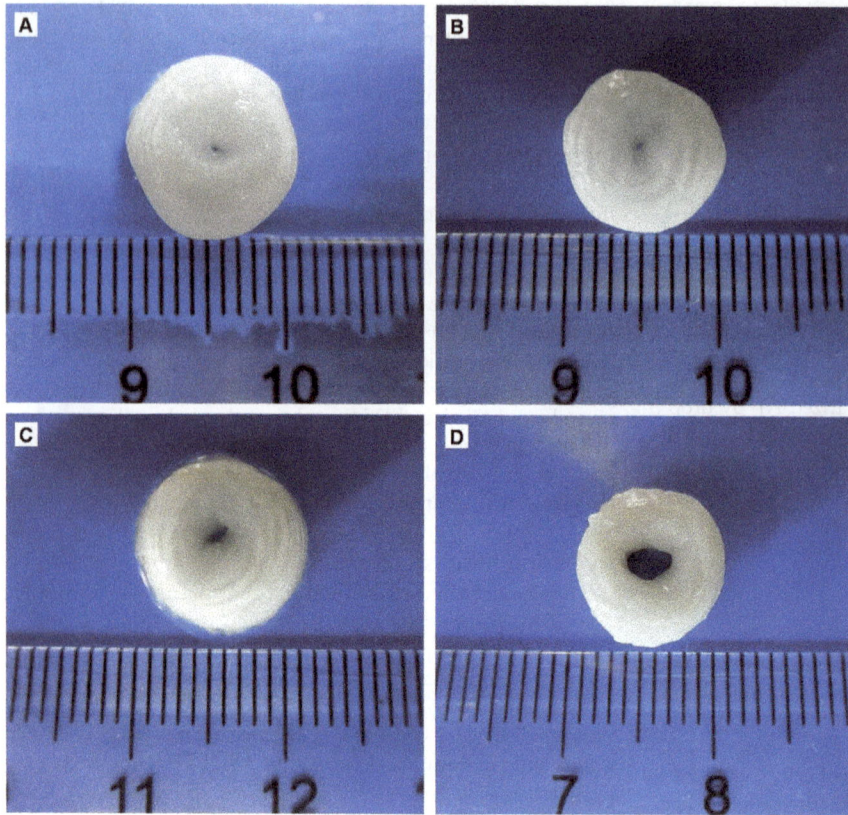

Figure 2. Representative macroscopic images of AF before and after decellularization. (A) Triton X-100, (B) SDS, (C) trypsin, (D) control.

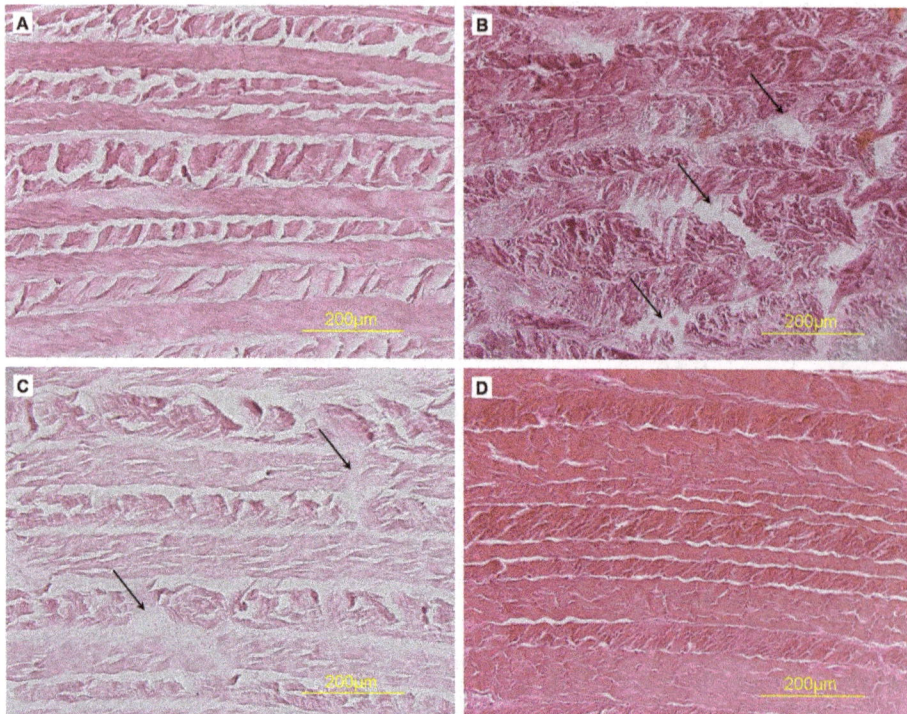

Figure 3. Hematoxylin and eosin (H&E) staining of cross-sections of AF samples. (A) Triton X-100, (B) SDS, (C) trypsin, (D) control. Collagen fiber fracture (arrows).

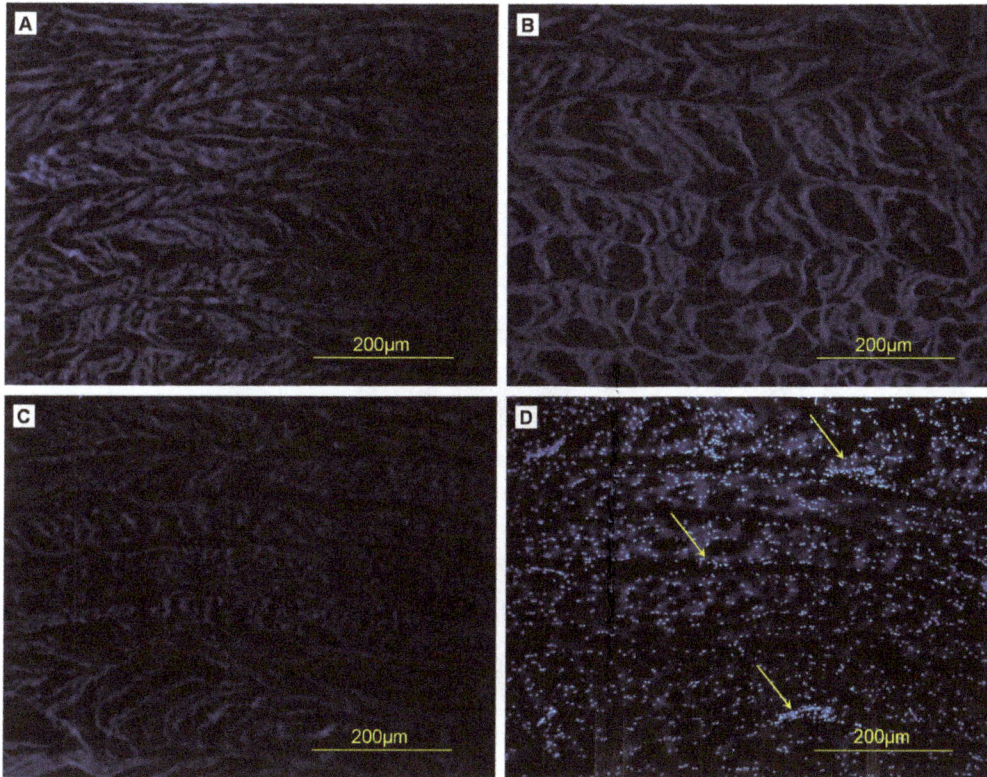

Figure 4. Hoechst 33258 staining of cross-sections of AF samples. (A) Triton X-100, (B) SDS, (C) trypsin, (D) control. DNA (arrows).

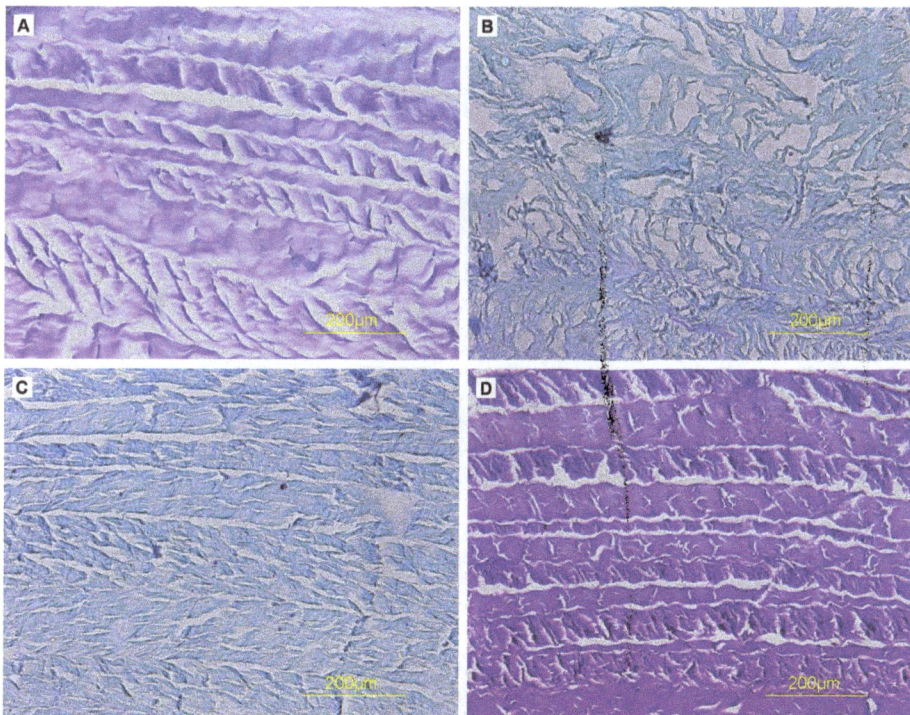

Figure 5. Toluidine blue staining of cross-sections of AF samples. (A) Triton X-100, (B) SDS, (C) trypsin, (D) control.

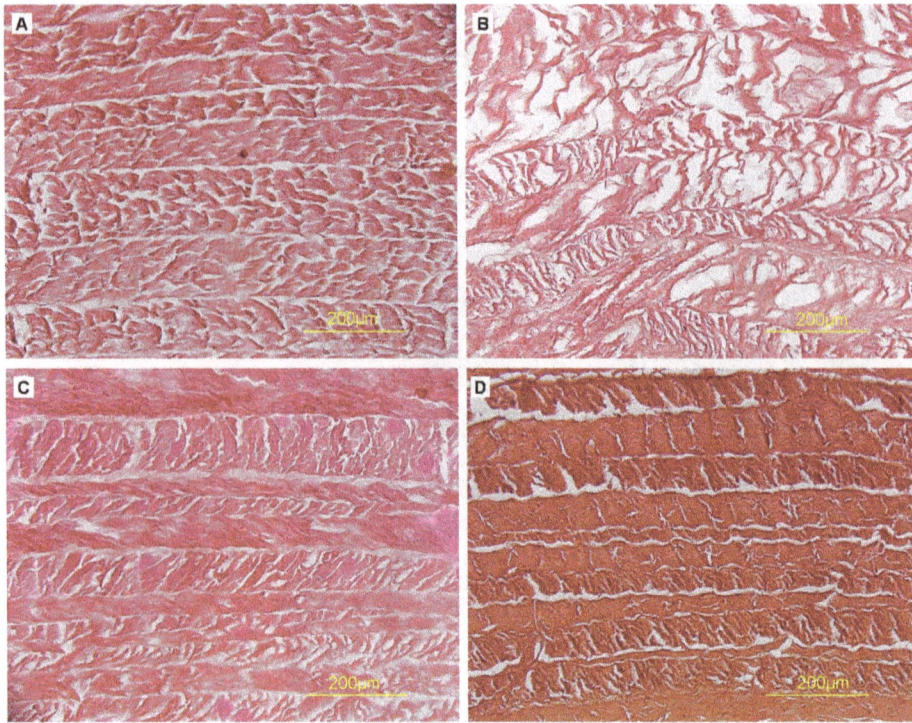

Figure 6. Safranin O staining of cross-sections of AF samples. (A) Triton X-100, (B) SDS, (C) trypsin, (D) control.

Cytotoxicity Assay

Different concentrations of extracts had no effect on cell proliferation, with no difference in OD values for the 4 groups at each time (P>0.05), so the decellularized AF were not cytotoxic (Fig. 11).

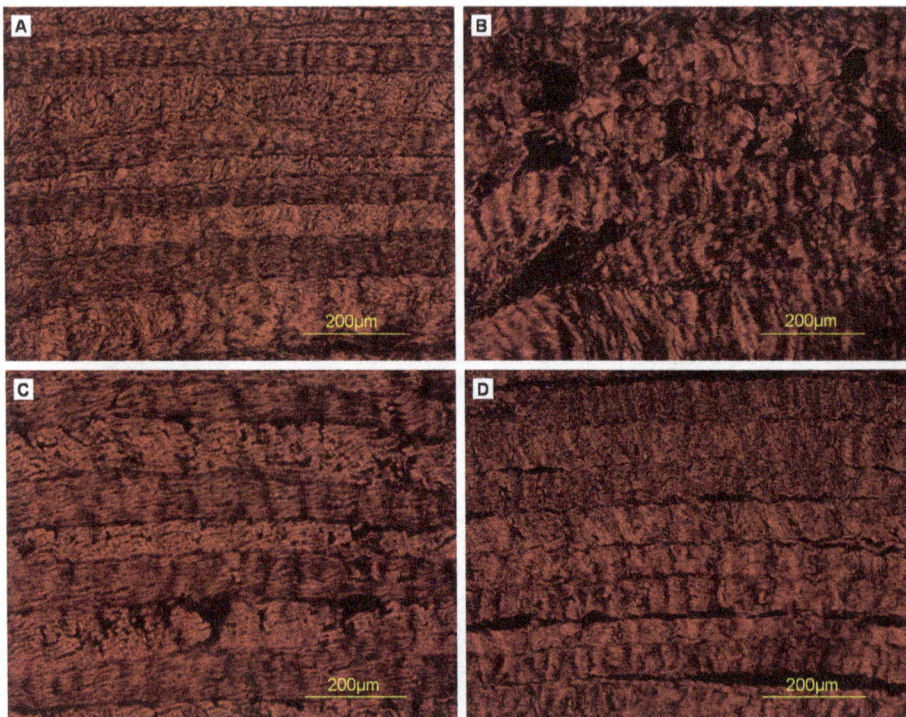

Figure 7. Sirius red stain of cross-sections of AF samples. (A) Triton X-100, (B) SDS, (C) trypsin, (D) control.

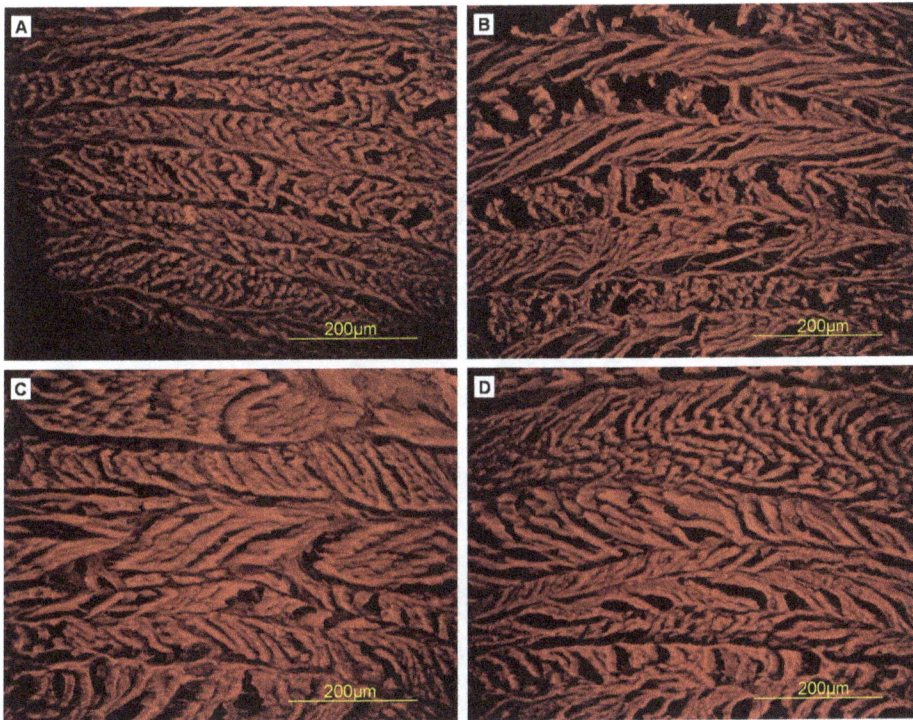

Figure 8. Collagen I immunouorescent staining of cross-sections of AF samples. (A) Triton X-100, (B) SDS, (C) trypsin, (D) control.

Cell Distribution and Viability Assessment

After 7 days of culture, AF cells infiltrated the mid-horizontal plane of decellularized AF (Fig. 12A). Live/dead staining showed live cells evenly distributed in decellularized AF, with no dead cells (Fig. 12B).

Discussion

In the present study, we explored the use of a non-ionic detergent (Triton X-100), an anionic detergent (SDS) and enzymatic agent (trypsin) to decellularize pig AF and compared the histological structure and biomechanical properties of decellularized AF as an ideal scaffold for AF tissue engineering. Triton X-100–treated AF retained the major ECM components after thorough cell removal, preserved the concentric lamellar structure and tensile mechanical properties, and possessed favorable biocompatibility, so it is a suitable candidate for producing scaffold material for AF tissue engineering.

The immunogenicity of acellular matrixes must be eliminated before they are used for tissue engineering. Cells are the main immunogenic factors in tissue. Histocompatibility antigens (human leukocyte antigen) are distributed on the surface of cell membranes in the form of lipoproteins or glycoproteins. They are genetically determined and differ among individuals within the same or different species. Histocompatibility antigens are recognized by T cells, and the tissue is attacked by the recipient host after transplantation of allogeneic cells. So, before ECM is used as scaffold, the cells must be removed to avoid immune rejection, inflammation, and potential transplant rejection [25]. Our H&E staining showed that all 3 decellularization agents removed cells. Furthermore, hochest 33258 staining, which emits blue fluorescence when bound to double-stranded DNA, showed no DNA in decellularized AF with the 3 agents. Therefore, use of the 3 agents was effective in AF decellularization. Previously, decellularization

with Triton X-100 completely removed nuclear material in nerve, pericardium and bone [11,16–17]; with SDS removed cells in meniscus, cornea and cartilage bone [12,14,18]; and with trypsin removed cells in dermal, aortic and aortic valve tissue [13,15,19,21]. However, the cell removal efficacy of Triton X-100 is controversial: nuclear material was observed in tendon, artery, and ligament after decellularization with Triton X-100 solution [26–28]. The decellularization effect of Triton X-100 is related to the organization of the material. As well, concentrations of detergents affect decellularization efficiency. Recently Chan et al. [24] decellularized bovine intervertebral disc to create a natural intervertebral disc scaffold with 0.1% SDS. Many dead cells were left in the intervertebral disc on live/dead staining, whereas in our study, 0.5% SDS produced no cells in decellularized AF.

Collagen and GAG are the main components of the AF ECM. They play an important role in guiding cellular attachment, survival, migration, proliferation, differentiation [29]. The ideal decellularized AF ECM should contain collagen and GAG content close to that of natural AF. We calculated collagen content by presence of hydroxyproline in the test samples and found no difference between decellularized AF and control samples, which indicates no collagen lost in the decellularization process with Triton X-100, SDS or trypsin. However, GAG content was reduced with decellularization, especially with trypsin, and the GAG content was closest to that of the control with Triton X-100. The preservation of collagen and loss of GAG may be related to their relative position. Within and between the lamellae is a proteoglycan-rich ground substance [30]. The orderly arranged collagen fibers are embedded in a matrix rich in proteoglycan and GAG, which are exposed to decellularization solution and more likely to be lost during decellularization as compared with collagen [31]. Especially, trypsin has the ability of disconnecting the interactions between the matrix proteins, thus creating a more

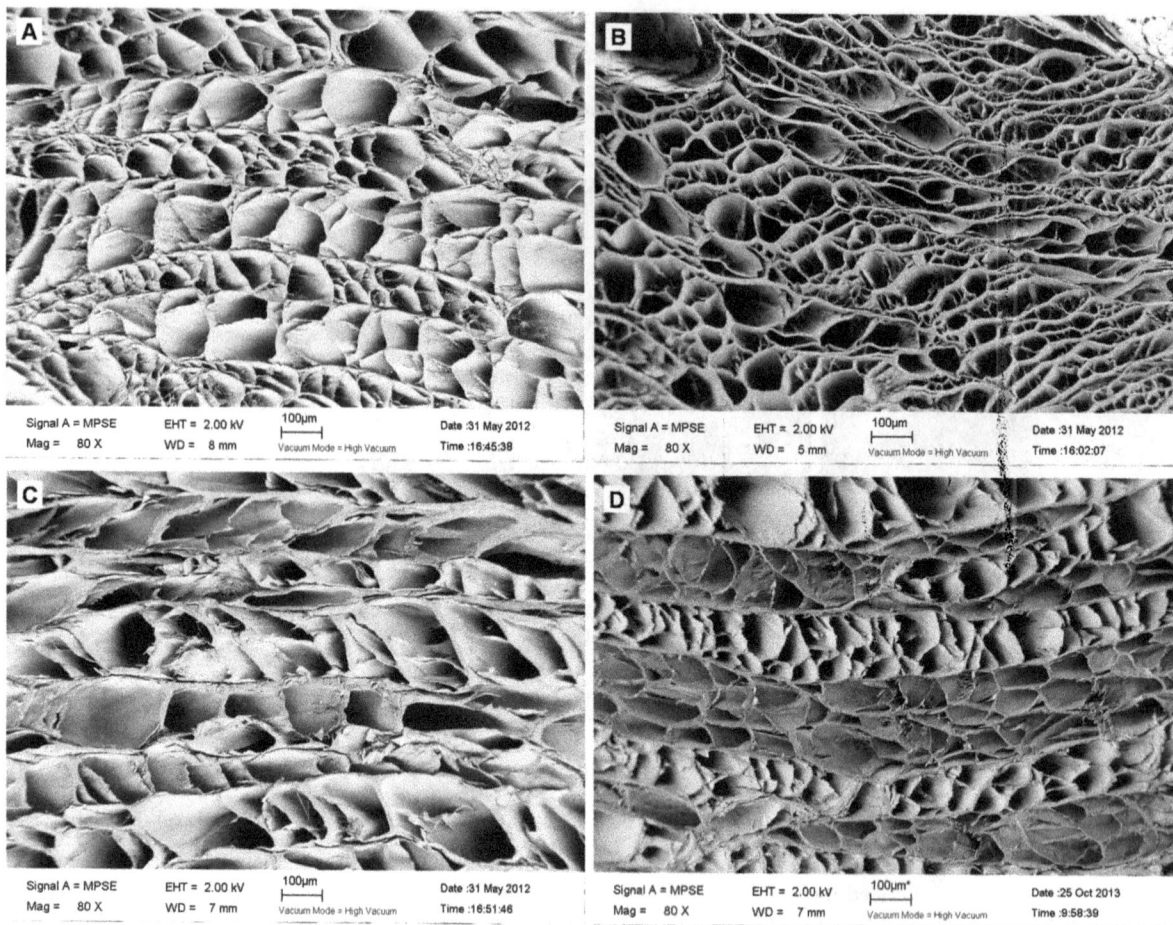

Figure 9. Scanning electron micrographs of cross-sections of AF samples. (A) Triton X-100, (B) SDS, (C) trypsin, (D) control.

open matrix, which results in more GAG lost. Triton X-100 was superior to the other treatments in retaining collagen and GAG content.

AF is a multi-lamellar fibro-cartilagenous ring. The unique angle-ply architecture of AF is critical for withstanding multi-axial physiologic loads for normal function of the spine. After decellularization, H&E staining and SEM revealed a well-preserved concentric lamellae structure with Triton X-100. With trypsin, the concentric lamellar structure was slightly disturbed, with some collagen fractures seen on H&E staining. With SDS, the concentric lamellar structure was severely destroyed, with large gaps between collagen fibers, as seen on H&E staining and SEM. This finding was consistent with the reported features of SDS treatment. SDS, which has a negatively charged head-group and belongs to anionic detergents, can bind and denature both soluble and membrane-bound proteins. It can disrupt non-covalent bands within proteins and cause them to lose their native conformation. So SDS tends to disrupt the native tissue structure and causes decreased GAG concentration and loss of collagen integrity [25]. Cartmell et al. [32] decellularized rat tail tendons with Triton X-100, TnBP, and SDS. Treatment with SDS resulted in a pronounced opening of the spaces between the aligned collagen fibers regardless of concentration or treatment time. Kasimir et al. [33] treated aortic and pulmonary porcine valves with 0.1%, 0.03% and 0.01% SDS for 24 and 48 h. All concentrations completely removed cells. However, the matrix fibers were markedly disintegrated after 24 and 48 h. Reports about the

effect of SDS differ. Liao et al. [15] processed porcine aortic valves with 0.1% SDS and preserved the trilayered structure of the native aortic valve. Therefore, the effects of SDS on tissue structure depend on the tissue substrate.

Mechanical property is an important parameter of the intervertebral disc. *In vivo*, intervertebral discs serve to support large spinal loads, which are combinations of tension, torsion, compression, and bending. The hydrostatic excess pressure in the nucleus pulposus caused by these loads generates large circumferential tensile stress in the surrounding AF [34]. The normal tensile mechanical properties of AF secure the nucleus pulposus in the right position and the intervertebral disc functions normally. AF exhibits regional variations in tensile mechanical properties [35–36]. The anterior AF has larger tensile values than the posterolateral annulus. Also, tensile values are larger in the outer than the inner regions of the annulus [8,37–38]. These variations are generally attributed to inhomogeneity in tissue structure and biochemical composition. In the current study, the mechanical samples were all dissected from the outer anterior section of AF to eliminate the regional variation caused by inhomogeneous biochemical composition and structural organization.

We found no significant difference in ultimate load and stress, toughness, elastic modulus and mechanical work to fracture between Triton X-100, trypsin and control treatment; however, these parameters were lower with SDS than control treatment. The mechanical results have much to do with the structure of decellularized AF. Tensile properties are closely related to collagen

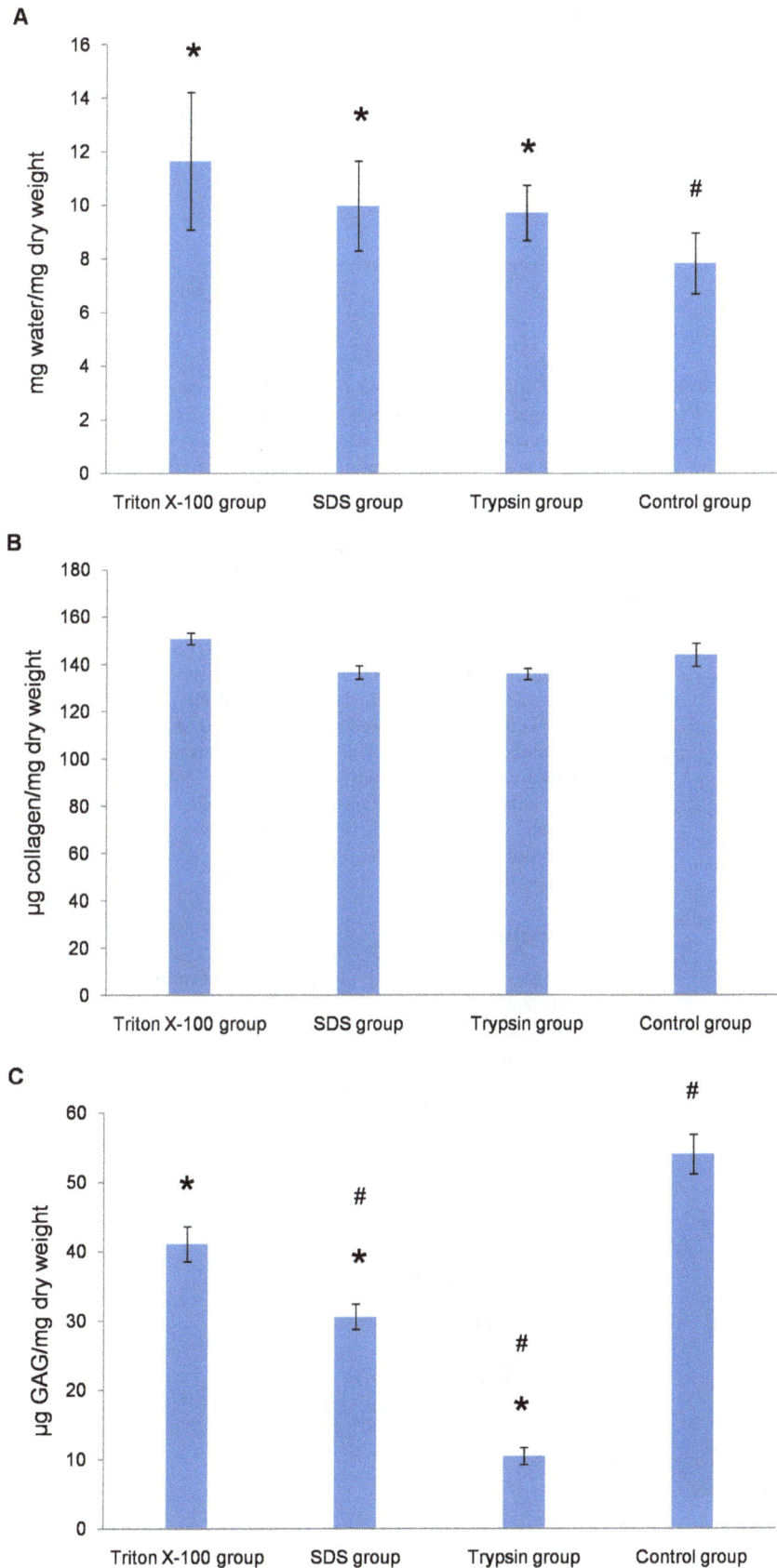

Figure 10. Water (A), collagen (B), and glycosaminoglycan (GAG) content (C) of AF. Data are mean ± SD. * = p<0.05 compared to control, # = p<0.05 compared to Triton X-100.

Table 1. The biomechanical properties of annulus fibrosus with decellularization treatments.

Group	Ultimate load (N)	Ultimate stress (MPa)	Ultimate strain (%)	Toughness (N/mm)	Elastic modulus (MPa)	Mechanical work to fracture ($\times 10^{-3}$ J)
Triton X-100	24.52±3.83	6.02±0.83	0.41±0.05	15.58±1.62	28.89±5.50	30.85±5.15
SDS	11.27±2.68*	2.86±0.34*	0.39±0.07	5.45±1.10*	14.71±1.19*	16.23±4.27*
Trypsin	20.18±3.31	4.94±0.58	0.28±0.06	17.67±3.28	34.94±3.53	35.14±4.93
Control	22.98±2.10	5.86±1.13	0.34±0.05	17.00±2.89	30.71±5.47	29.62±5.26

*$p<0.05$, vs. control.
Data are mean±SD, n = 10 in each group.

content and arrangement [8]. The specimens treated with SDS had a seriously disturbed structure and broken collagen fibers, so their mechanical properties were lower than those of natural AF. The collagen content and arrangement of specimens was similar with Triton X-100 or trypsin and natural AF, for no difference between these 2 groups and natural AF.

We tested the biocompatibility of treated specimens, the most important feature of decellularized scaffolds for tissue engineering. In the decellularization process, a wide variety of chemicals are used, including EDTA, RNase A, and DNase I. If the chemicals remain within the tissue after decellularization, they will be toxic to host cells when the scaffold is implanted *in vivo*. So, we extensively washed specimens in PBS at the end of decellularization to clear any residual reagents and detected the toxicity of scaffolds by MTT and live/dead staining. MTT assay showed that scaffold extracts had no effect on cell proliferation, so the residual reagents were successfully removed. As well, live/dead staining showed that live cells were evenly distributed in the scaffold, with no dead cells, which also inferred that the scaffolds were non-cytotoxic.

Recently, Chan et al. [24] decellularized bovine intervertebral disc as a natural scaffold for intervertebral disc tissue engineering. In his study, a protocol for decellularizing bovine disc was investigated, in which SDS combining with freeze–thaw cycles has been applied, but lots of dead cells remained in the disc after decellularization. As we mentioned above, the decellularization effect of detergents is related to the organization of tissue. Intervertebral disc as a new tissue proposed for decellularized

scaffold should be treated with different detergents to seek the optimal decellularization protocol. In 2011, the optimized decellularization procedure of NP tissue was studied by Mercuri JJ et al. [39]. To determine the optimal decellularization method suitable for AF, 3 protocols were applied in our study, including Triton X-100, SDS combined with freeze–thaw cycles and trypsin. The 3 protocols have been compared in cells removal, ECM content (collagen and GAG), microstructure (SEM) and tensile properties (ultimate load, stress, and strain; toughness; elastic modulus; and mechanical work to fracture). In our study, the concentric lamellar structure before and after decellularization was studied emphatically, for it is critical for withstanding multi-axial physiologic loads for normal function of the spine. We observed concentric lamellar structure of decellularized AF through history staining and SEM. While focus was concentrated on collagen fibril meshwork in Chan's study. Besides, we recellularized AF cells into decellularized AF and observed cell proliferation and viability, which showed a high survival rate over 7 days, with cell penetration. While Chan et al. have focused on recellularization of decellularized NP with bovine NP cells.

Conclusions

This study explored the possibility of using an AF matrix decellularlized with 3 agents as a tissue-engineered AF scaffold material. We compared decellularlized specimens with natural ones for cell removal efficiency, preservation of the matrix components, microstructure and mechanical function. Overall,

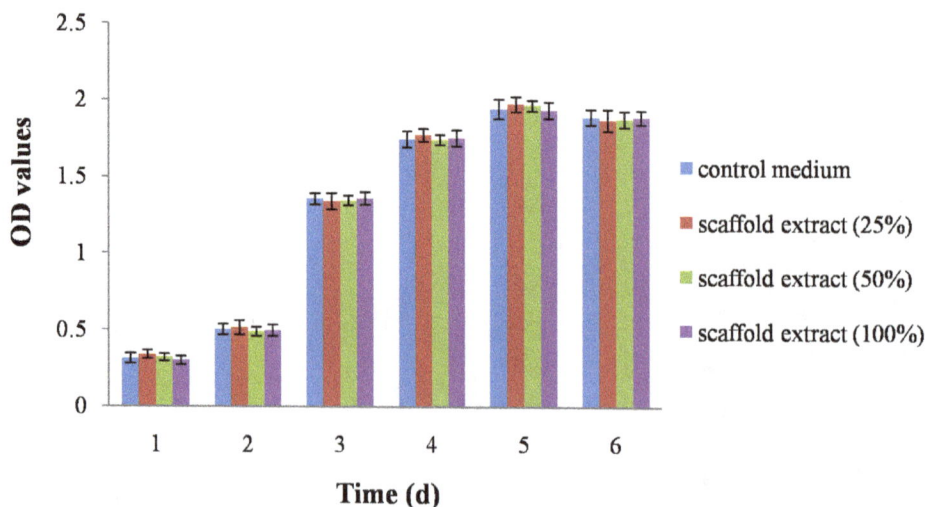

Figure 11. Cytotoxicity of decellularized AF. MTT assay of proliferation of AF cells cultured with different concentrations of scaffold extracts.

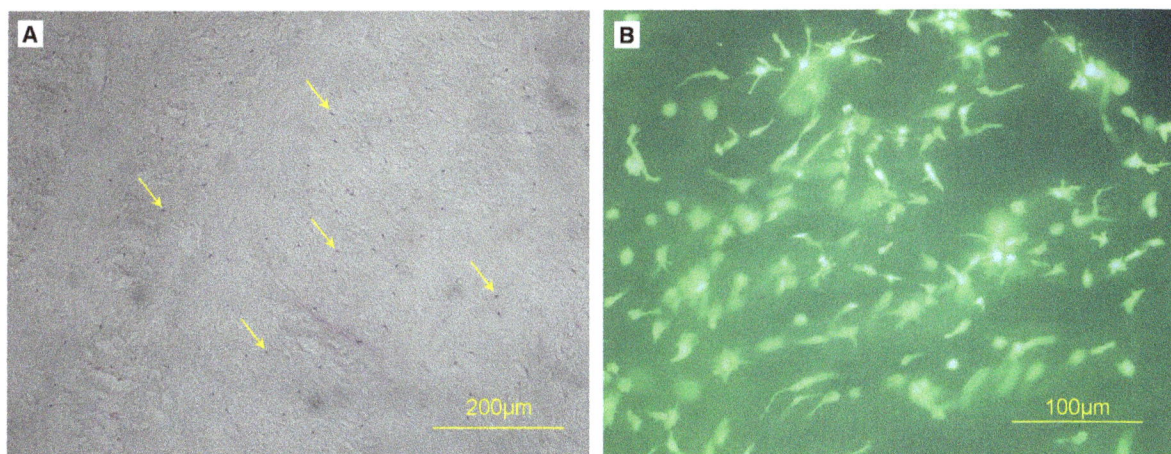

Figure 12. Recellularization of decellularized AF and Evaluation. (A) H&E staining of cell-containing constructs. AF cells (arrows). (B) Live/dead staining of cells seeded into decellularized AF. (Green: viable, red: necrotic).

Triton X-100–treated AF retained the major ECM components after thorough cell removal, preserved the concentric lamellar structure and tensile mechanical properties, and possessed favorable biocompatibility, so AF so treated would be a suitable candidate as a scaffold for AF tissue engineering. An *in vivo* study is still needed to determine whether the novel scaffold could have potential for intervertebral disc tissue engineering.

Acknowledgments

The authors thank technician Bin Zhao for help with the histology.

Author Contributions

Conceived and designed the experiments: HX BX QY. Performed the experiments: HX XL Yang Zhang. Analyzed the data: XM QX CZ. Contributed reagents/materials/analysis tools: YW. Wrote the paper: HX BX QY Yuanyuan Zhang.

References

1. Devereaux M (2009) Low back pain. Med Clin North Am 93: 477–501.
2. Wan Y, Feng G, Shen FH, Laurencin CT, Li X (2008) Biphasic scaffold for annulus fibrosus tissue regeneration. Biomaterials 29: 643–52.
3. Yang Q, Zhao YH, Xia Q, Xu BS, Ma XL, et al. (2013) Novel cartilage-derived biomimetic scaffold for human nucleus pulposus regeneration: a promising therapeutic strategy for symptomatic degenerative disc diseases. Orthop Surg 5: 60–3.
4. Shao X, Hunter CJ (2007) Developing an alginate/chitosan hybrid fiber scaffold for annulus fibrosus cells. J Biomed Mater Res A 82: 701–10.
5. Cassidy JJ, Hiltner A, Baer E (1989) Hierarchical structure of the intervertebral disc. Connect Tissue Res 23: 75–88.
6. Hsu EW, Setton LA (1999) Diffusion tensor microscopy of the intervertebral disc anulus fibrosus. Magn Reson Med 41: 992–9.
7. Eyre DR, Muir H (1976) Types I and II collagens in intervertebral disc. Interchanging radial distributions in annulus fibrosus. Biochem J 157: 267–70.
8. Han WM, Nerurkar NL, Smith LJ, Jacobs NT, Mauck RL, et al. (2012) Multi-scale structural and tensile mechanical response of annulus fibrosus to osmotic loading. Ann Biomed Eng 40: 1610–21.
9. Böer U, Lohrenz A, Klingenberg M, Pich A, Haverich A, et al. (2011) The effect of detergent-based decellularization procedures on cellular proteins and immunogenicity in equine carotid artery grafts. Biomaterials 32: 9730–7.
10. Cheng HL, Loai Y, Beaumont M, Farhat WA (2010) The acellular matrix (ACM) for bladder tissue engineering: A quantitative magnetic resonance imaging study. Magn Reson Med 64: 341–8.
11. Dong SW, Ying DJ, Duan XJ, Xie Z, Yu ZJ, et al. (2009) Bone regeneration using an acellular extracellular matrix and bone marrow mesenchymal stem cells expressing Cbfa1. Biosci Biotechnol Biochem 73: 2226–33.
12. Kheir E, Stapleton T, Shaw D, Jin Z, Fisher J, et al. (2011) Development and characterization of an acellular porcine cartilage bone matrix for use in tissue engineering. J Biomed Mater Res A 99: 283–94.
13. Zhao Y, Zhang Z, Wang J, Yin P, Zhou J, et al. (2012) Abdominal hernia repair with a decellularized dermal scaffold seeded with autologous bone marrow-derived mesenchymal stem cells. Artif Organs 36: 247–55.
14. Stapleton TW, Ingram J, Fisher J, Ingham E (2011) Investigation of the regenerative capacity of an acellular porcine medial meniscus for tissue engineering applications. Tissue Eng Part A 17: 231–42.
15. Liao J, Joyce EM, Sacks MS (2008) Effects of decellularization on the mechanical and structural properties of the porcine aortic valve leaflet. Biomaterials 29: 1065–74.
16. Dong X, Wei X, Yi W, Gu C, Kang X, et al. (2009) RGD-modified acellular bovine pericardium as a bioprosthetic scaffold for tissue engineering. J Mater Sci Mater Med 20: 2327–36.
17. Sun XH, Che YQ, Tong XJ, Zhang LX, Feng Y, et al. (2009) Improving nerve regeneration of acellular nerve allografts seeded with SCs bridging the sciatic nerve defects of rat. Cell Mol Neurobiol 29: 347–53.
18. Du L, Wu X (2011) Development and characterization of a full-thickness acellular porcine cornea matrix for tissue engineering. Artif Organs 35: 691–705.
19. Zhao Y, Zhang Z, Wang J, Yin P, Wang Y, et al. (2011) Preparation of decellularized and crosslinked artery patch for vascular tissue-engineering application. J Mater Sci Mater Med 22: 1407–17.
20. Cebotari S, Mertsching H, Kallenbach K, Kostin S, Repin O, et al. (2002) Construction of autologous human heart valves based on an acellular allograft matrix. Circulation 106: I63–I68.
21. Liu GF, He ZJ, Yang DP, Han XF, Guo TF, et al. (2008) Decellularized aorta of fetal pigs as a potential scaffold for small diameter tissue engineered vascular graft. Chin Med J (Engl) 121: 1398–406.
22. Edwards CA, O'Brien WD Jr (1980) Modified assay for determination of hydroxyproline in a tissue hydrolyzate. Clin Chim Acta 104: 161–7.
23. Farndale RW, Buttle DJ, Barrett AJ (1986) Improved quantitation and discrimination of sulphated glycosaminoglycans by use of dimethylmethylene blue. Biochim Biophys Acta 883: 173–7.
24. Chan LK, Leung VY, Tam V, Lu WW, Sze KY, et al. (2013) Decellularized bovine intervertebral disc as a natural scaffold for xenogenic cell studies. Acta Biomater 9: 5262–72.
25. Gilbert TW, Sellaro TL, Badylak SF (2006) Decellularization of tissues and organs. Biomaterials 27: 3675–83.
26. Badylak SF (2007) The extracellular matrix as a biologic scaffold material Biomaterials. 28: 3587–93.
27. Uriel S, Labay E, Francis-Sedlak M, Moya ML, Weichselbaum RR, et al. (2009) Extraction and assembly of tissue-derived gels for cell culture and tissue engineering. Tissue Eng Part C Methods 15: 309–21.
28. Leor J, Amsalem Y, Cohen S (2005) Cells, scaffolds, and molecules for myocardial tissue engineering. Pharmacol Ther 105: 151–63.
29. Macfelda K, Kapeller B, Wilbacher I, Losert UM (2007) Behavior of cardiomyocytes and skeletal muscle cells on different extracellular matrix components–relevance for cardiac tissue engineering. Artif Organs 31: 4–12.
30. Wagner DR, Lotz JC (2004) Theoretical model and experimental results for the nonlinear elastic behavior of human annulus fibrosus. J Orthop Res 22: 901–9.

31. Nerurkar NL, Elliott DM, Mauck RL (2007) Mechanics of oriented electrospun nanofibrous scaffolds for annulus fibrosus tissue engineering. J Orthop Res 25: 1018–28.

32. Cartmell JS, Dunn MG (2000) Effect of chemical treatments on tendon cellularity and mechanical properties. J Biomed Mater Res 49: 134–40.

33. Kasimir MT, Rieder E, Seebacher G, Silberhumer G, Wolner E, et al. (2003) Comparison of different decellularization procedures of porcine heart valves. Int J Artif Organs 26: 421–7.

34. Ambard D, Cherblanc F (2009) Mechanical behavior of annulus fibrosus: a microstructural model of fibers reorientation. Ann Biomed Eng 37: 2256–65.

35. Schroeder Y, Elliott DM, Wilson W, Baaijens FP, Huyghe JM (2008) Experimental and model determination of human intervertebral disc osmoviscoelasticity. J Orthop Res 26: 1141–6.

36. Yin L, Elliott DM (2005) A homogenization model of the annulus fibrosus. J Biomech 38: 1674–84.

37. Ebara S, Iatridis JC, Setton LA, Foster RJ, Mow VC, et al. (1996) Tensile properties of nondegenerate human lumbar anulus fibrosus. Spine (Phila Pa 1976) 21: 452–61.

38. Skaggs DL, Weidenbaum M, Iatridis JC, Ratcliffe A, Mow VC (1994) Regional variation in tensile properties and biochemical composition of the human lumbar anulus fibrosus. Spine (Phila Pa 1976) 19: 1310–9.

39. Mercuri JJ, Gill SS, Simionescu DT (2011) Novel tissue-derived biomimetic scaffold for regenerating the human nucleus pulposus. J Biomed Mater Res A 96: 422–35.

Tissue Engineered Skin Substitutes Created by Laser-Assisted Bioprinting Form Skin-Like Structures in the Dorsal Skin Fold Chamber in Mice

Stefanie Michael[1]*, Heiko Sorg[1], Claas-Tido Peck[1], Lothar Koch[2], Andrea Deiwick[2], Boris Chichkov[2], Peter M. Vogt[1], Kerstin Reimers[1]

1 Department of Plastic, Hand- and Reconstructive Surgery, Hannover Medical School, Hannover, Germany, **2** Laser Zentrum Hannover e.V., Hannover, Germany

Abstract

Tissue engineering plays an important role in the production of skin equivalents for the therapy of chronic and especially burn wounds. Actually, there exists no (cellularized) skin equivalent which might be able to satisfactorily mimic native skin. Here, we utilized a laser-assisted bioprinting (LaBP) technique to create a fully cellularized skin substitute. The unique feature of LaBP is the possibility to position different cell types in an exact three-dimensional (3D) spatial pattern. For the creation of the skin substitutes, we positioned fibroblasts and keratinocytes on top of a stabilizing matrix (Matriderm®). These skin constructs were subsequently tested *in vivo*, employing the dorsal skin fold chamber in nude mice. The transplants were placed into full-thickness skin wounds and were fully connected to the surrounding tissue when explanted after 11 days. The printed keratinocytes formed a multi-layered epidermis with beginning differentiation and *stratum corneum*. Proliferation of the keratinocytes was mainly detected in the suprabasal layers. *In vitro* controls, which were cultivated at the air-liquid-interface, also exhibited proliferative cells, but they were rather located in the whole epidermis. E-cadherin as a hint for adherens junctions and therefore tissue formation could be found in the epidermis *in vivo* as well as *in vitro*. In both conditions, the printed fibroblasts partly stayed on top of the underlying Matriderm® where they produced collagen, while part of them migrated into the Matriderm®. In the mice, some blood vessels could be found to grow from the wound bed and the wound edges in direction of the printed cells. In conclusion, we could show the successful 3D printing of a cell construct *via* LaBP and the subsequent tissue formation *in vivo*. These findings represent the prerequisite for the creation of a complex tissue like skin, consisting of different cell types in an intricate 3D pattern.

Editor: Andrzej T. Slominski, University of Tennessee, United States of America

Funding: This study has been supported by Deutsche Forschungsgemeinschaft, SFB TransRegio 37 and Rebirth Cluster of Excellence (Exc 62/1). The funders had no role in study design, data collection and analysis, decision to publish, or preparation of the manuscript.

Competing Interests: The authors have declared that no competing interests exist.

* E-mail: Michael.Stefanie@mh-hannover.de

Introduction

Major burn injuries often prove difficult in therapy due to their complexity, the high risk of infection, the large area which might be affected and the potential destruction of deeper skin layers including the dermis. Often, the availability of autologous split-thickness skin grafts and keratinocytes for wound coverage is limited, especially in case of large burned areas. Therefore, the need of skin substitutes for temporary or permanent wound coverage is high. Several skin substitutes like Integra® and Matriderm® are already employed in the clinical application, being complemented by the use of autologous split-thickness skin grafts [1–3]. While Integra® serves to prepare the wound bed in preparation for transplantation with autologous split-thickness skin three weeks later, Matriderm® is used in a single step procedure and must be covered immediately. Nevertheless, full success in burn wound regeneration has not been reached yet, neither under functional nor under aesthetic aspects. In nearly every case of treating large and deep burn injuries discolouring or scarring remains, the latter leading to undesirable contractions. Also, neither hair follicles nor sebaceous and perspiratory glands can be regenerated.

Tissue engineering promises to have high potential in the production of new skin. In this context, it remains a challenge to create a precise and complex new tissue comprising several cell types which are arranged in a specific 3D pattern. Furthermore, the different tissue functions strongly depend on its specific structure and on the cells which are influenced by their distinct microenvironment [4]. For example, the formation of vessels in a skin equivalent cultivated *in vitro* is thought to be dependent on the direct interaction of endothelial cells with fibroblasts and their secreted extracellular matrix proteins and growth factors [5]. Bioreactors are used for the *in vitro* cultivation of complex tissues offering the possibility to mimic and control the desired microenvironment [6].

One solution for the problem of creating complex 3D tissues might be the use of LaBP. It offers the possibility to produce specific high resolution two-dimensional (2D) as well as 3D patterns, incorporating different cell types like human osteosarcoma and mouse endothelial cells [7], human osteoprogenitor cells [8], rodent olfactory ensheathing cells [9], human endothelial cells

[10] and human adipose derived mesenchymal stem cells which can subsequently be differentiated to fat [11] as well as bone and cartilage [12]. Cells – including rat Schwann and astroglial cells, pig lens epithelial cells [13], Chinese hamster ovarian cells, human osteoblasts [14], murine embryonal carcinoma cells [15], and fibroblasts and kerationcytes [16] – survive the transfer without damage and alteration of cell phenotype. This represents a major prerequisite for the use of LaBP in tissue engineering. Commonly, also the terms cell printing or simply bioprinting are used. In advance of the *in vivo* testing of the here produced skin substitutes we could already show tissue formation and functional cell-cell contacts in corresponding 3D tissue constructs *in vitro* [17].

In this study, *via* the use of LaBP, we created a multi-layered, fully cellularized skin equivalent for the future treatment of burn patients. The transplanted skin equivalent was tested *in vivo* for its ability to form tissue as well as cellular behaviour of the printed cells, the differentiation of the keratinocytes and potential neovascularisation using the dorsal skin fold chamber in mice. *In vitro* controls supplemented the *in vivo* experiments.

Materials and Methods

Cell Culture

NIH3T3 fibroblasts (DSMZ, Braunschweig, Germany) and HaCaT keratinocytes (DKFZ, Heidelberg, Germany) have previously been labelled by stable transduction with lentiviral or gammaretroviral vectors encoding for either eGFP or mCherry [17]. In the following the four resulting cell lines are named accordingly: NIH3T3-eGFP, NIH3T3-mCherry, HaCaT-eGFP and HaCaT-mCherry. Fibroblasts were cultivated in Dulbecco's modified Eagle's medium (DMEM) with high glucose (4.5 g/L) (PAA, Pasching, Austria) supplemented with 10% fetal bovine serum (FBS) (Biochrom, Berlin, Germany), 1% of 100 mM sodium pyruvate (Biochrom), and 1% of penicillin/streptomycin (Biochrom) whereas keratinocytes were grown with DMEM/Ham's F12 medium (PAA) supplemented with 10% FBS and 1% of penicillin/streptomycin.

Cell Transfer and Production of the Transplant

Cells were arranged in 3D skin constructs using LaBP (as previously described in [16] [18–19]). Briefly, the setup consists of two co-planar glass slides. The upper one is coated with a thin layer of laser absorbing material (here 60 nm of gold) and a layer of biomaterial to be transferred (here 60 µm of cell containing collagen). This glass slide is mounted upside-down above a second (receiver) glass slide. The laser pulses are focused through the upper glass slide into the laser absorbing layer, which is evaporated locally. The vapor pressure propels a small amount of the subjacent biomaterial towards the receiver glass slide. By moving the glass slides relative to each other, arbitrary patterns of biomaterial can be produced. By repeating this procedure layer-by-layer also 3D patterns can be generated.

For the here presented skin substitutes the cells were trypsinized and centrifuged at 400 g. A pellet containing 1.5 million cells was resuspended in a mixture of 37 µl collagen (Collagen Type I, Rat Tail, BD Biosciences, Bedford, MA, USA), 5 µl phosphate buffered saline (10× PBS, Biochrom) and 0.85±0.5 µl sodium hydroxide (1N NaOH, Sigma-Aldrich) for neutralization (pH = 7.1±0.3) prior to the transfer. For the skin substitutes 20 layers of fibroblast-containing collagen and 20 layers of keratinocyte-containing collagen were printed subsequently onto a sheet of Matriderm® (2.3 cm×2.3 cm, Dr. Suwelack Skin & Health Care, Billerbeck, Germay), used as a stabilization matrix.

The printed cells were kept in the incubator under submerged conditions over night. The next day (defined as day 0), nine round pieces (diameter 6.0 mm) were removed from the large construct with a biopsy punch, three of which were implanted into the skin fold chambers *in vivo* (one per mouse). As four independent printing processes were conducted, altogether 12 animals were used. The remaining six pieces of each printing process served as *in vitro* controls. Two of them were directly fixed on day 0 to depict the situation at the beginning of the experiments, whereas the remaining four pieces were raised to the air-liquid-interface. *In vitro* controls were then fixed on days 5 and 11 (duplicates per time point) and *in vivo* specimen on day 11.

Cultivation of Constructs in vitro

Constructs were raised to the air-liquid-interface and cultivated with differentiation medium on top of plastic platforms. The latter consisted of cell strainers (BD Biosciences, Bedford, MA, USA) turned upside down. The medium was composed of DMEM high glucose medium (PAA) mixed with the same amount of DMEM/

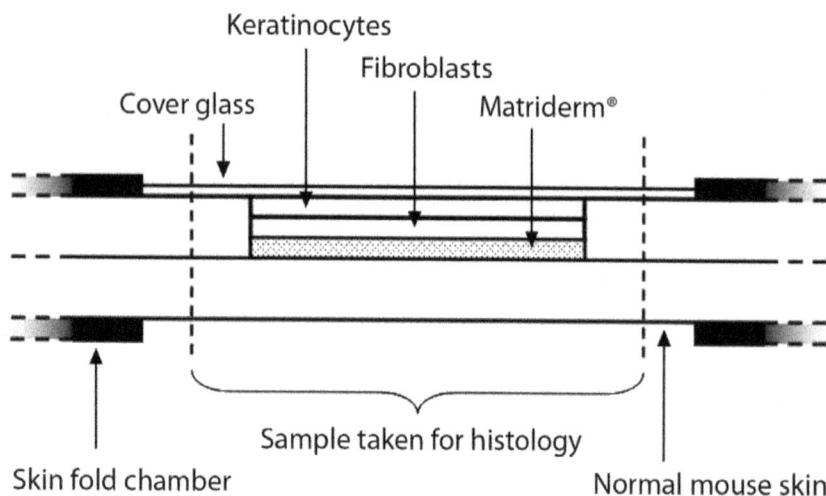

Figure 1. Scheme of the utilised dorsal skin fold chambers in mice. The chambers are attached to the back skin of the mice. The printed skin construct consisting of 20 layers of fibroblasts and 20 layers of keratinocytes on top of Matriderm® is placed into a round full-thickness wound in the mouse skin, while the opposite side remains intact. To close the chamber a cover glass is used.

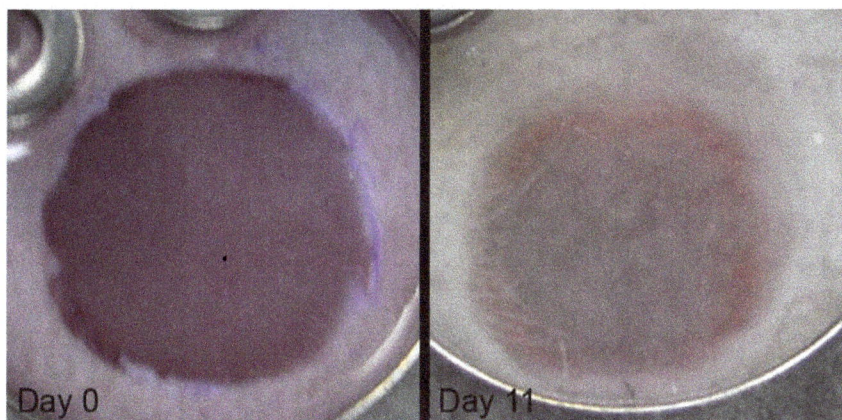

Figure 2. Tissue engineered skin construct in the dorsal skin fold chamber in nude mice. The pictures show a skin construct inserted into the wound directly after the implantation (left) and on day 11 (right). The implanted constructs were created *via* LaBP, consisting of 20 layers of fibroblasts and 20 layers of keratinocytes on top of Matriderm®. They fill the full-thickness wound completely.

Ham's F12 medium (PAA), supplemented with 1% FBS, 1% penicillin/streptomycin, 10^{-7} mM isoprenaline hydrochloride (Sigma-Aldrich, Steinheim, Germany), 10^{-7} mM hydrocortisone (Sigma-Aldrich) and 10^{-7} mM insulin (insulin bovine pancreas, Sigma-Aldrich). The resulting calcium chloride concentration of the culture medium is 158.3 mg/L.

Animals and Ethics Statement

All animal experiments were evaluated and approved by the responsible animal care committee (Nds. Landesamt für Verbraucherschutz und Lebensmittelsicherheit) and the Hannover Medical School (Institut für Versuchstierkunde). The animals (male BALB/c-Nude mice, 8 weeks) were purchased from Charles River and kept in the local animal care facility according to the institution guidelines. They received standardized food and water, living with a day-night cycle of 12 hours each. Animals were used for the experiments when they were at least 12 weeks old.

Skin Fold Chamber

The dorsal skin fold chamber was used for the evaluation of tissue engineered skin *in vivo* as published previously [20]. The skin constructs were placed into full-thickness wounds, while the skin on the other side of the chamber remained intact. All surgery was performed under isoflurane anesthesia, and all efforts were made to minimize suffering. The mice were sacrificed on day 11 after implantation, preparing the constructs surrounded by normal skin for histological analyses (Figure 1). Analogous constructs - without cells - have already been assessed *in vivo* [20] and are used as a comparison in this study.

Histology

Samples were fixed in 4% paraformaldehyde, embedded in paraffin and subsequently cut to sections of 5 μm thickness. Masson's trichrome stainings were conducted following standard procedures.

Immunohistochemistry

In order to detect the presence of e-cadherin, collagen IV, cytokeratin 14 and Ki67, immunohistochemistry was carried out. For Ki67 (Thermo Fisher, RM9106_S0, 1:200) and e-cadherin staining (Santa Cruz, SC 7870, 1:300) the deparaffinised and rehydrated paraffin sections were incubated in a 99°C heated water bath for 25 min (Ki67) and 15 min (e-cadherin), respectively, before blocking. For cytokeratin 14 (Biozol, DBB-DB099-1, 1:300) and collagen IV (Abcam, Ab6586, 1:2000) stainings the sections were incubated in a 37°C warm water bath for 10 min, using 100 ml 0.2 N HCl solution with 100 mg pepsin. All sections were blocked with 2% FBS in PBS at room temperature for 30 min, followed by incubation with first antibody in 1% FBS in PBS at 4°C over night. After washing with PBS for all but the collagen IV staining a goat anti rabbit secondary antibody (Sigma-Aldrich, A3687, 1:1000) coupled to an alkaline phosphatase was employed. The samples were incubated at 37°C for 1 h. As a substrate nitro-blue tetrazolium-5-bromo-4-chloro-3'-indolyphosphate (Roche, Mannheim, Germany) was used. In case of the collagen IV primary antibody, a biotin coupled secondary anti rabbit antibody (Dako, E0432, 1:400) was used for 90 min at room temperature. Subsequently, the Vectastain ABC Kit (Vector Laboratories, Peroxidase Standard PK-4000) was employed for 60 min at room temperature to enhance the signal. The latter was then visualised with diaminobenzidine tetrahydro chloride (ICN98068). Staining of all antibodies was detected with a light microscope (Olympus).

Results

Operations and Macroscopic Evaluation

To evaluate skin constructs *in vivo* generated *via* LaBP, we printed 20 layers of a keratinocyte cell line (HaCaT) on top of 20 layers of a fibroblast cell line (NIH3T3) by an established laser-assisted bioprinting procedure. Using stable transduction, cell lines were labelled with genes encoding for green or red fluorescent proteins, respectively: NIH3T3-eGFP, NIH3T3-mCherry, HaCaT-eGFP and HaCaT-mCherry. Matriderm® was used as a carrier matrix to enhance stability of the constructs for transplantation (Figure 1). The skin constructs were placed into full-thickness skin wounds in the dorsal skin fold chamber preparation in mice in such a way that the constructs and the surrounding skin laid in close contact to each other. Uninjured skin from the opposite side of the back fold served as a control for all experiments. Analogous constructs without the cells [20] serve as a comparison to this study.

All animals survived the surgical intervention and implantation procedure and tolerated the chambers well, showing no signs of discomfort or changes in sleeping and feeding habits. After 11

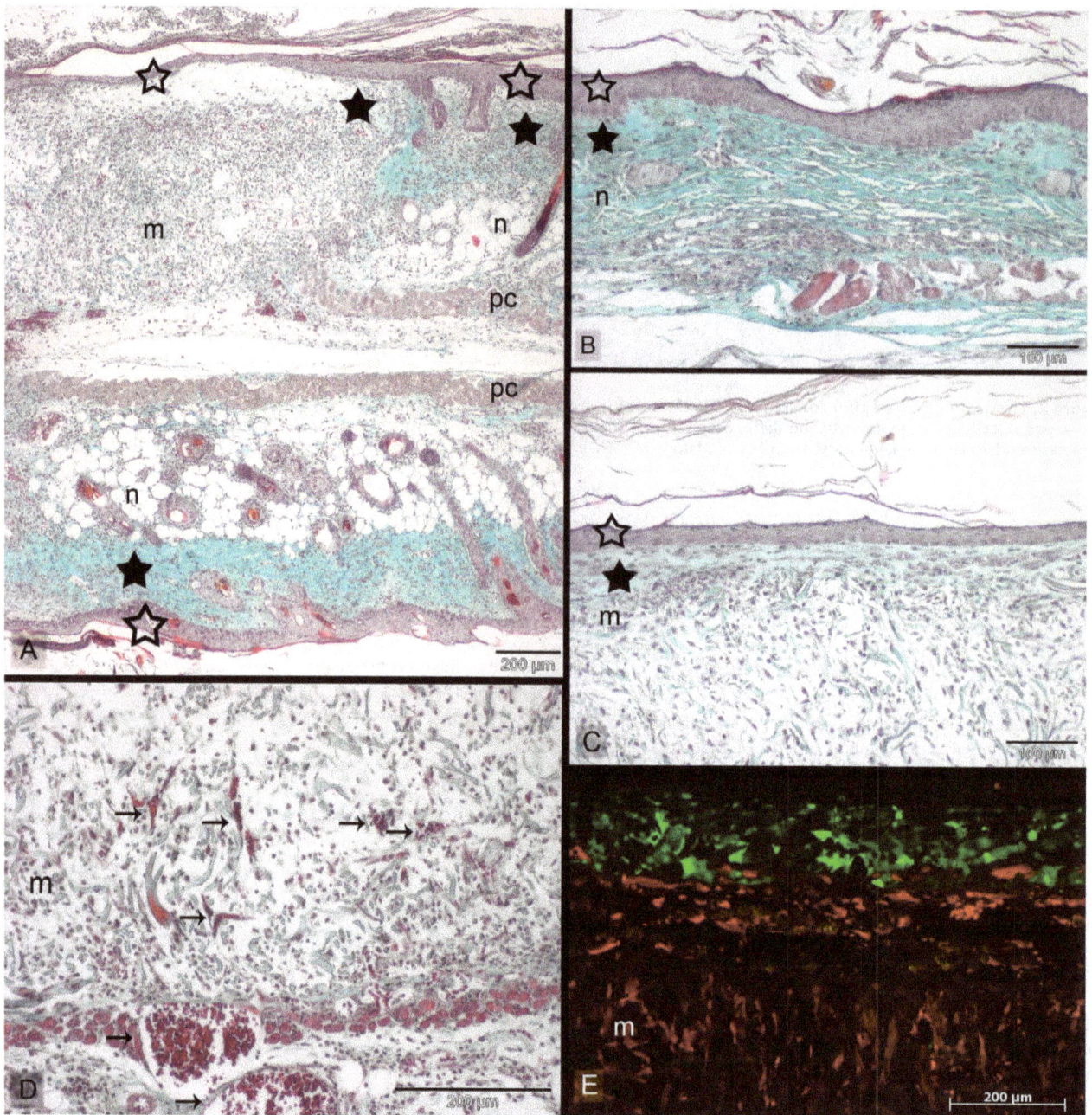

Figure 3. Histological sections of the tissue engineered skin constructs *in vivo.* Skin constructs were implanted in dorsal skin fold chambers in mice for 11 days. Sections were stained with Masson's trichrome (A–D) or analyzed with fluorescence microscopy (E), respectively. (A) illustrates an overview with the junction between the inserted skin construct (m = Matriderm®) and native mouse skin (n) at the wound edge after 11 days in the dorsal skin fold chamber in mice. The intact mouse skin opposite of the skin construct can be seen in the lower part of the picture (n) (see also Figure 1). The skin construct and the intact skin part in the sandwiched skin are separated by the *panniculus carnosus (pc)*. Both in native mouse skin (B) and the printed skin construct (C) a dense epidermis (empty asterisks) and a corneal layer can be observed. In case of the skin construct, the epidermis is formed by the printed keratinocytes (E). This can clearly be seen by the green fluorescence emitted by the used HaCaT-eGFP cells. The fibroblasts (NIH3T3-mCherry) partly migrated into the Matriderm® (yellowish fibres). The fibroblasts, which stayed on top of the Matriderm®, display an outstretched morphology (C), being accompanied by collagen deposition (filled asterisks). Blood vessels (arrows) can be detected in the skin constructs (D). Scale bars depict 200 μm (A, D, E) and 100 μm (B, C).

days, the borders of the skin constructs and the surrounding mouse skin were tightly grown together so that no sharp linings were visible between the two tissue types any more (Figure 2). During the course of time, the previously shining surface of the constructs became matt. Neither inflammatory/necrotic processes nor contraction of the wounds could be detected.

Formation of Skin-like Tissue in vivo by Printed Skin Constructs

First of all, the survival and tissue formation of the printed skin cells was of particular interest. On top of the fibroblasts and the Matriderm®, the keratinocytes (HaCaT-eGFP) developed a dense stratified tissue (Figure 3E), similar to normal epidermis as can be

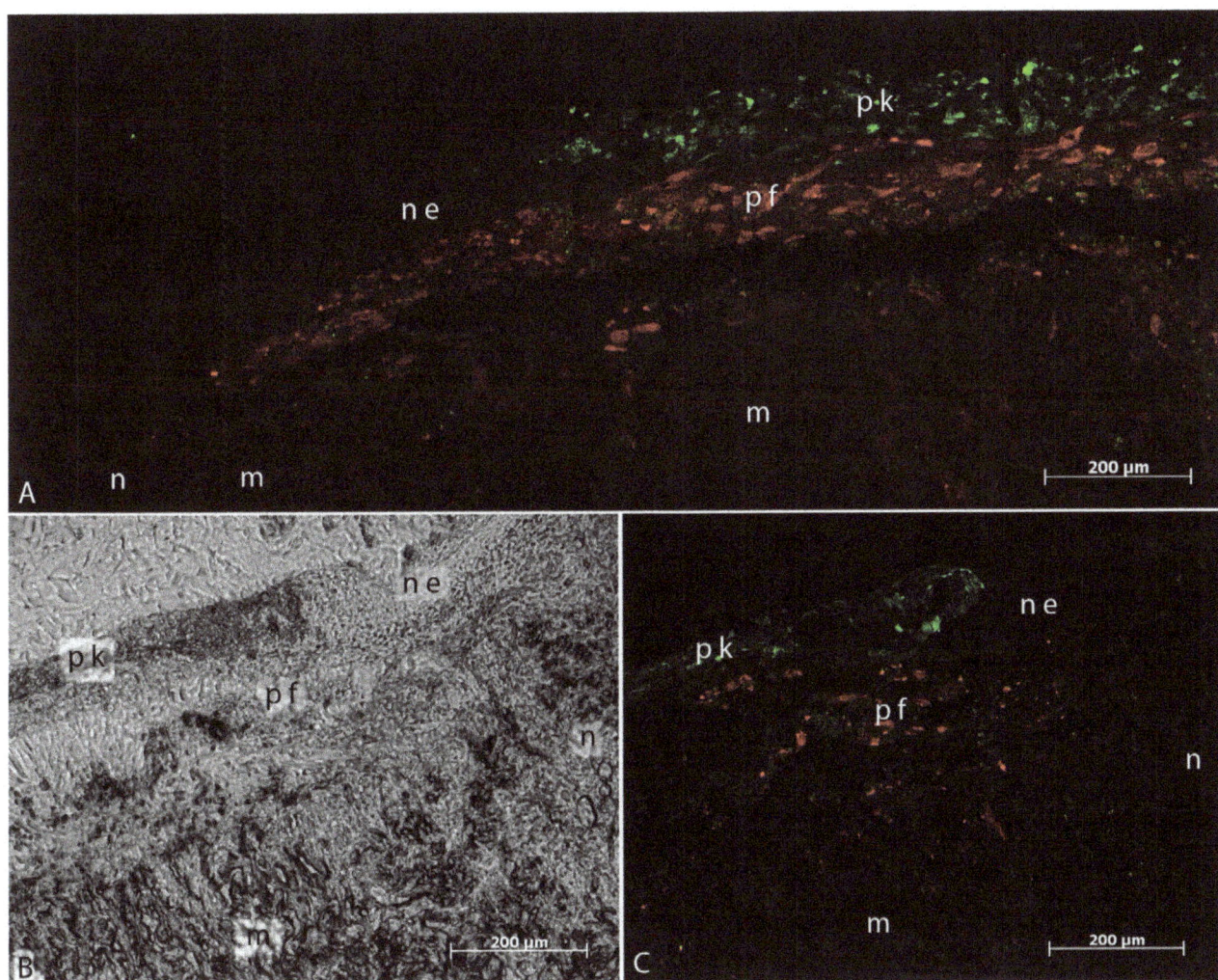

Figure 4. Fluorescent pictures of the tissue engineered skin constructs *in vivo*. Skin constructs were implanted in dorsal skin fold chambers in mice for 11. The sections show an overview of the junction between the inserted skin construct (m = Matriderm®) and native mouse skin (n) with either fluorescence microscopy (A, C) or transmitted light microscopy (B). Two different situations concerning the epidermis were observed during analysis of the junction zones: In some cases, as depicted in (A), the normal mouse epidermis (ne) started to grow on top of the Matriderm®, where it connected to the epidermis formed by the printed keratinocytes (pk). The latter were labelled in green by stable transduction (HaCaT-eGFP). In other cases, as depicted in (B) and (C), the epidermis formed by the printed keratinocytes ended at the border of the Matriderm®, synchronous to the presence of the printed fibroblasts (pf) labelled in red (NIH3T3-mCherry). As can be seen by comparing (B) and (C) - which depict the same location - the keratinocytes even partly grew on top of the normal mouse epidermis. In both cases, the printed fibroblasts formed a multi-layer tissue underneath the printed keratinocytes. Partly, they also migrated into the Matriderm®. All scale bars depict 200 µm.

seen in the Masson's trichrome staining (Figure 3A, C). In some samples, this was followed by a corneal layer (Figure 3A, C). However, the epidermis in the skin constructs was less thick than in native mouse skin. Besides, no rete ridges could be found in the skin constructs. After 11 days the tissue developed by the printed cells was connected to the surrounding native mouse skin tissue at the wound edges. Neither an interruption of the epidermis nor a gap between the dermis and the Matriderm® could be observed at the junction between the skin constructs and the mouse skin (Figure 3A, Figure 4). In this context, two different situations could be observed: In some cases, the normal mouse epidermis started to grow on top of the Matriderm®, becoming connected to the epidermis which was formed by the printed keratinocytes (Figure 4A). In other cases, the epidermis formed by the printed keratinocytes ended simultaneously with the printed fibroblasts (NIH3T3-mCherry), directly at the border of the Matriderm®.

Partly, the keratinocytes even grew on top of the normal mouse epidermis (Figure 4B, C).

Furthermore, the migration pattern of the printed fibroblasts was assessed. As can be seen in the histological sections, the fibroblasts partly migrated into the Matriderm®, closely following the fibres of the latter (Figure 3E, Figure 4). Some fibroblasts remained on top of the Matriderm®, composing a multi-layer sheet of tissue (Figure 3C) and secreting collagen as can be seen in the trichrome staining. As such, the printed skin cells survived well and formed a multi-layer, keratinized skin equivalent.

One important issue in respect of the use of skin substitutes is their vascularisation. Here, small blood vessels could be found in the skin constructs which seem to grow in from the depth of the wound bed as well as from wound edges into the Matriderm® in the direction of the printed cell layers (Figures 3D and 5).

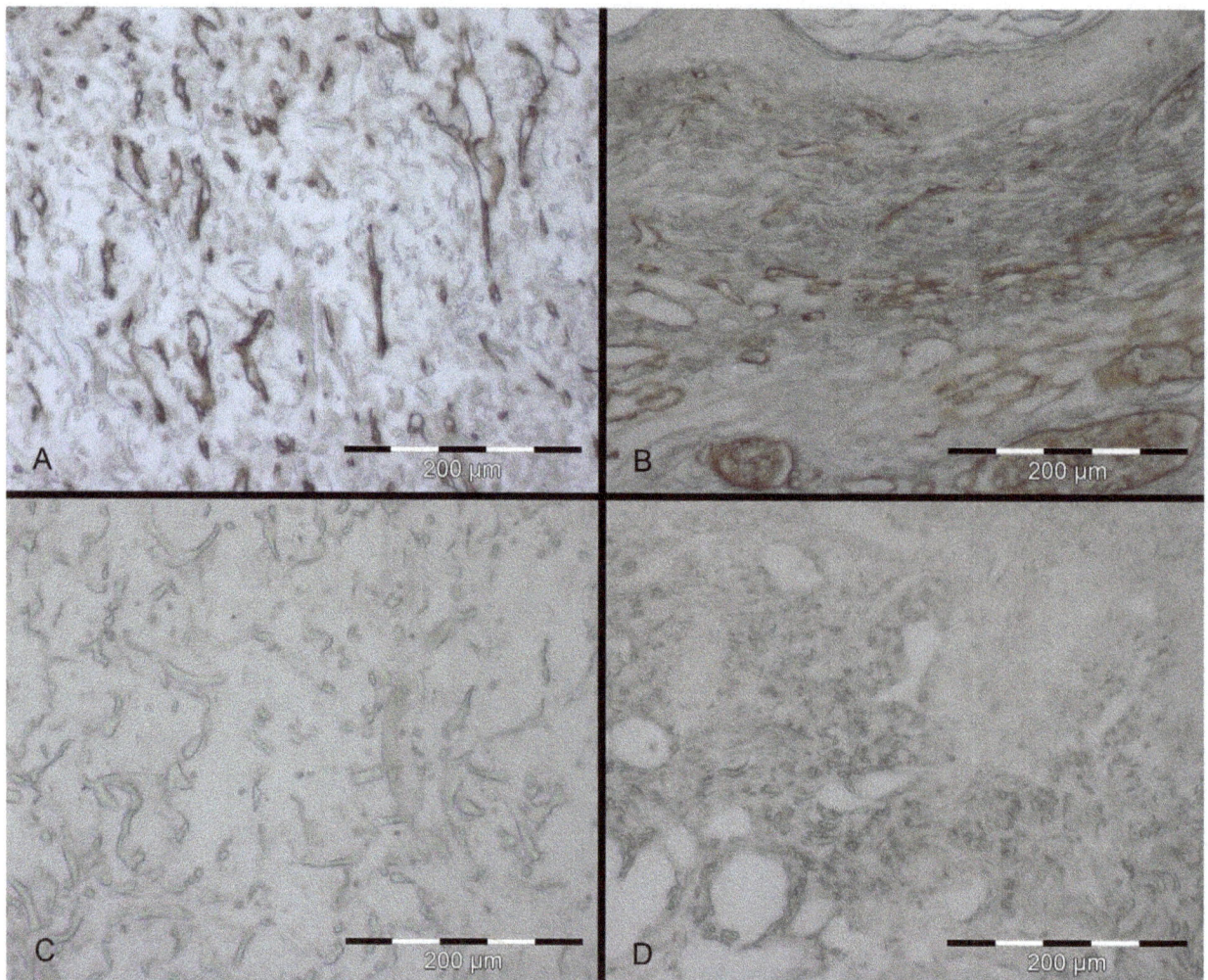

Figure 5. Blood vessel detection *via* immunohistochemistry in skin constructs cultivated *in vivo* for 11 days. Skin constructs were cultivated *in vivo* for 11 days in the dorsal skin fold chamber in mice. Collagen IV expression (brown) – indicating blood vessels/capillaries – can be detected in the Matriderm® as small tubes reaching from the wound bed in the direction of the cells (A). Small and large blood vessels are present in the normal mouse skin (B). Matriderm® (C) and normal mouse skin (D) without first antibody serve as the respective negative controls. Scale bars depict 200 μm each.

Adherens junctions – containing especially e-cadherin - are essential for stable cell-cell contact and can abundantly be found in epithelia like skin. Therefore, e-cadherin can be used as a hint for epithelia formation and consequently for tissue development. By means of immunostaining, e-cadherin could be detected between the keratinocytes of the skin constructs inserted into the wounds. Here, the pattern of the e-cadherin localisation is the same as in native skin and can be found in the whole epidermis (Figure 6).

One important characteristic of an epidermis is the differentiation of the keratinocytes. Cytokeratin 14 is a marker for undifferentiated keratinocytes. The corresponding immunostaining revealed the presence of cytokeratin 14 in the whole epidermis of the skin constructs (Figure 7A–C). In native skin cytokeratin 14 staining could be found in only the suprabasal layers.

In contrast, Ki67 as a proliferation marker, showed a signal mainly in the suprabasal layer of the skin constructs (Figure 7D–F), indicating that only those keratinocytes maintained their proliferating state. Note that in the skin constructs only a few cells showed a positive signal for Ki67 whereas in the normal mouse skin nearly all suprabasal cells were stained. Fibroblasts in the dermis showed proliferation in the skin constructs as well as in the native mouse skin.

In vitro Controls of Printed Skin Constructs

As a control, printed skin constructs were also cultivated *in vitro* with the addition of differentiation medium. Samples were taken for histology on day 0-at the same time point as the corresponding transplants were inserted into the chambers of the mice – to demonstrate the starting conditions of the experiments. Further on, samples were also secured on day 11, corresponding to the end of the *in vivo* experiments and in between (on day 5).

On day 0 the two multi-layers of different cell types on top of the Matriderm® could clearly be seen (Figure 8A, D, G). The keratinocytes were still round and embedded in the collagen gel without contact to each other and with quite large spacing between the cells. In contrast, the fibroblasts already began to stretch out and to migrate into the Matriderm®. While the keratinocytes formed a dense tissue during the course of time, part of the fibroblasts migrated into the Matriderm®, following the

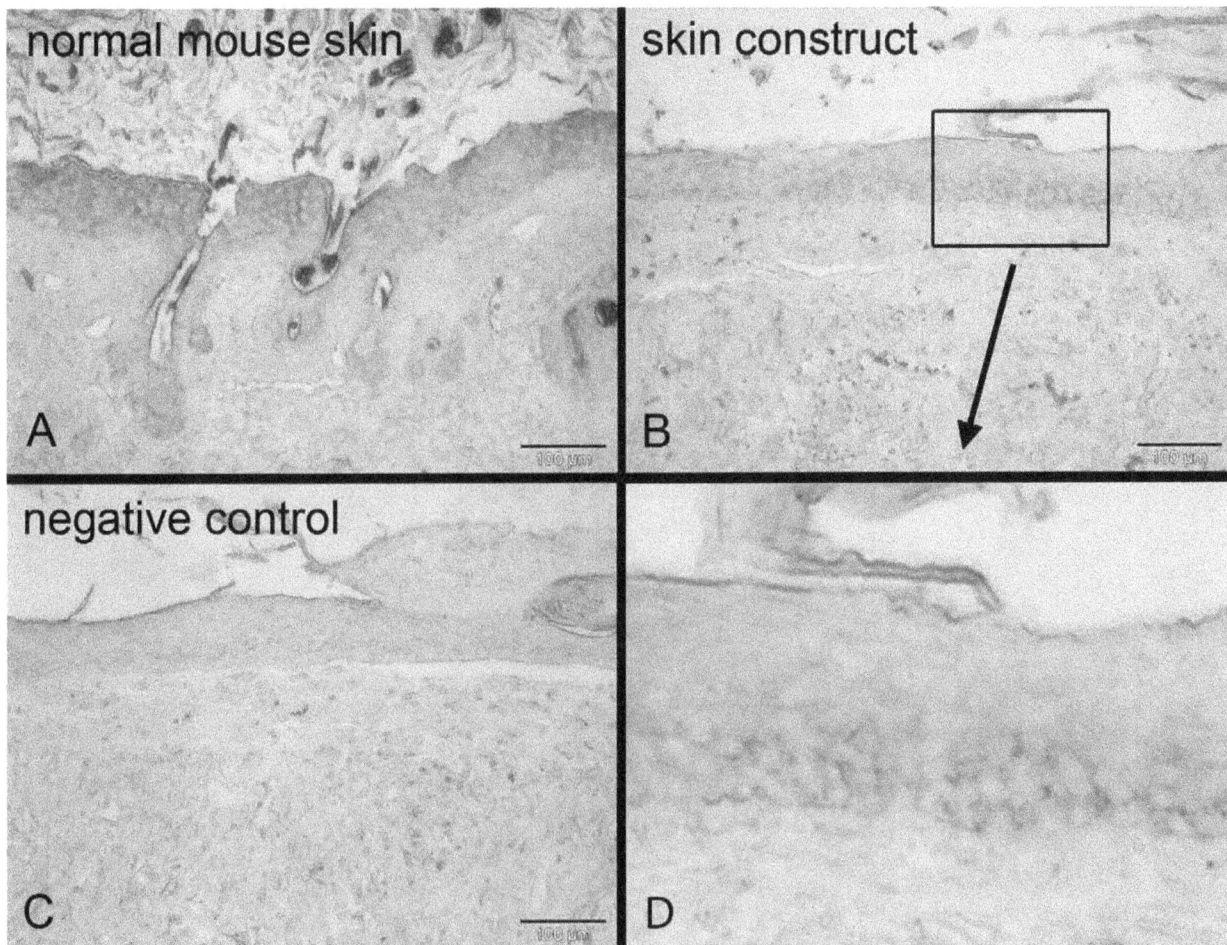

Figure 6. E-cadherin detection *via* immunohistochemistry in skin constructs cultivated *in vivo* for 11 days. Skin constructs were cultivated *in vivo* for 11 days in the dorsal skin fold chamber in mice. E-cadherin expression (dark brown) can be found throughout the epidermis in both normal mouse skin (A) and the skin constructs (B). Normal mouse skin without first antibody serves as a negative control (C). All scale bars depict 100 μm.

fibres closely (Figure 8, brown cells beside green Matriderm® fibres, especially clear in Figure 8H and I).

A thickening of the epidermis-like tissue formed by the keratinocytes could be observed comparing day 11 to day 5 (Figure 8). Surprisingly, trichrome staining revealed the presence of a horizontal line of collagen in the lower part of the epidermis-like tissue and some globular collagen accumulations in the upper part. These probably are the remnants of the collagen used for the printing process.

Also in the *in vitro* cultures, immunostaining was carried out. The presence of e-cadherin increased during time from none on day 0 to its detection in the whole epidermis-like tissue on day 11, (Figure 9A–C). This is in accordance with the trichrome stainings, where a dense tissue can be observed on days 5 and 11 (Figure 8).

Also *in vitro*, cytokeratin 14 could be detected evenly distributed throughout all layers of the epidermis-like tissue (Figure 9D–F). This indicates a lack of differentiation. Proliferation could be found in all examined stages of the *in vitro* cultures (Figure 9G–I). While nearly all of the just printed cells showed proliferation on day 0, only some of the cells did so on days 5 and 11. But in contrast to the *in vivo* situation, the proliferating cells were found in the whole epidermis-like structure, more or less evenly distributed. No special spatial pattern could be observed.

Discussion

The development of newly generated skin substitutes for burn therapy is very important. Here, we present the *in vivo* assessment of a simple skin equivalent created *via* LaBP. The printed cells form a tissue which is quite similar to native skin, including collagen producing fibroblasts and presumably differentiating keratinocytes, forming a dense epidermis. Although 11 days of cultivation is too short for a complete differentiation of keratinocytes, the distribution of Ki67 (as a marker for proliferation) mainly in the suprabasal layers hints at the beginning differentiation of the keratinocytes. In native skin, only the keratinocytes in the *stratum basale* maintain proliferation, whereas the differentiating keratinocytes in the other skin layers cease proliferating. Furthermore, the Matriderm® carrier becomes populated by the printed fibroblasts (presented in this work) as well as murine host fibroblasts (presented in previous work [20]). This leads to the integration and ingrowth of the skin construct into the wound. However, the absence of rete ridges and the thinner epidermis in the skin constructs may result in less stability of the constructs compared to native mouse skin. This may be solved by printing rete ridges and a thicker epidermis in future experiments, though.

As skin is a complex organ consisting of different cell types and substructures arranged in defined spatial patterns, LaBP is suited

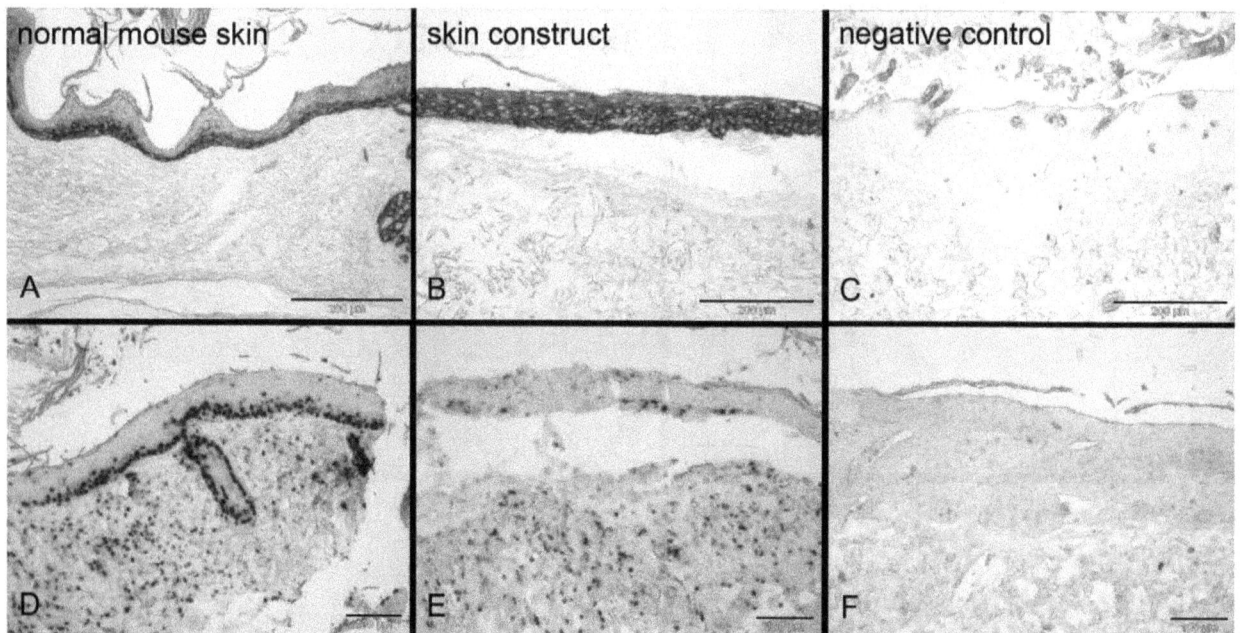

Figure 7. Detection of cytokeratin 14 and proliferation *via* immunohistochemistry in skin constructs cultivated *in vivo*. Skin constructs were cultivated *in vivo* in the dorsal skin fold chamber in mice for 11 days. The left column shows normal mouse skin, the middle column the skin construct and the right column the respective negative controls (normal mouse skin without first antibody) of the immunohistochemistry stainings. Cytokeratin 14 expression is limited to the suprabasal layers of the epidermis in mouse skin (A) but present in the whole epidermis in the skin constructs (B). Proliferation *via* Ki67 can be detected in the suprabasal layers of the epidermis and in the dermis in both normal mouse skin and skin constructs (D, E). Scale bars depict 200 μm (A–C) and 100 μm (D–F).

for the production of tissue engineered skin substitutes. It offers many possibilities and is a very promising technique for the fabrication of other kinds of tissues as e.g. bone or cartilage [8] [10]. [12]. Different levels of tissue generation have been investigated using bioprinting techniques. As we have previously shown *in vitro*, printed epidermal cells develop a dense epidermis including the expression of adherens and gap junctions [17]. In this manuscript, the next step has been carried out and tissue formation also *in vivo* could be documented. In a different setting, a pie-shaped multi-layered construct produced by inkjet printing (see below) and consisting of different cell types (stem cells, smooth muscle cells, endothelial cells) has already been analysed *in vivo* as a technical prerequisite to develop vascularised bone tissue in the future. The implanted cells could be detected several weeks after implantation indicating good survival rates. Interestingly, the used stem cells were able to differentiate into bone *in vivo* and endothelial cells formed a network of blood vessels in the implants after six weeks [21].

Concerning maintenance of cellular phenotype, human chondrocytes were found to express cartilage specific genes after being printed into cartilage lesions by inkjet printing and being cultivated *ex vivo*. They maintained their deposited positions due to simultaneous photopolymerization of a surrounding biomaterial scaffold, and attached firmly to the enclosing cartilage tissue [22]. These findings are in accordance with our own experiences [16] and that of others [7] [9] [13–14] as LaBP does not seem to impair cellular phenotype and behaviour. Even the differentiability of transferred mesenchymal stem cells [11–12] and pluripotent murine embryonal carcinoma cells [15] is maintained.

In the *in vitro* samples, a horizontal stripe of collagen without cells could be observed in the lower part of the epidermis-like structure. Obviously, the fibroblasts did not degrade the circumjacent collagen - which has been used as a hydrogel during the

printing process - but started to migrate into the Matriderm® right away. During the printing, the cells are mixed with a hydrogel (in general called printing matrix) to serve four different purposes. Firstly, the matrix is necessary to achieve a uniform coverage of the donor glass slide with the biomaterial to be transferred. This is important as only then a consistency and uniformity of the printing process is possible. Secondly, the matrix helps the cells to survive, providing a moist environment preventing drying. Thirdly, it presents a specific surrounding for the printed cells and thereby acts as a biomimetic gel to create the desired micro-environment. In our case, collagen is already present in physiological skin and therefore is very suited as a printing matrix. Fourthly, a matrix like collagen enables the formation of a 3D construct due to its gelling effect. While in the *in vitro* controls the collagen is left by the fibroblasts, in the *in vivo* situation outstretched fibroblasts can be found in the collagen, probably also expressing and producing collagen by themselves. As this is similar to the situation in physiological skin, this process is much desired.

Very important for the take of a grafted skin or skin substitute is its fast vascularisation. It is a major prerequisite for the successful clinical use of a skin substitute. In our experiments, blood vessels could be found to start growing into the Matriderm® from the wound bed and the wound edge mostly in the direction of the transplanted cells. In a previous study, in skin constructs consisting of Matriderm® covered with collagen type I but without cells [20] no vessels could be detected growing into the Matriderm®. This suggests that the neovascularisation might be induced or supported by the printed cells on top of the Matriderm®. Actually, keratinocytes were found to produce vascular endothelium growth factor (VEGF) [23], the expression of which is regulated by several growth factors and cytokines [24] as well as by insulin in a diabetic mouse model [25]. Furthermore, keratinocytes in a tissue engineered skin substitute regulate the size of newly growing

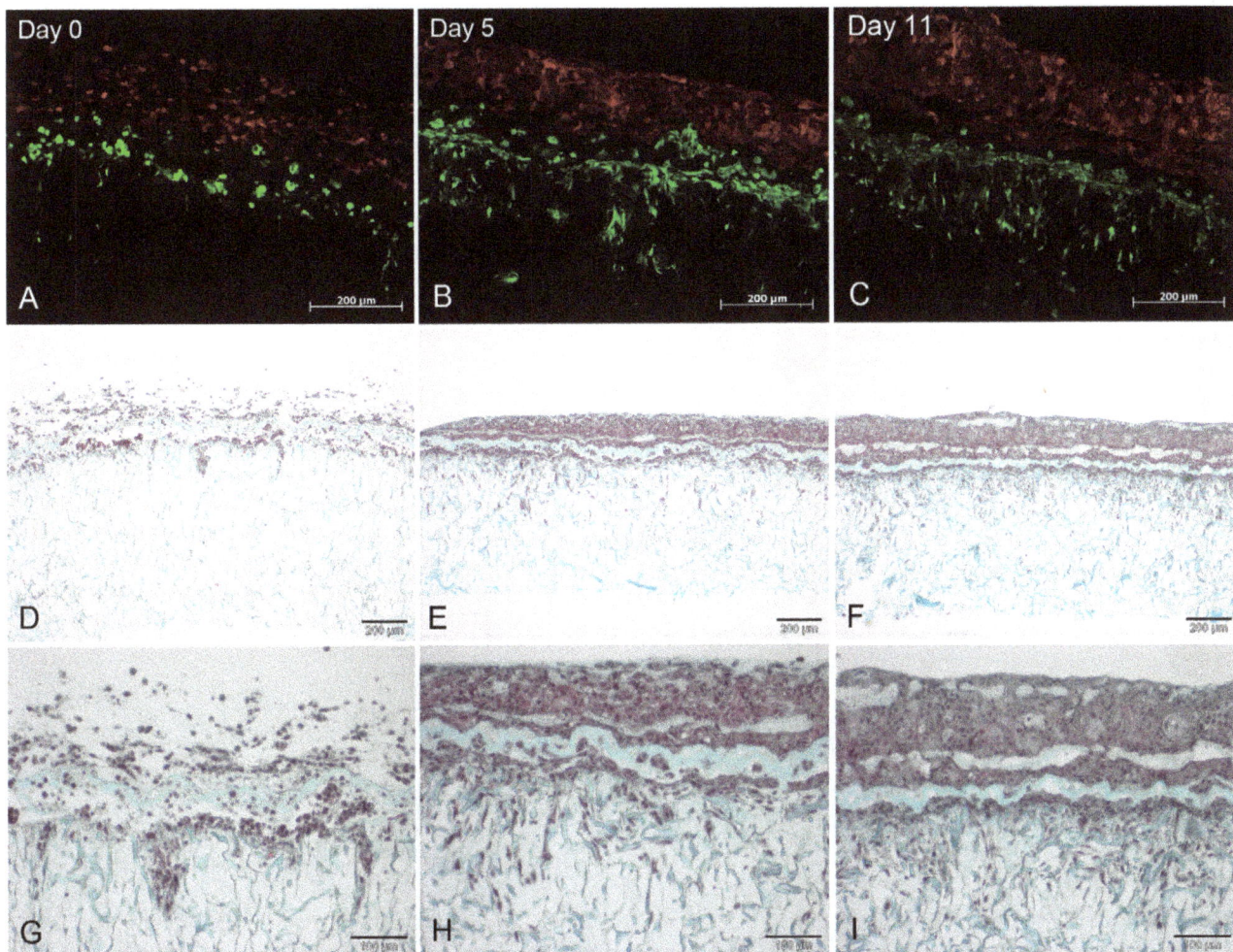

Figure 8. Histological sections of the tissue engineered skin constructs *in vitro*. Skin constructs were cultivated at the air-liquid-interface with differentiation medium for 11 days. Sections show cells using fluorescent microscopy and Masson's trichrome staining, respectively. The time points indicated in A–C are valid for the whole respective columns. The skin constructs were cultivated at the air-liquid-interface. The keratinocytes (HaCaT-mCherry) exhibit red fluorescence while the fibroblasts (NIH3T3-eGFP) appear in green (A–C). Masson's trichrome staining reveals the connective tissue containing collagen (green) and the cells (reddish) (D–I). The fibroblasts already start to grow into the Matriderm® underneath one day after printing (A, D, G). The keratinocytes, which still are rounded and are not connected to each other on day 0 (A, D, G), already form a dense tissue on day 5 (B, E, H). The thereby formed epidermis increases in height until day 11 (C, F, I). Scale bars depict 200 μm (A–F) and 100 μm (G–I).

vessels in the dermis *in vitro*, resulting in small vessels similar to capillaries present in the skin's microcirculation [26]. The same effect could be observed in the absence of keratinocytes when adding keratinocyte-conditioned medium or VEGF. Therefore, we assume that the printed keratinocytes in our skin substitutes might enhance vessel formation by VEGF production.

In our experiments no complete vascularisation of the printed skin equivalents could be achieved under the current conditions. Probably, the period was too short for complete vascularisation of the skin equivalent. Following the idea of improving graft vascularisation, Black *et al.* integrated human umbilical vein endothelial cells (HUVEC) into tissue engineered skin equivalents containing fibroblasts and keratinocytes in combination with a collagen chitosan scaffold [5] [27]. According to their study, the HUVEC were shown to form capillary-like tubules in the dermis *in vitro*. In a similar approach, skin equivalents constructed by seeding acellular dermis with keratinocytes and Bcl-2-transduced HUVEC showed perfusion through HUVEC-lined microvessels two weeks after implantation into mice [28]. This highlights the

necessity but also the probable success to incorporate endothelial cells into our printed constructs as a next step.

In contrast to the *in vivo* situation, our *in vitro* controls formed a multi-layered tissue with collagen producing fibroblasts, but did not show any differentiation of the keratinocytes (HaCaT). This might be due to the culturing method *in vitro*. Although differentiation aiding supplements were added to the culture medium and the skin constructs were raised to the air-liquid-interface, this might not have been appropriate enough to trigger the differentiation. Also, the culturing period of 11 days is quite short and induction of differentiation *in vitro* might have been observed at a later time point. However, the beginning differentiation of the keratinocytes *in vivo* could be due to the growth factors present in mice but absent *in vitro*.

Summing up, LaBP offers the possibility to place cells of a specific type wherever in the tissue they are needed. This is a unique feature of bioprinting techniques and it may be used to print skin supplemented with endothelial cells, hair follicle cells, peripheral nerve cells, Schwann cells, melanocytes or cells present

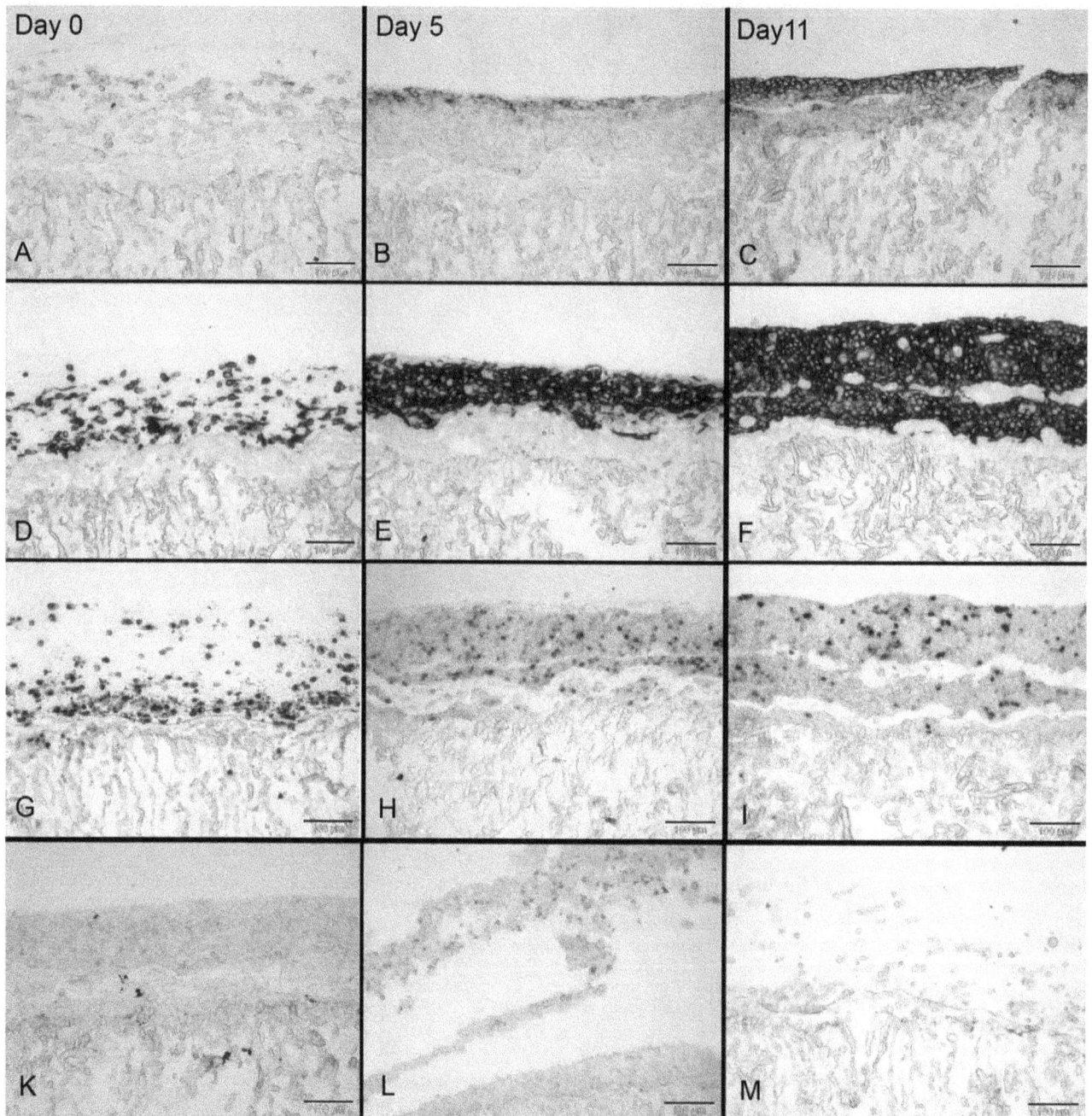

Figure 9. Sections of immunohistochemically stained skin constructs cultivated *in vitro*. Skin constructs were cultivated *in vitro* at the air-liquid-interface with differentiation medium for 11 days. The indicated time points in A–C are valid for the whole respective column. E-cadherin expression is absent on day 0 but can be detected on days 5 and 11 (A–C) while cytokeratin 14 expression is clearly visible at all time points in the whole epidermis (D–F). While nearly all cells exhibit Ki67 staining on day 0, only few cells do so at days 5 and 11 (G–I). The corresponding negative controls of the stainings (skin constructs without first antibody) are shown below (K – e-cadherin, L – cytokeratin 14, M – Ki67). All scale bars depict 100 μm.

in perspiratory and sebaceous glands. Therefore, we hope to be able to produce a much more similar skin construct to native skin compared to other current approaches. This is especially important for the future patients as they gain a much more functional and aesthetic skin substitute. This in turn would lead to a major increase of their quality of life, on the physical as well as on the mental level.

As an alternative to LaBP, a similar technique called inkjet printing is available, which has already been used to print 2D protein arrays [29], endothelial cells, smooth muscle cells [29–30], a 3D construct containing HeLa cells [31], or a 3D composite construct containing muscle cells, endothelial cells and stem cells [21]. Also, inkjet printing can be used to create antimicrobial assays [32] or to transfect cells with relatively large molecules [33]. The major disadvantage of this technique, however, is the high shear force of the nozzle, leading to severe cell impairment [34]. With LaBP – which is nozzle free – cells can be printed with a much higher density [10]. This is very important for the printing of skin, which is a tissue with a very high density of cells present. Therefore, for our purposes, LaBP remains the technique of choice.

To further improve the use of LaBP for the creation of (skin) tissue, an adaptation to automation would be of advantage. Our setup is not suited for the high throughput production of skin substitutes yet, but in principal an automation of the whole process - including the printing process as such as well as the cultivation of the skin substitutes - is conceivable.

We used the dorsal skin fold chamber in mice for the assessment of the tissue engineered skin constructs. This approach exhibits different advantages as well as drawbacks. The common approach of dorso-lateral full-thickness wounds without a chamber allows for the cultivation of a tissue engineered skin constructs for a long period of time. While two to eight weeks are the most frequently used time intervals [35–38], animals can be kept up to six months [39]. Using the chambers and the small nude mice, the observation period is quite limited as the mice would not be able to bear the weight of the chambers for several weeks. In our study, we demonstrated that skin constructs produced by LaBP are viable *in vivo*, forming a tissue similar to simple skin within the time frame of 11 days. As a further limitation, compared to the common approach of simple full-thickness wounds in the dorso-lateral region of mice with an area of 2 cm×2 cm to 2 cm×3 cm [35] [38–40], the wound area in the chambers is very small (round hole with 6 mm diameter). This, however, is partly compensated by the lack of wound contraction. The latter is the major way of wound healing in rodents [41], opposed to the main mechanisms in human wound healing, i.e. granulation tissue formation and subsequent reepithelialisation [3] [42]. In the chambers, the skin constructs are safely secured in the wound by the glass slide while the surrounding and opposite skin is firmly attached to the titanium frames [20]. Therefore, no contraction of the wound area occurs. This is very important since we aim at assessing the situation in humans and not in rodents. A further advantage of the skin fold chamber is the lack of customary wound dressings. Thereby, no changes of the dressings are necessary, which reduces the stress for the animals considerably [20]. Furthermore, the transparent glass slide allows for a continuous observation of the wound closure, without any stress for the animals due to the removal of wound dressings [43].

In conclusion, we could show LaBP to be an adequate technique for the creation and *in vivo* formation of a 3D tissue like skin. Therefore, LaBP represents a major promise for the improvement of burn therapy and thus for a raised quality of life for the patients.

Acknowledgments

The authors kindly thank Sabine Braun, Vincent Pritzel, Annika Köhler und Viktor Maurer for their excellent technical assistance and organizational help.

Author Contributions

Conceived and designed the experiments: SM LK AD HS BC PMV KR. Performed the experiments: SM HS CTP AD LK. Analyzed the data: SM HS CTP KR. Contributed reagents/materials/analysis tools: AD LK BC PMV KR. Wrote the paper: SM HS KR.

References

1. Vogt PM, Kolokythas P, Niederbichler A, Knobloch K, Reimers K, et al. (2007) Innovative wound therapy and skin substitutes for burns. Der Chirurg; Zeitschrift Fur Alle Gebiete Der Operativen Medizen 78: 335–342.
2. Supp DM, Boyce ST (2005) Engineered skin substitutes: Practices and potentials. Clinics in Dermatology 23: 403–412.
3. Singer AJ, Clark RA (1999) Cutaneous wound healing. The New England Journal of Medicine 341: 738–746.
4. Jakab K, Norotte C, Marga F, Murphy K, Vunjak-Novakovic G, et al. (2010) Tissue engineering by self-assembly and bio-printing of living cells. Biofabrication 2: 022001.
5. Black AF, Berthod F, L'heureux N, Germain L, Auger FA (1998) In vitro reconstruction of a human capillary-like network in a tissue-engineered skin equivalent. The FASEB Journal : Official Publication of the Federation of American Societies for Experimental Biology 12: 1331–1340.
6. Couet F, Mantovani D (2012) Perspectives on the advanced control of bioreactors for functional vascular tissue engineering in vitro. Expert Review of Medical Devices 9: 233–239.
7. Barron JA, Wu P, Ladouceur HD, Ringeisen BR (2004) Biological laser printing: A novel technique for creating heterogeneous 3-dimensional cell patterns. Biomedical Microdevices 6: 139–147.
8. Catros S, Fricain JC, Guillotin B, Pippenger B, Bareille R, et al. (2011) Laser-assisted bioprinting for creating on-demand patterns of human osteoprogenitor cells and nano-hydroxyapatite. Biofabrication 3: 025001.
9. Othon CM, Wu X, Anders JJ, Ringeisen BR (2008) Single-cell printing to form three-dimensional lines of olfactory ensheathing cells. Biomedical Materials (Bristol, England) 3: 034101.
10. Guillotin B, Souquet A, Catros S, Duocastella M, Pippenger B, et al. (2010) Laser assisted bioprinting of engineered tissue with high cell density and microscale organization. Biomaterials 31: 7250–7256.
11. Gruene M, Pflaum M, Deiwick A, Koch L, Schlie S, et al. (2011) Adipogenic differentiation of laser-printed 3D tissue grafts consisting of human adipose-derived stem cells. Biofabrication 3: 015005.
12. Gruene M, Deiwick A, Koch L, Schlie S, Unger C, et al. (2010) Laser printing of stem cells for biofabrication of scaffold-free autologous grafts. Tissue Engineering.Part C, Methods.
13. Hopp B, Smausz T, Kresz N, Barna N, Bor Z, et al. (2005) Survival and proliferative ability of various living cell types after laser-induced forward transfer. Tissue Engineering 11: 1817–1823.
14. Ringeisen BR, Kim H, Young HD, Spargo BJ, Auyeung RCY, et al. (2001) Cell-by-cell construction of living tissue. 698.
15. Ringeisen BR, Kim H, Barron JA, Krizman DB, Chrisey DB, et al. (2004) Laser printing of pluripotent embryonal carcinoma cells. Tissue Engineering 10: 483–491.
16. Koch L, Kuhn S, Sorg H, Gruene M, Schlie S, et al. (2010) Laser printing of skin cells and human stem cells. Tissue Engineering.Part C, Methods 16: 847–854.
17. Koch L, Deiwick A, Schlie S, Michael S, Gruene M, et al. (2012) Skin tissue generation by laser cell printing. Biotechnology and Bioengineering 109: 1855–1863.
18. Unger C, Gruene M, Koch L, Koch J, Chichkov BN (2011) Time-resolved imaging of hydrogel printing via laser-induced forward transfer. Applied Physics A: Materials Science & Processing 103: 271–277.
19. Gruene M, Unger C, Koch L, Deiwick A, Chichkov B (2011) Dispensing pico to nanolitre of a natural hydrogel by laser-assisted bioprinting. Biomedical Engineering Online 10: 19.
20. Michael S, Sorg H, Peck CT, Reimers K, Vogt PM (2012) The mouse dorsal skin fold chamber as a means for the analysis of tissue engineered skin. Burns : Journal of the International Society for Burn Injuries.
21. Xu T, Zhao W, Zhu JM, Albanna MZ, Yoo JJ, et al. (2013) Complex heterogeneous tissue constructs containing multiple cell types prepared by inkjet printing technology. Biomaterials 34: 130–139.
22. Cui X, Breitenkamp K, Finn MG, Lotz M, D'Lima DD (2012) Direct human cartilage repair using three-dimensional bioprinting technology. Tissue Engineering.Part A 18: 1304–1312.
23. Ballaun C, Weninger W, Uthman A, Weich H, Tschachler E (1995) Human keratinocytes express the three major splice forms of vascular endothelial growth factor. The Journal of Investigative Dermatology 104: 7–10.
24. Hacker C, Valchanova R, Adams S, Munz B (2010) ZFP36L1 is regulated by growth factors and cytokines in keratinocytes and influences their VEGF production. Growth Factors (Chur, Switzerland) 28: 178–190.
25. Goren I, Muller E, Schiefelbein D, Gutwein P, Seitz O, et al. (2009) Akt1 controls insulin-driven VEGF biosynthesis from keratinocytes: Implications for normal and diabetes-impaired skin repair in mice. The Journal of Investigative Dermatology 129: 752–764.
26. Rochon MH, Fradette J, Fortin V, Tomasetig F, Roberge CJ, et al. (2010) Normal human epithelial cells regulate the size and morphology of tissue-engineered capillaries. Tissue Engineering.Part A 16: 1457–1468.
27. Black AF, Hudon V, Damour O, Germain L, Auger FA (1999) A novel approach for studying angiogenesis: A human skin equivalent with a capillary-like network. Cell Biology and Toxicology 15: 81–90.
28. Schechner JS, Crane SK, Wang F, Szeglin AM, Tellides G, et al. (2003) Engraftment of a vascularized human skin equivalent. The FASEB Journal:

Official Publication of the Federation of American Societies for Experimental Biology 17: 2250–2256.

29. Wilson WC Jr, Boland T (2003) Cell and organ printing 1: Protein and cell printers. The Anatomical Record.Part A, Discoveries in Molecular, Cellular, and Evolutionary Biology 272: 491–496.

30. Boland T, Mironov V, Gutowska A, Roth EA, Markwald RR (2003) Cell and organ printing 2: Fusion of cell aggregates in three-dimensional gels. The Anatomical Record.Part A, Discoveries in Molecular, Cellular, and Evolutionary Biology 272: 497–502.

31. Arai K, Iwanaga S, Toda H, Genci C, Nishiyama Y, et al. (2011) Three-dimensional inkjet biofabrication based on designed images. Biofabrication 3: 034113.

32. Zheng Q, Lu J, Chen H, Huang L, Cai J, et al. (2011) Application of inkjet printing technique for biological material delivery and antimicrobial assays. Analytical Biochemistry 410: 171–176.

33. Owczarczak AB, Shuford SO, Wood ST, Deitch S, Dean D (2012) Creating transient cell membrane pores using a standard inkjet printer. Journal of Visualized Experiments : JoVE (61). pii: 3681. doi:10.3791/3681.

34. Born C, Zhang Z, Al-Rubeai M, Thomas CR (1992) Estimation of disruption of animal cells by laminar shear stress. Biotechnology and Bioengineering 40: 1004–1010.

35. Klingenberg JM, McFarland KL, Friedman AJ, Boyce ST, Aronow BJ, et al. (2010) Engineered human skin substitutes undergo large-scale genomic reprogramming and normal skin-like maturation after transplantation to athymic mice. The Journal of Investigative Dermatology 130: 587–601.

36. Xie JL, Li TZ, Qi SH, Huang B, Chen XG, et al. (2007) A study of using tissue-engineered skin reconstructed by candidate epidermal stem cells to cover the nude mice with full-thickness skin defect. Journal of Plastic, Reconstructive & Aesthetic Surgery : JPRAS 60: 983–990.

37. Egana JT, Fierro FA, Kruger S, Bornhauser M, Huss R, et al. (2009) Use of human mesenchymal cells to improve vascularization in a mouse model for scaffold-based dermal regeneration. Tissue Engineering.Part A 15: 1191–1200.

38. Kalyanaraman B, Boyce ST (2009) Wound healing on athymic mice with engineered skin substitutes fabricated with keratinocytes harvested from an automated bioreactor. The Journal of Surgical Research 152: 296–302.

39. Boyce ST, Medrano EE, Abdel-Malek Z, Supp AP, Dodick JM, et al. (1993) Pigmentation and inhibition of wound contraction by cultured skin substitutes with adult melanocytes after transplantation to athymic mice. The Journal of Investigative Dermatology 100: 360–365.

40. Rasmussen CA, Gibson AL, Schlosser SJ, Schurr MJ, Allen-Hoffmann BL (2010) Chimeric composite skin substitutes for delivery of autologous keratinocytes to promote tissue regeneration. Annals of Surgery 251: 368–376.

41. Galiano RD, Michaels J, Dobryansky M, Levine JP, Gurtner GC (2004) Quantitative and reproducible murine model of excisional wound healing. Wound Repair and Regeneration : Official Publication of the Wound Healing Society [and] the European Tissue Repair Society 12: 485–492.

42. Kirfel G, Herzog V (2004) Migration of epidermal keratinocytes: Mechanisms, regulation, and biological significance. Protoplasma 223: 67–78.

43. Laschke MW, Harder Y, Amon M, Martin I, Farhadi J, et al. (2006) Angiogenesis in tissue engineering: Breathing life into constructed tissue substitutes. Tissue Engineering 12: 2093–2104.

Tissue Engineering Bone Using Autologous Progenitor Cells in the Peritoneum

Jinhui Shen[1], Ashwin Nair[1], Ramesh Saxena[2], Cheng Cheng Zhang[3], Joseph Borrelli Jr.[4], Liping Tang[1,5]*

1 Department of Bioengineering, University of Texas at Arlington, Arlington, Texas, United States of America, **2** Division of Nephrology, University of Texas Southwestern Medical Center at Dallas, Dallas, Texas, United States of America, **3** Departments of Physiology and Developmental Biology, University of Texas Southwestern Medical Center at Dallas, Dallas, Texas, United States of America, **4** Texas Health Physicians Group, Texas Health Arlington Memorial Hospital, Arlington, Texas, United States of America, **5** Department of Biomedical Science and Environmental Biology, Kaohsiung Medical University, Kaohsiung, Taiwan

Abstract

Despite intensive research efforts, there remains a need for novel methods to improve the ossification of scaffolds for bone tissue engineering. Based on a common phenomenon and known pathological conditions of peritoneal membrane ossification following peritoneal dialysis, we have explored the possibility of regenerating ossified tissue in the peritoneum. Interestingly, in addition to inflammatory cells, we discovered a large number of multipotent mesenchymal stem cells (MSCs) in the peritoneal lavage fluid from mice with peritoneal catheter implants. The osteogenic potential of these peritoneal progenitor cells was demonstrated by their ability to easily infiltrate decalcified bone implants, produce osteocalcin and form mineralized bone in 8 weeks. Additionally, when poly(l-lactic acid) scaffolds loaded with bone morphogenetic protein-2 (a known osteogenic differentiation agent) were implanted into the peritoneum, signs of osteogenesis were seen within 8 weeks of implantation. The results of this investigation support the concept that scaffolds containing BMP-2 can stimulate the formation of bone in the peritoneum via directed autologous stem and progenitor cell responses.

Editor: Wan-Ju Li, University of Wisconsin-Madison, United States of America

Funding: This works was supported by National Institutes of Health grants EB007271 and CA172268. The funders had no role in study design, data collection and analysis, decision to publish, or preparation of the manuscript.

Competing Interests: The authors have declared that no competing interests exist.

* E-mail: ltang@uta.edu

Introduction

Bone loss often occurs as a result of open fractures, osteomyelitis, fractures which fail to heal, congenital malformations, tumors, and in a more general sense, osteoporosis. In recent years, significant progress has been made in the development of tissue engineered bone designed to replace or bridge large bone defects [1–3]. Typical tissue engineering strategies involve the use of an implant, usually in the form of a 3-dimensional scaffold, which integrates with existing bone tissue to restore bone and to some extent, the function of the damaged bone [4–6]. In recent years, adult mesenchymal stem cells (MSCs) have shown great promise in regenerating tissue engineered tissues and organs [7,8]. However, these approaches are still plagued by limitations associated with the recovery, differentiation ability and survival of autologous MSCs [9–13]. Therefore, there remains a need for better means to generate functional tissue by tissue engineering techniques, including bone.

Ossification of the peritoneum is a pathological condition often associated with patients undergoing peritoneal dialysis (PD) or as a result of traumatic splenic rupture or peritonitis [14–16]. Although inflammatory responses are believed to contribute to peritoneal ossification [15], the process(es) governing peritonitis-mediated ossification is not clear. Recent studies carried out in our laboratories have found that biomaterial-mediated inflammatory responses can prompt the recruitment of MSCs with multipotency, including osteogenic potential/activities [17]. In addition, we have

recently discovered that the implantation of peritoneal catheters prompts the immigration of MSCs to the peritoneal cavities in humans (unpublished results). Based on these observations, we assume that, with the localized release of an osteogenic agent, the recruited MSCs would differentiate into osteoblasts to promote mineralization of tissue engineered bone scaffolds *in situ*.

To test this hypothesis, we first analyzed the compositions and osteogenic properties of progenitor cells in lavage fluids of animals following intraperitoneal implantation of biomaterials. Using decalcified bone matrix as an osteoconductive material, we assessed the osteogenic activity of these cells in the peritoneum. Finally, to explore the applicability of this process for *in situ* bone tissue engineering, we developed and used poly l-lactic- acid (PLLA) scaffolds, combined with bone morphogenetic protein-2 (BMP-2), and evaluated the potential for stimulating the production of viable bone in the peritoneum by inducing osteogenic reactions from autologous progenitor cells.

Materials and Methods

Ethics Statement

The animal use protocols (A11-008, A07-030) were reviewed and approved by the Institutional Animal Care and Use Committee of the University of Texas at Arlington.

Materials

Goat anti-mouse SCF and Nanog antibodies and rabbit anti-mouse antibodies were obtained from Santa Cruz Biotechnology Inc. (Santa Cruz, CA). Anti-mouse CD45, Sca-1, c-kit, CD34, FLK2-, CD3, B220, TER-119-, antibodies (rat anti mouse) to various stem cell markers CD105, CD29 and CD44, along with secondary antibody streptavidin PE/Cy5.5, and donkey anti-rat – APC were obtained from eBioscience (San Diego, CA, USA). Rat anti mouse CD105 was obtained from Santa Cruz Biotechnology (Dallas, TX, USA). Biotin conjugated lineage antibody cocktail was obtained from Miltenyi Biotec (Miltenyi Biotec Inc, Auburn, CA). Bone morphogenetic protein-2 (BMP-2) was obtained from R&D Systems (R&D Systems, Minneapolis, MN). PLLA was obtained from Medisorb 100L 1A (Lakeshore Biomaterials, AL, USA) with inherent viscosity of 1.9 dL/g.

Methods

Mouse peritoneal cell collections. Balb/c mice (4~6 months) were used in this experiment. Mice were implanted with polyurethane umbilical vessel catheter (2 cm length, 5.0 FR, Sentry Medical Products, Lombard, IL, USA) based on modified procedure published earlier [18,19]. Briefly, the mice were sedated with Isoflurane inhalation. Following sterilization with 70% ethanol, a small incision (~5 mm) was made and two sections of catheter were implanted in the peritoneal cavities. The incisions were then closed with stainless steel wound clips. After implantation, the mice were euthanized at different times (0 h, 6 h, 12 h, 18 h, 1 d, 2 d, 4 d, 7 d, 10 d, 14 d) with carbon dioxide inhalation. The peritoneal cells were then recovered via peritoneal lavage with 5 ml of sterile saline twice. The isolated cells were then characterized by determining the expression of various cell markers and via cell differentiation studies.

Flow cytometry analyses and cell differentiation of peritoneal progenitor cells. For flow cytometry analysis, RBC lysing buffer (Sigma Chemical Co., St Louis, MO) was used to remove red blood cells from each sample following the manufacturer's instructions as previously described [20]. The cell density was adjusted to 5×10^6/ml and then stained with monoclonal antibodies including anti-mouse CD105, CD29, CD45, CD44, CD3, B220, Mac-1, TER-119, or biotin conjugated lineage antibody cocktail (CD3, B220, CD11b, CD14, Ter 119, Miltenyi Biotec), Streptavidin secondary antibody, Sca-1, c-kit, CD34, FLK2. Lin$^-$Sca-1$^+$Kit$^+$CD34$^-$FLK2- are widely used markers for long term hematopoietic stem cells (HSCs) [20,21], while CD105$^+$CD29$^+$CD44$^+$CD45$^-$ is well recognized as the marker set for MSCs [22,23]. Stained cells were analyzed on BD FACSCalibur (BD Bioscience, San Jose, USA) to determine the types and percentages of peritoneal cells. Osteogenic differentiation of peritoneal progenitor cells was performed on confluent cells in the presence of recombinant BMP-2 (R&D Systems, Minneapolis, MN) for 3 weeks as previously described [24]. Calcium-rich deposits by osteoblasts were then evaluated using Alizarin Red S staining [24]. Adipogenic, neurogenic and myogenic differentiation of peritoneal progenitor cells was performed and analyzed similar to earlier publications [17].

Induced bone formation in peritoneum. Both decalcified bone collagen scaffolds and porous PLLA scaffolds were used for triggering bone formation in the peritoneum. It is well established that decalcified bone scaffolds contain necessary bone morphogenetic protein for inducing bone formation [25,26]. Decalcified femur bone scaffolds (~1.5 mm×1 mm×15 mm in size) were prepared according to published procedures [27,28]. It is estimated that there are 3 mg of BMP-2 per gram of demineralized dentin [29], and ~50–100 ng of a combination of BMPs per 25 mg of bovine bone matrix [30]. PLLA scaffolds were fabricated following salt leaching technique [31]. PLLA scaffolds have been widely used in bone tissue engineering research [32,33]. For this study, PLLA scaffolds were fabricated based on optimal scaffold design criteria related to pore size and structure, as established previously [34–37]. To stimulate localized osteogenic differentiation, ethanol sterilized PLLA scaffolds (5 mm×5 mm×5 mm in size with a pore size of 150 to 300 μm) were immersed in osteogenic differentiation solution (complete medium supplemented with 50 μg/ml ascorbic acid-2-phosphate, 10 nM dexamethasone, 7 mM β-glycerolphosphate, and 1 μg/ml BMP-2) overnight, then lyophilized prior to implantation. BMP-2 is a known osteogenic differentiation agent and a few studies have shown high levels of osteogenic activities in various stem cells at 10–50 ng/ml [38,39]. Our pilot studies have shown that the *in vivo* release rates are approximately 5% (~50 ng) per scaffold per day for a period of 2 weeks. For *in vivo* testing, the scaffolds, including untreated scaffolds as control, were implanted in the peritoneal cavities [18,19]. The implants were isolated at 16 hours, 2, 6 and 12 weeks for histological evaluation.

Histological Evaluation of peritoneal implants. All explants were frozen, cryosectioned, fixed, and used for various histological evaluations as described earlier [17,40]. H&E stain was used for providing a general overview of tissue structures. Immunohistochemical analyses were performed for assessing the presence of stem and progenitor cell markers including; SCF and Nanog, and osteoblast markers - osteocalcin and alkaline phosphatase - based on previous studies [40]. Alkaline phosphatase activity (AP activity) was tested using a biochemical assay obtained from Sigma (St. Louis, MO, USA). Calcium content change was tested by Alizarin Red S staining and von Kossa staining. All stained sections were observed under light microscopy (Leica DM LB) and the images analyzed using ImageJ, as described earlier [17,40].

Statistical analyses. The extent of cell recruitment, stem cell marker expression and mineralization were analyzed using One way ANOVA. PLLA scaffold mineralization at the end of 8 weeks was evaluated using Student's t-test. The statistical significance was determined at $p < 0.05$.

Results

Recruitment and characterization of peritoneal fluid cells following dialysis in mice

A murine peritoneal implantation model was used to study the feasibility of using the peritoneum as a site for *in vivo* tissue regeneration. For that, it was essential to first investigate the cell populations that exist/arrive in the mice peritoneum following introduction of an implant. Polyurethane umbilical vessel catheters (2 cm in length) were implanted into the mouse (n = 5) peritoneal cavity to mimic the trauma and foreign body response caused by peritoneal dialysis procedures. After catheter implantation, peritoneal lavage fluid was collected at various time points up to day 14 from both catheter-implanted and control animals. Not surprisingly, after implantation for 2 days, we predominantly observed a number of inflammatory cells like T-cells (5.33±0.52%), B-cells (17.20±0.94%), myeloid cells (64.81±0.58%) and erythroid cells (4±0.41%) (Figure 1A). Interestingly, we also found that there were two unique sub-populations in the effluent cells that shared markers identical to those of MSCs (CD105$^+$CD44$^+$CD29$^+$CD45$^-$) and HSCs (Lin$^-$Sca-1$^+$Kit$^+$CD34$^-$FLK2$^-$) in nearly each of the animals, these cells accounted for 0.29±0.04% and 0.03±0.01% respectively (Figure 1A–B). Interestingly, peritoneal MSCs also express many progenitor cell markers, including CD8a, CD29, CD31, CD44, CD54, CD73, CD105, CD106, Stem Cell Factor (SCF), SH-3, Nanog, SSEA-3, and vimentin (Figure 1C). They also stained negative for CD10, CD11b, CD11c, CD13, CD14, CD19, CD30, CD34, CD45, CD49e, CD90, CD95, CD117, CD166, Nestin, Neurofilament-N, STRO-1, TRA-1-81, or alpha-smooth muscle actin. Similar cells were also identified from the peritoneal effluents of human patients with End Stage Renal Disease (ESRD) (unpublished observation). To test the functionality and plasticity of these peritoneal progenitor cells, we assessed their multipotency by culturing them *in vitro* in the presence of various differentiation-inducing media. Specifically, the undifferentiated cells exhibited fibroblast like morphology; the culture of these cells in specific differentiation mediums led to differentiation into osteogenic, adipogenic, neurogenic and myogenic phenotypes (Figure 1D). Collectively, these results suggest that the implantation of catheters in the peritoneum of adult mice prompted the migration of multipotent MSCs and other progenitor cells that express markers similar to those expressed on bone marrow stem cells into the area.

Calcification of decalcified bone implants

To determine the ability of these MSCs cells to form bone tissue in the peritoneum, we initially used decalcified bone scaffolds (approximately 1.5 mm×1 mm×15 mm in size). By doing so, we were able to study the ability of intraperitoneally implant-recruited MSCs to differentiate into bone forming cells and to produce mineralized bone tissue. Since peritoneal MSCs were found to express SCF and Nanog, both markers were used to assess the extent of MSC recruitment following scaffold implantation. H&E staining of the bone scaffolds showed an increase in eosinic staining from 16 hours to 12 weeks along with high cell infiltration by 12 weeks (Figure 2A). Interestingly, shortly after implantation (16 hour), there were many recruited cells including SCF+ and Nanog+ progenitor cells and an increased number on the surfaces of scaffold implants (Figure 2A). As expected, implant-associated osteoblast activity remained low as reflected by slight osteocalcin production on the tissue: scaffold interface of all the evaluated time points. After two weeks, SCF+/Nanog+ progenitor cells (stained in green) were found to be almost 2 times that found at 16 hours, indicating the recruitment and infiltration of progenitor cells into bone scaffolds (Figure 2B). The numbers of SCF+/Nanog+ cells remained the same through week 6 and were substantially reduced in number by week 12. By week 2, osteocalcin expression increased almost 2 times compared to that at the end of 16 hours. By 6 weeks this increase was almost 3 times, and by week 12 osteocalcin expression was almost 4 times that at 16 hours. Interestingly, over the 6 to 12 week period as the levels of osteocalcin increased, the expression of SCF and Nanog returned to the same levels as that at the end of 16 hours, suggesting the differentiation of the SCF+/Nanog+ progenitor cells into osteocalcin+ osteoblasts.

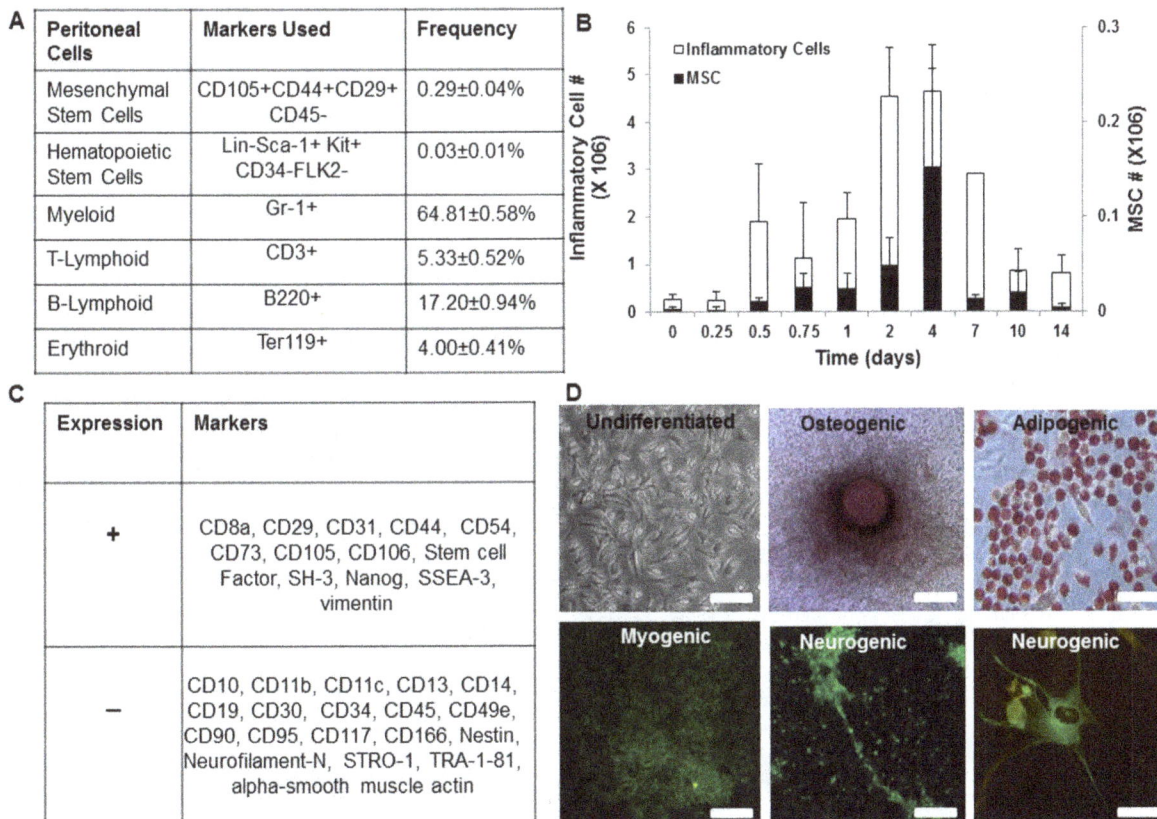

Peritoneal Cells	Markers Used	Frequency
Mesenchymal Stem Cells	CD105+CD44+CD29+ CD45-	0.29±0.04%
Hematopoietic Stem Cells	Lin-Sca-1+ Kit+ CD34-FLK2-	0.03±0.01%
Myeloid	Gr-1+	64.81±0.58%
T-Lymphoid	CD3+	5.33±0.52%
B-Lymphoid	B220+	17.20±0.94%
Erythroid	Ter119+	4.00±0.41%

Expression	Markers
+	CD8a, CD29, CD31, CD44, CD54, CD73, CD105, CD106, Stem cell Factor, SH-3, Nanog, SSEA-3, vimentin
−	CD10, CD11b, CD11c, CD13, CD14, CD19, CD30, CD34, CD45, CD49e, CD90, CD95, CD117, CD166, Nestin, Neurofilament-N, STRO-1, TRA-1-81, alpha-smooth muscle actin

Figure 1. Multipotent MSCs exist in Balb/c mice peritoneal effluents. (A) Peritoneal cell population from animals with 2 day implants. (B) Recruitment of inflammatory cells and MSCs in mice peritoneal cavity after catheter implantation. (C) Expression of surface markers on peritoneal MSCs. (D) Peritoneal cells differentiated into specific lineages appropriate conditions. Morphology of undifferentiated cells is compared with cells differentiating into specific lineages like osteogenic (Alizarin Red S stain), adipogenic (Oil Red O stain), myogenic (α-smooth muscle actin) and neurogenic (NF-M and Class III β-tubulin stain). (Mag 200×, scale bar 100 μm, statistical significance of cell numbers at various time points tested using One Way ANOVA, *p<0.05).

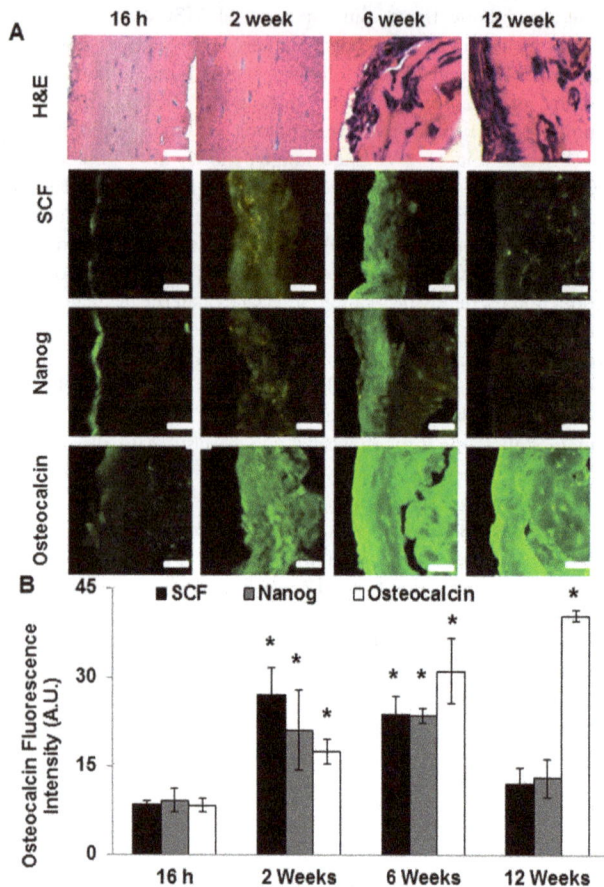

Figure 2. Progenitor cells are found around decalcified peritoneal bone implants. (A) Hematoxylin & Eosin (H&E) staining and immunohistochemical staining of SCF, Nanog and Osteocalcin on decalcified bone scaffold 16 hours, 2 weeks, 6 weeks and 12 weeks after implantation, positive stained as green. (B) Quantification of SCF, Nanog and osteocalcin expression was performed using NIH ImageJ. (Mag 200×, scale bar 200 μm, statistical significance of cell surface marker expression at various time points tested using One Way ANOVA, * $p < 0.05$).

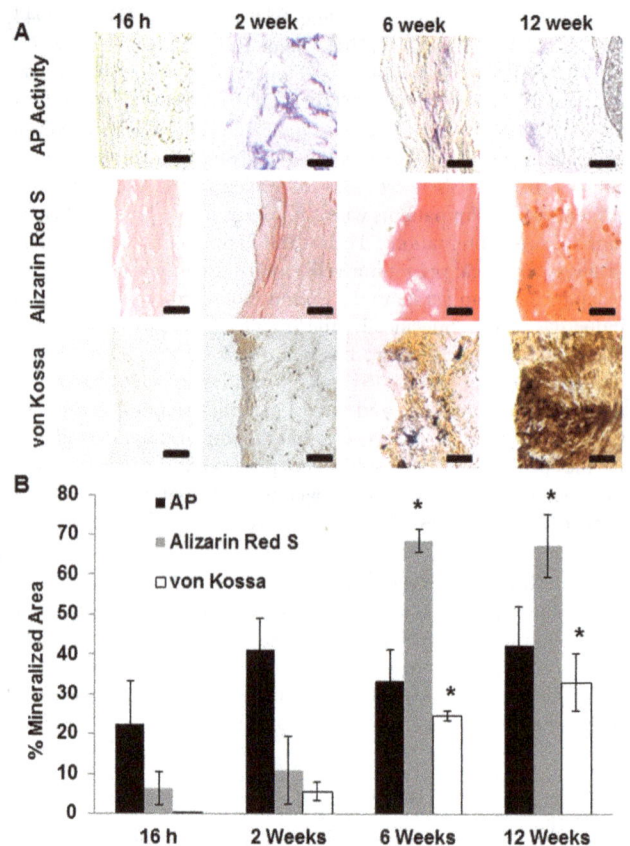

Figure 3. Mineralization of decalcified bone implants. Alkaline phosphatase (AP) assay and calcification staining of implanted decalcified bone implants at 16 hours, 2 weeks, 6 weeks, and 12 weeks (A). Quantification of AP, and calcification based on Alizarin Red S and von Kossa staining was performed using NIH ImageJ (B) (Mag 200×, scale bar 200 μm, statistical significance of mineralization at various time points tested using One Way ANOVA, * $p < 0.05$).

Mineralization in decalcified bone implants

Mineralization of the decalcified bone scaffolds in the peritoneum was determined based on alkaline phosphatase (AP) activity, Alizarin Red S staining and von Kossa staining (Figure 3A). The area fraction of the implants that were mineralized was determined using ImageJ (Figure 3B). We found that AP activity increased at week 2 and that the tissue associated AP activity remained stable from weeks 2 to 12. Alizarin Red S stain (orange or red) and von Kossa stain (brown or black deposits) were also carried out to assess scaffold mineralization. There was no significant increase in calcium content between 16 hours to 2 weeks. However, scaffold mineralization dramatically increased during the period between week 6 and 12 by more than 7 fold compared to the 16 hours to 2 week period.

Ossification of scaffolds implanted within the peritoneum

Use of decalcified bone scaffolds to induce osteogenic differentiation of stem and progenitor cells in the peritoneum is not a clinically relevant approach. To find an alternative bone tissue engineering strategy, subsequent studies were carried out using poly-l- lactic acid (PLLA) polymer scaffolds that were loaded with BMP-2, which is an osteogenic differentiation agent and has been used clinically for bone tissue engineering [41,42]. Interestingly, even without stem cell pre-seeding, PLLA scaffolds promote the recruitment of MSCs ($\sim 0.35 \pm 0.10\%$) by day 3. By week 8, we found substantial infiltration of cells into the scaffolds (as seen by the H&E staining). Coincidentally, many infiltrated cells expressed osteocalcin, indicating osteoblast activity (Figure 4A). There was a nearly 5 fold increase in osteocalcin expression (Figure 4B). Signs of mineralization in the form of calcium and phosphate deposits were observed within the scaffold as well (Figure 4C). Quantification of these markers of osteogenic activity showed a 5 fold increase in mineralization at the end of 8 weeks in PLLA scaffolds loaded with osteogenic differentiation agent BMP-2 (Figure 4D).

Discussion

For years, peritoneum has been used for evaluating host responses to biomaterials that have subsequently been employed in human subjects to meet clinical needs [18,43–49]. Recently, a few studies have even explored the possibility of using the peritoneum to grow visceral organs like bladder, uterus and vas deferens [50,51]. Most of these organ/tissue regeneration strategies involve the transplantation of cell-seeded scaffolds. Additional studies attempted to grow blood vessels by implanting biomaterials into

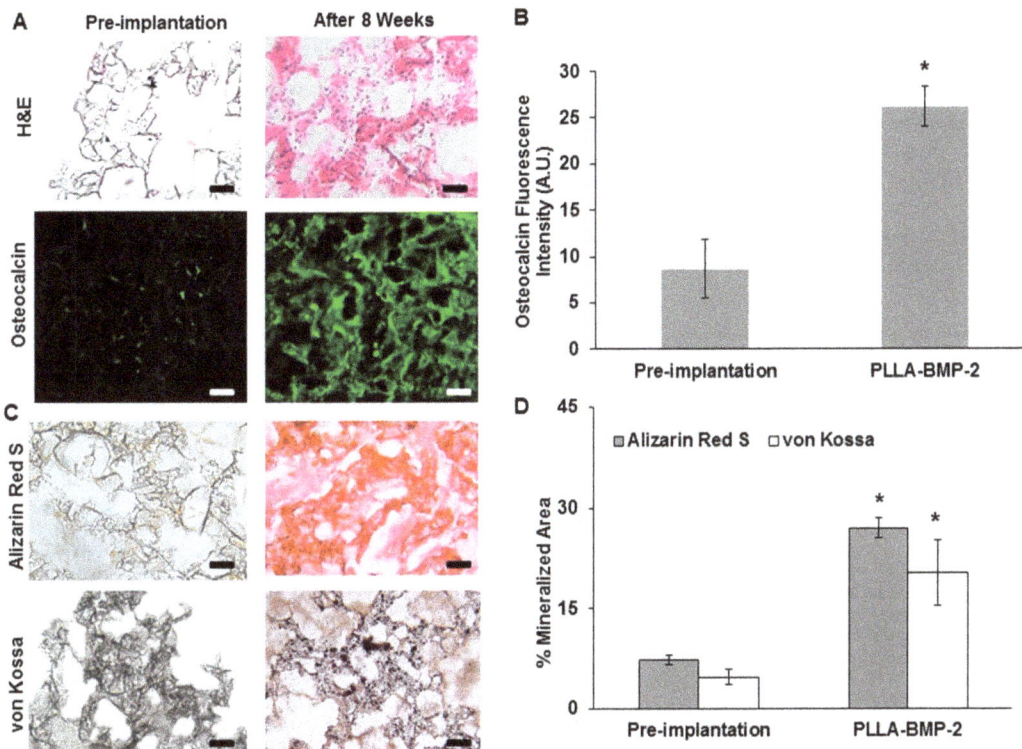

Figure 4. Mineralization and pro-osteogenic activity in PLLA-BMP2 scaffold implants in Balb/c mice peritoneal cavity after 8 weeks. (A) H&E staining and osteocalcin staining of PLLA-BMP2 scaffold after 8 weeks. (B) Quantification of osteocalcin intensity was performed using NIH ImageJ. (C) Examination of calcium deposits on scaffold based on Alizarin Red S and von Kossa staining. (D) Quantification of calcified area was performed using NIH ImageJ. (Mag 200×, scale bar 200 μm, significance of scaffold implant after vs. before implantation using Student's t-test, *$p < 0.05$).

the peritoneum [52,53]. These studies showed that free-floating implants within the peritoneum acquired layers of macrophage derived myofibroblasts and collagen matrix along with mesothelial cells and undifferentiated cells bearing markers similar to those expressed on bone marrow stem cells [52,53]. Although several recent works, including ours, indicated the presence of progenitor cells in the peritoneum [53–55], the potential of using peritoneal stem cells for tissue engineering application has not been explored. Since tissue calcification has been observed following peritoneal dialysis, and splenic rupture [18,44,46–48,56], it is possible that pathogenic processes might lead to recruitment and osteogenic differentiation of progenitor cells in the peritoneum. To support these observations, we found that progenitor cells expressing various MSC surface markers were recruited to peritoneal implants with their numbers increasing dramatically until day 4, post implantation. Based on the expression of cell surface markers and lineage specific differentiation, we found that the peritoneal stem cells were similar to bone marrow stem cells [17,57]. Such cells were multipotent, as they had the ability to differentiate into osteogenic, adipogenic, neurogenic and myogenic phenotypes. The mechanism or process for stem and progenitor cell recruitment to the peritoneal cavity following biomaterial implantation has yet to be determined. It is likely that foreign body reaction-associated inflammatory signals are responsible, since anti-inflammatory agents have been shown to reduce stem cell recruitment to the subcutaneous implants [17]. In addition, inflammatory chemokines, such as CCL2, CCL3, CCL4 and CXCL12 have been shown to promote mesenchymal stem cell migration [58–63].

Osteogenic differentiation is essential for stem cell-mediated bone regeneration [64]. To promote osteogenic differentiation of peritoneal stem cells, we first used decalcified bone scaffolds and then BMP-2-loaded PLLA scaffolds. It is well established that decalcified bone scaffolds contain osteogenic agents essential for osteogenic differentiation [65,66]. We found the presence of SCF+/Nanog+ progenitor cells reached their peak within 2 weeks of implantation and the cell density remained stable for up to 6 weeks, post implant. However, after week 6, the reduction of SCF+/Nanog+ progenitor cell number coincided with an increase of osteoblast osteocalcin activities and bone mineralization. A recent study has reported elevated expression of Nanog during undifferentiated state of stem cells followed by a drop in the levels upon osteoblastic differentiation [67]. Our results support that scaffold-containing osteogenic agents prompted the differentiation of progenitor cells into osteoblasts.

The formation of mineralized bone tissue in the peritoneum is a unique phenomenon that may be related to a disease - peritoneal ossification (PO). The cause of PO is not fully understood. However, it is generally believed that chronic inflammatory responses associated with peritoneal dialysis may be responsible [68–70]. It is possible that mineralized scaffolds and PO share a similar mechanism [16]. In fact, a few studies have suggested that an inflammatory stimulus, as that associated with implantation in the peritoneum, could lead to osteogenic cell migration from surrounding bone into the peritoneum or the inflammatory stimulus acts on stem cells to produce mesoblasts and osteoblasts [69,71,72].

While this phenomenon opens up a new vista in tissue regenerative strategies, there are additional opportunities for further advancement. For example, better designed scaffolds which can release bioactive chemokines at a more desirable rate could possibly enhance this bone formation phenomenon. We have made further advancement in scaffold development and investigated bone regeneration in critical sized defect models [2,64]. The results from this study indicates the suitability of the peritoneal environment for autologous progenitor cell recruitment and differentiation into bone. These findings can pave the way for future research by exploring smarter materials to facilitate the

development of more efficient strategies that engineer bone *in vivo* in the peritoneum so it can be readily transplanted to the defective sites, such as critical size defects, as needed.

Author Contributions

Conceived and designed the experiments: LT JS. Performed the experiments: JS. Analyzed the data: JS AN. Contributed reagents/materials/analysis tools: RS CCZ JB LT. Wrote the paper: JS AN.

References

1. Holzwarth JM, Ma PX (2011) Biomimetic nanofibrous scaffolds for bone tissue engineering. Biomaterials 32: 9622–9629.
2. Patel ZS, Young S, Tabata Y, Jansen JA, Wong ME, et al. (2008) Dual delivery of an angiogenic and an osteogenic growth factor for bone regeneration in a critical size defect model. Bone 43: 931–940.
3. Huang YC, Kaigler D, Rice KG, Krebsbach PH, Mooney DJ (2005) Combined angiogenic and osteogenic factor delivery enhances bone marrow stromal cell-driven bone regeneration. J Bone Miner Res 20: 848–857.
4. Diao H, Wang J, Shen C, Xia S, Guo T, et al. (2009) Improved cartilage regeneration utilizing mesenchymal stem cells in TGF-beta1 gene-activated scaffolds. Tissue Eng Part A 15: 2687–2698.
5. Shin M, Yoshimoto H, Vacanti JP (2004) In vivo bone tissue engineering using mesenchymal stem cells on a novel electrospun nanofibrous scaffold. Tissue Eng 10: 33–41.
6. Wayne JS, McDowell CL, Shields KJ, Tuan RS (2005) In vivo response of polylactic acid-alginate scaffolds and bone marrow-derived cells for cartilage tissue engineering. Tissue Eng 11: 953–963.
7. Caplan AI (2007) Adult mesenchymal stem cells for tissue engineering versus regenerative medicine. J Cell Physiol 213: 341–347.
8. Meinel L, Karageorgiou V, Fajardo R, Snyder B, Shinde-Patil V, et al. (2004) Bone tissue engineering using human mesenchymal stem cells: effects of scaffold material and medium flow. Ann Biomed Eng 32: 112–122.
9. Stocum DL, Zupanc GK (2008) Stretching the limits: stem cells in regeneration science. Dev Dyn 237: 3648–3671.
10. Ikada Y (2006) Challenges in tissue engineering. J R Soc Interface 3: 589–601.
11. Laschke MW, Harder Y, Amon M, Martin I, Farhadi J, et al. (2006) Angiogenesis in tissue engineering: breathing life into constructed tissue substitutes. Tissue Eng 12: 2093–2104.
12. Bonab MM, Alimoghaddam K, Talebian F, Ghaffari SH, Ghavamzadeh A, et al. (2006) Aging of mesenchymal stem cell in vitro. BMC Cell Biol 7: 14.
13. Crisostomo PR, Wang M, Wairiuko GM, Morrell ED, Terrell AM, et al. (2006) High passage number of stem cells adversely affects stem cell activation and myocardial protection. Shock 26: 575–580.
14. Ioannidis O, Sekouli A, Paraskevas G, Kotronis A, Chatzopoulos S, et al. (2012) Intra-abdominal heterotopic ossification of the peritoneum following traumatic splenic rupture. J Res Med Sci 17: 92–95.
15. Agarwal A, Yeh BM, Breiman RS, Qayyum A, Coakley FV (2004) Peritoneal calcification: causes and distinguishing features on CT. AJR Am J Roentgenol 182: 441–445.
16. Di Paolo N, Sacchi G, Lorenzoni P, Sansoni E, Gaggiotti E (2004) Ossification of the peritoneal membrane. Perit Dial Int 24: 471–477.
17. Nair A, Shen J, Lotfi P, Ko CY, Zhang CC, et al. (2011) Biomaterial implants mediate autologous stem cell recruitment in mice. Acta Biomater 7: 3887–3895.
18. Tang L, Eaton JW (1993) Fibrin(ogen) mediates acute inflammatory responses to biomaterials. J Exp Med 178: 2147–2156.
19. Hu WJ, Eaton JW, Ugarova TP, Tang L (2001) Molecular basis of biomaterial-mediated foreign body reactions. Blood 98: 1231–1238.
20. Zheng J, Umikawa M, Cui C, Li J, Chen X, et al. (2012) Inhibitory receptors bind ANGPTLs and support blood stem cells and leukaemia development. Nature 485: 656–660.
21. Christensen JL, Weissman IL (2001) Flk-2 is a marker in hematopoietic stem cell differentiation: a simple method to isolate long-term stem cells. Proc Natl Acad Sci U S A 98: 14541–14546.
22. Chartoff EH, Damez-Werno D, Sonntag KC, Hassinger L, Kaufmann DE, et al. (2011) Detection of intranasally delivered bone marrow-derived mesenchymal stromal cells in the lesioned mouse brain: A cautionary report. Stem Cells International 2011: 5686586.
23. Soleimani M, Nadri S (2009) A protocol for isolation and culture of mesenchymal stem cells from mouse bone marrow. Nature Protocols 4: 102–106.
24. Spinella-Jaegle S, Roman-Roman S, Faucheu C, Dunn FW, Kawai S, et al. (2001) Opposite effects of bone morphogenetic protein-2 and transforming growth factor-beta1 on osteoblast differentiation. Bone 29: 323–330.
25. Takahashi S, Iwata H, Hanamura H (1987) Nature of bone morphogenetic protein (BMP) from decalcified rabbit bone matrix. Nihon Seikeigeka Gakkai Zasshi 61: 197–204.
26. Urist MR, Strates BS (1971) Bone morphogenetic protein. J Dent Res 50: 1392–1406.
27. King RD, Brown B, Hwang M, Jeon T, George AT (2010) Fractal dimension analysis of the cortical ribbon in mild Alzheimer's disease. Neuroimage 53: 471–479.
28. Yaccoby S, Ling W, Zhan F, Walker R, Barlogie B, et al. (2007) Antibody-based inhibition of DKK1 suppresses tumor-induced bone resorption and multiple myeloma growth in vivo. Blood 109: 2106–2111.
29. Mizutani H, Mera K, Ueda M, Iwata H (1996) A study of the bone morphogenetic protein derived from bovine demineralized dentin matrix. Nagoya Journal of Medical Sciences 59: 37–47.
30. Sampath TK, Maliakal JC, Hauschka PV, Jones WK, Sasak H, et al. (1992) Recombinant human osteogenic protein-1 (hOP-1) induces new bone formation in vivo with a specific activity comparable with natural bovine osteogenic protein and stimulates osteoblast proliferation and differentiation in vitro. J Biol Chem 267: 20352–20362.
31. Thevenot P, Nair A, Dey J, Yang J, Tang L (2008) Method to analyze three-dimensional cell distribution and infiltration in degradable scaffolds. Tissue Eng Part C Methods 14: 319–331.
32. Ho MH, Kuo PY, Hsieh HJ, Hsien TY, Hou LT, et al. (2004) Preparation of porous scaffolds by using freeze-extraction and freeze-gelation methods. Biomaterials 25: 129–138.
33. Nam YS, Park TG (1999) Porous biodegradable polymeric scaffolds prepared by thermally induced phase separation. J Biomed Mater Res 47: 8–17.
34. Gupta B, Revagade N, Hilborn J (2007) Poly(lactic acid) fiber: an overview. Prog Polym Sci 32: 455–482.
35. Hofmann S, Hagenmuller H, Koch AM, Muller R, Vunjak-Novakovic G, et al. (2007) Control of in vitro tissue-engineered bone-like structures using human mesenchymal stem cells and porous silk scaffolds. Biomaterials 28: 1152–1162.
36. Murphy CM, Haugh MG, O'Brien FJ (2010) The effect of mean pore size on cell attachment, proliferation and migration in collagen–glycosaminoglycan scaffolds for bone tissue engineering. Biomaterials 31: 461–466.
37. Rezwan K, Chen QZ, Blaker JJ, Boccaccini AR (2006) Biodegradable and bioactive porous polymer/inorganic composite scaffolds for bone tissue engineering. Biomaterials 27: 3413–3431.
38. Hakki SS, Bozkurt B, Hakki EE, Kayis SA, Turac G, et al. (2014) Bone morphogenetic protein-2, -6, and -7 differently regulate osteogenic differenti-ation of human periodontal ligament stem cells. Journal of Biomedical Materials Research Part B: Applied Biomaterials 102: 119–130.
39. Song I, Kim BS, Kim CS, Im GI (2011) Effects of BMP-2 and vitamin D3 on the osteogenic differentiation of adipose stem cells. Biochem Biophys Res Commun 408: 126–131.
40. Shen J, Tsai YT, Dimarco NM, Long MA, Sun X, et al. (2011) Transplantation of mesenchymal stem cells from young donors delays aging in mice. Sci Rep 1: 67.
41. Bessa PC, Casal M, Reis RL (2008) Bone morphogenetic proteins in tissue engineering: the road from laboratory to clinic, part II (BMP delivery). J Tissue Eng Regen Med 2: 81–96.
42. Bessa PC, Casal M, Reis RL (2008) Bone morphogenetic proteins in tissue engineering: the road from the laboratory to the clinic, part I (basic concepts). J Tissue Eng Regen Med 2: 1–13.
43. Bellón JM, Jurado F, García-Honduvilla N, López R, Carrera-San Martín A, et al. (2002) The structure of a biomaterial rather than its chemical composition modulates the repair process at the peritoneal level. The American Journal of Surgery 184: 154–159.
44. Guo W, Willen R, Andersson R, Parsson H, Liu X, et al. (1993) Morphological response of the peritoneum and spleen to intraperitoneal biomaterials. Int J Artif Organs 16: 276–284.
45. Matthews BD, Mostafa G, Carbonell AM, Joels CS, Kercher KW, et al. (2005) Evaluation of adhesion formation and host tissue response to intra-abdominal polytetrafluoroethylene mesh and composite prosthetic mesh. Journal of Surgical Research 123: 227–234.
46. Riser BL, Barreto FC, Rezg R, Valaitis PW, Cook CS, et al. (2011) Daily peritoneal administration of sodium pyrophosphate in a dialysis solution prevents the development of vascular calcification in a mouse model of uraemia. Nephrol Dial Transplant 26: 3349–3357.
47. Tang L, Jennings TA, Eaton JW (1998) Mast cells mediate acute inflammatory responses to implanted biomaterials. Proc Natl Acad Sci U S A 95: 8841–8846.
48. Tang L, Ugarova TP, Plow EF, Eaton JW (1996) Molecular determinants of acute inflammatory responses to biomaterials. J Clin Invest 97: 1329–1334.

49. Najman S, Savic V, Djordjevic L, Ignjatovic N, Uskokovic D (2004) Biological evaluation of hydroxyapatite/poly-L-lactide (HAp/PLLA) composite biomaterials with poly-L-lactide of different molecular weights intraperitoneally implanted into mice. Bio-Medical Materials and Engineering 14: 61–70.

50. Campbell GR, Turnbull G, Xiang L, Haines M, Armstrong S, et al. (2008) The peritoneal cavity as a bioreactor for tissue engineering visceral organs: bladder, uterus and vas deferens. J Tissue Eng Regen Med 2: 50–60.

51. Cao Y, Zhang B, Croll T, Rolfe BE, Campbell JH, et al. (2008) Engineering tissue tubes using novel multilayered scaffolds in the rat peritoneal cavity. J Biomed Mater Res A 87: 719–727.

52. Campbell JH, Efendy JL, Han CL, Campbell GR (2000) Blood vessels from bone marrow. Ann N Y Acad Sci 902: 224–229.

53. Chue WL, Campbell GR, Caplice N, Muhammed A, Berry CL, et al. (2004) Dog peritoneal and pleural cavities as bioreactors to grow autologous vascular grafts. J Vasc Surg 39: 859–867.

54. Vranken I, De Visscher G, Lebacq A, Verbeken E, Flameng W (2008) The recruitment of primitive Lin(−) Sca-1(+), CD34(+), c-kit(+) and CD271(+) cells during the early intraperitoneal foreign body reaction. Biomaterials 29: 797–808.

55. Le SJ, Gongora M, Zhang B, Grimmond S, Campbell GR, et al. (2010) Gene expression profile of the fibrotic response in the peritoneal cavity. Differentiation 79: 232–243.

56. Matthews BD, Mostafa G, Carbonell AM, Joels CS, Kercher KW, et al. (2005) Evaluation of adhesion formation and host tissue response to intra-abdominal polytetrafluoroethylene mesh and composite prosthetic mesh. J Surg Res 123: 227–234.

57. Dominici M, Le Blanc K, Mueller I, Slaper-Cortenbach I, Marini F, et al. (2006) Minimal criteria for defining multipotent mesenchymal stromal cells. The International Society for Cellular Therapy position statement. Cytotherapy 8: 315–317.

58. Dudek AZ, Nesmelova I, Mayo K, Verfaillie CM, Pitchford S, et al. (2003) Platelet factor 4 promotes adhesion of hematopoietic progenitor cells and binds IL-8: novel mechanisms for modulation of hematopoiesis. Blood 101: 4687–4694.

59. Lord BI, Woolford LB, Wood LM, Czaplewski LG, McCourt M, et al. (1995) Mobilization of early hematopoietic progenitor cells with BB-10010: a genetically engineered variant of human macrophage inflammatory protein-1 alpha. Blood 85: 3412–3415.

60. Pelus LM, Fukuda S (2006) Peripheral blood stem cell mobilization: the CXCR2 ligand GRObeta rapidly mobilizes hematopoietic stem cells with enhanced engraftment properties. Exp Hematol 34: 1010–1020.

61. Thevenot PT, Nair AM, Shen J, Lotfi P, Ko CY, et al. (2010) The effect of incorporation of SDF-1alpha into PLGA scaffolds on stem cell recruitment and the inflammatory response. Biomaterials 31: 3997–4008.

62. Zhang F, Tsai S, Kato K, Yamanouchi D, Wang C, et al. (2009) Transforming growth factor-beta promotes recruitment of bone marrow cells and bone marrow-derived mesenchymal stem cells through stimulation of MCP-1 production in vascular smooth muscle cells. J Biol Chem 284: 17564–17574.

63. Zhang J, Lu SH, Liu YJ, Feng Y, Han ZC (2004) Platelet factor 4 enhances the adhesion of normal and leukemic hematopoietic stem/progenitor cells to endothelial cells. Leuk Res 28: 631–638.

64. Nair AM, Tsai YT, Shah KM, Shen J, Weng H, et al. (2013) The effect of erythropoietin on autologous stem cell-mediated bone regeneration. Biomaterials 34: 7364–7371.

65. Gruskin E, Doll BA, Futrell FW, Schmitz JP, Hollinger JO (2012) Demineralized bone matrix in bone repair: history and use. Adv Drug Deliv Rev 64: 1063–1077.

66. Tuli SM, Singh AD (1978) The osteoinductive property of decalcified bone matrix. An experimental study. J Bone Joint Surg Br 60: 116–123.

67. Arpornmaeklong P, Brown SE, Wang Z, Krebsbach PH (2009) Phenotypic characterization, osteoblastic differentiation, and bone regeneration capacity of human embryonic stem cell-derived mesenchymal stem cells. Stem Cells Dev 18: 955–968.

68. Gandhi VC, Humayun HM, Ing TS, Daugirdas JT, Jablokow VR, et al. (1980) Sclerotic thickening of the peritoneal membrane in maintenance peritoneal dialysis patients. Arch Intern Med 140: 1201–1203.

69. Garosi G, Di Paolo N (2001) Inflammation and gross vascular alterations are characteristic histological features of sclerosing peritonitis. Perit Dial Int 21: 417–418.

70. Libetta C, De Nicola L, Rampino T, De Simone W, Memoli B (1996) Inflammatory effects of peritoneal dialysis: evidence of systemic monocyte activation. Kidney Int 49: 506–511.

71. Fadare O, Bifulco C, Carter D, Parkash V (2002) Cartilaginous differentiation in peritoneal tissues: a report of two cases and a review of the literature. Mod Pathol 15: 777–780.

72. Wlodarski KH (1991) Bone histogenesis mediated by nonosteogenic cells. Clin Orthop Relat Res: 8–15.

13

Altering the Architecture of Tissue Engineered Hypertrophic Cartilaginous Grafts Facilitates Vascularisation and Accelerates Mineralisation

Eamon J. Sheehy[1,2], **Tatiana Vinardell**[1,2,4], **Mary E. Toner**[5], **Conor T. Buckley**[1,2], **Daniel J. Kelly**[1,2,3]*

1 Trinity Centre for Bioengineering, Trinity Biomedical Sciences Institute, Trinity College Dublin, Dublin, Ireland, **2** Department of Mechanical and Manufacturing Engineering, School of Engineering, Trinity College Dublin, Dublin, Ireland, **3** Advanced Materials and Bioengineering Research Centre (AMBER), Trinity College Dublin, Dublin, Ireland, **4** School of Agriculture and Food Science, University College Dublin, Belfield, Dublin, Ireland, **5** Department of Pathology, School of Dental Science, Trinity College Dublin, Dublin, Ireland

Abstract

Cartilaginous tissues engineered using mesenchymal stem cells (MSCs) can be leveraged to generate bone *in vivo* by executing an endochondral program, leading to increased interest in the use of such hypertrophic grafts for the regeneration of osseous defects. During normal skeletogenesis, canals within the developing hypertrophic cartilage play a key role in facilitating endochondral ossification. Inspired by this developmental feature, the objective of this study was to promote endochondral ossification of an engineered cartilaginous construct through modification of scaffold architecture. Our hypothesis was that the introduction of channels into MSC-seeded hydrogels would firstly facilitate the *in vitro* development of scaled-up hypertrophic cartilaginous tissues, and secondly would accelerate vascularisation and mineralisation of the graft *in vivo*. MSCs were encapsulated into hydrogels containing either an array of micro-channels, or into non-channelled 'solid' controls, and maintained in culture conditions known to promote a hypertrophic cartilaginous phenotype. Solid constructs accumulated significantly more sGAG and collagen *in vitro*, while channelled constructs accumulated significantly more calcium. *In vivo*, the channels acted as conduits for vascularisation and accelerated mineralisation of the engineered graft. Cartilaginous tissue within the channels underwent endochondral ossification, producing lamellar bone surrounding a hematopoietic marrow component. This study highlights the potential of utilising engineering methodologies, inspired by developmental skeletal processes, in order to enhance endochondral bone regeneration strategies.

Editor: Kent Leach, University of California at Davis, United States of America

Funding: This work was supported by Science Foundation Ireland under the President of Ireland Young Researcher Award (Grant No. SFI/08/Y15/B1336) and a starter grant from the European Research Council (Stem Repair Project No. 258463). The funders had no role in study design, data collection and analysis, decision to publish, or preparation of the manuscript.

Competing Interests: The authors have declared that no competing interests exist.

* E-mail: kellyd9@tcd.ie

Introduction

The goal of tissue engineering is to replace or regenerate damaged tissues through the combination of cells, three-dimensional scaffolds, and signalling molecules [1,2]. Bone tissue engineering has, thus far, generally focussed on the direct osteoblastic differentiation of mesenchymal stem cells (MSCs), in a process resembling intramembranous ossification [3]. However, the success of this approach to bone regeneration has been hampered by insufficient blood vessel infiltration, preventing the necessary delivery of oxygen and nutrients to the engineered graft [4,5]. Recently an endochondral approach to bone tissue engineering, which involves remodelling of an intermediary hypertrophic cartilaginous template [6–14], has been proposed as an alternative to direct intramembranous ossification for bone regeneration using MSCs. Chondrogenically primed bone marrow derived MSCs have an inherent tendency to undergo hypertrophy [15], an undesirable attribute in MSC-based articular cartilage repair therapies, but one which may be harnessed for use in endochondral bone tissue engineering strategies. Hypertrophic

chondrocytes are equipped to survive the hypoxic environment a tissue engineered graft will experience once implanted *in vivo* [14]. Furthermore, cells undergoing hypertrophy release pro-angiogenic factors, such as vascular endothelial growth factor, to facilitate the conversion of avascular tissue to vascularised tissue [16]. The endochondral approach to bone regeneration may therefore circumvent many of the issues associated with the traditional intramembranous method; however, in order to be used to repair large bone defects, this approach first requires strategies to engineer scaled-up hypertrophic cartilage. This may be challenging, as recent attempts to generate large hypertrophic constructs using MSC seeded collagen scaffolds have demonstrated the formation of a core region devoid of cells and matrix [12].

Such 'scaling up' of engineered grafts, and the associated issue of core degradation, is a well-documented challenge in the field of tissue engineering [17]. Strategies to promote nutrient supply and waste removal include the use of bioreactors [18–21], modification of scaffold architectures [22,23], or a combination of both [24–26]. An alternative approach might be to recapitulate the mechanisms adopted during skeletogenesis to provide nutrients to the cartilaginous templates. During normal bone development, canals within the developing hypertrophic cartilage have been identified [27,28]. These canals play an important role in supplying nutrients to the developing cartilage [29]. Furthermore, endochondral ossification is dependent on neo-vascularisation of the cartilaginous template, with the in-growth of blood vessels via these cartilage canals identified as a key event during bone development [29]. Finally, these canals act as conduits for the migration of osteogenic cells into the cartilaginous template which in turn lay down new bone matrix [30]. Executing a developmental engineering paradigm [31], directed at mimicking the structure and function of the cartilage canal network, may therefore be a powerful tool in endochondral bone tissue engineering strategies.

Inspired by the cartilage canal network observed during endochondral skeletogenesis, we hypothesised that the introduction of channels into hypertrophic cartilaginous constructs would firstly facilitate extracellular matrix synthesis and the *in vitro* development of the construct, and secondly would accelerate mineralisation and vascularisation of the graft once implanted *in vivo*. To assess the influence of scaffold architecture on *in vitro* graft development, bone marrow derived MSCs were encapsulated in agarose hydrogels either containing an array of microchannels (termed 'channelled'), or in non-channelled (termed 'solid') controls, and cultured long-term (up to 10 weeks) to undergo hypertrophic chondrogenic differentiation. Furthermore, to test the efficacy of channelled architectures to enhance mineralisation and vascularisation *in vivo*, channelled and solid constructs were also subjected to a shorter period of hypertrophic priming (6 weeks) prior to subcutaneous implantation in nude mice. Constructs were harvested at 4 and 8 weeks post-implantation to test our hypothesis that graft architecture would influence mineralisation and vascularisation of the engineered hypertrophic tissue.

Methods

Experimental Design

The first phase of this study investigated the *in vitro* development of engineered cartilaginous constructs undergoing long- term hypertrophic chondrogenesis. MSCs were encapsulated in solid and channelled hydrogels, cultured in chondrogenic conditions (further details below), for a period of 5 weeks. Thereafter constructs were switched to hypertrophic conditions for an additional 5 weeks, resulting in a total *in vitro* culture period of 10 weeks. The second phase of this study investigated the potential of engineered hypertrophic cartilaginous constructs to undergo mineralisation and vascularisation *in vivo*. MSCs were encapsulated in solid and channelled constructs and cultured in chondrogenic conditions for a period of 5 weeks, as per phase 1. Thereafter constructs received an additional week in hypertrophic conditions (6 weeks total *in vitro* priming) prior to subcutaneous implantation in nude mice. Constructs were harvested at 4 and 8 weeks post-implantation.

Cell Isolation and Expansion

Porcine bone marrow derived MSCs (4 months old) from the femoral shaft were isolated aseptically and expanded according to a modified method for human MSCs [32] in high glucose Dulbecco's modified eagle's medium (DMEM) GlutaMAX supplemented with 10% v/v foetal bovine serum and 100 U/mL penicillin –100 µg/mL streptomycin (all Gibco, Biosciences, Dublin Ireland) at 20% pO_2. Following colony formation, MSCs were trypsinised, counted, seeded at density of 5×10^3 cells/cm^2,

and expanded to passage 1 (P1). MSCs underwent 15 population doublings prior to encapsulation within hydrogels.

Cell Encapsulation in Solid and Channelled Agarose Hydrogels

At the end of P1 MSCs were suspended in 2% agarose at a density of 20×10^6 cells/mL. The agarose cell suspension was cast in a stainless steel mould to produce regular solid (non-channelled) cylindrical constructs (Ø5 mm×3 mm). Channelled cylindrical constructs were fabricated utilising a pillared polydimethylsiloxane array structure [24] inserted into a cylindrical Tufset mould to produce a unidirectional 4×3 channelled array in the longitudinal direction with diameters of 500 µm and a centre-centre spacing of 1 mm.

Chondrogenic and Hypertrophic Culture Conditions

The chondrogenic conditions applied in this study are defined as culture at 5% pO_2 in a chondrogenic medium consisting of high glucose DMEM GlutaMAX supplemented with 100 U/mL penicillin/streptomycin (both Gibco), 100 µg/mL sodium pyruvate, 40 µg/mL L-proline, 50 µg/mL L-ascorbic acid-2-phosphate, 4.7 µg/mL linoleic acid, 1.5 mg/mL bovine serum albumine, 1×insulin–transferrin–selenium, 100 nM dexamethasone (all from Sigma-Aldrich) and 10 ng/mL of human transforming growth factor-β3 (TGF-β3) (Prospec-Tany TechnoGene Ltd., Israel) [33]. The hypertrophic conditions applied are defined as culture at 20% pO_2 in a hypertrophic medium consisting of high glucose DMEM GlutaMAX supplemented with 100 U/mL penicillin/streptomycin, 100 µg/mL sodium pyruvate, 40 µg/mL L-proline, 50 µg/mL L-ascorbic acid-2-phosphate, 4.7 µg/mL linoleic acid, 1.5 mg/mL bovine serum albumine, 1×insulin–transferrin–selenium, 1 nM dexamethasone, 1 nM L-thyroxine and 20 µg/mL β-glycerophosphate (both Sigma-Aldrich) [13,34].

In vivo Subcutaneous Transplantation

In phase 2 of the study, following 6 weeks *in vitro* priming, MSC-seeded solid and channelled constructs were implanted subcutaneously into the back of nude mice (Balb/c; Harlan, UK) as previously described [34]. Briefly, two subcutaneous pockets were made along the central line of the spine, one at the shoulders and the other at the hips, and into each pocket three constructs were inserted. Nine constructs were implanted per each experimental group. Mice were sacrificed at 4 and 8 weeks post-implantation by CO_2 inhalation. In addition, three acelluar solid and three acelluar channelled hydrogels were implanted as controls and harvested at 8 weeks post-implantation. These controls were not primed in chondrogenic and hypertrophic culture conditions before transplantation. A total of 7 mice were used for the experiment. The animal protocol was reviewed and approved by the ethics committee of Trinity College Dublin.

Biochemical Analysis

The biochemical content of constructs was analysed at each time point. Prior to biochemical analysis, constructs were sliced in half, washed in phosphate buffered saline (PBS), weighed, and frozen for subsequent assessment. Half of each construct was digested with papain (125 µg/mL) in 0.1 M sodium acetate, 5 mM L-cysteine-HCL, 0.05 M ethylenediaminetetraacetic acid (EDTA), pH 6.0 (all from Sigma-Aldrich) at 60°C and 10 rpm for 18 h. DNA content was quantified using the Hoechst Bisbenzimide 33258 dye assay, with a calf thymus DNA standard. Proteoglycan content was estimated by quantifying the amount of sulphated glycosaminoglycan (sGAG) using the dimethylmethylene blue dye-binding assay (Blyscan, Biocolor Ltd., Northern Ireland), with a chondroitin sulphate standard. Total collagen content was determined by measuring the hydroxyproline content, using a hydroxyproline-to-collagen ratio of 1:7.69 [35,36]. The other half was digested in 1 M hydrochloric acid (HCL) (Sigma-Aldrich) at 60°C and 10 rpm for 18 h. The calcium content was determined using a Sentinel Calcium kit (Alpha Laboratories Ltd, Uk). 3–4 constructs were analysed biochemically per each *in vitro* experimental group. 4–6 constructs were analysed biochemically per each *in vivo* experimental group.

Histology and Immunohistochemistry

At each time point samples were fixed in 4% paraformaldehyde overnight, dehydrated in a graded series of ethanols, embedded in paraffin wax, sectioned at 8 μm and affixed to microscope slides. Samples harvested at 8 weeks post- *in vivo* implantation were decalcified in EDTA for 3–4 days prior to wax embedding. The sections were stained with haematoxylin and eosin (H&E), 1% alcian blue 8GX in 0.1 M HCL to assess sGAG content with a counter stain of nuclear fast red to assess cellular distribution, picro-sirius red to assess collagen distribution, and 1% alizarin red to assess calcium accumulation (all Sigma-Aldrich). Collagen types I, II and X were evaluated using a standard immunohistochemical technique; briefly, sections were treated with peroxidase, followed by treatment with chondroitinase ABC (Sigma-Aldrich) in a humidified environment at 37°C to enhance permeability of the extracellular matrix. Sections were incubated with goat serum to block non-specific sites and collagen type I (ab6308, 1:400; 1 mg/mL), collagen type II (ab3092, 1:100; 1 mg/mL) or collagen type X (ab49945, 1:200; 1.4 mg/mL) primary antibodies (mouse monoclonal, Abcam, Cambridge, UK) were applied for 1 h at room temperature. Next, the secondary antibody (Anti-Mouse IgG biotin conjugate, 1:200; 2.1 mg/mL) (Sigma-Aldrich) was added for 1 h followed by incubation with ABC reagent (Vectastain PK-400, Vector Labs, Peterborough, UK) for 45 min. Finally sections were developed with DAB peroxidase (Vector Labs) for 5 min. Positive and negative controls were included in the immunohis-tochemistry staining protocol for each batch. Terminal deoxynu-cleotidyl Transferase (TdT) cells were identified on the fully automated IHC Leica BOND-MAX (Leica biosystems). Briefly, sections were incubated with a mouse monoclonal TdT Clone SEN28 ready to use primary antibody (PA0339) (Leica biosys-tems). Heat mediated antigen retrieval was performed with epitope retrieval solution 2 (AR9640) for 10 min (Leica biosystems). Thereafter, visualisation of the antibody was per-formed with the Bond Polymer Refine Detection kit (DS9800) (Leica biosystems). Heat mediated antigen retrieval was performed on CD31 sections with sodium citrate solution (Sigma-Alrich) for 20 min, and incubated with goat serum, biotin, and avidin, to block non-specific sites, and the CD31 (ab28364, 1:50; 0.2 mg/mL) primary antibody (Rabbit polyclonal, Abcam, Cambridge, UK) was applied overnight at 4°C. Thereafter sections were treated with peroxidase followed by the secondary antibody (ab97051, goat polyclonal secondary antibody to Rabbit IgG, 1:200, 1 mg/mL, Abcam, Cambridge, UK) for 1 h, ABC reagent for 45 min, and developed with DAB peroxidase for 5 min. (both Vector Labs). Histological images are taken from two represen-tative constructs.

Micro-computed Tomography

Micro-computed tomography (μCT) scans were carried out on constructs using a Scanco Medical 40 μCT system (Scanco Medical, Bassersdorf, Switzerland) in order to quantify mineral content and assess mineral distribution. In phase 1 of the experiment, constructs were scanned at the end of the 10 week *in vitro* culture period. In phase 2, constructs were scanned at 4 and 8 weeks post-implantation. Constructs were scanned in PBS, at a voxel resolution of 12 μm, a voltage of 70 kVp, and a current of 114 μA. Circular contours were drawn in an anti-clockwise direction around the periphery of the constructs. Additionally, circular constructs were drawn in a clockwise direction around the perimeter of the channels, so as to exclude the area with the channels from the analysis. A Gaussian filter (sigma = 0.8, support = 1) was used to suppress noise and a global threshold corresponding to a density of 172.6 mg hydroxyapatite/cm^3 was applied. This threshold was selected by visual inspection of individual scan slices so as to include mineralised tissue and exclude non-mineralised tissue. 3D evaluation was carried out on the segmented images to determine mineral volume and to reconstruct a 3D image. The variance of mineralisation with depth through the constructs was analysed qualitatively by examining sections ~0.75 mm of depth from the top of the construct (quarter section), and ~1.5 mm of depth from the top of the construct (centre section). 10 slices were compiled per section, corresponding to a thickness of 120 μm.

Statistical Analysis

All statistical analyses were carried out using Minitab 15.1. Results are reported as mean ± standard deviation. The number of constructs analysed per group are provided in the figure legends. Groups were analysed by a general linear model for analysis of variance with groups of factors. Tukey's test was used to compare conditions. Anderson-Darling normality tests were conducted on residuals to confirm a normal distribution. Non-normal data was transformed using the Box-Cox procedure. Any non-normal data which the Box-Cox procedure could not find a suitable transformation for was transformed using the Johnson procedure. Significance was accepted at a level of $p < 0.05$.

Results

Chondrogenically Primed MSCs can be Stimulated in vitro to Produce a Calcified Cartilaginous Tissue within Solid and Channelled Hydrogels

MSCs were encapsulated in solid and channelled hydrogels and were subjected to chondrogenic culture conditions for a period of 5 weeks, followed by an additional 5 weeks in hypertrophic culture conditions, resulting in a total *in vitro* culture period of 10 weeks. Solid MSC-seeded hydrogels accumulated significantly more sGAG compared to channelled MSC- seeded hydrogels after 5 weeks (1.31±0.06 vs. 0.94±0.08%ww, $p < 0.0001$) and 10 weeks (0.70±0.01 vs. 0.58±0.04%ww, $p = 0.0304$) of *in vitro* culture, see Fig. 1(A). Both solid and channelled constructs show a significant reduction in sGAG after 10 weeks compared to their levels at 5 weeks. Both collagen and calcium accumulation increased with time in culture for both solid and channelled constructs, with greater levels of collagen observed in solid constructs after 10 weeks of culture (1.58±0.09 vs. 1.27±0.15%ww, $p = 0.03$), see Fig. 1(B). Channelled constructs accumulated significantly more calcium than solid constructs after 10 weeks in culture (3.72±0.61 vs. 2.58±0.15%ww, $p = 0.0035$), see Fig. 1(C). μCT analysis confirmed the enhancement of mineralisation in channelled constructs, see Fig. 1(D). Both solid and channelled constructs mineralised preferentially around their periphery, with a reduction in mineralisation observed with depth through the constructs.

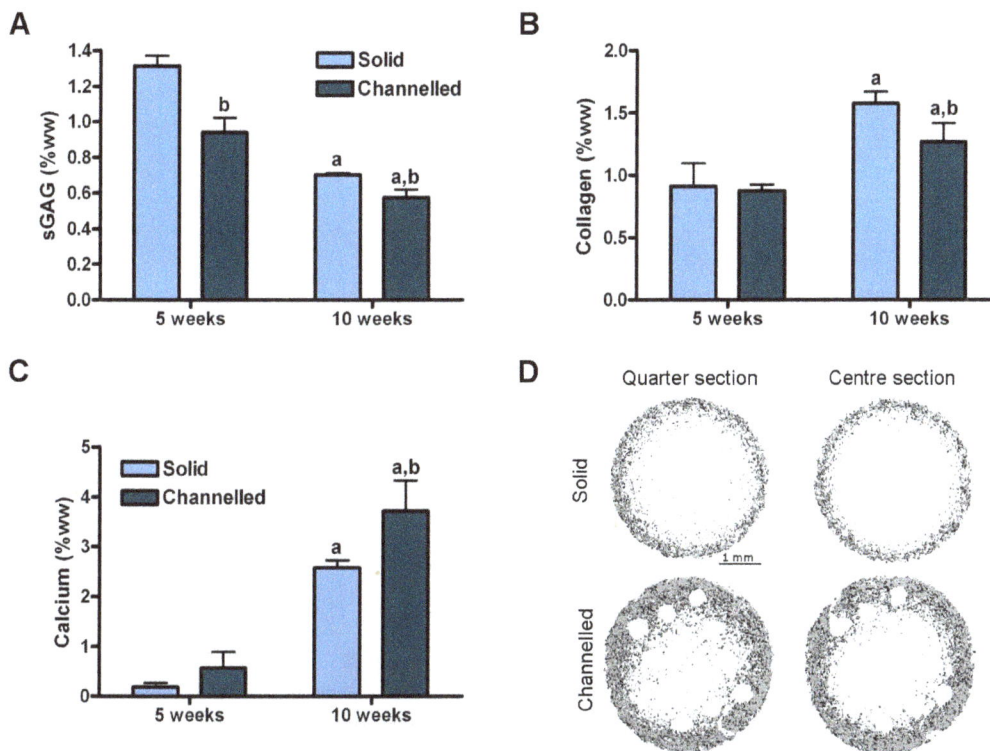

Figure 1. Biochemical and μCT analysis of constructs after 10 weeks of *in vitro* culture. Solid and channelled constructs were subjected to 5 weeks culture in chondrogenic conditions, followed by an additional 5 weeks culture in hypertrophic conditions. (A) sGAG, (B) collagen, (C) calcium (% wet weight) accumulation of solid and channelled constructs after 5 and 10 weeks of *in vitro* culture. 3–4 constructs per group were analysed. Significance $p<0.05$: a vs. 5 weeks, b vs. solid constructs. (D) μCT analysis of solid and channelled constructs after 10 weeks *in vitro* culture. Quarter section corresponds to a region ~0.75 mm into the depth of the construct. Centre section corresponds to a region ~1.5 mm into the depth of the construct. Sections correspond to a thickness of 120 μm. Scale bar is consistent across all images. Images are representative of 3 constructs analysed.

Channelled Architectures Accelerate in vivo Mineralisation of MSC-seeded Hydrogels

MSC- seeded solid and channelled hydrogels underwent 5 weeks of culture in chondrogenic conditions, followed by an additional week of culture in hypertrophic conditions, prior to subcutaneous implantation in nude mice. Constructs were harvested at 4 and 8 weeks post-implantation. The DNA content of solid constructs 4 and 8 weeks post-implantation were lower than pre- implantation levels, whereas the DNA content of channelled constructs was higher after 4 weeks *in vivo* compared to pre-implantation levels, and further increased after 8 weeks *in vivo*, see Fig. 2(A). The sGAG content of both solid and channelled constructs was lower 4 and 8 weeks post-implantation compared to pre-implantation levels, with solid constructs maintaining significantly higher levels of sGAG at both post-implantation time points compared to channelled constructs (0.34 ± 0.1 vs. $0.18\pm0.07\%$ww at 8 weeks post-implantation, $p = 0.0001$), see Fig. 2(B). Prior to implantation, collagen accumulation was similar in solid and channelled constructs. Collagen levels did not change in solid constructs post-implantation, whereas collagen accumulation was significantly higher in channelled constructs at 4 and 8 weeks post implantation compared to pre-implantation levels (2.12 ± 0.53 vs. $1.2\pm0.39\%$ww, 8 weeks post-implantation vs. pre-implantation, $p = 0.0005$), see Fig. 2(C). The calcium content of solid and channelled constructs increased at 4 weeks post-implantation compared to pre-implantation, with a further increase evident at 8 weeks post-implantation compared to 4 weeks post-implantation, see Fig. 2(D). Calcium accumulation in channelled constructs

showed a trend towards a significant increase, as compared to solid constructs, at 4 weeks post-implantation (4.37 ± 0.61 vs. $3.27\pm0.33\%$ww, $p = 0.0593$), with this difference becoming significant at 8 weeks post-implantation (7.97 ± 1.15 vs. $5.12\pm0.53\%$ww, $p<0.0001$).

Histological analysis of constructs, pre- and post- implantation, was carried out using alcian blue, picro-sirius red and alizarin red staining to assess the spatial distribution of sGAG, collagen, and calcium deposition respectively, see Fig. 3. Prior to implantation, solid and channelled constructs stained positively and homo-genously for alcian blue and picro-sirius red, whereas both constructs only stained positively for alizarin red around their periphery. At 4 weeks post-implantation, solid and channelled constructs stained weakly for alcian blue with channels beginning to show positive staining for nuclear fast red. An increased number of channels stained positive for nuclear fast red at 8 weeks post-implantation. Tissue that had filled the channels also stained intensely for picro-sirius red at 8 weeks post-implantation. At 4 weeks post-implantation, channelled constructs stained homoge-nously for alizarin red, whereas the staining in solid constructs was heterogeneous, with the core region of the engineered tissue staining negatively, indicating that it was devoid of mineral. H&E staining of acellular (i.e. MSC-free) hydrogels 8 weeks post-implantation demonstrated no bone or cartilage tissue formation in either solid or channelled constructs, though small levels of cellular infiltration were apparent within the channels of channelled constructs (data not shown).

μCT analysis was performed on MSC-seeded hydrogels at 4 and 8 weeks post-implantation to assess the spatial distribution of

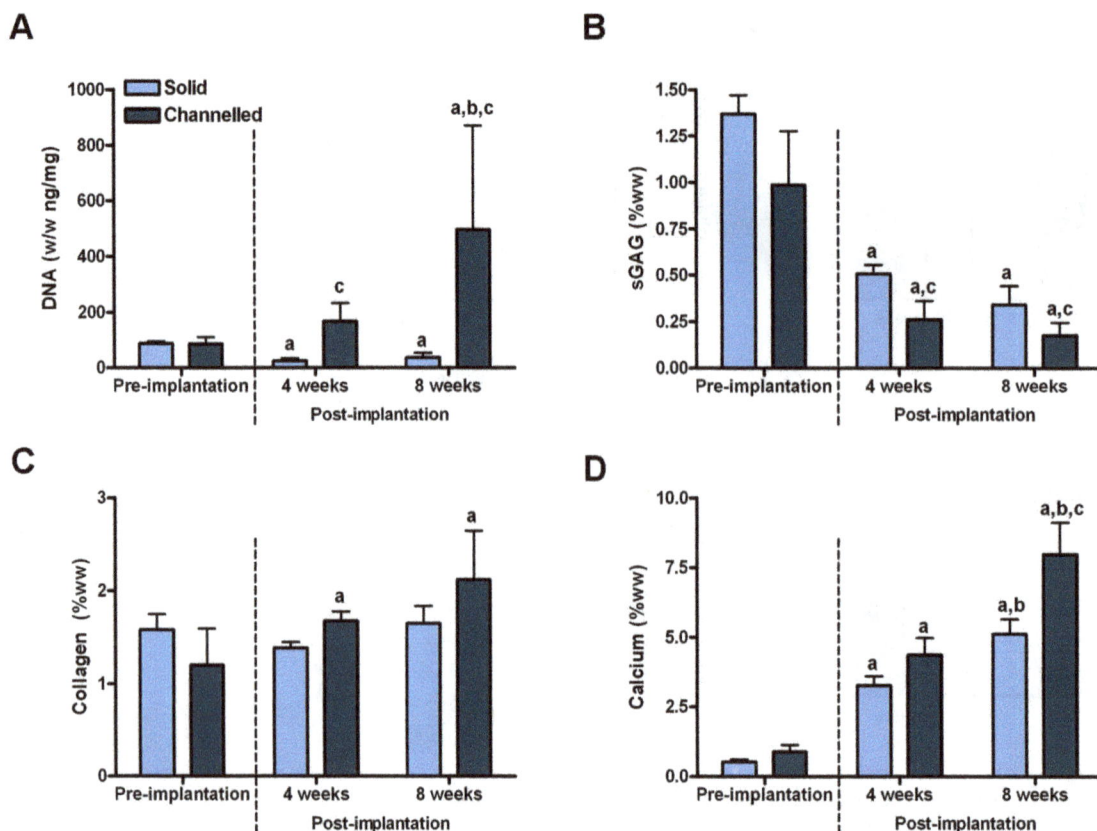

Figure 2. Biochemical analysis of constructs pre-implantation and post-implantation. Solid and channelled constructs were subjected to 5 weeks culture in chondrogenic conditions, followed by 1 week of culture in hypertrophic conditions, and then implanted subcutaneously into nude mice to be harvested at 4 weeks and 8 weeks post-implantation. (A) DNA content, normalised to mg wet weight. (B) sGAG, (C) collagen and (D) calcium content (% wet weight). Significance $p<0.05$: a vs pre-implantation, b vs. 4 weeks post-implantation, c vs. solid constructs. 3–4 constructs per group were analysed pre-implantation. 4–6 constructs per group were analysed post-implantation.

mineral and it's variance with depth through the construct, and also to quantify the volume of mineral in the construct. Mineral volume, as quantified by μCT, was significantly higher for channelled constructs as compared to solid constructs at both 4 and 8 weeks post-implantation (44.99 ± 6.54 vs. $28.54\pm1.53\%$ MV/TV, channelled constructs vs. solid constructs 8 weeks post-implantation, $p=0.014$), see Fig. 4(A). At 8 weeks post-implantation, at a depth of ~0.75 mm from the tissue surface, solid constructs had a core region devoid of mineral, whereas a homogenous deposition of mineral was observed in the channelled constructs, see Fig. 4(B). At a depth of ~1.5 mm from the surface, the core region devoid of mineral in solid constructs had increased in area. While there was also evidence of a non-mineralised core region developing in channelled constructs, this area appeared smaller than the corresponding region in the solid constructs.

Channels Act as Conduits for Vascularisation and Provide a Milieu for Endochondral Bone and Marrow Formation

Upon retrieval from the back of nude mice, macroscopically the channels appeared to the filled with a reddish, well vascularised tissue, see Fig. 5(A,B). This vascularised tissue spread through the depth of the channels, see Fig. 5(C). In contrast, macroscopically the solid constructs did not appear to be vascularised, see Fig. 5(D). Histological and immunohistochemical analysis was carried out on constructs to further investigate vascularisation and *de novo* tissue formation within the channels, and to determine the pathway

through which this tissue formation occurs. At 4 weeks post-implantation channels were stained for H&E and for CD31 to examine the presence of blood vessels and endothelial cells. TdT staining was also carried out to assess the proportion of primitive marrow cells in the cellular marrow. Blood vessel structures were identified by H&E and CD31 staining, see Fig. 5(E,F). Vascularisation did not appear to progress significantly from the channels into the calcified cartilage within the hydrogel. TdT staining at 4 weeks post implantation indicated a scanty positive staining as expected for normal marrow, see Fig. 5(G).

H&E staining of solid constructs did not demonstrate bone formation 4 weeks post-implantation, see Fig. 6(A). H&E staining of channelled constructs at 4 weeks post-implantation demonstrated woven bone formation within the channels, surrounding a marrow component consisting of a mixture of hematopoietic foci and marrow adipose tissue, see Fig. 6(B). At 8 weeks post-implantation no bone formation was evident in solid constructs, see Fig. 6(C), whereas within the channels of channelled constructs lamellar-like bone was evident, with the appearance of osteocyte-like cells embedded within the bone matrix, which again surrounded a marrow component, see Fig. 6(D).

Prior to subcutaneous implantation (i.e. during the *in vitro* priming phase), channels had partially filled with a cartilaginous matrix containing MSCs which stained intensely for alcian blue and collagen type II and negatively for collagens type I and X, see Fig. 7(A-D). At 4 weeks post-implantation channels stained very

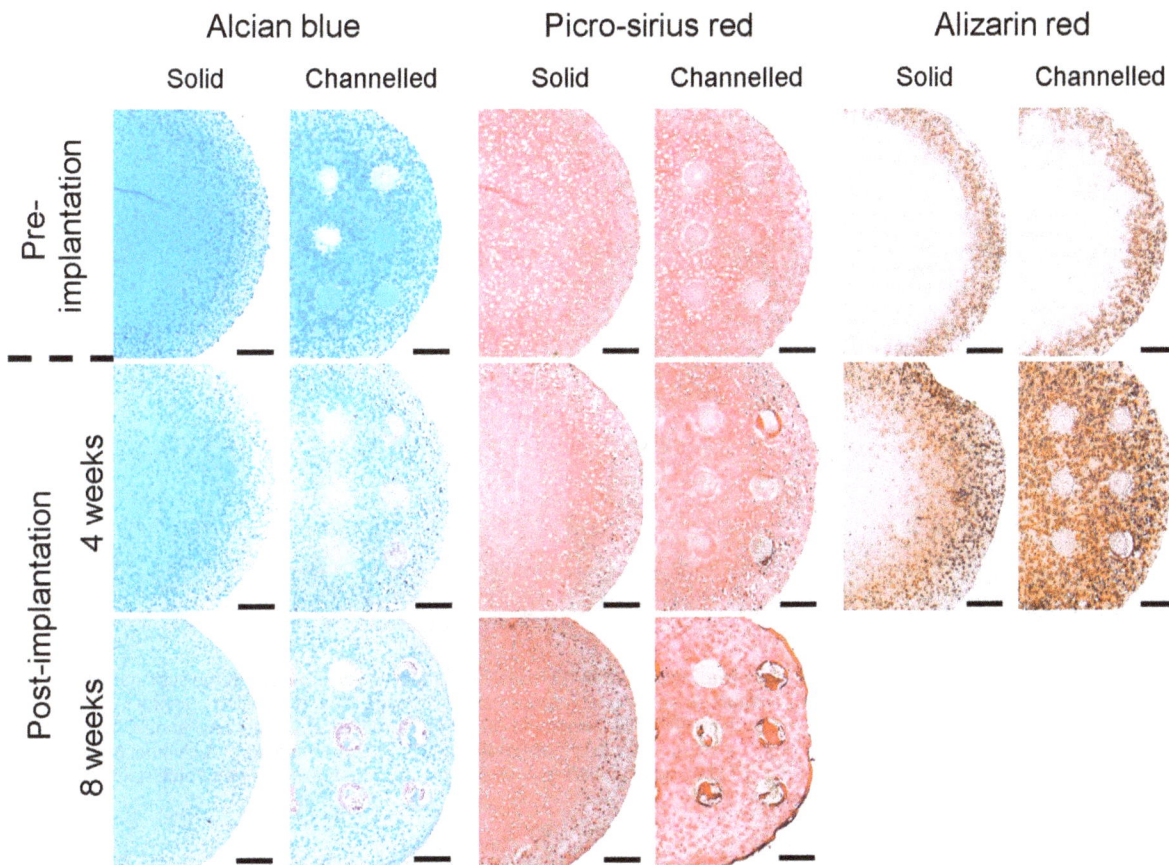

Figure 3. Histology of constructs pre-implantation and post-implantation. Solid and channelled constructs were subjected to 5 weeks culture in chondrogenic conditions, followed by 1 week of culture in hypertrophic conditions, and then implanted subcutaneously into nude mice to be harvested at 4 weeks and 8 weeks post-implantation. Constructs were stained with alcian blue, picro-sirius red and alizarin red to assess sGAG, collagen, and calcium accumulation respectively. 8 weeks post-implantation samples were decalcified prior to histological analysis, hence alizarin red staining was not undertaken at this time point. Images show half the construct. Scale bars are 500 μm.

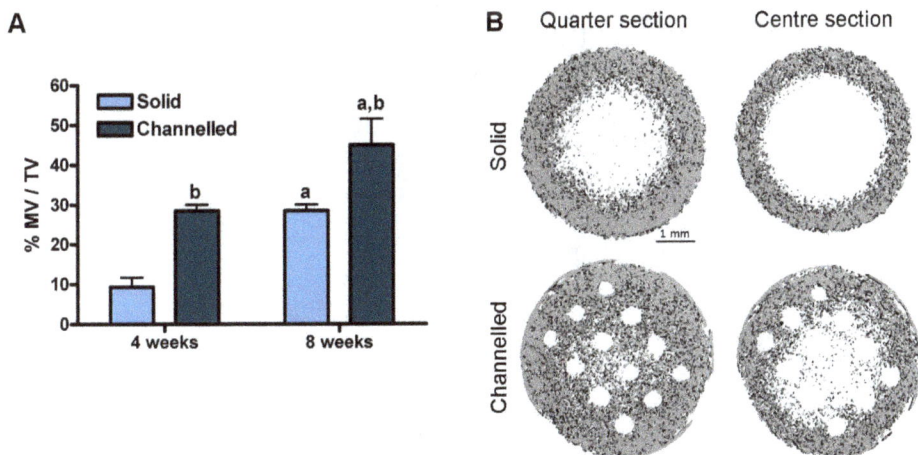

Figure 4. μCT analysis of constructs post-implantation. Constructs were analysed by μCT at 4 weeks and 8 weeks post-implantation to examine mineralisation. (A) % Mineral volume per total volume of construct (% MV/TV). 3 constructs per group were analysed. Significance $p < 0.05$: a vs. 4 weeks post-implantation, b vs. solid constructs. (B) μCT images of constructs 8 weeks post-implantation showing spatial variance of mineralisation with depth through the construct. Quarter section corresponds to a region ~0.75 mm into the depth of the construct. Centre section corresponds to a region ~1.5 mm into the depth of the construct. Sections correspond to a thickness of 120 μm. Images are representative of 3 constructs analysed.

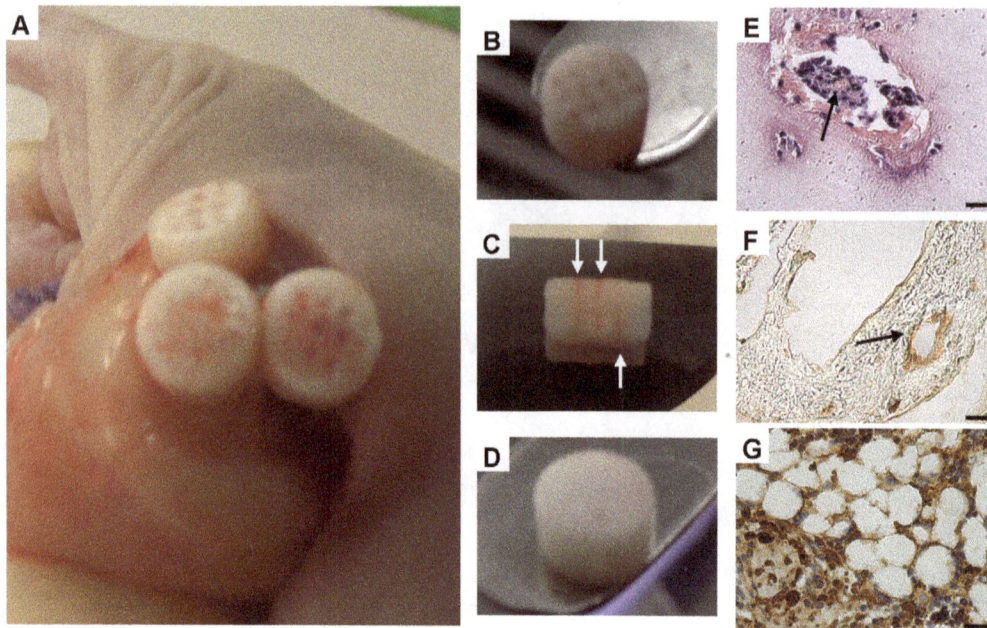

Figure 5. Vascularisation of channelled constructs post-implantation. (A) Retrieval of channelled constructs from the subcutaneous pocket 8 weeks post-implantation. (B) Image of a channelled construct 8 weeks post-implantation. (C) Image of a channelled construct sliced in half longitudinally to enable visualisation through the depth of the channels. White arrows indicate vascularised tissue within the channels. (D) Image of a solid construct 8 weeks post-implantation. (E) H&E staining of a channel 4 weeks post-implantation. (F) CD31 staining of a channel 4 weeks post-implantation. Black arrows indicate vessel-like structures. (G) TdT staining of a channel 4 weeks post-implantation. Brown and blue staining indicates cells positive and negative for TdT respectively. Scale bars are 25 μm.

weakly for alcian blue, and very intensely for collagen type I, see Fig. 7(E,F). At this time point a decrease in the intensity of staining for collagen type II, as well as an increase in the intensity of staining for collagen type X, was also observed within channels as compared to pre-implantation levels, see Fig. 7(G,H). This suggests that the cartilage that had filled the channels *in vitro* supported endochondral bone formation *in vivo*.

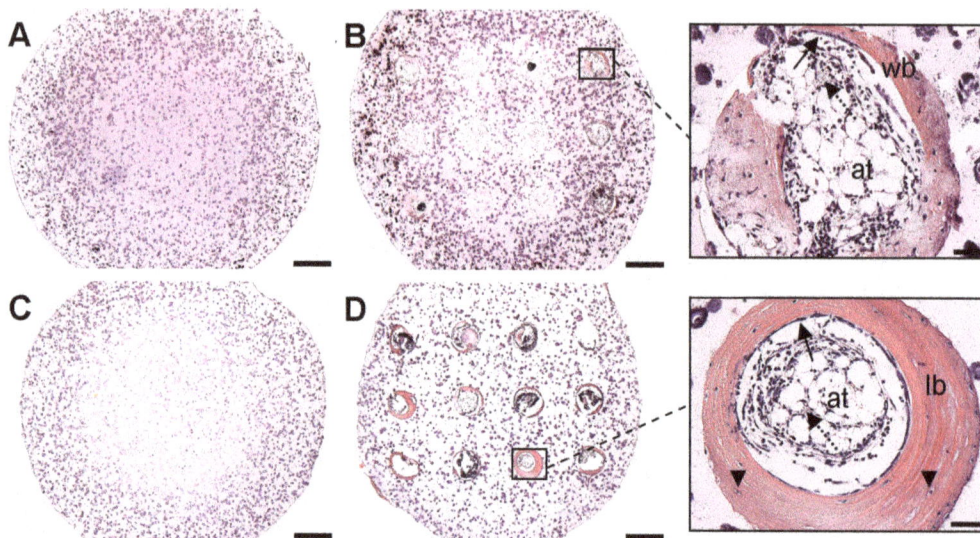

Figure 6. H&E staining of constructs post-implantation. Constructs were stained with H&E at 4 weeks and 8 weeks post-implantation to examine bone formation. (A) Solid construct 4 weeks post-implantation. (B) Channelled construct 4 weeks post-implantation. (C) Solid construct 8 weeks post-implantation. (D) Channelled construct 8 weeks post-implantation. Arrows show lining of osteoblasts laying down new bone, arrowheads show osteocytes embedded within bone matrix, dotted arrows show hematopoietic elements. 'at' – adipose tissue, 'wb' – woven bone, 'lb' – lamellar bone. Main image scale bars are 500 μm. Inset scale bars are 50 μm.

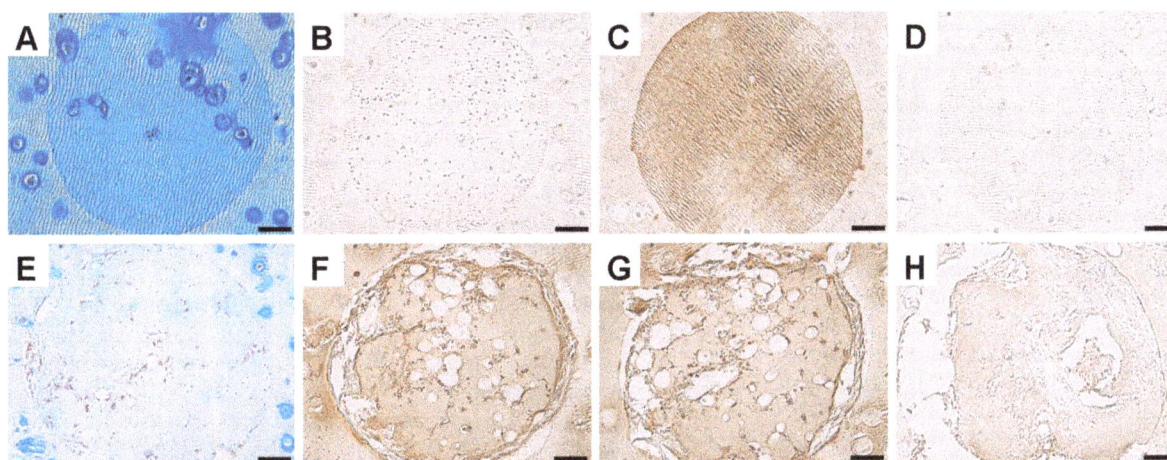

Figure 7. Histology and immunohistochemistry of the channels within channelled constructs pre-implantation and 4 weeks post-implantation. Channels were examined to determine the pathway through which bone formation occurs. (A,E) Alcian blue staining. (B,F) Collagen I staining. (C,G) Collagen II staining. (D,H) Collagen X staining. (A–D) Pre-implantation. (E–H) 4 weeks post-implantation. Scale bars are 50 µm.

Discussion

New vessel formation is critical for bone tissue regeneration. During long bone growth, canals within the developing hypertrophic cartilage facilitate vascular invasion and the migration of osteogenic cells, thus playing a key role in endochondral ossification. The aim of this study was to mimic this function in tissue engineered hypertrophic cartilaginous constructs. We hypothesised that introducing an array of channels into MSC-seeded constructs would firstly facilitate the *in vitro* engineering of scaled-up hypertrophic cartilaginous tissues, and secondly, would accelerate the vascularisation and mineralisation of the engineered graft following implantation *in vivo*. During long-term *in vitro* culture, the presence of this array of channels lead to a reduction in total sGAG and collagen accumulation, whereas calcium deposition was enhanced in channelled constructs. In support of our hypothesis, the channels acted as conduits for blood vessel infiltration and their presence accelerated mineralisation of the engineered tissue *in vivo*. Finally, channels themselves provided a milieu for bone and marrow formation, with *de novo* bone being generated within these channels via the endochondral pathway.

Previous studies have investigated introducing nutrient channels into cartilaginous constructs in order to enhance matrix synthesis and the functional properties of tissue engineered cartilage [24]. For example, the introduction of a single nutrient channel into chondrocyte seeded agarose hydrogels results in superior mechanical properties and the formation of a more uniform fibrillar network whilst maintaining similar levels of sGAG and collagen content [22]. In the present study, scaffold architecture differentially regulated matrix synthesis, with solid constructs accumulating significantly more sGAG and collagen, and channelled constructs accumulating significantly more calcium. It may be that the channels provide a pathway for the diffusion of sGAGs and collagen out of the engineered tissue and into the media, and that calcium deposition is less influenced by such phenomena. Alternatively, or perhaps in conjunction, it may be that the altered rates of sGAG, collagen and calcium deposition in solid and channelled constructs is due to differences in the spatial gradients in regulatory cues that develop in these tissues. In solid hydrogels, the cellular consumption of oxygen at the periphery of the construct will result in the development of a low oxygen microenvironment within the core of the engineered tissue

[37,38], which has been shown to enhance sGAG and collagen synthesis [33,39]. The introduction of channels into scaffolds may increase core oxygen levels [40], which has been shown to enhance hypertrophy and mineralisation of chondrogenically primed MSCs [33,41].

Previous attempts to engineer scaled-up cartilaginous grafts for endochondral bone repair have reported the development of a core region devoid of cells and cartilaginous matrix [12]. This phenomenon was not observed in our study. However, if the endochondral approach is to be implemented to treat critically sized bone defects, further scaling-up of *in vitro* cartilaginous grafts will be required, and at these very large scales nutrient diffusion limitations are likely to develop within the construct. It is perhaps only at these dimensions, and where core degradation would otherwise occur, that the introduction of channels would benefit the *in vitro* development of an engineered cartilaginous tissue.

Following *in vivo* implantation, biochemical and histological analysis of MSC- seeded constructs revealed a reduction in sGAG content and an increase in calcium accumulation, indicating a loss of the chondrogenic phenotype and progression down the endochondral route. Interestingly, while solid constructs maintained their pre-implantation collagen levels during the course of the *in vivo* implantation period, channelled constructs continued to accumulate collagen, which may be due in part to the formation of new tissue within the channels themselves. Both histological data and the DNA assay suggested that channels became highly cellularised once implanted *in vivo*. While we cannot rule out a role played by donor cells, as there were MSCs present within channels pre-implantation, previous studies investigating bone tissue engineering *via* the endochondral pathway have highlighted an important role played by recruited host derived cells in driving endochondral ossification [10,14]. It would appear that the channels in our tissue engineered hypertrophic cartilaginous constructs may be acting as conduits for cellular infiltration from the host, which are then playing a role in laying down new tissue within the construct.

In bone tissue engineering strategies, the prevention of core necrosis during *in vitro* development and following *in vivo* implantation is a significant challenge [5]. In the present study we utilised µCT to investigate the spatial variance of *in vivo* mineralisation with depth through the tissue and found that channelled architectures accelerated mineralisation, and furthermore, promoted the development of a more homogenous mineralised construct and limited the development of a core region devoid of viable cells and mineral. In addition to the inherent increase in nutrients and oxygen due to presence of the channels, they also facilitate blood vessel infiltration as evident by staining for CD31 positive endothelial cells, which in turn further increases oxygen levels and provides soluble factors and signals to promote osteogenesis [42].

Subcutaneous implantation of chondrogenically primed MSCs has been shown to produce endochondral bone containing hematopoietic marrow at 8 weeks post-implantation [8,9]. In the present study, high magnification images of channels at 4 weeks post implantation demonstrated deposition of an immature woven bone, being laid down by osteoblasts, which at 8 weeks post-implantation had developed into lamellar-like bone with osteocyte-like cells evident within the bone matrix. Temporal analysis of chondrogenic extracellular matrix progression within channels indicated degradation of the sGAG/collagen type II rich-matrix following *in vivo* implantation, and a corresponding increase in the accumulation of collagen type I and collagen type X, suggesting that *de novo* bone found in the channels was being generated *via* the endochondral pathway. Moreover, at both *in vivo* time points there was evidence of hematopoietic foci and marrow adipose tissue, enclosed within the developing bone matrix. Given that it has previously been demonstrated that endochondral ossification is required for hematopoietic stem cell niche formation [43], this provides further validation that the mechanism through which bone is formed within the channels is indeed endochondral.

In the current study cartilaginous tissues were engineered using 4 month old, skeletally immature, porcine MSCs. Skeletally immature MSC seeded hydrogels have been shown to generate cartilaginous tissues with superior matrix deposition and mechanical properties as compared to skeletally mature MSC seeded hydrogels [44]. Further studies are required to confirm if the beneficial effect of channels, as shown in this study, would occur with the use of engineered tissues generated from a skeletally mature donor where the inherent regenerative capacity is not as high.

Bone tissue engineering *via* endochondral ossification has recently been demonstrated with collagen mesh scaffolds [12], on three-dimensional electrospun fibers [45], and with the bone void filler NuOss [46]. Utilising hydrogels as scaffolds for endochondral bone regeneration may be a particularly powerful tool in 'scaling up' tissue engineered grafts in order to treat defects

of a clinically relevant size [38]. However, the chondrogenically primed MSC-seeded agarose hydrogels utilised in this study appeared to only partially support progression towards endochondral ossification, with the engineered tissue laid down in the main body of the hydrogel apparently locked within a hypertrophic, calcified cartilage state at 8 weeks post- implantation. Since the cartilaginous matrix within channels, which had filled up at least partially *in vitro* with scaffold-free tissue, had the ability to undergo full endochondral ossification *in vivo* it would appear that the agarose hydrogel acted as a barrier for vascularisation and the transition from mineralised cartilage to endochondral bone. This may be due to the agarose hydrogel undergoing minimal degradation during the 8 week *in vivo* period. A previous study investigating alginate hydrogels as scaffolds for bone tissue engineering demonstrated enhanced bone formation by accelerating the degradation properties of the hydrogel [47]. Future studies in our lab will examine the capacity of different naturally derived biodegradable hydrogels to produce endochondral bone *in vivo*, with the ultimate goal of scaling up anatomically accurate cartilaginous grafts as a paradigm for whole bone tissue engineering via endochondral ossification. However, all scaffolds or hydrogels will likely impede vascularisation to some degree, and adopting the channel strategy developed in this study is expected to accelerate the process of endochondral ossification regardless of the hydrogel used to tissue engineer hypertrophic cartilage grafts for bone regeneration.

Conclusion

Inspired by the cartilage canal network observed during endochondral skeletogenesis, this study demonstrated that tissue engineering channelled hypertrophic cartilaginous constructs accelerates mineralisation and vascularisation of the graft *in vivo* and shows promise for use in future endochondral bone regeneration strategies. The study reinforces the importance of optimising the architecture of engineered constructs targeting bone tissue regeneration, even if this is achieved via an endochondral pathway as opposed to the traditional intramembranous route.

Author Contributions

Conceived and designed the experiments: ES TV MT CB DK. Performed the experiments: ES TV. Analyzed the data: ES MT. Contributed reagents/materials/analysis tools: ES CB. Wrote the paper: ES DK.

References

1. Langer R (2000) Tissue Engineering. Molecular Therapy 1: 12–15.
2. Koh CJ, Atala A (2004) Tissue Engineering, Stem Cells, and Cloning: Opportunities for Regenerative Medicine. Journal of the American Society of Nephrology 15: 1113–1125.
3. Meijer GJ, De Bruijn JD, Koole R, Van Blitterswijk CA (2007) Cell-based bone tissue engineering. PLoS Medicine 4: 0260–0264.
4. Santos MI, Reis RL (2010) Vascularization in bone tissue engineering: Physiology, current strategies, major hurdles and future challenges. Macromolecular Bioscience 10: 12–27.
5. Lyons FG, Al-Munajjed AA, Kieran SM, Toner ME, Murphy CM, et al. (2010) The healing of bony defects by cell-free collagen-based scaffolds compared to stem cell-seeded tissue engineered constructs. Biomaterials 31: 9232–9243.
6. Jukes JM, Both SK, Leusink A, Sterk LMT, Van Blitterswijk CA, et al. (2008) Endochondral bone tissue engineering using embryonic stem cells. Proceedings of the National Academy of Sciences of the United States of America 105: 6840–6845.
7. Farrell E, Van Der Jagt OP, Koevoet W, Kops N, Van Manen CJ, et al. (2009) Chondrogenic priming of human bone marrow stromal cells: A better route to bone repair? Tissue Eng Part C Methods 15: 285–295.
8. Scotti C, Tonnarelli B, Papadimitropoulos A, Scherberich A, Schaeren S, et al. (2010) Recapitulation of endochondral bone formation using human adult mesenchymal stem cells as a paradigm for developmental engineering.

Proceedings of the National Academy of Sciences of the United States of America 107: 7251–7256.
9. Janicki P, Kasten P, Kleinschmidt K, Luginbuehl R, Richter W (2010) Chondrogenic pre-induction of human mesenchymal stem cells on β-TCP: Enhanced bone quality by endochondral heterotopic bone formation. Acta Biomaterialia 6: 3292–3301.
10. Tortelli F, Tasso R, Loiacono F, Cancedda R (2010) The development of tissue-engineered bone of different origin through endochondral and intramembranous ossification following the implantation of mesenchymal stem cells and osteoblasts in a murine model. Biomaterials 31: 242–249.
11. Lau TT, Lee LQP, Vo BN, Su K, Wang DA (2012) Inducing ossification in an engineered 3D scaffold-free living cartilage template. Biomaterials 33: 8406–8417.
12. Scotti C, Piccinini E, Takizawa H, Todorov A, Bourgine P, et al. (2013) Engineering of a functional bone organ through endochondral ossification. Proceedings of the National Academy of Sciences of the United States of America 110: 3997–4002.
13. Sheehy EJ, Vinardell T, Buckley CT, Kelly DJ (2013) Engineering osteochondral constructs through spatial regulation of endochondral ossification. Acta Biomaterialia 9: 5484–5492.
14. Farrell E, Both SK, Odörfer KI, Koevoet W, Kops N, et al. (2011) In-vivo generation of bone via endochondral ossification by in-vitro chondrogenic

priming of adult human and rat mesenchymal stem cells. BMC Musculoskeletal Disorders 12.

15. Pelttari K, Winter A, Steck E, Goetzke K, Hennig T, et al. (2006) Premature induction of hypertrophy during in vitro chondrogenesis of human mesenchymal stem cells correlates with calcification and vascular invasion after ectopic transplantation in SCID mice. Arthritis and Rheumatism 54: 3254–3266.

16. Gerber HP, Vu TH, Ryan AM, Kowalski J, Werb Z, et al. (1999) VEGF couples hypertrophic cartilage remodeling, ossification and angiogenesis during endochondral bone formation. Nat Med 5: 623–628.

17. Lee CH, Marion NW, Hollister S, Mao JJ (2009) Tissue formation and vascularization in anatomically shaped human joint condyle ectopically in vivo. Tissue Engineering - Part A 15: 3923–3930.

18. Martin I, Wendt D, Heberer M (2004) The role of bioreactors in tissue engineering. Trends Biotechnol 22: 80–86.

19. Mauck RL, Soltz MA, Wang CCB, Wong DD, Chao PHG, et al. (2000) Functional tissue engineering of articular cartilage through dynamic loading of chondrocyte-seeded agarose gels. Journal of Biomechanical Engineering 122: 252–260.

20. Mauck RL, Nicoll SB, Seyhan SL, Ateshian GA, Hung CT (2003) Synergistic action of growth factors and dynamic loading for articular cartilage tissue engineering. Tissue Eng 9: 597–611.

21. Thorpe SD, Buckley CT, Vinardell T, O'Brien FJ, Campbell VA, et al. (2010) The response of bone marrow-derived mesenchymal stem cells to dynamic compression following tgf-β3 induced chondrogenic differentiation. Annals of Biomedical Engineering 38: 2896–2909.

22. Bian L, Angione SL, Ng KW, Lima EG, Williams DY, et al. (2009) Influence of decreasing nutrient path length on the development of engineered cartilage. Osteoarthritis Cartilage 17: 677–685.

23. Zhang Y, Yang F, Liu K, Shen H, Zhu Y, et al. (2012) The impact of PLGA scaffold orientation on invitro cartilage regeneration. Biomaterials 33: 2926–2935.

24. Buckley CT, Thorpe SD, Kelly DJ (2009) Engineering of large cartilaginous tissues through the use of microchanneled hydrogels and rotational culture. Tissue Engineering - Part A 15: 3213–3220.

25. Sheehy EJ, Buckley CT, Kelly DJ (2011) Chondrocytes and bone marrow-derived mesenchymal stem cells undergoing chondrogenesis in agarose hydrogels of solid and channelled architectures respond differentially to dynamic culture conditions. Journal of Tissue Engineering and Regenerative Medicine 5: 747–758.

26. Mesallati T, Buckley CT, Nagel T, Kelly DJ (2012) Scaffold architecture determines chondrocyte response to externally applied dynamic compression. Biomechanics and Modeling in Mechanobiology: 1–11.

27. Ganey TM, Ogden JA, Sasse J, Neame PJ, Hilbelink DR (1995) Basement membrane composition of cartilage canals during development and ossification of the epiphysis. Anatomical Record 241: 425–437.

28. Ytrehus B, Carlson CS, Lundeheim N, Mathisen L, Reinholt FP, et al. (2004) Vascularisation and osteochondrosis of the epiphyseal growth cartilage of the distal femur in pigs - Development with age, growth rate, weight and joint shape. Bone 34: 454–465.

29. Blumer MJF, Longato S, Fritsch H (2008) Structure, formation and role of cartilage canals in the developing bone. Annals of Anatomy 190: 305–315.

30. Blumer MJF, Schwarzer C, Pérez MT, Konakci KZ, Fritsch H (2006) Identification and location of bone-forming cells within cartilage canals on

their course into the secondary ossification centre. Journal of Anatomy 208: 695–707.

31. Lenas P, Moos M, Luyten FP (2009) Developmental engineering: A new paradigm for the design and manufacturing of cell-based products. Part I: From three-dimensional cell growth to biomimetics of in Vivo development. Tissue Engineering - Part B: Reviews 15: 381–394.

32. Lennon DP, Caplan AI (2006) Isolation of human marrow-derived mesenchymal stem cells. Exp Hematol 34: 1604–1605.

33. Sheehy EJ, Buckley CT, Kelly DJ (2012) Oxygen tension regulates the osteogenic, chondrogenic and endochondral phenotype of bone marrow derived mesenchymal stem cells. Biochemical and Biophysical Research Communications 417: 305–310.

34. Vinardell T, Sheehy EJ, Buckley CT, Kelly DJ (2012) A comparison of the functionality and in vivo phenotypic stability of cartilaginous tissues engineered from different stem cells sources. Tissue Engineering Part A 18: 1161–1170.

35. Kafienah W, Sims TJ (2004) Biochemical methods for the analysis of tissue-engineered cartilage. Methods Mol Biol 238: 217–230.

36. Ignat'eva NY, Danilov NA, Averkiev SV, Obrezkova MV, Lunin VV, et al. (2007) Determination of hydroxyproline in tissues and the evaluation of the collagen content of the tissues. J Anal Chem 62: 51–57.

37. Thorpe SD, Nagel T, Carroll SF, Kelly DJ (2013) Modulating Gradients in Regulatory Signals within Mesenchymal Stem Cell Seeded Hydrogels: A Novel Strategy to Engineer Zonal Articular Cartilage. PLoS ONE 8.

38. Buckley CT, Meyer EG, Kelly DJ (2012) The influence of construct scale on the composition and functional properties of cartilaginous tissues engineered using bone marrow-derived mesenchymal stem cells. Tissue Engineering - Part A 18: 382–396.

39. Meyer EG, Buckley CT, Thorpe SD, Kelly DJ (2010) Low oxygen tension is a more potent promoter of chondrogenic differentiation than dynamic compression. Journal of Biomechanics 43: 2516–2523.

40. Buckley CT, O'Kelly KU (2010) Fabrication and characterization of a porous multidomain hydroxyapatite scaffold for bone tissue engineering investigations. Journal of Biomedical Materials Research - Part B Applied Biomaterials 93: 459–467.

41. Gawlitta D, Van Rijen MHP, Schrijver EJM, Alblas J, Dhert WJA (2012) Hypoxia impedes hypertrophic chondrogenesis of human multipotent stromal cells. Tissue Engineering - Part A 18: 1957–1966.

42. Brandi ML, Collin-Osdoby P (2006) Vascular biology and the skeleton. Journal of Bone and Mineral Research 21: 183–192.

43. Chan CKF, Chen CC, Luppen CA, Kim JB, DeBoer AT, et al. (2009) Endochondral ossification is required for haematopoietic stem-cell niche formation. Nature 457: 490–494.

44. Erickson IE, Van Veen SC, Sengupta S, Kestle SR, Mauck RL (2011) Cartilage matrix formation by bovine mesenchymal stem cells in three-dimensional culture is age-dependent. Clinical Orthopaedics and Related Research 469: 2744–2753.

45. Yang W, Yang F, Wang Y, Both SK, Jansen JA (2013) In vivo bone generation via the endochondral pathway on three-dimensional electrospun fibers. Acta Biomaterialia 9: 4505–4512.

46. Weiss HE, Roberts SJ, Schrooten J, Luyten FP (2012) A semi-autonomous model of endochondral ossification for developmental tissue engineering. Tissue Engineering - Part A 18: 1334–1343.

47. Simmons CA, Alsberg E, Hsiong S, Kim WJ, Mooney DJ (2004) Dual growth factor delivery and controlled scaffold degradation enhance in vivo bone formation by transplanted bone marrow stromal cells. Bone 35: 562–569.

14

Induced Collagen Cross-Links Enhance Cartilage Integration

Aristos A. Athens[1,2], Eleftherios A. Makris[1,3], Jerry C. Hu[1]*

1 Department of Biomedical Engineering, University of California Davis, Davis, California, United States of America, **2** Davis Senior High School, Davis, California, United States of America, **3** Department of Orthopedic Surgery and Musculoskeletal Trauma, University of Thessaly, Larisa, Greece

Abstract

Articular cartilage does not integrate due primarily to a scarcity of cross-links and viable cells at the interface. The objective of this study was to test the hypothesis that lysyl-oxidase, a metalloenzyme that forms collagen cross-links, would be effective in improving integration between native-to-native, as well as tissue engineered-to-native cartilage surfaces. To examine these hypotheses, engineered cartilage constructs, synthesized via the self-assembling process, as well as native cartilage, were implanted into native cartilage rings and treated with lysyl-oxidase for varying amounts of time. For both groups, lysyl-oxidase application resulted in greater apparent stiffness across the cartilage interface 2–2.2 times greater than control. The construct-to-native lysyl-oxidase group also exhibited a statistically significant increase in the apparent strength, here defined as the highest observed peak stress during tensile testing. Histology indicated a narrowing gap at the cartilage interface in lysyl-oxidase treated groups, though this alone is not sufficient to indicate annealing. However, when the morphological and mechanical data are taken together, the longer the duration of lysyl-oxidase treatment, the more integrated the interface appeared. Though further data are needed to confirm the mechanism of action, the enhancement of integration may be due to lysyl-oxidase-induced pyridinoline cross-links. This study demonstrates that lysyl-oxidase is a potent agent for enhancing integration between both native-to-native and native-to-engineered cartilages. The fact that interfacial strength increased manifold suggests that cross-linking agents should play a significant role in solving the difficult problem of cartilage integration. Future studies must examine dose, dosing regimen, and cellular responses to lysyl-oxidase to optimize its application.

Editor: Alejandro Almarza, University of Pittsburgh, United States of America

Funding: This work was supported by the DHS Blue and White Foundation. The funders had no role in study design, data collection and analysis, decision to publish, or preparation of the manuscript.

Competing Interests: The authors have declared that no competing interests exist.

* E-mail: jcyhu@ucdavis.edu

Introduction

Because of articular cartilage's lack of inherent healing potential, lesions tend to degenerate to osteoarthritis (OA), a significant problem affecting over a third of adults aged 65 and over [1]. Currently, there are no cartilage treatments that offer long-term functionality. Mosaicplasty and microfracture require defect site preparation via cartilage removal. Subsequently, the defect is filled by either cartilage plugs or a "super clot" [2]. Autografts and allografts are also options. For these and other procedures, success is predicated upon the fill tissue's integration with native cartilage. Various strategies and materials have been proposed to integrate cartilage and bone [3–6]. However, cartilage-to-cartilage integration has proven to be notoriously difficult, even when using tissue engineering approaches [7,8]. To achieve long-term, durable repair, grafts and engineered articular cartilage alike need to be integrated with native cartilage. Without proper integration, the implant will fall out of place or degrade rapidly [9], likely due to the high stress concentrations that occur at cartilage interfaces *in vivo*.

The general consensus regarding the main factors that hinder integration are: 1) Cell death, at the wound edge [8] and in surgically prepared defects, leads to metabolically inactive tissue, which prevents cell adhesion and migration to the injury site [10–14]. 2) Cell migration to the wound edge is hindered by the dense collagen network [10–14]; in native cartilage, cells are locked into lacunae and are not observed to migrate [15]. 3) Lack of cross-links between matrices of native and implant tissues [16,17]. In short, the insufficiency of viable cells at the wound edge prevents synthesis of integrative matrix between the two surfaces to be joined [12–14,18,19], in part by lack of matrix synthesis. Even when viable cells are present, the newly synthesized matrix may not be sufficiently cross-linked to the native tissue. This study aims to overcome all of these factors by supplying viable cells to the interface via engineered neocartilage to mitigate the issues of cell death and lack of cell migration at the wound edge by exogenously inducing cross-links.

One way to deliver cells at an interface may be via the use of constructs engineered using the self-assembling process, which is an established method for generating tissue with abundant cells at the construct edge [20]. This method has also generated neocartilage with properties approaching those of native tissue [20]. Maintenance of cartilage with normal functional properties requires sustaining cell density; large areas of cell death would undoubtedly result in biomechanically inferior matrix or none at all [21]. Thus, this study seeks to use tissue engineered constructs created via chondrocyte self-assembly to deliver a higher cell density to the wound edge to enhance integration.

Another suggested mechanism for the enhancement of integration is collagen pyridinoline (PYR) cross-links [22]. PYR cross-links have been shown to be a major factor in determining the stiffness of connective tissues. PYR naturally forms within cartilage and other musculoskeletal tissues during development and aging via the enzyme lysyl oxidase (LOX), a metalloenzyme that converts amine side-chains of lysine and hydroxylysine into aldehydes. *In vivo*, LOX is most active at sites of growing collagen fibrils [1]. A potential method for inducing collagen cross-linking across cartilage interfaces is thus the exogenous application of this enzyme. Since LOX is a small-sized molecule, at roughly ~50 kDa, and since cross-link formation occurs over several weeks, exogenous LOX can be applied to *in vitro* cultures on a continuous basis to ensure full penetration via diffusion and to allow sufficient time for cross-link formation. By employing LOX, one would expect the formation of "anchoring" sites, composed of PYR cross-links in the collagen network of the engineered tissue as well as of the native tissue, to bridge the two tissues together. Thus, LOX application combined with the delivery of high cell numbers to the wound edge are expected to promote tissue integration.

Using the self-assembling process, the objective of this study was to determine whether LOX can alter the integration of native-to-construct and native-to-native tissue systems through two experiments. It was hypothesized that application of LOX would enhance integration, as evidenced through tensile measurements. The first experiment sought to examine whether LOX would promote integration between native cartilage and neocartilage and to determine time and duration of application. The second experiment sought to determine whether the results from the first experiment can be replicated in a native-to-native cartilage system.

Materials and Methods

Cell and tissue harvest

Articular cartilage was harvested from distal femurs of one-week old male calves (Research 87 Inc., Boston, MA) less than 36 hr after sacrifice. To obtain the cells, following harvest, the tissue was digested in 0.2% collagenase type II (Worthington, Lakewood, NJ) in culture medium for 24 hr as previously described [23]. Culture medium formulation is as follows: DMEM with 4.5 mg/mL glucose and L-glutamine, 100 nM dexamethasone, 1% fungizone, 1% penicillin/streptomycin, 1% ITS+, 50 µg/mL ascorbate-2-phosphate, 40 µg/mL L-proline, and 100 µg/mL sodium pyruvate. Cell viability was assessed using trypan blue exclusion, and cells were frozen at $-80°C$ using DMEM containing 20% fetal bovine serum (Atlanta Biologicals, Lawrenceville, GA) and 10% dimethyl sulfoxide until use. To reduce animal variability, cells from four animals were pooled together for cell seeding.

Self-assembly of constructs

Cylindrical, non-adherent, agarose wells were prepared by placing 5 mm diameter stainless steel posts in 48 well plates filled with 1 ml of 2% molten agarose, as previously described [24]. After the agarose gelled, posts were removed. The resultant wells were saturated with two exchanges of medium. After thawing, cells were counted, viability was assessed using trypan blue exclusion, and cells were seeded into the agarose wells at a concentration of 5.5 million/100 µl medium. After 4 hr, an additional 400 µl of medium was added per well. To prevent disruption of the construct, complete medium change did not occur until after 24 hr. Constructs were cultured at 10% CO_2, 37°C, in a humidified incubator for a total of t = 28 d (t = 1 d defined as 24 hr post seeding). Medium was changed daily (500 µl).

Tissue integration

To examine the study's hypotheses, two separate, but concurrent, experiments were conducted. First, the use of LOX was examined for the construct-to-native interface. At t = 28 d, engineered constructs were removed from culture and prepared for integration with native articular cartilage. The t = 28 d culture time was chosen to coincide with prior work in self-assembled cartilage and with other cartilage tissue engineering efforts. Bovine articular cartilage explants, measuring 6 mm×1 mm, were harvested using biopsy punches. A concentric, 4 mm diameter defect was punched from the explant. From the engineered constructs, 4 mm diameter biopsies were obtained and press-fitted into the defect in the explant (Fig. 1). To ensure that all constructs were in firm contact with the explants, cyanoacrylate was applied; a penetration depth of 25 µm (~2.5% of thickness) and degradation within the culture period were verified using histology. These construct/explant assemblies were cultured for an additional 14 d, at which point they were removed for assessments. The second experiment consisted entirely of explants instead of constructs. Native-to-native tissue assemblies were formed using the same methods as described above.

Collagen cross-linking via lysyl oxidase

The LOX medium contained a concentration of 0.15 µg/ml LOX (GenWay Biotech, Inc., San Diego, CA). This concentration is based a pilot study in which three concentrations, 0.0015, 0.015, and 0.15 µg/ml of LOX, were examined. The results showed that only 0.15 µg/ml of LOX improved pyridinoline content over the culture duration employed; at this concentration, neither the collagen nor glycosaminoglycan per wet weight (collagen/ww and GAG/ww, respectively) was altered when LOX was applied to

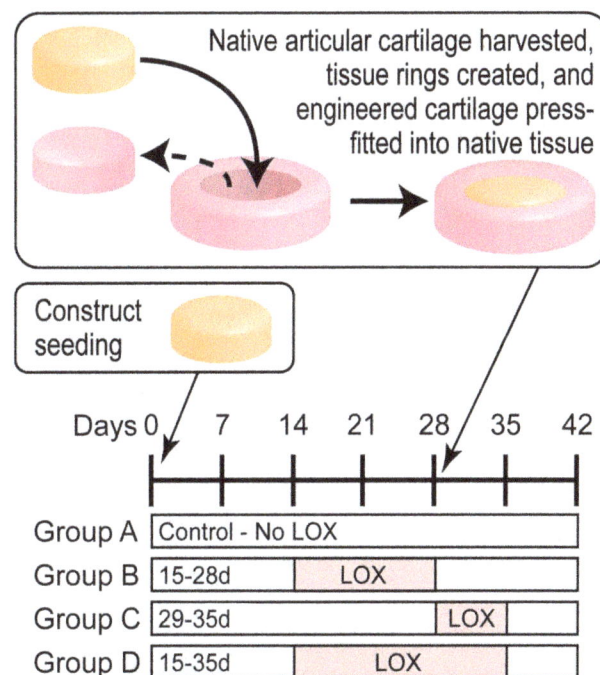

Figure 1. Schematic of the experiment examining integration of tissue engineered cartilage to native cartilage. For Group B, LOX was applied during construct formation, t = 15–28 d. For Group C, LOX was applied after forming the construct-to-native assemblies, t = 29–35 d. For Group D, LOX was applied both before and after the formation of the construct-to-native assemblies, t = 15–35 d.

either native or engineered cartilage separately. For the construct-to-native study, four groups were examined: The control (Group A) consisted of construct/explant assemblies maintained in culture medium only. Group B was treated with the LOX medium during $t = 15–28$ d. Group C was treated during $t = 29–35$ d. Group D was treated during $t = 15–35$ d. These groups were chosen to examine LOX treatment prior (Group B), after (Group C), or throughout (Group D) integration with the native tissue. Assemblies were assessed at $t = 42$ d to allow for a total of 14 d of integration time. For the native-to-native study, two groups were examined. The control group was allowed to integrate for 14 d in culture medium, while the LOX Group was maintained in a LOX medium during the same time.

Histology

Frozen sections were collected at 14 µm on positively charged slides to promote maximal adherence. These were fixed in 10% neutral buffered formalin and stained. Picrosirius red was used to demonstrate collagen distribution, as previously described [20]. After staining, slides were dehydrated through ascending alcohol percentages (50%, 60%, 70%, 80%, 90%, 95%, 100%) to minimize dehydration artifacts, and coverslips were applied using Permount.

Biochemistry

Total collagen was assessed using a hydroxyproline assay, and glycosaminoglycan (GAG) content was measured using a Blyscan kit, both as previously described [20].

Tensile testing

Assemblies were cut into strips 1 mm wide. Thickness and width were verified photographically using ImageJ (National Institutes of Health, Bethesda, MD). Specimens were glued onto test strips separated by a pre-defined spacing of 1 mm, and the strips were clamped and exposed to constant uniaxial strain of 1% of the 1 mm spacing per second until failure using a uniaxial materials testing machine (Instron 5565). Force-deformation data were collected and then normalized with respect to the cross-sectional area and initial spacing length of the specimens. From this, an apparent "stiffness" was derived by calculating the slope of the linear region of the graph. The ultimate tensile apparent strength (UTS) was defined as the maximum stress attained by the specimen before failure.

Statistical analysis

Based on prior data used to determine LOX concentration, application time, and effects on cellular activity, a power analysis was performed to determine an $n = 6$ required to discern differences in tensile properties at $p < 0.05$. Data were compiled as mean ± standard deviation and analyzed using a single factor ANOVA. If the F-test was statistically significant, a Tukey's *post hoc* test was employed to identify significant groups. Significance was defined as $p < 0.05$.

Results

Integration of engineered constructs to native articular cartilage

For all treatment durations, LOX-treated construct-to-native assemblies displayed better integration as compared to controls using gross morphology, histology, and biomechanical evaluations. Prior to histological processing, the assemblies were evaluated straight from culture for gross morphology. Although LOX

addition increased the stiffness of the assemblies, it did not affect the size and dimensions of the samples. Grossly, gaps were seen between the construct and native tissue in 33% of the controls (Fig. 2). Gaps were not seen for any of the LOX-treated specimens. Similarly, histological evaluation showed gaps in the controls, while LOX-treated samples showed construct adherence to the native tissue. Tensile testing across the integration interface showed significantly higher apparent stiffness when LOX was applied during $t = 15–35$ d (Group D) (1.6 ± 0.6 MPa, versus control values of 0.7 ± 0.2 MPa (Fig. 3, top)). Significantly higher apparent strength values were observed for both Groups B and D (0.42 ± 0.07 MPa and 0.39 ± 0.06 MPa, respectively), where LOX was applied before formation of the construct-to-native assembly (Fig. 3, bottom). Control and Group C values were 0.23 ± 0.08 MPa and 0.28 ± 0.1 MPa. No significant differences were detected in the GAG/ww or collagen/ww among the construct or the explant portions of the assemblies. Specifically, no significant differences were detected in the GAG/ww content among the constructs ($4.5 \pm 1.4\%$, $2.9 \pm 1.4\%$, $3.1 \pm 0.5\%$, and $4.5 \pm 0.6\%$ for Groups A–D, respectively) or among the explant rings ($8.6 \pm 1.1\%$, $10.9 \pm 1.9\%$, $9.6 \pm 01.2\%$, and $10.5 \pm 2.8\%$ for Groups A-D, respectively). Additionally, no significant differences were detected in the collagen/ww content among the constructs ($5.4 \pm 2.2\%$, $5.7 \pm 1.7\%$, $5.7 \pm 0.7\%$, and $7.2 \pm 3.1\%$ for Groups A-D, respectively) or among the explant rings ($5.6 \pm 3\%$, $10.6 \pm 7.5\%$, $6.2 \pm 3\%$, and $10.5 \pm 3.2\%$ for Groups A-D, respectively).

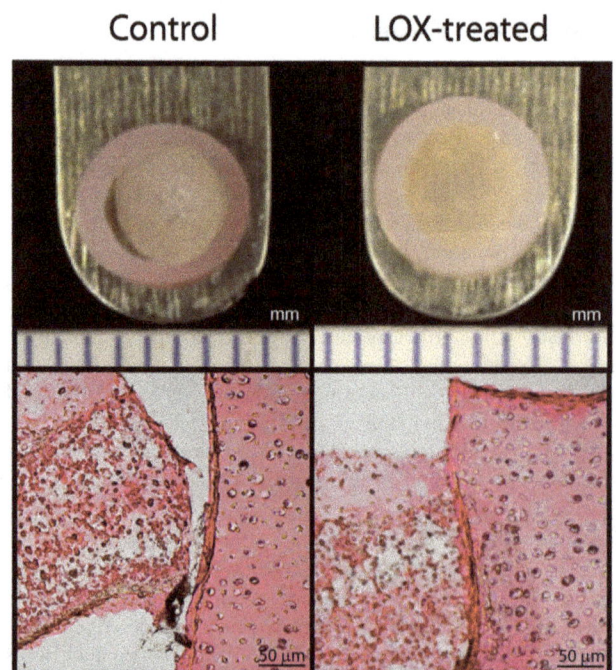

Figure 2. Gross morphology and histology of constructs/explant assemblies. Straight from culture, most controls resembled LOX-treated samples, though gaps were seen in one-third of the controls (upper left panel). None of the LOX-treated samples displayed gaps that were grossly visible; a representative sample (Group D) is shown (upper right). Gaps in the controls were also seen after histological processing using picrosirius red (lower left) versus LOX-treated samples (lower right, Group D).

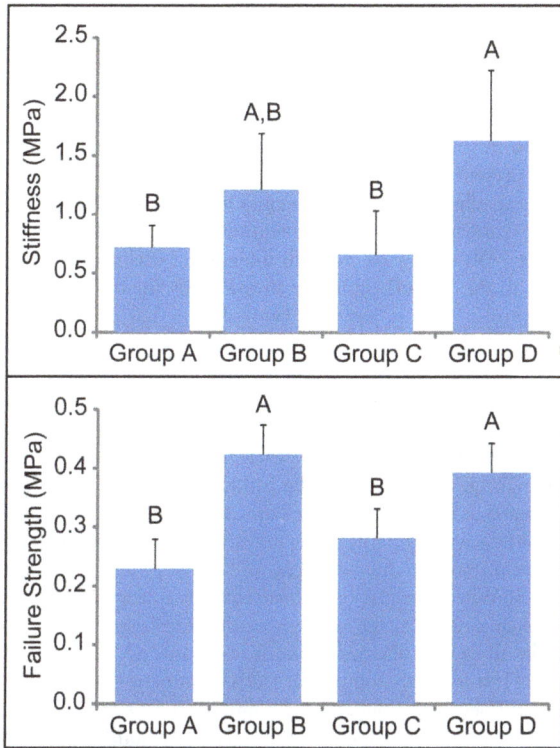

Figure 3. Tensile mechanical data of construct/explant interface. Significantly higher apparent stiffness (top) was seen when LOX was applied during t = 15–35 d (Group D) than controls (Group A). Significantly higher apparent strength was obtained across the integration interface when engineered cartilage was treated with LOX before being press-fitted into the native cartilage (bottom). Bars with different letters are significantly different (p<0.05).

Integration between native cartilage tissues

Qualitatively, the only difference between control and LOX-treated native-to-native assemblies was seen by histology (Fig. 4). The apparent stiffness of the LOX-treated group was more than twice that of the control (1.5±1.1 MPa versus 0.7±0.4 MPa), though neither this property nor the apparent strength were statistically significant (Fig. 5). This is potentially due to biological variations, since, in contrast with the engineered tissues which were formed using cells pooled from multiple animals, each native-to-native assembly is derived from a different animal. GAG and collagen content were not different between the two groups.

Discussion

Motivated by the as-of-yet unsolved issue of cartilage integration, the objective of this study was to examine the hypothesis that LOX would induce cartilage integration. This enzyme naturally occurs in cartilage and promotes PYR cross-links in collagen, thereby holding potential for strengthening cartilage-to-cartilage interfaces. The hypothesis was proven to be correct as evidenced by the biomechanical and histological data. At the dosage applied, this naturally occurring enzyme did not alter cellular response with respect to collagen and GAG production. Engineered tissues, formed using a self-assembling process, were integrated to native tissue explants by applying LOX to a ring-and-implant assembly (Fig. 1). Additionally, LOX was applied to native-to-native cartilage interfaces to examine whether this novel integration method can also be applicable to cases where there is not an

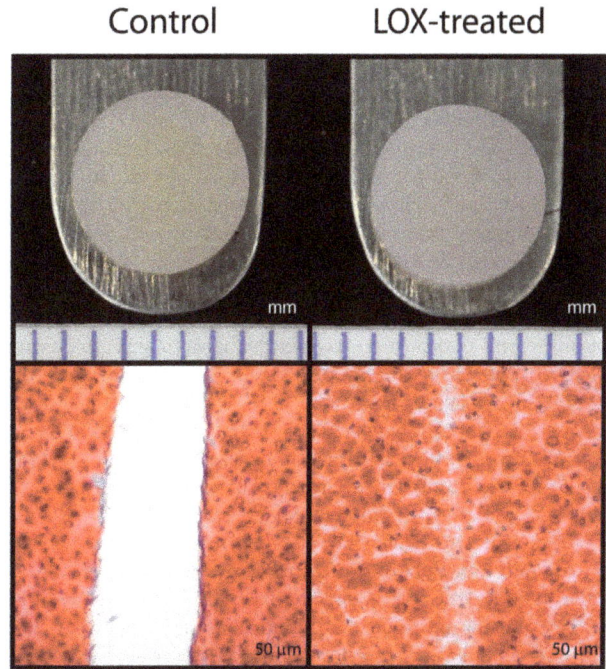

Figure 4. Gross morphology and histology of explant/explant assemblies. Neither control nor LOX-treated native-to-native assemblies displayed grossly visible gaps when removed from culture (top row). However, gaps can be seen after histological processing using picrosirius red in the control group, unlike the LOX-treated group (bottom).

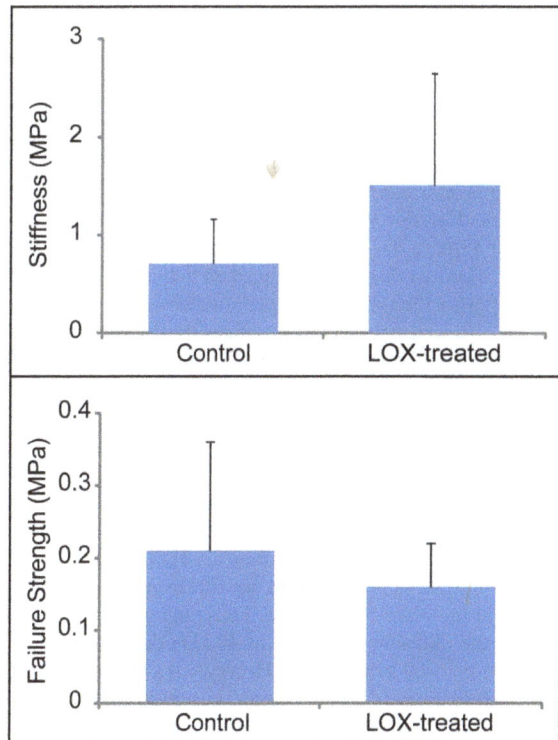

Figure 5. Tensile mechanical data of explant/explant interface. No differences in apparent stiffness (top) or apparent strength (bottom) were seen when LOX was applied to native-to-native integration.

abundance of cells at the wound edge. The results showed that cartilage integration can be enhanced if the interface is stocked with metabolically active cells and PYR cross-links simultaneously. Enhanced interfacial properties were observed for construct-to-native but not for the native-to-native case. Group D, which was treated with LOX for the longest period of time, had statistically higher tensile properties at the interface than did the other three groups. Specifically, Group D had approximately 2.2 times the tensile strength of controls. This was confirmed with morphological and histological data. The results of this study are significant for both current and prospective cartilage regeneration and repair methods.

It is worth noting that, despite the lack of any significant differences in the collagen and GAG content in either the construct or explant groups, there were significant LOX-induced increases in interface biomechanics. The fact that interfacial mechanical properties (apparent stiffness and apparent strength) increased significantly in the absence of increases in the main extracellular matrix (ECM) components suggests that cross-links play a central role in integration. Unfortunately, a relationship between the strength of the interface and the number of cross-links at the interface cannot be directly assessed. This is because the interface cannot be isolated without adjacent tissues that, too, contain cross-links. It is, therefore, difficult to ascertain the fraction of cross-links belonging to the interface alone. The same can be said of the collagen and GAG production by chondrocytes at the interface. Since the interface consists of a thin layer, minute changes in the ECM of this area would be masked by the comparatively greater ECM content of the cartilages undergoing integration. The mechanism of LOX-induced collagen cross-linking is well-established [22] and a strong candidate for explaining the results obtained in this study, though this was not directly proven here. Bolstering this hypothesis, recent studies have shown induced collagen crosslinks in engineered cartilage improves tensile stiffness [25,26]. Future studies may consider techniques such as time-resolved fluorescence spectroscopy [27] to quantify PYR at the interface.

It was observed during tensile testing that samples always broke at the interface, indicating that the interface is not as strong as either neocartilage or native cartilage. The mechanical property of the interface is, thus, due to newly synthesized matrix that has had relatively little time to develop cross-links in contrast to the rest of the tissues. It is known that LOX-induced PYR formation requires 7 to 30 days [28], which may also explain why increased apparent strength was observed only for groups whose LOX treatment was initiated 28 d prior to tensile assessment. In these cases, PYR cross-link precursors were allowed to accumulate within the constructs prior to their being press-fitted into the native tissue, at which time these precursors readily bridged the construct and native tissue together by maturing into cross-links. In terms of native-to-native interfaces, it may be prudent to consider longer durations of LOX application in future studies.

It is worthy to note that, for the timescale applied, diffusion of LOX should not be a bottleneck to its effectiveness. LOX is a relatively small molecule of ~50 kDa. For comparison, BMP-1 is 30 kDa, and the diffusion coefficient of 40 kDa dextran has been determined to be ~60 $\mu m^2/s$ [29]. However, LOX may require time to act before it promotes integration since it can take weeks to complete the final PYR product. This can be seen with Group C, which consists of LOX applied at t = 25–39 d only. This treatment did not result in significant increases in tensile properties. It is unclear whether this is due 1) to the short duration of LOX application or 2) to the late initiation of application. These two variables should be examined in a future study at greater detail.

For example, a variety of initiation and culture times can be examined, extending the total time of culture up to 8 weeks to identify the "ceiling" of effectiveness. Once this saturation level has been determined, one can then optimize not only the time of initial application, but also the total duration of application.

Aside from the dependence on *in vitro* culture time [30], cytokines present *in vivo* can also influence integration. A study examined the effects of steroid hormones in bovine cartilage that is lacking a known inhibitor to integration, interleukin-1β. An increase of ~50 kPa in mechanical integration was seen [31], as compared to the 700 kPa obtained in this study for the native-to-native controls. Also, it has also been shown that, without the assistance of exogenous agents, strength of half that which is seen in intact cartilage can be achieved in an equine model for chondrocyte transplantation [32]. It is worth noting that, in the present study, by delivering cells to the interface in concert with LOX, integration strength can be increased to 1.6 MPa (Group D). Comparing this result to the stiffness of fibrin, which is clinically used as tissue glue and sealant, the stiffness of the LOX-treated interface is roughly fifty times higher. Averaged over various strain rates, the stiffness of fibrin alone is under 30 kPa.[33] When fibrin is combined with chondrocytes to serve as a cartilage adhesive, the stiffness of the interface is increased over fibrin alone and also with time *in vivo*, to 0.645 MPa after 8 months [11]. It is worth noting that LOX-treatment achieves two-times the stiffness in a fraction of the time. It is expected that the stiffness of interfaces enhanced with LOX and chondrocytes will continue to improve *in vivo* as the cells remodel the matrix over time. Chondrocyte transplantation is a current therapy that, similar to the self-assembled constructs employed in this study, delivers metabolic cells to the wound edge using fibrin. It is conceivable for LOX to assist this clinical procedure, especially since the LOX treatment produces comparable results to fibrin at a shorter time. Of course, additional studies on 1) optimal dosing time, 2) cross-linker concentration, and 3) activity profile as related to not only the chondrocytes but also other cell types surrounding cartilage would need to be completed to ensure safety and efficacy, prior to deploying this technique clinically.

A major component of articular cartilage ECM is the electronegative aggrecan. This electric charge is an obstacle to integration because the similar charges in two pieces of tissue would cause them to repel [7]. Further studies need to be completed to fully understand the role which aggrecan's electronegativity may play in blocking integration. Future studies might also include the combination of LOX with other bioactive agents that are known to influence cartilage behavior. Already, transforming growth factor β1 (TFG-β1) has shown efficacy when combined with a biomaterial [34], and it would be interesting to examine how LOX can assist this case. TFG-β1 may work in synergism with LOX, the cytokine and enzyme working in tandem to effect greater collagen production and cross-linking.

It should be mentioned that, for this study, LOX concentration was based on a pilot study that examined LOX on native and engineered cartilages separately (described in "Materials and Methods"). Following this study's exciting results, it may be prudent to conduct a systematic examination of various LOX concentrations to identify a minimum, yet effective, concentration between 0.015 and 0.15 μg/ml that enhances interfacial stiffness and strength to the levels of the engineered or native cartilages, or even for other tissues where cross-linking plays important functional roles. For instance, integrating engineered knee meniscus to native knee meniscus has shown dependence on maturation state [35], and therefore the extent of collagen cross-links, and LOX may be used similarly for this tissue. Finding this minimum dose will be significant in not only reducing cost but also in mitigating any potential for this enzyme to interfere with other cellular processes, despite this being a naturally-occurring enzyme. Already, it has been shown here that LOX does not interfere with chondrocyte metabolism with respect to collagen and GAG synthesis, but, for its use *in vivo*, the effects of LOX on other cells may need to be elucidated prior to conducting animal studies with this enzyme. A similar process would allow the identification of an optimal LOX concentration for maximizing the native-to-native integration strength; this will be immensely useful from a clinical perspective, once the safety and efficacy of exogenous LOX has been shown.

While other cross-linkers such as ribose, glutaraldehyde, genipin, and methylglyoxal have all been investigated in conjunction with engineered articular cartilage [36,37], these agents have all been shown to alter cellular activity. Some of these agents are even cytotoxic and thus preclude their use with live cells in influencing integration. Furthermore, unnatural cross-linkers such as glutaraldehyde have been shown to elicit a foreign body giant cell reaction [38], in contrast to LOX, which is found naturally in multiple musculoskeletal tissues. This study demonstrates that LOX is a potent agent for enhancing integration between native and tissue engineered cartilage. It also paves the way for the use of LOX in improving native cartilage integration. These results could potentially be used to solve the problem of large cartilage defects by allowing tissue engineered cartilage implants to be integrated into the surrounding tissue.

Author Contributions

Conceived and designed the experiments: AAA EM JCH. Performed the experiments: AAA EM. Analyzed the data: AAA EM JCH. Contributed reagents/materials/analysis tools: JCH. Wrote the paper: AAA EM JCH.

References

1. Athanasiou KA, Darling EM, Hu JC (2009) Articular cartilage tissue engineering; Athanasiou KA, editor. Electronic reference: Morgan & Claypool.

2. Smith GD, Knutsen G, Richardson JB (2005) A clinical review of cartilage repair techniques. J Bone Joint Surg Br 87: 445–449.

3. Wang W, Li B, Li Y, Jiang Y, Ouyang H, et al. (2010) In vivo restoration of full-thickness cartilage defects by poly(lactide-co-glycolide) sponges filled with fibrin gel, bone marrow mesenchymal stem cells and DNA complexes. Biomaterials 31: 5953–5965.

4. Kon E, Delcogliano M, Filardo G, Fini M, Giavaresi G, et al. (2010) Orderly osteochondral regeneration in a sheep model using a novel nano-composite multilayered biomaterial. J Orthop Res 28: 116–124.

5. St-Pierre JP, Gan L, Wang J, Pilliar RM, Grynpas MD, et al. (2012) The incorporation of a zone of calcified cartilage improves the interfacial shear strength between in vitro-formed cartilage and the underlying substrate. Acta Biomater 8: 1603–1615.

6. Augst A, Marolt D, Freed LE, Vepari C, Meinel L, et al. (2008) Effects of chondrogenic and osteogenic regulatory factors on composite constructs grown using human mesenchymal stem cells, silk scaffolds and bioreactors. J R Soc Interface 5: 929–939.

7. Khan IM, Gilbert SJ, Singhrao SK, Duance VC, Archer CW (2008) Cartilage integration: evaluation of the reasons for failure of integration during cartilage repair. A review. Eur Cell Mater 16: 26–39.

8. Hunziker EB, Quinn TM (2003) Surgical removal of articular cartilage leads to loss of chondrocytes from cartilage bordering the wound edge. J Bone Joint Surg Am 85-A Suppl 2: 85–92.

9. Hunziker EB (1999) Biologic repair of articular cartilage. Defect models in experimental animals and matrix requirements. Clin Orthop Relat Res: S135–146.

10. van de Breevaart Bravenboer J, In der Maur CD, Bos PK, Feenstra L, Verhaar JA, et al. (2004) Improved cartilage integration and interfacial strength after enzymatic treatment in a cartilage transplantation model. Arthritis Res Ther 6: R469–476.

11. Peretti GM, Zaporojan V, Spangenberg KM, Randolph MA, Fellers J, et al. (2003) Cell-based bonding of articular cartilage: An extended study. J Biomed Mater Res A 64: 517–524.

12. Bos PK, DeGroot J, Budde M, Verhaar JA, van Osch GJ (2002) Specific enzymatic treatment of bovine and human articular cartilage: implications for integrative cartilage repair. Arthritis Rheum 46: 976–985.

13. Hunziker EB, Driesang IM, Saager C (2001) Structural barrier principle for growth factor-based articular cartilage repair. Clin Orthop Relat Res: S182–189.

14. Tew SR, Kwan AP, Hann A, Thomson BM, Archer CW (2000) The reactions of articular cartilage to experimental wounding: role of apoptosis. Arthritis Rheum 43: 215–225.

15. Minns RJ, Steven FS (1977) The collagen fibril organization in human articular cartilage. J Anat 123: 437–457.

16. Ahsan T, Lottman LM, Harwood F, Amiel D, Sah RL (1999) Integrative cartilage repair: inhibition by beta-aminopropionitrile. J Orthop Res 17: 850–857.

17. McGowan KB, Sah RL (2005) Treatment of cartilage with beta-aminopropionitrile accelerates subsequent collagen maturation and modulates integrative repair. J Orthop Res 23: 594–601.

18. Obradovic B, Martin I, Padera RF, Treppo S, Freed LE, et al. (2001) Integration of engineered cartilage. J Orthop Res 19: 1089–1097.

19. Reindel ES, Ayroso AM, Chen AC, Chun DM, Schinagl RM, et al. (1995) Integrative repair of articular cartilage in vitro: adhesive strength of the interface region. J Orthop Res 13: 751–760.

20. Hu JC, Athanasiou KA (2006) A self-assembling process in articular cartilage tissue engineering. Tissue Eng 12: 969–979.

21. Archer CW, Redman S, Khan I, Bishop J, Richardson K (2006) Enhancing tissue integration in cartilage repair procedures. J Anat 209: 481–493.

22. Eyre DR, Paz MA, Gallop PM (1984) Cross-linking in collagen and elastin. Annu Rev Biochem 53: 717–748.

23. Natoli RM, Responte DJ, Lu BY, Athanasiou KA (2009) Effects of multiple chondroitinase ABC applications on tissue engineered articular cartilage. J Orthop Res 27: 949–956.

24. Responte DJ, Arzi B, Natoli RM, Hu JC, Athanasiou KA (2012) Mechanisms underlying the synergistic enhancement of self-assembled neocartilage treated with chondroitinase-ABC and TGF-beta1. Biomaterials 33: 3187–3194.

25. Makris EA, Hu JC, Athanasiou KA (2013) Hypoxia-induced collagen crosslinking as a mechanism for enhancing mechanical properties of engineered articular cartilage. Osteoarthritis Cartilage.

26. Makris EA, MacBarb RF, Responte DJ, Hu JC, Athanasiou KA (2013) A copper sulfate and hydroxylysine treatment regimen for enhancing collagen crosslinking and biomechanical properties in engineered neocartilage. FASEB Accepted.

27. Sun Y, Responte D, Xie H, Liu J, Fatakdawala H, et al. (2012) Nondestructive evaluation of tissue engineered articular cartilage using time-resolved fluorescence spectroscopy and ultrasound backscatter microscopy. Tissue Eng Part C Methods 18: 215–226.

28. Ahsan T, Harwood F, McGowan KB, Amiel D, Sah RL (2005) Kinetics of collagen crosslinking in adult bovine articular cartilage. Osteoarthritis Cartilage 13: 709–715.

29. Leddy HA, Guilak F (2003) Site-specific molecular diffusion in articular cartilage measured using fluorescence recovery after photobleaching. Ann Biomed Eng 31: 753–760.

30. Theodoropoulos JS, De Croos JN, Park SS, Pilliar R, Kandel RA (2011) Integration of tissue-engineered cartilage with host cartilage: an in vitro model. Clin Orthop Relat Res 469: 2785–2795.

31. Englert C, Blunk T, Fierlbeck J, Kaiser J, Stosiek W, et al. (2006) Steroid hormones strongly support bovine articular cartilage integration in the absence of interleukin-1beta. Arthritis Rheum 54: 3890–3897.

32. Gratz KR, Wong VW, Chen AC, Fortier LA, Nixon AJ, et al. (2006) Biomechanical assessment of tissue retrieved after in vivo cartilage defect repair: tensile modulus of repair tissue and integration with host cartilage. J Biomech 39: 138–146.

33. Sierra DH, Eberhardt AW, Lemons JE (2002) Failure characteristics of multiple-component fibrin-based adhesives. J Biomed Mater Res 59: 1–11.

34. Maher SA, Mauck RL, Rackwitz L, Tuan RS (2010) A nanofibrous cell-seeded hydrogel promotes integration in a cartilage gap model. J Tissue Eng Regen Med 4: 25–29.

35. Ionescu LC, Lee GC, Garcia GH, Zachry TL, Shah RP, et al. (2011) Maturation state-dependent alterations in meniscus integration: implications for scaffold design and tissue engineering. Tissue Eng Part A 17: 193–204.

36. Elder BD, Mohan A, Athanasiou KA (2010) Beneficial effects of exogenous crosslinking agents on self-assembled tissue engineered cartilage construct biomechanical properties. Journal of Mechanics in Medicine and Biology 11: 433–443.

37. Eleswarapu SV, Chen JA, Athanasiou KA (2011) Temporal assessment of ribose treatment on self-assembled articular cartilage constructs. Biochem Biophys Res Commun 414: 431–436.

38. Speer DP, Chvapil M, Eskelson CD, Ulreich J (1980) Biological Effects of Residual Glutaraldehyde in Glutaraldehyde-Tanned Collagen Biomaterials. J Biomed Mater Res 14: 753–764.

15

Modulating Gradients in Regulatory Signals within Mesenchymal Stem Cell Seeded Hydrogels: A Novel Strategy to Engineer Zonal Articular Cartilage

Stephen D. Thorpe[1,2]**, Thomas Nagel**[1,2]**, Simon F. Carroll**[1,2]**, Daniel J. Kelly**[1,2]*

1 Trinity Centre for Bioengineering, Trinity Biomedical Sciences Institute, Trinity College Dublin, Dublin, Ireland, **2** Department of Mechanical and Manufacturing Engineering, School of Engineering, Trinity College Dublin, Dublin, Ireland

Abstract

Engineering organs and tissues with the spatial composition and organisation of their native equivalents remains a major challenge. One approach to engineer such spatial complexity is to recapitulate the gradients in regulatory signals that during development and maturation are believed to drive spatial changes in stem cell differentiation. Mesenchymal stem cell (MSC) differentiation is known to be influenced by both soluble factors and mechanical cues present in the local microenvironment. The objective of this study was to engineer a cartilaginous tissue with a native zonal composition by modulating both the oxygen tension and mechanical environment thorough the depth of MSC seeded hydrogels. To this end, constructs were radially confined to half their thickness and subjected to dynamic compression (DC). Confinement reduced oxygen levels in the bottom of the construct and with the application of DC, increased strains across the top of the construct. These spatial changes correlated with increased glycosaminoglycan accumulation in the bottom of constructs, increased collagen accumulation in the top of constructs, and a suppression of hypertrophy and calcification throughout the construct. Matrix accumulation increased for higher hydrogel cell seeding densities; with DC further enhancing both glycosaminoglycan accumulation and construct stiffness. The combination of spatial confinement and DC was also found to increase proteoglycan-4 (lubricin) deposition toward the top surface of these tissues. In conclusion, by modulating the environment through the depth of developing constructs, it is possible to suppress MSC endochondral progression and to engineer tissues with zonal gradients mimicking certain aspects of articular cartilage.

Editor: Hani A. Awad, University of Rochester, United States of America

Funding: Funding was provided by Science Foundation Ireland (President of Ireland Young Researcher Award: 08/Y15/B1336) and the European Research Council (StemRepair – Project number 258463). The funders had no role in study design, data collection and analysis, decision to publish, or preparation of the manuscript.

Competing Interests: The authors have declared that no competing interests exist.

* E-mail: kellyd9@tcd.ie

Introduction

Adult articular cartilage consists of three separate structural zones; the superficial tangential, middle and deep zones. This depth dependent composition and organisation is fundamental to the normal physiological function of articular cartilage [1,2]. Not only does cell morphology and arrangement change with depth, but each zone has distinct extra-cellular matrix (ECM) composition, architecture and mechanical properties. The dominant load carrying structural components of the ECM are collagen ($\sim$75% tissue by dry weight) and proteoglycan (20%–30% tissue by dry weight), the concentrations of which vary with depth from the articular surface [1,3,4]. Collagen content is highest in the superficial zone, decreasing by $\sim$20% in the middle and deep zones [1,3]. Proteoglycan content is lowest at the surface, increasing by as much as 50% into the middle and deep zones [3,4]. The zonal composition and structural organisation of the ECM determine the biomechanical properties which also vary through the tissue depth; such that the compressive modulus increases from the superficial zone to the deep zone [2,5,6], while the tensile modulus decreases from the superficial surface to the deep zone [7].

An on-going challenge in the field of articular cartilage regeneration is the attainment of this stratified zonal structure. Classical tissue engineering approaches focus primarily on forming homogeneous tissues by embedding chondrocytes or stem cells in various scaffolds and do not attempt to mimic the organised zonal architecture of articular cartilage. One approach toward this aim is to utilise chondrocytes from specific zones of articular cartilage in the corresponding regions of an engineered construct [8–10]. It has been shown that chondrocytes from different zones demonstrate different biosynthetic activities [9,10]. Layering such zonal chondrocytes in a photo-polymerising hydrogel has been shown to result in increased sulphated glycosaminoglycan (sGAG) accumulation in the bottom of the construct, although collagen content was also significantly higher in the bottom when compared to the top [10]. Another approach to engineering zonal cartilage is to vary biomaterial properties such as pore size [11], stiffness [12,13] or composition [14,15] through the depth of the scaffold or hydrogel. For example, combining layers of 2% and 3% agarose leads to zonal differences in the initial mechanical properties of the construct, however chondrocyte matrix elaboration in such bi-layered constructs was inferior to that in uniform 2% agarose [12]. Further improvements were observed with the application of dynamic compressive strain, or seeding zone specific chondrocytes into layered agarose hydrogels [13,16]. While promising, there are potential limitations associated with such an approach, including the development of a distinct boundary between gel layers which could delineate as a result of shear stress [17]. Furthermore, isolation of chondrocytes from separate zones of articular cartilage can be difficult, particularly in damaged and diseased human tissue where the zonal differences are less distinct and potential biopsies are limited in size.

An alternative strategy to engineering zonal cartilage is to attempt to recapitulate aspects of the tissue micro-environment which may be responsible for the creation of depth-dependent properties in articular cartilage during development and maturation. In this respect, undifferentiated mesenchymal stem cells (MSCs) could prove to be a more appropriate cell source, as when provided with suitable stimuli they may differentiate into chondrocytes with specific zonal phenotypes. For example, low oxygen tension, a characteristic of avascular articular cartilage, has been shown to enhance chondrogenesis of mesenchymal stem cells (MSCs) [18–25]. In addition to oxygen, it has long been proposed that mechanical signals guide the differentiation of mesenchymal stem cells [26–30]. Hydrostatic pressure has been shown to promote a chondrogenic phenotype [31–34], enhancing collagen and sulphated glycosaminoglycan (sGAG) accumulation [35] and increasing the mechanical stiffness [36] of cartilaginous grafts engineered using MSCs encapsulated in agarose hydrogels. Dynamic compressive strain, another key component of the mechanical environment of articular cartilage, has also been shown to promote MSC chondrogenesis [30,37–42]; positively modulating the functional development of cartilaginous constructs engineered using MSCs [43]. It has also been demonstrated that intermittent cyclic tensile strain applied to MSC seeded constructs increases collagen accumulation [44,45].

The objective of this study was to engineer a cartilaginous construct with native-like zonal composition using MSCs by controlling both the oxygen tension and mechanical environment thorough the depth of the developing tissue. In an attempt to create a gradient in oxygen tension through the depth of the construct mimicking that in normal articular cartilage [46], the bottom halves of MSC seeded agarose constructs were radially confined; limiting oxygen transport into this region of the construct. Furthermore, by subjecting these radially confined constructs to dynamic compression it is possible to modulate the mechanical environment throughout the depth of the tissue, with higher levels of fluid pressure in the bottom of the construct and greater strains across the top of the construct. It is hypothesised that such a depth dependent microenvironment will lead to the development of zonal cartilage tissues with a composition mimicking that of normal articular cartilage.

Materials and Methods

Cell Isolation and Expansion

Porcine MSCs were isolated and maintained as previously described [42]. Animals were bred and raised for food and not research purposes and were not subject to any procedures prior to their sacrifice, hence no specific ethical approval was required for this study. Briefly, mononuclear cells were isolated from the femora of 4 month old pigs (~50 kg) within 2 hours of sacrifice and plated at 10×10^6 mononuclear cells per 75 cm^2 culture flask (Nunclon; Nunc, VWR, Dublin, Ireland) allowing colony formation. MSCs were maintained in high-glucose Dulbecco's modified eagle medium (4.5 mg/mL D-Glucose; hgDMEM) supplemented with 10% foetal bovine serum (FBS), penicillin (100 U/mL)-

streptomycin (100 µg/mL) (all Gibco, Invitrogen, Dublin, Ireland) and amphotericin B (0.25 µg/mL; Sigma-Aldrich, Arklow, Ireland). Cultures were washed in Dulbecco's phosphate buffered saline (PBS) after 72 hrs. When ~75% confluent, MSCs were replated at 5×10^3 cells/cm^2 and expanded to passage two in a humidified atmosphere at 37°C and 5% CO$_2$.

Agarose Hydrogel Encapsulation and Construct Confinement

MSCs were suspended in defined chondrogenic medium consisting of hgDMEM supplemented with penicillin (100 U/mL)-streptomycin (100 µg/mL) (both Gibco), 0.25 µg/mL amphotericin B, 100 µg/ml sodium pyruvate, 40 µg/mL L-proline, 1.5 mg/mL bovine serum albumin, 4.7 µg/mL linoleic acid, 1× insulin–transferrin–selenium, 50 µg/mL L-ascorbic acid-2-phosphate, 100 nM dexamethasone (all Sigma-Aldrich) and 10 ng/mL TGF-β3 (Pro Spec-Tany TechnoGene Ltd., Rehovot, Israel). This cell suspension was mixed with agarose (Type VII; Sigma-Aldrich) in PBS at a ratio of 1:1 at approx. 40°C, to yield a final agarose concentration of 2% and a cell density of either 20×10^6 cells/mL or 50×10^6 cells/mL. The agarose-cell suspension was cast between stainless steel plates, one of which was overlaid with a patterned PDMS layer, allowed cool to 21°C for 30 min., and cored to produce cylindrical constructs (Ø 6 mm×4 mm thickness) which were patterned on one surface. Constructs remained patterned side up throughout culture and were maintained in ~1 mL chondrogenic medium per 1×10^6 cells/day with medium exchanged every 3 or 4 days and sampled for biochemical analysis. Either directly after fabrication or at day 21 of culture, constructs were press-fitted into custom made PTFE confinement chambers (Fig. 1) where they remained for the outstanding culture duration.

Dynamic Compression Application

Dynamic compressive loading was applied as described previously [42] to constructs from day 21 to day 42 of culture. Unconfined intermittent dynamic compression (DC) was carried out in an incubator-housed, dynamic compression bioreactor and consisted of a sine wave of 10% strain amplitude superimposed upon a 1% pre-strain, with a 0.01 N per construct preload at a frequency of 1 Hz for 4 hours/day, 5 days/week.

Mechanical Testing and Analysis of Physical Parameters

On removal from culture, construct diameter and wet weight (ww) were recorded. Constructs were mechanically tested in unconfined compression between impermeable platens using a standard materials testing machine (Bose Electroforce 3100;

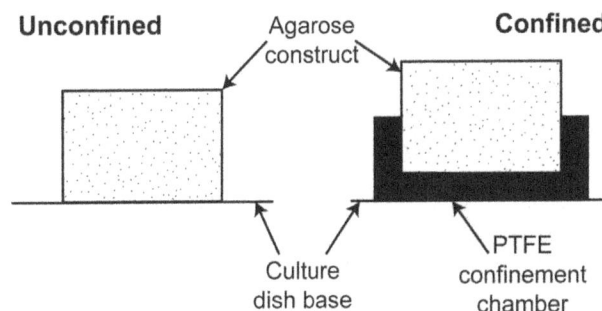

Figure 1. Experimental design. Constructs were press-fitted into custom made PTFE wells such that the bottom 2 mm of the construct thickness was confined.

Bose Corporation, Gillingham, UK) as previously described [47]. A preload of 0.01 N was applied to ensure that the construct surface was in direct contact with the impermeable loading platens. Stress relaxation tests were performed consisting of a ramp displacement of 0.025%/s up to 10% strain, which was maintained until equilibrium was reached (~30 minutes). This was followed by a dynamic test where cyclic strain amplitude of 1% (10%–11% total strain) was applied for 10 cycles at 1 Hz. Samples were subsequently sliced into top and bottom regions using a custom built rig and each region was tested simultaneously on separate, randomly assigned material testing machines (Zwick Roell Z005; Zwick Testing Machines Ltd., Herefordshire, UK) as above.

Biochemical Constituents

After mechanical testing of top and bottom construct regions, the wet weight (ww) of each was recorded and the constructs frozen at −85°C for further analyses. The biochemical content of constructs was assessed as previously described [42]. Samples were digested with papain (125 μg/ml) in 0.1 M sodium acetate, 5 mM L-cysteine HCl, 0.05 M EDTA (all Sigma-Aldrich), pH 6.0 at 60°C under constant rotation for 18 hours. DNA content was quantified using the Hoechst Bisbenzimide 33258 dye assay (Sigma-Aldrich) as previously described [48]. The sulphated glycosaminoglycan (sGAG) content was quantified using the dimethylmethylene blue dye-binding assay (Blyscan; Biocolor Ltd., Carrickfergus, Northern Ireland). Total collagen content was determined by measuring orthohydroxyproline via the dimethylaminobenzaldehyde and chloramine T assay [49]. A hydroxyproline-to-collagen ratio of 1:7.69 was used [50]. Cell culture media was analysed for sGAG and collagen secreted.

Histology and Immunohistochemistry

Constructs were fixed in 4% paraformaldehyde, paraffin embedded and sectioned at 5 μm to produce a cross section perpendicular to the disc face. Sections were stained for sGAG with 1% alcian blue 8 GX in 0.1 M HCl, for collagen with picro-sirius red and for calcific deposition with 1% alizarin red (all Sigma-Aldrich). Immunohistochemistry was performed as described previously [42]. Sections were enzymatically treated with chondroitinase ABC (Sigma-Aldrich) in a humidified environment at 37°C. Slides were blocked with goat serum (Sigma-Aldrich) and sections were incubated for 1 hour with a primary antibody diluted in blocking buffer specific to either collagen type I (1:400, 1 mg/mL), collagen type II (1:100; 1 mg/mL), collagen type X (1:200; 1.4 mg/mL) (all Abcam, Cambridge, UK) or proteoglycan-4 (PRG4; 1:200; 1 mg/mL) (Sigma-Aldrich). After washing in PBS, sections were incubated for 1 hour in the secondary antibody, anti-mouse IgG biotin antibody produced in goat (1:200; 2 mg/mL; Sigma-Aldrich). Colour was developed using the Vectastain ABC kit followed by exposure to peroxidase DAB substrate kit (both Vector Laboratories, Peterborough, UK). Immunofluorescent staining was employed for collagen type X, which involved permeabilisation of the cell membrane with 0.1% Triton-X100. In place of colour development, sections were incubated with ExtrAvidin-FITC (1:100; Sigma-Aldrich) for 1 hour, washed several times in PBS, nuclei counterstained with DAPI (1:500; 1 mg/mL; VWR), and sections mounted using Vectashield (Vector Laboratories). Sections were imaged with an Olympus IX51 inverted fluorescent microscope fitted with an Olympus DP70 camera. Sections of porcine cartilage, ligament and/or growth plate were included as controls.

Theoretical Prediction of Mechanical Environment within Agarose Constructs

To estimate the effect of semi-confinement on the mechanical environment within the construct, pore pressure and maximum principal strain were predicted for day 0 constructs. The loading protocol of the bioreactor was simulated with material properties derived from a sample specific fit to constructs mechanically tested at day 0. Cell-seeded constructs were modelled as fluid saturated porous media using a finite strain formulation applied previously to agarose in compression [51]. Surfaces in contact with media were modelled as free draining, while fluid flow was prohibited across surfaces in contact with the loading platens or the confinement well. Nonlinear permeability was modelled following Gu et al. [52]. Briefly, the solid matrix was modelled as nonlinearly viscoelastic based on a multiplicative decomposition of the deformation gradient into elastic and viscous parts $F = F_e F_v$. An evolution equation was defined for the viscous right Cauchy-Green tensor C_v:

$$\dot{C}_v = \frac{1}{\eta_v} C_v T_{ov} C$$

where η_v is the viscosity function, T_{ov} is the viscoelastic overstress and C is the right Cauchy-Green tensor [51,53]. While the equilibrium free Helmholtz energy potentials where based on a Neo-Hookean formulation

$$\psi_{eq} = C_1[I_1(C) - \ln(I_3(C)) - 3] + D_2(\ln I_3(C))^2$$

the viscoelastic overstresses where derived from the exponential potential

$$\psi_{ov} = \frac{C_{1v}}{\alpha_v}\left[e^{\alpha_v[I_1(CC_v^{-1}) - \ln(I_3(CC_v^{-1})) - 3]} - 1\right] + D_{2v}\left(\ln I_3(CC_v^{-1})\right)^2$$

The equilibrium properties C_1 and D_2 were determined analytically from the relaxed part of the ramp and hold test. D_{2v} was set to $C_{1v}D_2/C_1$. The remaining parameters in the viscoelastic potential and the viscosity were fit to unconfined ramp and hold force relaxation curves using a differential evolution algorithm developed by Storn and Price [54]. For further details see Görke et al. [51].

Measurement and Prediction of Oxygen Concentration within Agarose Constructs

Local oxygen concentration was assessed using implantable fibre optic oxygen micro-sensors (Microx TX; PreSens – Precision Sensing GmbH, Regensburg, Germany). Media was added to day 5 constructs seeded with 20×10^6 cells/mL in confined and unconfined configurations so that the surface of the media was 1 mm above the top surface of the construct. The sensor tip was positioned at the media surface above the centre of the construct. Constructs were allowed equilibrate for ≥24 hours in an incubator at 18.5% oxygen prior to oxygen measurement. A linear actuator (NA0830; Zaber Technologies Inc., Vancouver, Canada) was used to control the movement of the sensor tip which was moved vertically downward along the construct axis at 1.8 μm/s to within ~1 mm of the construct base; to avoid collision of the probe with the culture dish or confinement chamber. Oxygen concentration was sampled every 5 s.

Oxygen concentration in the constructs was modelled using a diffusion-reaction type equation. The reaction term followed Michaelis-Menten kinetics:

$$\frac{\partial c}{\partial t} = D\nabla^2 c - n\frac{Q_m c}{K_m + c}$$

Here, c is the oxygen concentration, D the diffusion coefficient, n the cell density, Q_m the maximum consumption rate and K_m the concentration at half the maximum consumption rate. The diffusion coefficient in 2% agarose was determined using the Mackie and Meares relation

$$\frac{D_{ag}}{D_{H_2O}} = \frac{\vartheta_f^2}{(2-\vartheta_f)}$$

so that $D_{ag} = 2.77 \times 10^{-3}$ mm^2/s and φ_f is the fluid phase volume fraction [55]. The cellular consumption value was set to $Q_m = 13.5 \times 10^{-18}$ mol/cell/s as determined by the model fit to the experimental data for unconfined constructs with 20×10^6 cells/mL (Fig. 2A). This value was similar to previously reported values for MSC oxygen consumption while undergoing chondrogenic differentiation [56]. The Michaelis-Menten constant, K_m was set to 5×10^{-5} μmol/mm^3. The sensitivity of the simulation outcome to this value is very low. The construct was modelled as axisymmetric. Oxygen diffusion through the culture media was accounted for by setting $D_{media} = 3.0 \times 10^{-3}$ mm^2/s. The oxygen concentration at the media surface was prescribed as 185 μM, while at surfaces in contact with the bottom of the well, the confining chamber and the symmetry axis the flux was set to zero. The simulations were performed for cell concentrations of $n = 20 \times 10^6$ cells/mL and $n = 50 \times 10^6$ cells/mL.

Statistical Analysis

Presented are results from one of two replicate studies with unique donors where n refers to the number of constructs analysed for each assay within a given replicate (n numbers provided in figure legends). Statistics were performed using MINITAB 15.1 software package (Minitab Ltd., Coventry, UK). Where necessary, a Box-Cox transformation was used to normalise data sets. Construct groups were analysed for significant differences using a general linear model for analysis of variance with factors of group, confinement, dynamic compression, construct region and interactions between these factors examined. Tukey's test for multiple comparisons was used to compare conditions. Significance was accepted at a level of $p \leq 0.05$. Numerical and graphical results are presented as mean ± standard error. Statistical results displayed in figures are from the post hoc tests and represent differences between specific treatment groups. p values in the text may refer to either main effects or post hoc tests.

Results

Radial Confinement Spatially Alters the Pore Pressure and Tensile Strain within Agarose Hydrogels during Dynamic Compression

The mechanical environment within an agarose hydrogel during dynamic compression was predicted for both unconfined and confined configurations (Fig. 1). A relatively homogenous strain environment is predicted within the unconfined construct while the confined constructs experience higher tensile strains in the top of the construct with predominantly compressive strains of lower magnitude present in the bottom (Fig. 3). The model predicts pore pressures in the bottom of the confined constructs which are about an order of magnitude greater than that in the unconfined configuration where a relatively homogenous pressure environment exists.

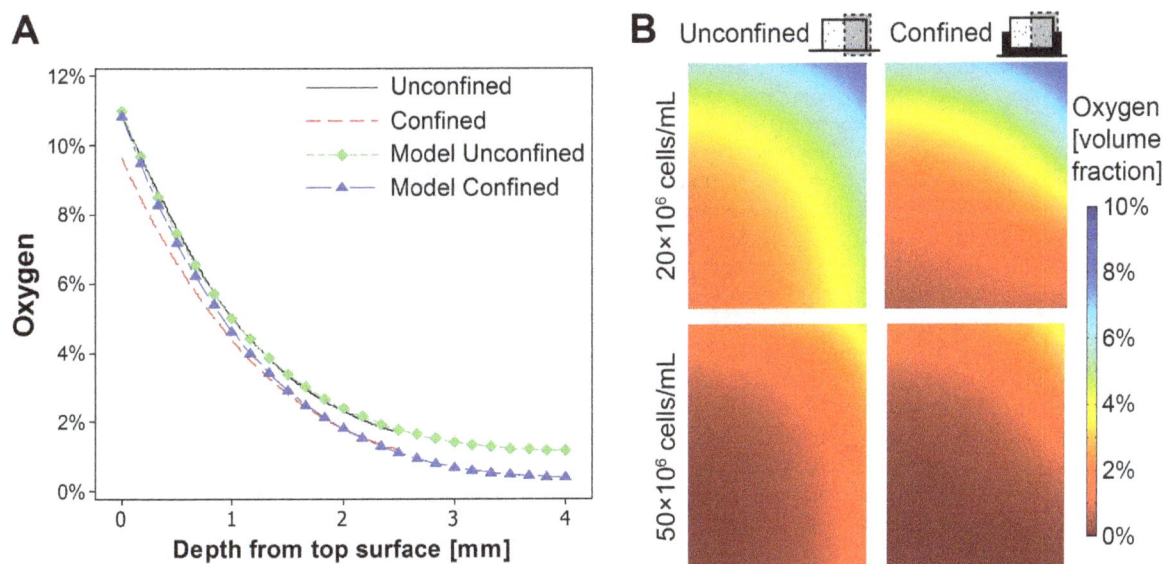

Figure 2. Oxygen tension is modulated by the cell seeding density and radial confinement. MSCs encapsulated in agarose at 20×10^6 cells/mL were cultured for 4 days, at which point constructs were confined and allowed to equilibrate for 24 hours. (A) Oxygen concentration through the media and along the construct axis was measured for unconfined and confined constructs with representative samples of each presented (Exp). Model predictions were fit to experimental data obtained from the surface of the culture media through the depth of the construct. For clarity, only the fit through the construct is shown (Model). (B) Predicted gradients in oxygen volume fraction for unconfined and confined constructs with seeding densities of 20×10^6 and 50×10^6 cells/mL. Oxygen volume fraction corresponds to molar fraction ($\times 10$ pmol/mm^3).

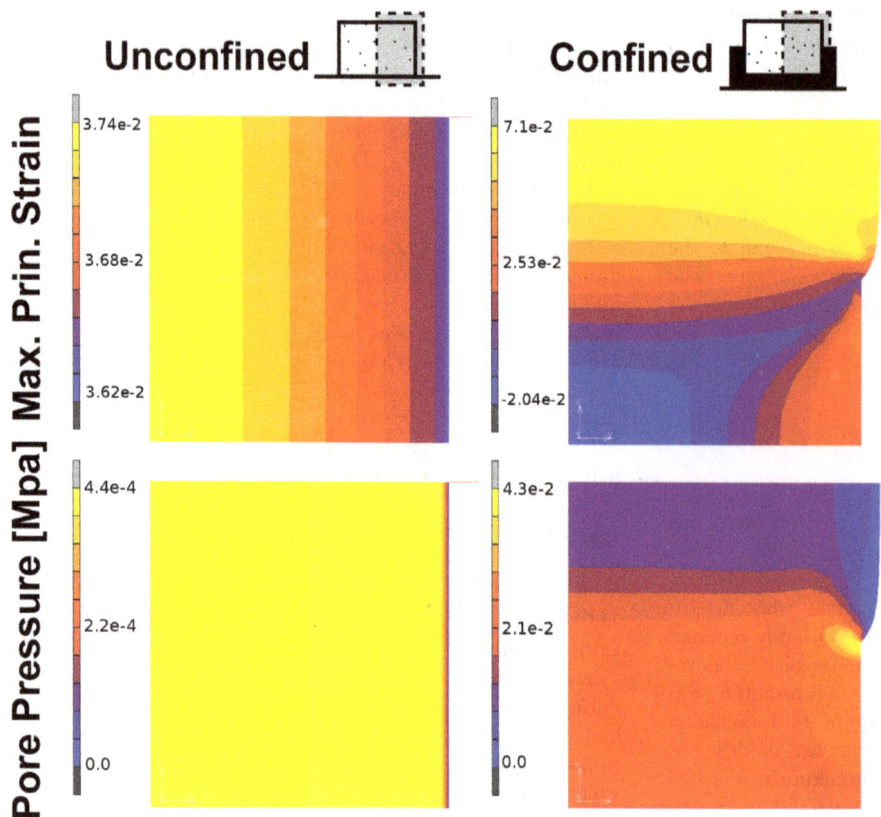

Figure 3. Partial radial confinement alters pore pressure and strain distribution under dynamic compression. Theoretical predictions of the maximum principle strain and pore pressure [MPa] for both unconfined and confined configurations during steady state dynamic compression at day 0.

Oxygen Tension within Engineered Tissues is Modulated by the Cell Seeding Density and Radial Confinement

The oxygen concentration along the construct axis was measured as a function of depth for constructs seeded at 20×10^6 cells/mL (Fig. 2A). Due to cellular consumption, an oxygen gradient develops over time. For unconfined constructs sitting on the base of a dish (i.e. the unconfined free swelling constructs in this study), the oxygen concentration decreases towards the bottom centre; from 10.82% at the top surface to 1.70% at a depth of 2.5 mm from this top surface. Oxygen concentration at a depth of 2.5 mm from the top of the construct was lower in confined constructs when compared to unconfined controls; 1.20% vs. 1.70% respectively. This data was fit to a computational model to predict spatial gradients in oxygen concentration throughout constructs seeded with both 20×10^6 cells/mL and 50×10^6 cells/mL (Fig. 2B). Increasing the cell density to 50×10^6 cells/mL accentuated these gradients in oxygen tension. Confinement led to further reductions in oxygen concentration and enlargement of the low oxygen region across the bottom of the construct; culminating in the development of a low oxygen region (with a minimum predicted value of 0.39%) at 20×10^6 cells/mL and an anoxic region (approaching 0%) predicted at the 50×10^6 cells/mL seeding density.

Radial Confinement Enhances sGAG Accumulation in the Bottom of Engineered Cartilaginous Constructs

MSCs were encapsulated in agarose at 20×10^6 cells/mL, confined, and cultured in free-swelling (unloaded) conditions to day 21 at which point the biochemical content was independently assessed within the top and bottom regions of the construct (Fig. 4A). No differences in DNA content were observed between the top and bottom of either confined or unconfined constructs. For unconfined constructs, sGAG content at day 21 did not change significantly with construct depth. However confinement led to an increase in sGAG in the construct bottom when compared to unconfined controls (0.470 ± 0.009%ww vs. 0.283 ± 0.024%ww; $p = 0.0002$); resulting in depth-dependent sGAG accumulation by day 21 ($p < 0.00005$). When the media was analysed, it was revealed that confinement served to decrease sGAG secretion to the media ($p = 0.004$; Fig. 4B); in spite of this, total sGAG produced (accumulated plus secreted to media) was still greatest in confined constructs (527.95 ± 4.536 μg vs. 485.348 ± 13.595 μg; $p = 0.0154$). Collagen accumulation was also depth dependent in unconfined constructs at day 21 with greater collagen accumulation in the bottom ($p = 0.0247$; Fig. 4A). Confinement acted to nullify this non-cartilaginous zonal variation in collagen concentration, with no difference between the two regions. Analysis of the media revealed that while confinement did not affect collagen accumulated, it did reduce collagen secreted to the media, such that more collagen was synthesised in unconfined constructs (929.45 ± 15.426 μg vs. 753.852 ± 12.597 μg; $p < 0.0001$; Fig. 4B).

Histological staining confirmed that spatial variations exist in the distribution of sGAG and collagen within constructs (Fig. 4C). In confined constructs, more intense alcian blue and collagen type II staining is evident towards the edge in the bottom (confined) region of the construct when compared to unconfined controls.

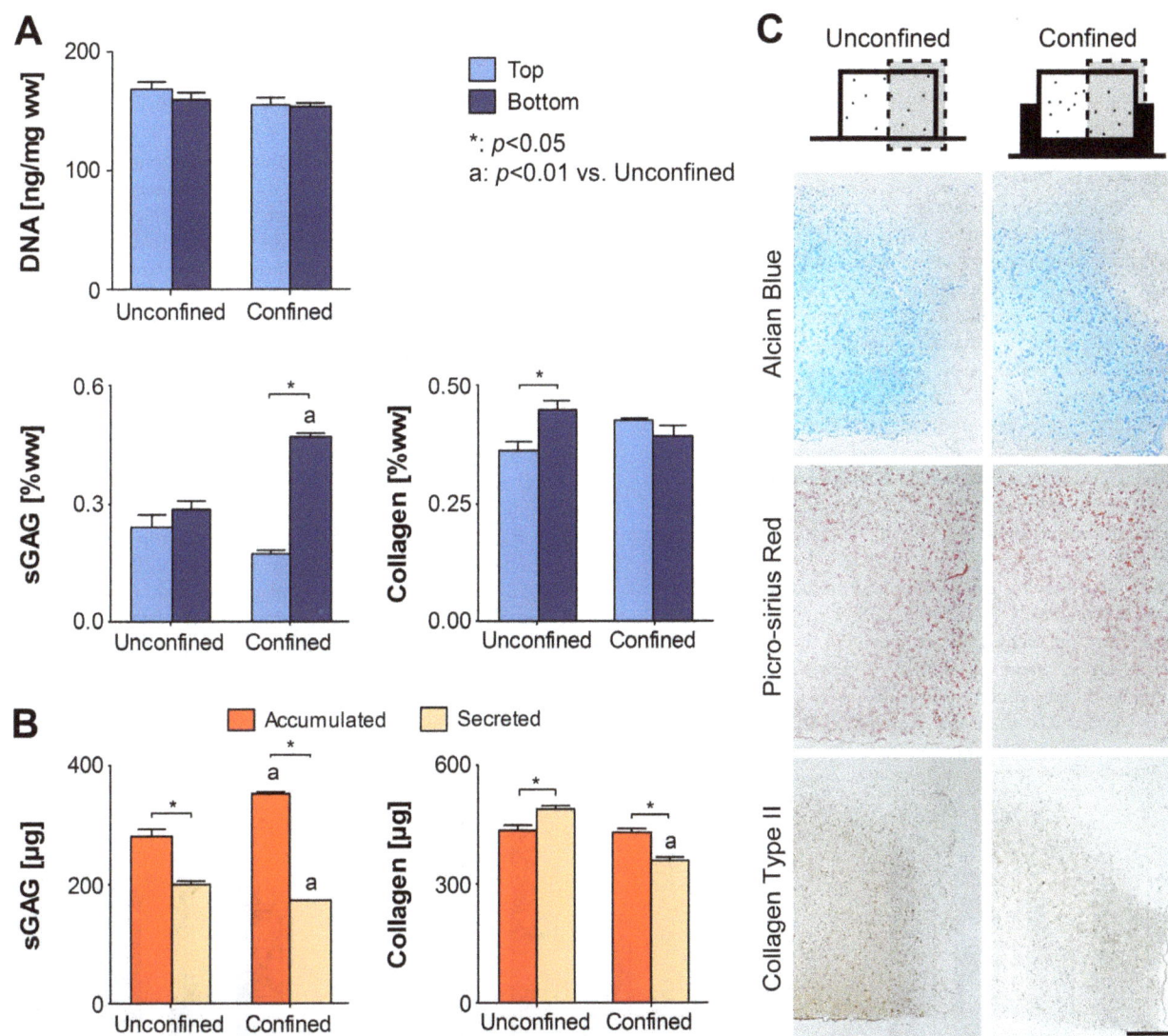

Figure 4. Radial confinement enhances sGAG accumulation in the bottom of engineered cartilaginous constructs. MSCs encapsulated in agarose at 20×10^6 cells/mL were cultured for 21 days in unconfined or confined conditions. (A) The top and bottom regions of unconfined and confined free-swelling constructs were analysed for DNA, sGAG and collagen contents. ($n=4$) (B) Total sGAG and collagen accumulated in the constructs (sum of top and bottom) and secreted to the media. Media data is presented as μg accumulated and secreted ($n=4$) (C) Unconfined and confined constructs were stained with alcian blue for sulphated mucins, picro-sirius red for collagen and immunohistochemically for collagen type II. Representative full-depth half construct sections are shown as indicated. ($n=2$) Scale bar 500 μm.

Radial Confinement Coupled with Dynamic Compression Enhances Collagen Accumulation in the Top of the Construct

MSCs were encapsulated in agarose at 20×10^6 cells/mL and cultured in unconfined free-swelling (unloaded) conditions for 21 days, at which point constructs were confined and dynamic compression applied from day 21 to 42. This delayed application of dynamic compression was motivated by our previous finding that dynamic compression application from day 0 inhibited chondrogenesis of MSCs [30,42,57,58].

While there was no difference in sGAG content between top and bottom construct regions for unconfined constructs at day 21, ensuring constructs remained the same way up over 42 days of culture did eventually lead to greater sGAG accumulation in the construct bottom when compared to the top ($p<0.0001$; Fig. 5). Confinement from day 21 did not further enhance sGAG

accumulation in either region. When normalised to DNA content, only unconfined constructs exhibited a significant difference in sGAG/DNA with depth (Fig. S1). Neither dynamic compression nor confinement had a significant effect on total sGAG accumulation, although compression did increase sGAG secretion to the media ($p=0.0344$; Fig. S1). By day 42 unconfined constructs also exhibited zonal variation in collagen content with greater accumulation in the bottom of the construct ($p=0.0016$; Fig. 5). However, construct confinement again acted to reduce this zonal variation such that there was no longer a significant difference between top and bottom. Moreover, when confinement was combined with dynamic compression, this zonal gradient was reversed such that collagen content in the top of confined compressed constructs was greater than that in unconfined controls (0.868 ± 0.033 vs. 0.736 ± 0.011; $p=0.0252$; Fig. 5). When normalised to DNA, collagen synthesis in the top of confined constructs was significantly higher than that in the bottom

($p = 0.0382$; Fig. S1). This change in relative collagen accumulation with depth was due both to an increase in collagen accumulation in the top of the construct, and a decrease in collagen accumulation in the bottom. Approximately half the total collagen produced was secreted to the media, with neither confinement nor dynamic compression having any significant effect (Fig. S1).

Confinement acted to increase the intensity of collagen staining in the top half of the construct when compared to unconfined controls; particularly in constructs subjected to dynamic compression (Fig. 6). Similar staining patterns were evident for collagen type II (Fig. 6). The combination of dynamic compression and confinement also led to a reduction in staining for collagen type I (Fig. 6). Dynamic compression reduced calcific deposition in unconfined constructs, as evident by reduced alizarin red staining (Fig. 6). A further reduction in alizarin red staining was observed when constructs were both confined and subjected to dynamic compression. Although collagen type X immunofluorescence was present in confined constructs, the combination of confinement and dynamic compression also led to a reduction in staining intensity (Fig. 6).

Though the equilibrium and dynamic moduli increased with time for all conditions ($p < 0.001$), bulk construct mechanical properties were unaffected by confinement or dynamic compression at day 42 (data not shown). In agreement with sGAG zonal variation, the equilibrium modulus was greater in the bottom of constructs than the top ($p < 0.0001$; Fig. 5). This region-specific increase in equilibrium modulus was greater in unconfined controls ($p = 0.0008$; Fig. 5).

MSC Response to Extrinsic Signals is Dependent on the Cell Seeding Density

While cartilaginous constructs with a varying zonal composition can be engineered by controlling the environment through the depth of the developing construct, absolute levels of matrix accumulation were lower than native values. In an attempt to address this issue, MSCs were encapsulated in agarose at a higher seeding density of 50×10^6 cells/mL and cultured in unconfined free-swelling (unloaded) conditions for 21 days, at which point constructs were confined and dynamic compression applied from day 21 to 42.

Dynamic compression increased total sGAG content for both confined and unconfined constructs compared to free swelling controls at day 42 ($p = 0.0005$; Fig. 7A). However this increase in sGAG occurred in the top of the construct ($p = 0.0001$); such that confinement combined with dynamic compression led to greater sGAG accumulation in the construct top compared to the bottom ($1.538 \pm 0.040\%$ww vs. $1.170 \pm 0.092\%$ww; $p = 0.0013$; Fig. 7A). On analysis of culture media, confinement was seen to inhibit total sGAG produced ($p = 0.0094$; Fig. S2). Notably more sGAG was secreted to the media than retained within the construct ($p = 0.0020$). Confinement acted to increase collagen accumulation by day 42 in the construct top when compared with unconfined controls irrespective of the applica-

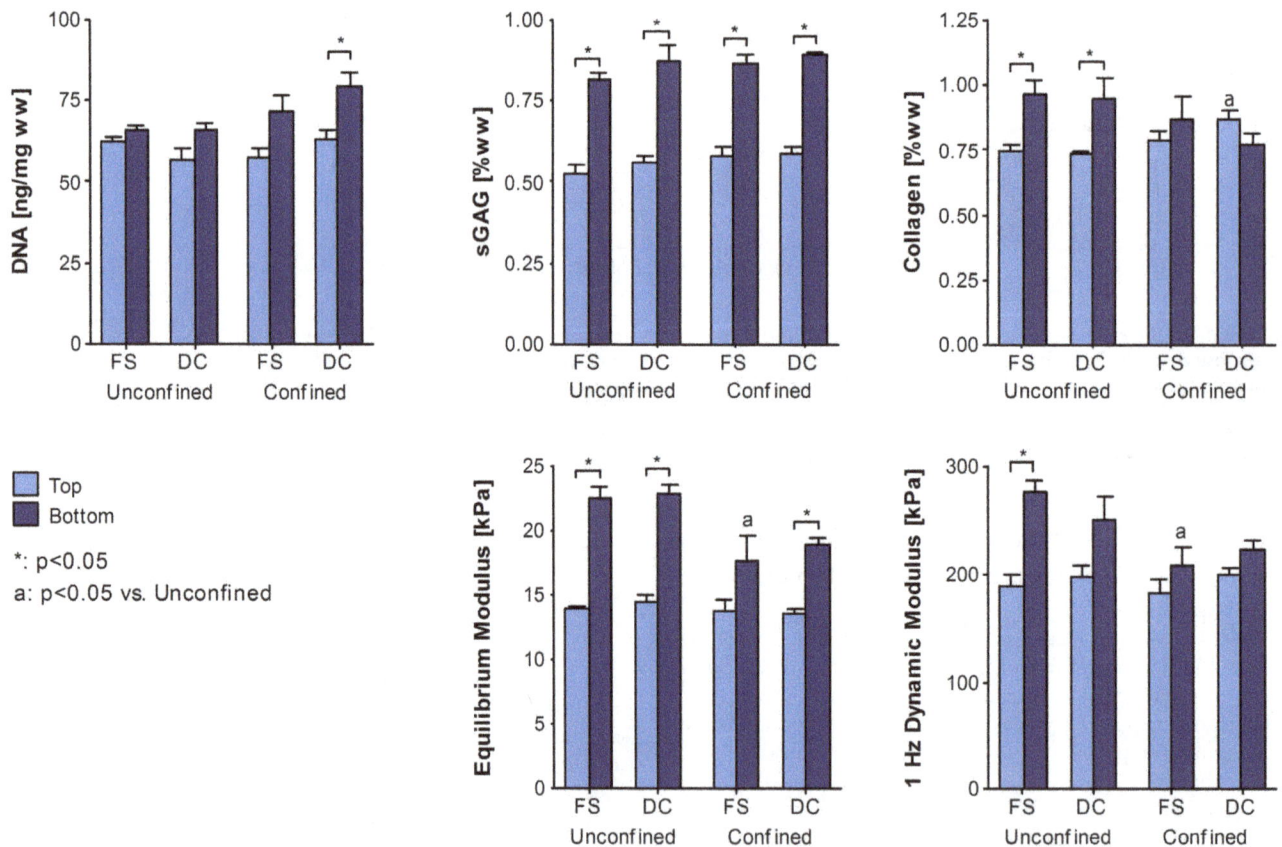

Figure 5. Radial confinement coupled with dynamic compression enhances collagen accumulation in the top of the construct. Agarose constructs containing MSCs at 20×10^6 cells/mL were confined from day 21 to day 42 of culture while 10% dynamic compression was applied. The top and bottom regions of constructs were analysed for DNA, sGAG and collagen contents. Top and bottom regions of constructs were also mechanically tested for both the equilibrium modulus and the dynamic modulus at 1 Hz. FS: free-swelling; DC: dynamic compression. ($n = 4$).

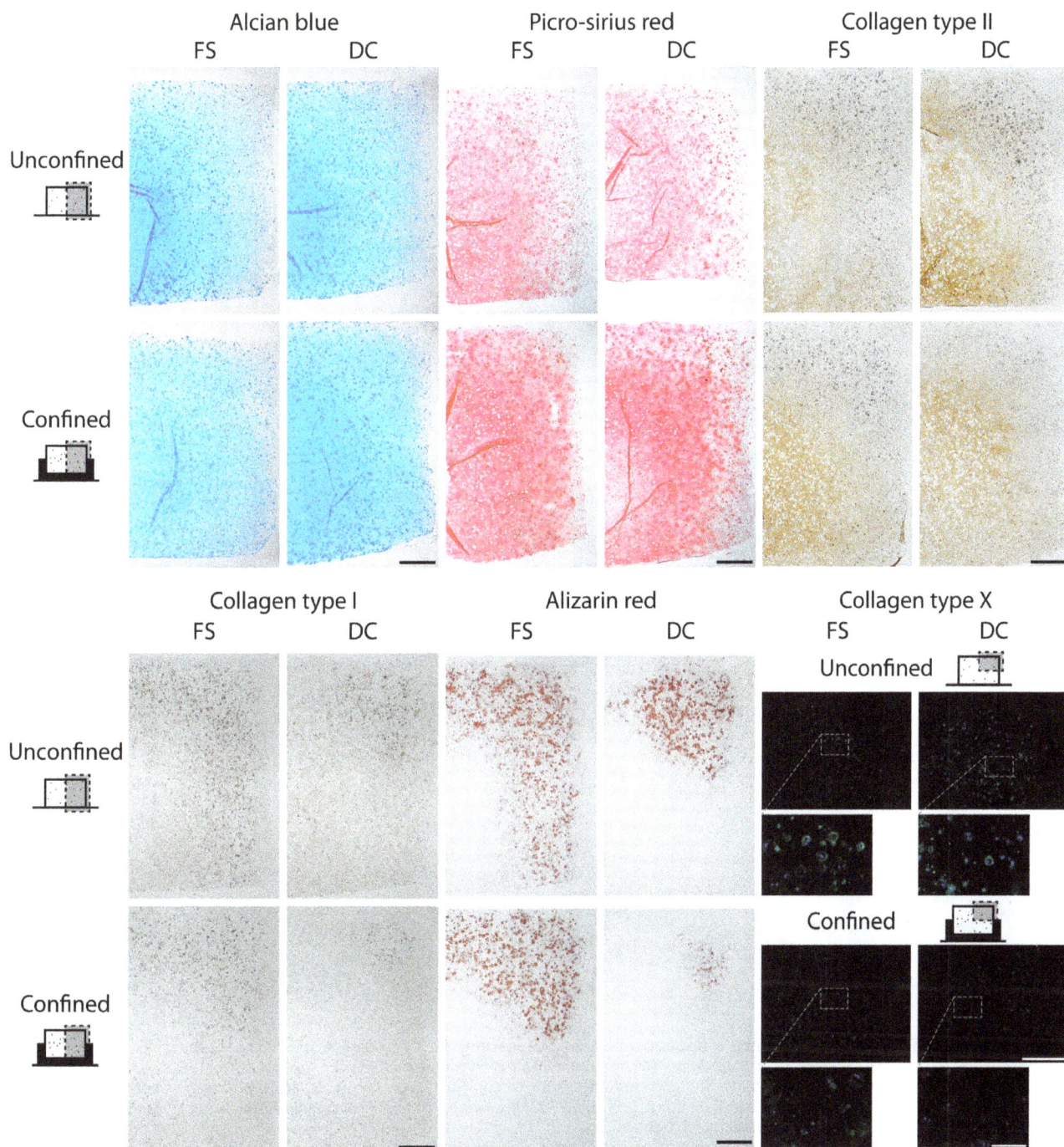

Figure 6. Radial confinement coupled with dynamic compression suppresses endochondral progression. Agarose constructs containing MSCs at 20×10^6 cells/mL were confined from day 21 to day 42 of culture while 10% dynamic compression was applied. Constructs at day 42 were stained with alcian blue for sulphated mucins, picro-sirius red for total collagen, alizarin red for calcific deposition and immunohistochemically for collagen type I, collagen type II and collagen type X. Representative full-depth half construct sections are shown for all but collagen type X where a representative quarter section is shown in immunofluorescence. Collagen type I and collagen type II staining is indicated in brown, with calcific deposits evident in black. FS: free-swelling; DC: dynamic compression. ($n = 2$) Scale bar 500 μm. Inset scale bar 100 μm.

tion of dynamic compression ($p = 0.0137$; Fig. 7). However, confinement led to an overall decrease in collagen content in the construct bottom in comparison to unconfined controls ($p = 0.0026$; Fig. 7, Fig. S2). On inclusion of collagen secreted to the media, it was evident that confinement at 50×10^6 cells/mL also inhibited total collagen production ($p = 0.0407$; Fig. S2).

Although proteoglycan-4 (PRG4) staining was weak in constructs seeded with 20×10^6 cells/mL, at 50×10^6 cells/mL the combination of confinement and dynamic compression was observed to increase PRG4 staining at the top surface of constructs (Fig. 7B). Collagen staining decreased toward the core region of all

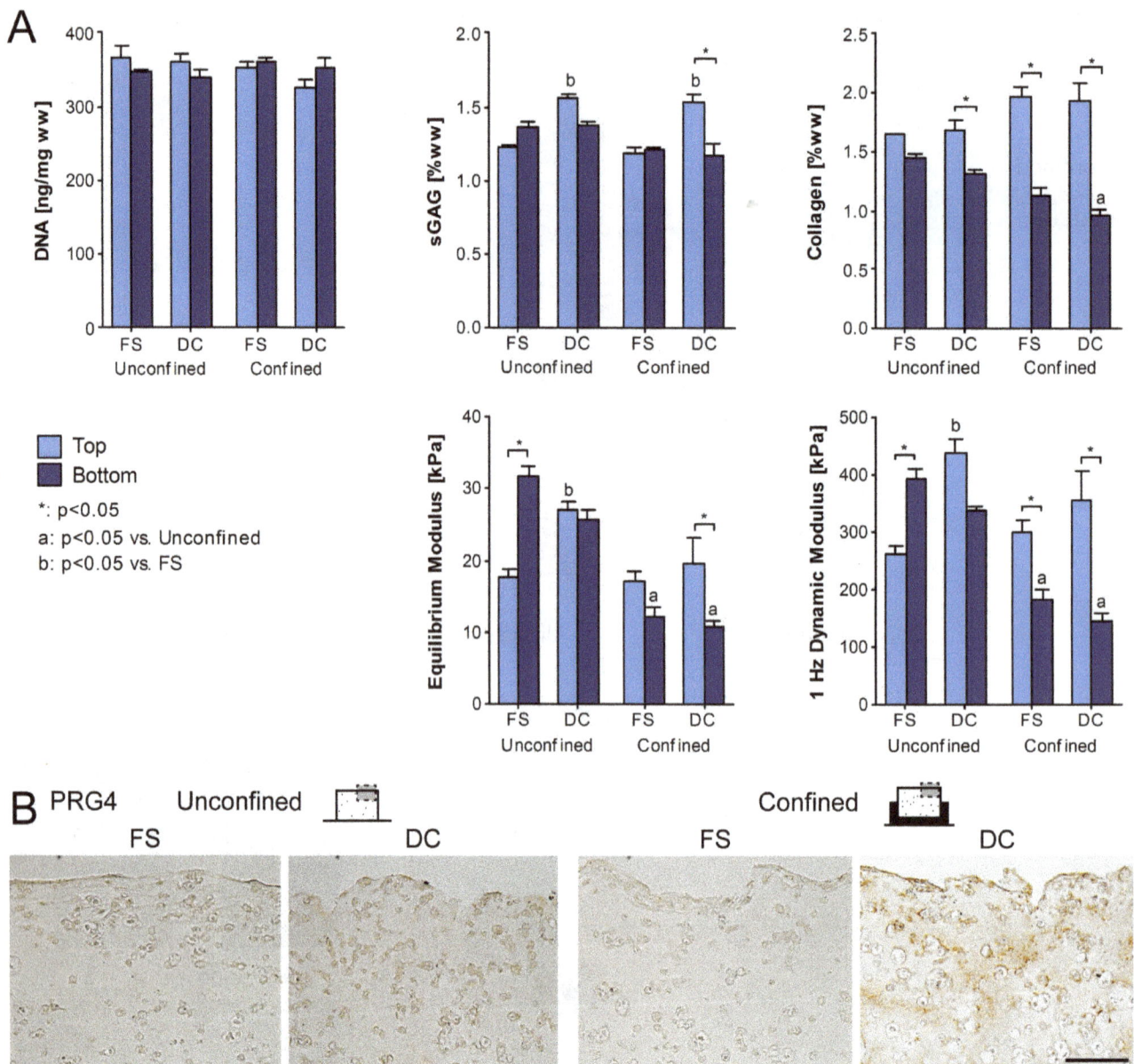

Figure 7. MSC response to extrinsic signals is dependent on the cell seeding density. Agarose constructs containing MSCs at 50×10^6 cells/mL were confined from day 21 to day 42 of culture while 10% dynamic compression was applied. (A) The top and bottom regions of constructs were analysed for DNA, sGAG and collagen contents. Top and bottom regions of constructs were also mechanically tested for both the equilibrium modulus and the dynamic modulus at 1 Hz. ($n = 4$) (B) Constructs at day 42 were stained immunohistochemically for proteoglycan-4 (PRG4). FS: free-swelling; DC: dynamic compression. ($n = 2$) Scale bar 100 μm.

constructs and was more heterogeneous than alcian blue staining (Fig. S3).

Dynamic compression acted to enhance both the equilibrium and dynamic modulus for the whole construct ($p < 0.05$; data not shown). Regionally, this increase was only evident in the construct top, which was stiffer than the corresponding region in FS controls ($p < 0.05$; Fig. 7). Confinement had a negative effect on whole construct stiffness ($p < 0.05$); attributed to a decrease in both equilibrium and dynamic moduli in the bottom of confined constructs corresponding to lower matrix accumulation in this region ($p < 0.01$; Fig. 7).

Discussion

Significant developments have been made regarding the use of MSCs for functional cartilage tissue engineering [59–61]. However, attempts to engineer grafts with a zonal composition and organisation mimicking normal articular cartilage have been limited. Creating such zonal variations in tissue composition within engineered grafts may be critical to regenerating hyaline cartilage with a normal Benninghoff architecture [62]. Here we show that the depth dependent properties of cartilaginous grafts engineered using MSCs can be modulated through spatial alteration of the oxygen tension and the mechanical environment within the developing construct. MSCs were encapsulated in agarose hydrogel and confined up to half their thickness. This reduced oxygen levels in the bottom of the construct and when combined with dynamic compression, increased tensile strains across the top of the construct. These spatial changes in the local environment correlated with increased sGAG accumulation in the bottom of constructs, increased collagen accumulation in the top of the construct, and a near complete suppression of MSC hypertrophy and tissue calcification throughout the depth of the engineered tissue. Consequently, a tissue with depth dependent variation in sGAG, collagen and mechanical properties similar, but not identical to native articular cartilage was established. In an effort to increase extracellular matrix (ECM) accumulation and mechanical properties, constructs were seeded at a higher cell density (50×10^6 cells/mL). This failed to instigate the creation of a native-like zonal composition in biochemical constituents and mechanical properties, demonstrating that the spatial environment within constructs seeded at very high densities was not conducive to engineering zonal constructs with cartilage-like properties.

Implantable fibre-optic oxygen micro-sensors and computational modelling were used to assess the spatial alterations in oxygen tension and the mechanical environment induced by confinement and dynamic compression respectively. Due to the uncertainty associated with material parameter identification for constructs later in culture where elaborated extracellular matrix may alter nutrient transfer and inhomogeneously alter construct mechanics, the simulations were performed for the initial cell seeded agarose constructs only. Despite this limitation, the measurement of oxygen availability and the finite element simulations provided insight into the spatial gradients in the mechanical environment and the oxygen tension throughout the construct and its dependence on cell density.

Initially, the effect of confinement was examined over 21 days. Confinement enhanced sGAG accumulation in the bottom (confined) region of the construct. One explanation for this was that confinement was simply acting to reduce secretion of ECM components into the media. While less sGAG was secreted from confined constructs, the total produced remained significantly higher with confinement (Fig. 4B), supporting the hypothesis that the low oxygen microenvironment predicted within the bottom of confined constructs may be enhancing chondrogenesis. Regions of faint alcian blue and collagen type II staining (Fig. 4C) in both confined and unconfined constructs also appeared to correlate with predicted regions of higher oxygen availability (Fig. 2B). While confinement lowers the availability of oxygen in the bottom of confined constructs, it could also act to reduce the availability of other regulatory factors such as TGF-β3 or glucose. However, in contrast to the abundance of data available in the literature supporting the argument that lower oxygen levels enhance chondrogenesis of MSCs [18–25], to the author's knowledge, there is no data available in the literature demonstrating that lower levels of other factors such as TGF-β3 or glucose will enhance cartilage matrix specific ECM synthesis in bone marrow derived MSCs. While we believe the lower levels of oxygen within the bottom of confined constructs are acting to enhance chondrogenesis in this region of the construct, we cannot rule out the possibility that gradients in other factors may also be playing a role.

In contrast to the findings over the first three weeks of culture, confinement from day 21 to day 42 did not lead to enhanced sGAG accumulation. In fact, sGAG accumulation was concentrated in the bottom of both confined and unconfined constructs at day 42. Given that care was taken to prevent constructs from flipping over during culture (thereby ensuring that the top surface of the construct remained face upwards for the culture duration), a low oxygen region will naturally develop in the bottom of the construct due to cellular consumption (Fig. 2), which may explain

why sGAG accumulation in the long term is higher in the bottom of both confined and unconfined constructs. It should be noted that the oxygen levels within the engineered tissue will also theoretically depend on the depth of media above the construct, and this should be considered when designing culture strategies to engineer zonal tissues.

When confinement was combined with dynamic compression, collagen accumulation increased in the top of the construct resulting in a depth-dependent zonal variation in both the biochemical composition and compressive properties of the engineered tissue. The increased levels of collagen synthesis in the top of mechanically stimulated confined constructs may be due to the higher levels of strain within this region of the engineered tissue (Fig. 3). Intermittent cyclic tension has specifically been shown to increase collagen content within MSC seeded constructs [44,45]. In the present study, tensile strains across the top of dynamically compressed constructs were predicted to approximately double due to radial confinement (Fig. 3); perhaps implicating this stimulus in promoting the higher levels of collagen accumulation in this region of the construct. Interestingly, the combination of confinement and dynamic compression also acted to inhibit hypertrophy (as evident by reduced type X collagen staining) and calcification (as evident by reduced alizarin red staining) of the engineered tissue. We have previously demonstrated that a low oxygen tension supresses the endochondral phenotype of chondrogenically primed MSCs [25], and here we provide further support for the role of dynamic compression in ensuring the development of a stable chondrogenic phenotype [30,63], suggesting that gradients in oxygen levels and mechanical cues may together regulate hypertrophy in developing articular cartilage. Although fluid pressures in the bottom of confined constructs were increased due to dynamic compression, this did not lead to a corresponding increase in sGAG accumulation in this region of the construct at either cell seeding density. This may be due to the fact that the pressure generated remains over an order of magnitude lower than that reported to elicit increases in chondrogenic gene expression for MSCs [31,32,35].

The effect of dynamic compression was also dependent on the cell seeding density. At a seeding density of 50×10^6 cells/mL dynamic compression acted to enhance sGAG accumulation and the bulk construct mechanical properties in both unconfined and confined conditions. Long-term dynamic compression of MSC seeded agarose applied after 21 days of unloaded pre-culture has previously been shown to augment sGAG accumulation [42] and enhance construct stiffness [43]. This finding that MSC response to dynamic compression is dependent on cell seeding density has also recently been reported [63]. At the higher seeding density of 50×10^6 cells/mL, it is possible that greater cell-cell contact and signalling may enhance MSC response to dynamic compression. It is also possible that a threshold level of ECM accumulation is required for dynamic compression to stimulate additional synthesis, as increasing the cell density was seen to increase total matrix accumulation. Démarteau et al. demonstrated that chondrocyte response to dynamic compression was positively correlated to the level of ECM accumulation [64]. Additionally, dynamic compression may facilitate enhanced transport of nutrients and other regulatory factors within the construct [65–67]. Increasing cell density acts to increase the severity of nutrient gradients due to cellular consumption in the construct periphery, which may be somewhat overcome through the application of dynamic compression.

At 50×10^6 cells/mL, the combination of confinement and loading also led to an increase in PRG4 staining across the top of the engineered tissue, which may be due to the high levels of deformation present in this region of the construct. This protein is known to play an important role in joint lubrication, and its expression has previously been shown to increase with mechanical stimulation [68,69].

While a construct with a zonal variation in mechanical properties and biochemical composition mimicking certain aspects of normal articular cartilage was achieved with a seeding density of 20×10^6 cells/mL, this was not the case at the higher seeding density of 50×10^6 cells/mL. Increasing the cell seeding density presumably increased consumption of oxygen and other solutes within the construct, resulting in a more acute decrease in nutrient availability away from the periphery. Confinement of constructs added to the severity of this gradient, such that an anoxic region was predicted in the bottom of confined constructs at the 50×10^6 cells/mL seeding density (Fig. 2B); lower than that predicted within the deep zone of native articular cartilage [46]. It is expected that gradients similar to that predicted for oxygen will develop for other solutes such as ascorbate and glucose and that one, or a combination of these may compromise cell metabolism culminating in the reduced ECM production observed at 50×10^6 cells/mL. While matrix accumulation was significantly reduced in regions of nutrient limitation, DNA content did not follow the same trend, with no differences between construct regions or conditions (Fig. 7A), suggesting that cells were viable but in a quiescent state. When cultured at low oxygen, MSCs have demonstrated a robust glycolytic potential [56,70] which may explain the survival of these cells in this potentially anoxic environment. It has recently been shown that optimal nutrient supply is crucial to the production of functional cartilage matrix using MSCs [71]. Vascularised cartilage canals are present in developing cartilage [72,73] and may provide a route for oxygen and nutrient supply. Mimicking such nutrient paths in engineered constructs [74,75] may provide a means to overcome transport limitations in such tissues.

Recently, through incorporation of specific natural and synthetic components into polyethylene glycol hydrogels, it has been demonstrated that it is possible to direct MSC differentiation into zone-specific phenotypes [15]; and through layering of these zone-specific hydrogels, engineer a depth-dependent tissue [76]. This approach was successful in the induction of cartilage-like zonal variations in collagen type II, type X, and sGAG production. The strategy adopted in this study, namely to modulate gradients in biomechanical and biochemical signals within the developing tissue, also has the potential to induce cartilage-like zonal variations in ECM content in one cohesive construct. Furthermore, this system may better recapitulate the spatial patterns of regulatory cues determining articular cartilage organisation during postnatal development, and may prime the construct for *in vivo* implantation as it will be subject to similar environmental cues within a load bearing defect. By coupling this zonal approach with strategies that attempt to spatially regulate endochondral ossification within MSC seeded hydrogels [77], it may be possible to also engineer functional osteochondral grafts.

Conclusion

Engineering cartilaginous grafts with structural composition and organisation is crucial to the long term repair of cartilage lesions.

By controlling the oxygen tension and mechanical environment through the depth of the developing tissue, MSC differentiation was modulated such that a construct with depth-dependent sGAG and collagen content somewhat akin to that of articular cartilage was engineered. This paper represents a novel approach toward engineering an organ or tissue with zonal variations in biochemical composition and mechanical properties. While there are still challenges to be overcome in order to engineer a native-like zonal articular cartilage tissue, and whether such bioreactor systems are ultimately used to engineer cartilaginous grafts for the clinic remains an open question [78], the results of this study help us to elucidate how environmental factors regulate MSC differentiation and in this case, the development and organisation of articular cartilage; knowledge that will be central to developing new therapies for damaged and diseased joints.

Supporting Information

Figure S1 Radial confinement coupled with dynamic compression enhances collagen accumulation in the top of the construct. Agarose constructs containing MSCs at 20×10^6 cells/mL were confined from day 21 to day 42 of culture while 10% dynamic compressive strain was applied. The top and bottom regions of constructs were analysed for sGAG and collagen contents which were normalised to DNA content. sGAG and collagen accumulated within the construct and secreted to culture media was also measured. FS: free-swelling; DC: dynamic compression. Dynamic compression as a main effect led to enhanced sGAG secretion to the media; $p = 0.0344$. $n = 4$. *: $p < 0.05$; **a**: $p < 0.05$ vs. Unconfined.

Figure S2 MSC response to extrinsic signals is dependent on the cell seeding density. Agarose constructs containing MSCs at 50×10^6 cells/mL were confined from day 21 to day 42 of culture while 10% dynamic compressive strain was applied. The top and bottom regions of constructs were analysed for sGAG and collagen contents which were normalised to DNA content. sGAG and collagen accumulated within the construct and secreted to culture media was measured. FS: free-swelling; DC: dynamic compression. $n = 4$. *: $p < 0.01$; **a**: $p < 0.05$ vs. Unconfined; **b**: $p < 0.05$ vs. FS.

Figure S3 MSC response to extrinsic signals is dependent on the cell seeding density. Constructs at day 42 were stained with alcian blue for sulphated mucins, picro-sirius red for total collagen, and immunohistochemically for collagen type I and type II. Representative full-depth half construct sections are shown. FS: free-swelling; DC: dynamic compression. ($n = 2$) Scale bar 500 μm.

Author Contributions

Conceived and designed the experiments: ST TN DK. Performed the experiments: ST TN SC. Analyzed the data: ST TN SC DK. Wrote the paper: ST DK.

References

1. Mow VC, Hung CT (2001) Biomechanics of articular cartilage. In: Nordin M, Frankel VH, editors. Basic biomechanics of the musculoskeletal system. 3rd ed. Philadelphia: Lippincott Williams & Wilkins. 60–101.

2. Gannon AR, Nagel T, Kelly DJ (2012) The role of the superficial region in determining the dynamic properties of articular cartilage. Osteoarthritis Cartilage 20: 1417–1425.

3. Mow VC, Guo XE (2002) Mechano-electrochemical properties of articular cartilage: their inhomogeneities and anisotropies. Annu Rev Biomed Eng 4: 175–209.

4. Brocklehurst R, Bayliss MT, Maroudas A, Coysh HL, Freeman MA, et al. (1984) The composition of normal and osteoarthritic articular cartilage from

human knee joints. With special reference to unicompartmental replacement and osteotomy of the knee. J Bone Joint Surg Am 66: 95–106.

5. Schinagl RM, Gurskis D, Chen AC, Sah RL (1997) Depth-dependent confined compression modulus of full-thickness bovine articular cartilage. J Orthop Res 15: 499–506.

6. Laasanen MS, Toyras J, Korhonen RK, Rieppo J, Saarakkala S, et al. (2003) Biomechanical properties of knee articular cartilage. Biorheology 40: 133–140.

7. Akizuki S, Mow VC, Muller F (1986) Tensile properties of human knee joint cartilage: I. Influence of ionic conditions, weight bearing, and fibrillation on the tensile modulus. J Orthop Res 4: 379–392.

8. Kim TK, Sharma B, Williams CG, Ruffner MA, Malik A, et al. (2003) Experimental model for cartilage tissue engineering to regenerate the zonal organization of articular cartilage. Osteoarthritis Cartilage 11: 653–664.

9. Klein TJ, Schumacher BL, Schmidt TA, Li KW, Voegtline MS, et al. (2003) Tissue engineering of stratified articular cartilage from chondrocyte subpopulations. Osteoarthritis Cartilage 11: 595–602.

10. Sharma B, Williams CG, Kim TK, Sun D, Malik A, et al. (2007) Designing zonal organization into tissue-engineered cartilage. Tissue Eng 13: 405–414.

11. Woodfield TB, Van Blitterswijk CA, De Wijn J, Sims TJ, Hollander AP, et al. (2005) Polymer scaffolds fabricated with pore-size gradients as a model for studying the zonal organization within tissue-engineered cartilage constructs. Tissue Eng 11: 1297–1311.

12. Ng KW, Wang CCB, Mauck RL, Kelly TAN, Chahine NO, et al. (2005) A layered agarose approach to fabricate depth-dependent inhomogeneity in chondrocyte-seeded constructs. J Orthop Res 23: 134–141.

13. Ng KW, Mauck RL, Statman LY, Lin EY, Ateshian GA, et al. (2006) Dynamic deformational loading results in selective application of mechanical stimulation in a layered, tissue-engineered cartilage construct. Biorheology 43: 497–507.

14. Hwang NS, Varghese S, Lee HJ, Theprungsirikul P, Canver A, et al. (2007) Response of zonal chondrocytes to extracellular matrix-hydrogels. FEBS Lett 581: 4172–4178.

15. Nguyen LH, Kudva AK, Guckert NL, Linse KD, Roy K (2011) Unique biomaterial compositions direct bone marrow stem cells into specific chondrocytic phenotypes corresponding to the various zones of articular cartilage. Biomaterials 32: 1327–1338.

16. Ng KW, Ateshian GA, Hung CT (2009) Zonal chondrocytes seeded in a layered agarose hydrogel create engineered cartilage with depth-dependent cellular and mechanical inhomogeneity. Tissue Eng Part A 15: 2315–2324.

17. Lee CS, Gleghorn JP, Won Choi N, Cabodi M, Stroock AD, et al. (2007) Integration of layered chondrocyte-seeded alginate hydrogel scaffolds. Biomaterials 28: 2987–2993.

18. Kanichai M, Ferguson D, Prendergast PJ, Campbell VA (2008) Hypoxia promotes chondrogenesis in rat mesenchymal stem cells: a role for AKT and hypoxia-inducible factor (HIF)-1alpha. J Cell Physiol 216: 708–715.

19. Krinner A, Zscharnack M, Bader A, Drasdo D, Galle J (2009) Impact of oxygen environment on mesenchymal stem cell expansion and chondrogenic differentiation. Cell Prolif 42: 471–484.

20. Meyer EG, Buckley CT, Thorpe SD, Kelly DJ (2010) Low oxygen tension is a more potent promoter of chondrogenic differentiation than dynamic compression. J Biomech 43: 2516–2523.

21. Buckley CT, Meyer EG, Kelly DJ (2012) The influence of construct scale on the composition and functional properties of cartilaginous tissues engineered using bone marrow-derived MSCs. Tissue Eng Part A 18: 382–396.

22. Baumgartner L, Arnhold S, Brixius K, Addicks K, Bloch W (2010) Human mesenchymal stem cells: Influence of oxygen pressure on proliferation and chondrogenic differentiation in fibrin glue in vitro. J Biomed Mater Res A 93: 930–940.

23. Buckley CT, Vinardell T, Kelly DJ (2010) Oxygen tension differentially regulates the functional properties of cartilaginous tissues engineered from infrapatellar fat pad derived MSCs and articular chondrocytes. Osteoarthritis Cartilage 18: 1345–1354.

24. Li J, He F, Pei M (2011) Creation of an in vitro microenvironment to enhance human fetal synovium-derived stem cell chondrogenesis. Cell Tissue Res 345: 357–365.

25. Sheehy EJ, Buckley CT, Kelly DJ (2012) Oxygen tension regulates the osteogenic, chondrogenic and endochondral phenotype of bone marrow derived mesenchymal stem cells. Biochem Biophys Res Commun 417: 305–310.

26. Pauwels F (1960) A new theory on the influence of mechanical stimuli on the differentiation of supporting tissue. The tenth contribution to the functional anatomy and causal morphology of the supporting structure. Z Anat Entwicklungsgesch 121: 478–515.

27. Carter DR, Blenman PR, Beaupre GS (1988) Correlations between mechanical stress history and tissue differentiation in initial fracture healing. J Orthop Res 6: 736–748.

28. Prendergast PJ, Huiskes R, Soballe K (1997) ESB Research Award 1996. Biophysical stimuli on cells during tissue differentiation at implant interfaces. J Biomech 30: 539–548.

29. Kelly DJ, Jacobs CR (2010) The role of mechanical signals in regulating chondrogenesis and osteogenesis of mesenchymal stem cells. Birth Defects Res C Embryo Today 90: 75–85.

30. Thorpe SD, Buckley CT, Steward AJ, Kelly DJ (2012) European Society of Biomechanics S.M. Perren Award 2012: The external mechanical environment can override the influence of local substrate in determining stem cell fate. J Biomech 45: 2483–2492.

31. Angele P, Yoo JU, Smith C, Mansour J, Jepsen KJ, et al. (2003) Cyclic hydrostatic pressure enhances the chondrogenic phenotype of human mesenchymal progenitor cells differentiated in vitro. J Orthop Res 21: 451–457.

32. Miyanishi K, Trindade MC, Lindsey DP, Beaupre GS, Carter DR, et al. (2006) Effects of hydrostatic pressure and transforming growth factor-beta 3 on adult human mesenchymal stem cell chondrogenesis in vitro. Tissue Eng 12: 1419–1428.

33. Steward AJ, Thorpe SD, Vinardell T, Buckley CT, Wagner DR, et al. (2012) Cell-matrix interactions regulate mesenchymal stem cell response to hydrostatic pressure. Acta Biomater 8: 2153–2159.

34. Vinardell T, Rolfe RA, Buckley CT, Meyer EG, Ahearne M, et al. (2012) Hydrostatic pressure acts to stabilise a chondrogenic phenotype in porcine joint tissue derived stem cells. Eur Cell Mater 23: 121–132; discussion 133–124.

35. Meyer EG, Buckley CT, Steward AJ, Kelly DJ (2011) The effect of cyclic hydrostatic pressure on the functional development of cartilaginous tissues engineered using bone marrow derived mesenchymal stem cells. J Mech Behav Biomed Mater 4: 1257–1265.

36. Liu Y, Buckley CT, Downey R, Mulhall KJ, Kelly DJ (2012) The role of environmental factors in regulating the development of cartilaginous grafts engineered using osteoarthritic human infrapatellar fat pad-derived stem cells. Tissue Eng Part A 18: 1531–1541.

37. Angele P, Schumann D, Angele M, Kinner B, Englert C, et al. (2004) Cyclic, mechanical compression enhances chondrogenesis of mesenchymal progenitor cells in tissue engineering scaffolds. Biorheology 41: 335–346.

38. Campbell JJ, Lee DA, Bader DL (2006) Dynamic compressive strain influences chondrogenic gene expression in human mesenchymal stem cells. Biorheology 43: 455–470.

39. Huang CY, Hagar KL, Frost LE, Sun Y, Cheung HS (2004) Effects of cyclic compressive loading on chondrogenesis of rabbit bone-marrow derived mesenchymal stem cells. Stem Cells 22: 313–323.

40. Kisiday JD, Frisbie DD, McIlwraith CW, Grodzinsky AJ (2009) Dynamic compression stimulates proteoglycan synthesis by mesenchymal stem cells in the absence of chondrogenic cytokines. Tissue Eng Part A 15: 2817–2824.

41. Mauck RL, Byers BA, Yuan X, Tuan RS (2007) Regulation of cartilaginous ECM gene transcription by chondrocytes and MSCs in 3D culture in response to dynamic loading. Biomech Model Mechanobiol 6: 113–125.

42. Thorpe SD, Buckley CT, Vinardell T, O'Brien FJ, Campbell VA, et al. (2010) The response of bone marrow-derived mesenchymal stem cells to dynamic compression following tgf-β3 induced chondrogenic differentiation. Ann Biomed Eng 38: 2896–2909.

43. Huang AH, Farrell MJ, Kim M, Mauck RL (2010) Long-term dynamic loading improves the mechanical properties of chondrogenic mesenchymal stem cell-laden hydrogel. Eur Cell Mater 19: 72–85.

44. Connelly JT, Vanderploeg EJ, Mouw JK, Wilson CG, Levenston ME (2010) Tensile loading modulates bone marrow stromal cell differentiation and the development of engineered fibrocartilage constructs. Tissue Eng Part A 16: 1913–1923.

45. Baker BM, Shah RP, Huang AH, Mauck RL (2011) Dynamic tensile loading improves the functional properties of mesenchymal stem cell-laden nanofiber-based fibrocartilage. Tissue Eng Part A 17: 1445–1455.

46. Zhou S, Cui Z, Urban JP (2004) Factors influencing the oxygen concentration gradient from the synovial surface of articular cartilage to the cartilage-bone interface: a modeling study. Arthritis Rheum 50: 3915–3924.

47. Buckley CT, Thorpe SD, O'Brien FJ, Robinson AJ, Kelly DJ (2009) The effect of concentration, thermal history and cell seeding density on the initial mechanical properties of agarose hydrogels. J Mech Behav Biomed Mater 2: 512–521.

48. Kim YJ, Sah RL, Doong JY, Grodzinsky AJ (1988) Fluorometric assay of DNA in cartilage explants using Hoechst 33258. Anal Biochem 174: 168–176.

49. Kafienah W, Sims TJ (2004) Biochemical methods for the analysis of tissue-engineered cartilage. Methods Mol Biol 238: 217–230.

50. Ignat'eva NY, Danilov NA, Averkiev SV, Obrezkova MV, Lunin VV, et al. (2007) Determination of hydroxyproline in tissues and the evaluation of the collagen content of the tissues. J Anal Chem 62: 51–57.

51. Görke U-J, Günther H, Nagel T, Wimmer MA (2010) A Large Strain Material Model for Soft Tissues With Functionally Graded Properties. J Biomech Eng 132: 074502.

52. Gu WY, Yao H, Huang CY, Cheung HS (2003) New insight into deformation-dependent hydraulic permeability of gels and cartilage, and dynamic behavior of agarose gels in confined compression. J Biomech 36: 593–598.

53. Lion A (1997) A physically based method to represent the thermo-mechanical behaviour of elastomers. Acta Mechanica 123: 1–25.

54. Storn R, Price K (1997) Differential Evolution – A Simple and Efficient Heuristic for Global Optimization over Continuous Spaces. J of Global Optimization 11: 341–359.

55. Sengers BG, Heywood HK, Lee DA, Oomens CWJ, Bader DL (2005) Nutrient Utilization by Bovine Articular Chondrocytes: A Combined Experimental and Theoretical Approach. J Biomech Eng 127: 758–766.

56. Pattappa G, Heywood HK, de Bruijn JD, Lee DA (2011) The metabolism of human mesenchymal stem cells during proliferation and differentiation. J Cell Physiol 226: 2562–2570.

57. Thorpe SD, Buckley CT, Vinardell T, O'Brien FJ, Campbell VA, et al. (2008) Dynamic compression can inhibit chondrogenesis of mesenchymal stem cells. Biochem Biophys Res Commun 377: 458–462.

58. Haugh MG, Meyer EG, Thorpe SD, Vinardell T, Duffy GP, et al. (2011) Temporal and Spatial Changes in Cartilage-Matrix-Specific Gene Expression in Mesenchymal Stem Cells in Response to Dynamic Compression. Tissue Eng Part A 17: 3085–3093.

59. Khan WS, Johnson DS, Hardingham TE (2010) The potential of stem cells in the treatment of knee cartilage defects. Knee 17: 369–374.

60. Vinatier C, Bouffi C, Merceron C, Gordeladze J, Brondello JM, et al. (2009) Cartilage tissue engineering: towards a biomaterial-assisted mesenchymal stem cell therapy. Curr Stem Cell Res Ther 4: 318–329.

61. Huang AH, Farrell MJ, Mauck RL (2010) Mechanics and mechanobiology of mesenchymal stem cell-based engineered cartilage. J Biomech 43: 128–136.

62. Nagel T, Kelly DJ (2012) The Composition of Engineered Cartilage at the Time of Implantation Determines the Likelihood of Regenerating Tissue with a Normal Collagen Architecture. Tissue Eng Part A: In press.

63. Bian L, Zhai DY, Zhang EC, Mauck RL, Burdick JA (2012) Dynamic compressive loading enhances cartilage matrix synthesis and distribution and suppresses hypertrophy in hMSC-laden hyaluronic acid hydrogels. Tissue Eng Part A 18: 715–724.

64. Demarteau O, Wendt D, Braccini A, Jakob M, Schafer D, et al. (2003) Dynamic compression of cartilage constructs engineered from expanded human articular chondrocytes. Biochem Biophys Res Commun 310: 580–588.

65. Albro MB, Chahine NO, Li R, Yeager K, Hung CT, et al. (2008) Dynamic loading of deformable porous media can induce active solute transport. J Biomech 41: 3152–3157.

66. Mauck RL, Hung CT, Ateshian GA (2003) Modeling of neutral solute transport in a dynamically loaded porous permeable gel: implications for articular cartilage biosynthesis and tissue engineering. J Biomech Eng 125: 602–614.

67. Albro MB, Banerjee RE, Li R, Oungoulian SR, Chen B, et al. (2011) Dynamic loading of immature epiphyseal cartilage pumps nutrients out of vascular canals. J Biomech 44: 1654–1659.

68. Kupcsik L, Stoddart MJ, Li Z, Benneker LM, Alini M (2010) Improving chondrogenesis: potential and limitations of SOX9 gene transfer and mechanical stimulation for cartilage tissue engineering. Tissue Eng Part A 16: 1845–1855.

69. Candrian C, Vonwil D, Barbero A, Bonacina E, Miot S, et al. (2008) Engineered cartilage generated by nasal chondrocytes is responsive to physical forces resembling joint loading. Arthritis Rheum 58: 197–208.

70. Pattappa G, Thorpe SD, Jegard NC, Heywood H, de Bruijn JD, et al. (2013) Continuous and uninterrupted oxygen tension influences the colony formation and oxidative metabolism of human mesenchymal stem cells. Tissue Eng Part C Methods 19: 68–79.

71. Farrell MJ, Comeau ES, Mauck RL (2012) Mesenchymal stem cells produce functional cartilage matrix in three-dimensional culture in regions of optimal nutrient supply. Eur Cell Mater 23: 425–440.

72. Hall BK (2005) Bones and cartilage: developmental and evolutionary skeletal biology. London: Elsevier Academic. xxviii, 760 p.

73. Lecocq M, Girard CA, Fogarty U, Beauchamp G, Richard H, et al. (2008) Cartilage matrix changes in the developing epiphysis: early events on the pathway to equine osteochondrosis? Equine Vet J 40: 442–454.

74. Buckley CT, Thorpe SD, Kelly DJ (2009) Engineering of large cartilaginous tissues through the use of microchanneled hydrogels and rotational culture. Tissue Eng Part A 15: 3213–3220.

75. Sheehy EJ, Buckley CT, Kelly DJ (2011) Chondrocytes and bone marrow-derived mesenchymal stem cells undergoing chondrogenesis in agarose hydrogels of solid and channelled architectures respond differentially to dynamic culture conditions. J Tissue Eng Regen Med 5: 747–758.

76. Nguyen LH, Kudva AK, Saxena NS, Roy K (2011) Engineering articular cartilage with spatially-varying matrix composition and mechanical properties from a single stem cell population using a multi-layered hydrogel. Biomaterials 32: 6946–6952.

77. Sheehy EJ, Vinardell T, Buckley CT, Kelly DJ (2013) Engineering osteochondral constructs through spatial regulation of endochondral ossification. Acta Biomater 9: 5484–5492.

78. Pelttari K, Wixmerten A, Martin I (2009) Do we really need cartilage tissue engineering? Swiss Med Wkly 139: 602–609.

Noninvasive Quantification of *In Vitro* Osteoblastic Differentiation in 3D Engineered Tissue Constructs Using Spectral Ultrasound Imaging

Madhu Sudhan Reddy Gudur[9]**, Rameshwar R. Rao**[9]**, Alexis W. Peterson, David J. Caldwell, Jan P. Stegemann**[¶]*****, Cheri X. Deng**[¶]*****

Department of Biomedical Engineering, University of Michigan, Ann Arbor, Michigan, United States of America

Abstract

Non-destructive monitoring of engineered tissues is needed for translation of these products from the lab to the clinic. In this study, non-invasive, high resolution spectral ultrasound imaging (SUSI) was used to monitor the differentiation of MC3T3 pre-osteoblasts seeded within collagen hydrogels. SUSI was used to measure the diameter, concentration and acoustic attenuation of scatterers within such constructs cultured in either control or osteogenic medium over 21 days. Conventional biochemical assays were used on parallel samples to determine DNA content and calcium deposition. Construct volume and morphology were accurately imaged using ultrasound. Cell diameter was estimated to be approximately 12.5–15.5 μm using SUSI, which corresponded well to measurements of fluorescently stained cells. The total number of cells per construct assessed by quantitation of DNA content decreased from $5.6 \pm 2.4 \times 10^4$ at day 1 to $0.9 \pm 0.2 \times 10^4$ at day 21. SUSI estimation of the equivalent number of acoustic scatters showed a similar decreasing trend, except at day 21 in the osteogenic samples, which showed a marked increase in both scatterer number and acoustic impedance, suggestive of mineral deposition by the differentiating MC3T3 cells. Estimation of calcium content by SUSI was 41.7 ± 11.4 μg/ml, which agreed well with the biochemical measurement of 38.7 ± 16.7 μg/ml. Color coded maps of parameter values were overlaid on B-mode images to show spatiotemporal changes in cell diameter and calcium deposition. This study demonstrates the use of non-destructive ultrasound imaging to provide quantitative information on the number and differentiated state of cells embedded within 3D engineered constructs, and therefore presents a valuable tool for longitudinal monitoring of engineered tissue development.

Editor: Hani A. Awad, University of Rochester, United States of America

Funding: This work was supported in part by a National Science Foundation Graduate Research Fellowship, Grant # DGE 1256260 (to RRR) http://www.nsfgrfp.org/, and by the National Institutes of Health through R21-AR064041 (to CXD and JPS). The funders had no role in study design, data collection and analysis, decision to publish, or preparation of the manuscript.

Competing Interests: The authors have declared that no competing interests exist.

* E-mail: jpsteg@umich.edu (JPS); cxdeng@umich.edu (CXD)

¶ These authors also contributed equally to this work.

Introduction

Bone tissue engineering approaches combine cells, biomaterials, and growth factors to recreate native bone tissue [1]. Traditionally, biochemical and histological assays are performed to monitor cell function and development in these engineered tissues. However, these techniques require sample processing and are destructive in nature, and therefore do not allow for an individual sample to be tracked as it develops. For example, traditional methods for characterizing cell number include manual counting chambers, automated cell counters, spectrophotometers, and flow cytometers [2,3,4]. These methods require a variety of sample processing steps including disruption of the tissue construct into constituents for counting (destructive in nature) and sample dilution. In addition, they may require specialized equipment and reagents that can be expensive. Importantly, most currently used measurement techniques describe only single timepoint, aggregate characteristics of the sample, and do not provide three dimensional (3D) spatial and temporal information.

Non-destructive approaches based on confocal microscopic imaging to count cell nuclei have been used to provide 3D assessment of cell numbers [5]. However, such techniques require high quality microscopy images, are time consuming, and involve complex processing algorithms to acquire entire spatially registered 3D images of the construct. Magnetic resonance imaging (MRI) and micro-computed tomography (μ-CT) techniques have been used to estimate bone mineral densities [6,7]. However, these methods require the use of calibration phantoms and involve long data acquisition times [8]. Long exposures to X-ray may affect cell-seeded constructs in terms of the structure, viability, and cellular development of the constructs. Conventional MRI imaging systems do not provide the ability to study the microstructural details of 3D engineered tissue constructs due to their low resolution. Therefore there is a need for non-destructive imaging and characterization modalities, capable of providing both spatial and temporal information of engineered tissues as they develop *in vitro*. Such methods would greatly facilitate the translation of tissue engineering products from the lab to the clinic.

Ultrasound imaging is a widely used non-invasive and non-destructive method that has the potential for quantitative evaluation of tissue development both *in vitro* and *in vivo*. It has been reported recently that ultrasound can be used to quantify cell number in BMSC/β-TCP composites using a grayscale equivalent parameter [9]. Fite et al. used an ultrasound method to monitor the chondrogenic differentiation of equine adipose stem cells in 3D poly(lactide-co-glycolide) scaffolds [10] by correlating signal attenuation measured through gray scale image analysis to extracellular matrix (ECM) deposition, which was considered to be a marker of cell differentiation. Kreitz et al. tracked collagen deposition by myofibroblasts in fibrin tissue constructs over an 18 day culture period [11]. Their quantitative analysis correlated observed gray scale values to ECM deposition as measured by hydroxyproline content. Ultrasound has also been used as a tool to measure the mechanical properties of agarose hydrogels as they develop over time [12], by correlating material properties such as elastic modulus with obtained acoustic properties.

Ultrasound propagation and acoustic scattering in a tissue volume depend on tissue microstructure, composition, and physical properties such as density and compressibility. Therefore, backscattered ultrasound signals may be used to extract information about the structure and composition of the tissue under investigation, as well as its mechanical and physical properties. Although tissue properties such as speed of sound, acoustic attenuation and the tissue volume can be calculated directly from the backscattered radiofrequency (RF) data, tissue microstructural details are not apparent from the raw RF signals. Tunis *et al.* [13] studied the envelope statistics of ultrasound backscatter signals from cisplatin-treated aggregated acute myeloid leukemia (AML) cells and evaluated the applicability of various statistical distribution functions to model the envelope histograms. They reported that shape parameters of the generalized gamma distribution function were sensitive to structural changes within cells induced by the drug.

Quantitative ultrasound imaging methods using spectral analysis of the RF signals have been developed to extract additional parameters for enhanced tissue characterization. The power spectrum of the backscattered RF data includes information about tissue microstructure, and the spectral regression parameters can be related to scatterer properties such as effective sizes, concentrations and acoustic impedances [14,15]. Spectral slope has been shown to depend on the scatterer size, whereas mid-band fit (MBF) relates to the size, concentration and relative acoustic impedances of the scattering elements [14]. Spectral analysis has been used in various applications, including characterization of plaque composition by intravascular ultrasound (IVUS) [16,17], lesions induced by high intensity focused ultrasound (HIFU) [18,19] and RF ablation [20]. Spectral parameters have also shown the ability to identify changes in tissue state for prostate, breast, pancreas, lymph node, and other cancer types [21,22,23,24,25,26,27]. Oelze *et al.* [22] developed methods to differentiate and characterize rat mammary fibroadenomas and 4T1 mouse carcinomas by estimating scatterer properties from backscatter RF signals in the spectral domain with a Gaussian form factor model [28].

The use of high frequency (20–60 MHz) ultrasound imaging has provided higher spatial resolution than conventional ultrasound imaging (5–15 MHz) in diagnostic radiology [23]. Kolios *et al.* have developed spectral analysis technique to characterize the properties of cell aggregates that were used as simplified models of tumors [23,29], to detect cellular changes with high spatial resolution and sensitivity after exposure to chemotherapy drug treatments [30]. They found that ultrasound backscatter intensity and spectral slope increased due to treatment, which was

interpreted as a consequence of the decrease in effective scatterer size of cell aggregates. The use of higher ultrasound frequency imaging, with corresponding ultrasound wavelengths in the order of 100 μm, permits sensing of changes in cell nuclei and cell structure.

In order to achieve non-invasive and quantitative assessment of engineered tissue constructs with high spatial resolution, we have implemented a high frequency spectral ultrasound imaging (SUSI) technique, and have validated its use to characterize the composition and structure engineered tissue constructs. In previous work, we used the spectral MBF and slope parameters to measure the quantity and spatial distribution of particulate hydroxyapatite in acellular collagen hydrogels [31]. We observed a strong correlation between MBF and mineral concentration, and between the spectral slope and particle size. The amount of mineral deposited from simulated body fluid on acellular collagen constructs over a period of 3 weeks was also studied using spectral parameters, and showed strong correlation with MBF.

The work presented here extends our previous study using high resolution SUSI to quantitatively characterize the osteogenic differentiation of MC3T3 mouse pre-osteoblast cells seeded within 3D collagen-based engineered tissues. MC3T3 cells are a well characterized mineralizing cell type that can be induced towards the osteogenic lineage with the addition of a defined set of supplements in the culture medium [32]. Collagen is a widely used biomaterial in orthopaedic tissue engineering due to its ability to support cell attachment and proliferation, as well as to serve as an osteoconductive and osteoinductive matrix [33]. In this study we non-destructively quantified the bulk properties of the hydrogels, including speed of sound, acoustic attenuation and volume compaction, and tracked these parameters over time. Microstructural properties of the cell-seeded constructs, including cell size, cell number, and cell differentiation, were also assessed using spectral ultrasound and were compared to data generated by traditional biochemical assays and confocal fluorescence imaging. This study demonstrates that SUSI can be used to non-destructively characterize cell-seeded engineered tissue constructs longitudinally over time with high spatial resolution.

Materials and Methods

Cell Culture

Mouse pre-osteoblast MC3T3-E1 (generously provided by Dr. R.T. Franchesci, University of Michigan) were cultured in α-MEM without ascorbic acid (Life Technologies, Grand Island, NY) supplemented with 10% fetal bovine serum (FBS; Life Technologies) and 1% penicillin and streptomycin (PS: Life Technologies) and used at passage 8. Media was changed every other day.

Collagen Hydrogel Synthesis

Three-dimensional (3D) collagen hydrogels were created as previously described [31]. Briefly, collagen type I (MP Biomedicals, Solon, OH) was prepared at 4.0 mg/ml in 0.02 N acetic acid. Collagen hydrogels (2.0 mg/ml final concentration) were formed by mixing 10% Dulbecco's modified Eagle's medium (DMEM; Life Technologies), 10% FBS, 20% 5X-concentrated DMEM (stock concentration), 10% 0.1 N NaOH (Sigma Aldrich, St. Louis, MO), and 50% collagen stock solution. 500 μL of the mixture was then pipetted into a 24-well plate and allowed to gel for 30 mins at 37°C. Cells were encapsulated within the hydrogels at the time of gelation at a concentration of 1.0×10^6 cells/ml.

After gelation, hydrogels were moved into a 6-well culture plate containing α-MEM supplemented with 10% FBS and 1% PS to allow cells to compact their matrices. After 24 hours, the media was changed with either a control media or osteogenic media containing 10 mM beta-glycero phosphate (β-GP; Sigma) and 50 μg/ml ascorbic acid 2-phosphate (Sigma).

Cell Viability

Cell viability was visualized and quantified as previously described [34]. At days 1 and 21, cell-seeded hydrogels were washed 3X in phosphate buffered saline (PBS; Life Technologies) for 5 mins and then incubated in 4 μM calcein-AM (EMD Millipore, Billerica, MA) and 4 μM ethidium homodimer-1 (Sigma) in PBS for 45 mins. Constructs were washed 3X in PBS prior to imaging on a Nikon A1 Confocal Microscope (Nikon Instruments, Melville, NY). Cell viability was quantified using ImageJ software (National Institute of Health, Bethesda, MD).

Fluorescence Staining

At days 1 and 21, cells were stained for their actin cytoskeleton and nuclei. Hydrogels were washed 2X in PBS for 5 mins/wash and then fixed in zinc-buffered formalin (Z-Fix; Battle Creek, MI) for 10 mins at 4°C. Gels were washed another 2X in PBS and then permeabilized using 0.5% Triton-X 100 (Sigma) in PBS for 20 mins at room temperature. Constructs were washed again 2X, and then incubated in a solution containing 165 nM AlexaFluor 488 phalloidin (Life Technologies) and 10 nM fluorescent DAPI (Life Technologies) in 1% bovine serum albumin (BSA; Sigma) in PBS for 45 mins. Hydrogels were washed again prior to imaging.

Biochemical Assays

Cellular DNA content and calcium were quantified as previously described [35]. For DNA quantification, hydrogels were collected and degraded overnight in 10 mM Tris-HCl (Sigma) containing 0.6 mg/mL collagenase type I (MP Biomedicals), 0.2% IGEPAL (Sigma), and 2 mM phenylmethanesulfonyl-fluoride (Sigma). DNA was then measured using the PicoGreen DNA assay (Life Technologies). Calcium secretion was assayed by first dissolving the constructs in 1 N acetic acid (Sigma) overnight. The cell-hydrogel lysate was then assayed using the ortho-cresolphthalein (OCPC) method [31].

Phantom Studies

Agar phantoms embedded with Polybead® microspheres (Polysciences, Inc., Warrington, PA) of different diameters at various concentrations were used to validate the size and concentration estimation from SUSI. Polybeads were added to the 2% agar solution at 45°C and thoroughly mixed to disperse them uniformly throughout the phantom. Four polybead diameters of interest were chosen: 6, 10, 16 and 25 μm. For each polybead size, phantoms with four different concentrations of polybeads were made. Each phantom was approximately 2000 mm^3 in size.

Ultrasound Imaging and Backscattered Signal Acquisition

Figure 1 shows a schematic diagram of the ultrasound imaging setup. A gel slab with 8% agarose (Sigma) was placed at the bottom of a 60 mm Petri dish to reduce ultrasound reflection from the bottom of the dish. The dish was then filled with α-MEM at room temperature and the constructs were placed on top of the agarose gel pad. Ultrasound imaging was performed using a Vevo 770 (VisualSonics Inc., Toronto, Canada) and an RMV 708 imaging probe with a nominal 55 MHz center frequency, 20–75 MHz bandwidth (−6 dB), 4.5 mm focal distance, and 1.5 mm depth of focus (−6 dB). Ultrasound B-mode imaging was

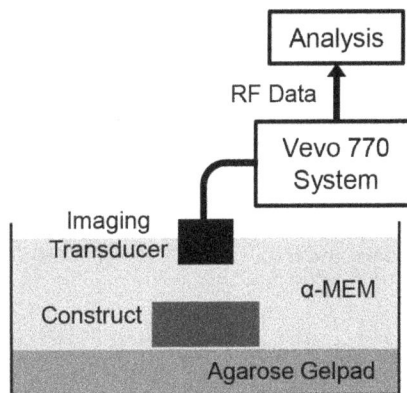

Figure 1. Schematic of experimental setup used for spectral ultrasound imaging (SUSI) of engineered tissue constructs.

performed with the ultrasound beam focus placed 0.5 mm below the top surface of each sample. The interval between adjacent A-lines in the B-scans was set at 31 μm. 3D imaging of the construct was performed by acquiring a series of B-scans with 200 μm interval between adjacent scans across the tissue construct using a computer-controlled automatic translational stage. The 3D image data were used to estimate the volume of the construct. The backscattered radiofrequency (RF) signals of all ultrasound images were acquired at a sampling rate of 420 million samples/s. For estimating the speed of sound and acoustic attenuation of the construct, backscattered RF data with ultrasound focus placed at the gel pad surface were collected with and without the presence of a construct.

Ultrasound Imaging Analysis

1) Construct Volume. A semi-automated segmentation procedure and edge detection algorithm from the Vevo 770 system were used to detect the contour of the construct in a B-image. The volume of the construct was then calculated as the volume within the contours defined from each of B-mode images separated by 200 μm in 3D image data.

2) Speed of Sound in Construct. A grayscale parameter [31] was computed from the RF data of a B-scan. The time of travel of the ultrasound pulse from the imaging transducer to the construct top surface (t_{top}), bottom surface (t_{bottom}), and the agar gel pad surface ($t_{pad}^{construct}$) was determined based on grayscale thresholding using an automated algorithm. The time of travel to the agar gel pad without the construct (t_{pad}^{ref}) was also determined and used as the reference. Assuming the speed of sound in the surrounding fluid medium (C_f) to be 1480 m/s, the thickness of the construct (L) was determined as

$$L = 0.5 C_f \left[\left(t_{bottom} - t_{top} \right) + \left(t_{pad}^{ref} - t_{pad}^{construct} \right) \right], \quad (1)$$

and speed of sound in the tissue construct C_{tc} as

$$C_{tc} = \frac{2L}{t_{bottom} - t_{top}}. \quad (2)$$

3) Attenuation. Frequency dependent attenuation in dB/cm was calculated as:

$$\alpha(f)_{tc} = \frac{20}{2L} \log_{10} \frac{|A(f)|}{|A_0(f)|}, \quad (3)$$

where $|A(f)|$ and $|A_0(f)|$ are the spectral magnitudes of the RF signal from the gel pad surface with and without (reference) the presence of construct respectively. The slope of α against f was estimated by a linear fit between 20–55 MHz to yield the attenuation coefficient in the construct in dB/(cm-MHz).

SUSI analysis

1) Scatterer Size. The calibrated power spectrum of the RF signals for each A-line was obtained using linear regression to find the spectral parameters, i.e., the slope (m') and the mid-band fit (MBF') within a $-$ 9 dB bandwidth [31]. The spectral parameters were corrected for the attenuation (α, dB/(cm-MHz)) of the tissue construct as $m = m'+2\alpha z$ and $MBF = MBF'+2\alpha z f_c$ where z is the ultrasound propagation distance in the tissue construct. Ultrasonic spectral parameters have been related to the system factors and the physical properties of effective acoustic scatterers in tissue [14]. As described previously [14], spectral slope represents a parameter associated with scatterer radius (a), its geometry (n) and the center frequency of the imaging transducer (f_c) and bandwidth (b), and is given by:

$$m = 26.06 \frac{\left[b - \left(1 - \frac{b^2}{4} \ln\left(\frac{2+b}{2-b}\right)\right)\right]}{b^3 f_c} n - (105.5 f_c) a^4. \quad (4)$$

Thus, the scatterer radius can be calculated as:

$$a = \sqrt{0.25 \frac{\left[b - \left(1 - \frac{b^2}{4}\right)\right] \ln\left(\frac{2+b}{2-b}\right)}{b^3 f_c^2} n - \frac{m}{105.5 f_c}}. \quad (5)$$

2) Acoustic Concentration. The spectral MBF depends on an additional parameter, the acoustic concentration (CQ^2),

$$MBF = 4.34 \ln\left(E a^{2(n-1)} CQ^2\right) + g_1(f_c,b)n + g_2(f_c,b)a^2, \quad (6a)$$

$$g_1(f_c,b) = 4.34 \left[\ln\left(f_c\left(1 - \frac{b^2}{4}\right)^{1/2}\left(\frac{2+b}{2-b}\right)^{1/b}\right) - 1\right], \quad (6b)$$

$$g_2(f_c,b) = -76.9 f_c^2\left(3 + \frac{b^2}{4}\right). \quad (6c)$$

where C is the number concentration of the acoustic scatterers (/ mm³), Q is the relative acoustic impedance, and E is a shape-dependent parameter. Eq. (6) can be rearranged to obtain C as

$$C = \frac{e^{0.23(MBF - g_1 n - g_2 a^2)}}{E a^{2(n-1)} Q^2}. \quad (7)$$

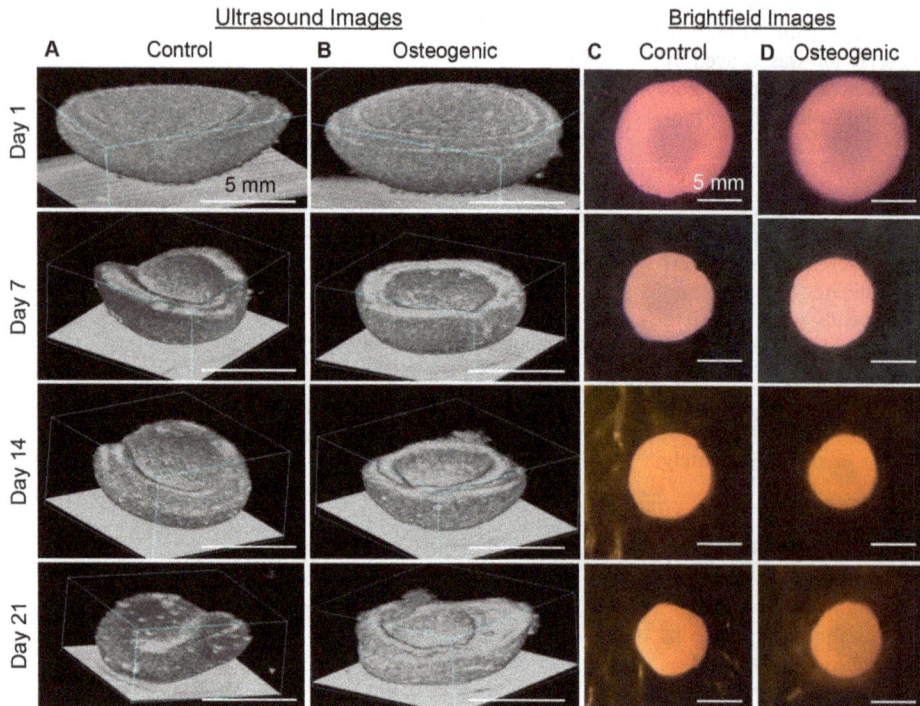

Figure 2. Longitudinal monitoring of MC3T3 cells seeded in collagen constructs. 3D rendered ultrasound backscattered images of the constructs in (A) control and (B) osteogenic media on day 1, 7, 14 and 21 of the development process. Brightfield images of corresponding constructs are shown in (C) and (D).

The relative acoustic impedance, Q, of MC3T3 cells was estimated from a known number concentration, C, of cells in constructs.

3) Deposited calcium by cells. Differentiation of MC3T3 cells was assessed by detecting and quantifying the mass of the calcium deposited [36]. The mass of calcium at day 21 was calculated by comparing the relative acoustic impedance of the scatterers at day 0 to day 21. The relative acoustic impedance is defined as

$$Q_i = \frac{\rho_i c_i - \rho c}{\rho c}, \qquad (8)$$

where ρ and ρ_i are the mass densities of the ECM and scatterers on the ith day, c and c_i are the speed of sound in the ECM and scatterers on i^{th} day respectively. On day 21, the presence of deposited calcium around the cells will increase relative acoustic impedance of the scatterer. With known relative acoustic impedance of the scatterer on day 0 (cell alone without calcium) and day i (cell and calcium), the mass of secreted calcium can be calculated as (derivation in Appendix S1.)

$$M_i^{cal} = N_i \rho c \left(\frac{V_i}{c_i}(1+Q_i) - \frac{V_0}{c_0}(1+Q_0) \right), \qquad (9)$$

where N_i is the total number of cells on day i and V_i is the volume of the net scatterer on day i and is approximately $\frac{4}{3}\pi a_i^3$.

4) SUSI Parametric Images. The spatial distribution of scatterer features (scatter size and calcium concentration) within a construct was represented as parametric images where each pixel within a B-mode image was marked with a color that corresponded to the values of the scatterer size or calcium concentration.

Statistical Analysis

Analysis of the scatterer size and concentration was carried out in an element volume with dimensions of 0.6 mm×5.0 mm× thickness (~1 mm) throughout the construct. Results are presented as mean ± standard deviation. Normalization test was carried out successfully with $p<0.01$ on the data wherever statistical analysis was made. Statistical comparisons between any two parameters were performed using Student's t-test for paired samples and the differences were considered significant at a level of $p<0.05$.

Results

Validation of SUSI Estimation of Scatterer Size and Concentration

Verification of SUSI technique was performed using agarose phantoms embedded with polystyrene microspheres of known size and concentration. As shown in Table S1, experiments were

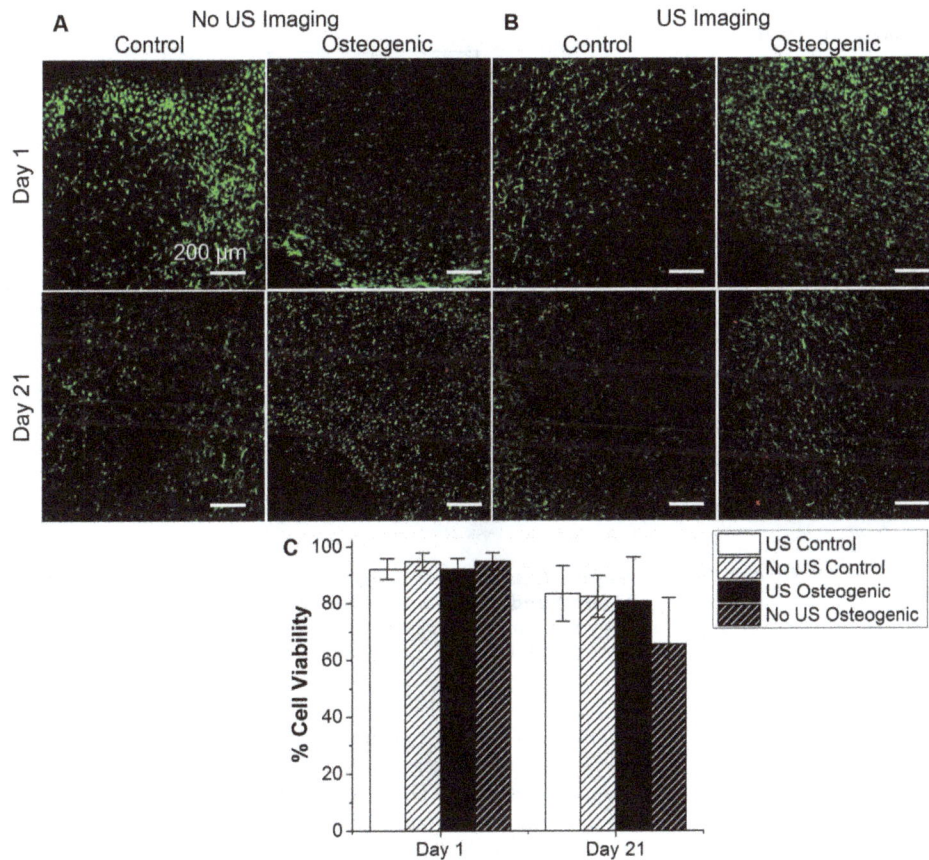

Figure 3. Comparison of cell viability at day 1 and 21 of MC3T3 cells seeded in collagen constructs, in either control or osteogenic media (cytoplasm of living cells is stained green, nuclei of dead cells is stained red). Constructs in (A) were not imaged using ultrasound, while those in (B) were imaged using ultrasound. Bar plot in (C) shows quantification of cell viability calculated from the images. Scale bar = 200 µm. Best viewed in color.

performed on phantoms with 4 different concentrations of Polybead® polystyrene microsphere (Polysciences Inc.) of different sizes (6, 10, 16 and 25 µm diameter). The SUSI method was able to detect the scatterer (polystyrene spheres) size and concentration in these phantoms, validating the SUSI estimation protocol. Additional validation of SUSI for estimation of cell size was performed using MC3T3-seeded engineered tissue constructs on day 0 with four known concentrations of cells (Supplemental Figure S1). Constructs with a known concentration of cells (2×10^6 cells/ml) were used to estimate the relative acoustic impedance of the MC3T3 cells via SUSI analysis, providing a value of approximately 0.6. SUSI was also used to estimate cell size, and the estimated cell diameter was approximately 14 µm ($n = 9$, range $= \approx 13.5$–15.5 µm). The estimated diameter decreased slightly with increasing cell concentration (Figure S1E), possibly due to increased compaction of the constructs at higher cell concentration. The estimated relative acoustic impedance was then used to estimate the cell concentration of other constructs with prepared at different cell concentration (0.5, 1 and 5×10^6 cells/ml) at day 0. These data are shown in Figures S1E, F, and show a good linear fit between the estimated concentration and the true concentration ($R^2 = 0.92$). These data confirmed the ability of SUSI to detect particle size and concentration in hydrogel constructs.

Virtual Histology for Longitudinal Monitoring of Tissue Constructs

Imaging using the Vevo 770 system at 55 MHz achieved rapid and non-invasive 3D ultrasound imaging of MC3T3-seeded collagen constructs. During ultrasound imaging, the constructs in α–MEM media were removed from the incubator for less than 20 minutes. Figure 2 shows an example of non-destructive longitudinal monitoring of the progression and development of cell seeded collagen constructs incubated in control and osteogenic media. The 3D rendered ultrasound images, constructed on Visualsonics Vevo 770 imaging system, at different time-points (day 1, 7, 14 and 21) clearly show a reduction in the volume of the constructs over time ("gel compaction") in both control and osteogenic media. Constructs adopted a symmetrical concave shape from day 1, which is a typical result of cell-mediated gel compaction. In addition, constructs in osteogenic medium became more echogenic than those in control medium, indicating changes occurring during development in osteogenic medium.

Cell viability in the constructs with and without ultrasound imaging was compared to assess possible effects of ultrasound imaging. As shown in Figure 3, viability was greater than 90% at day 1 in all of the samples and greater than 70% in all samples at day 21. There were no statistical differences between the samples with and without ultrasound imaging at either time point, indicating that exposure to ultrasound imaging did not affect cell viability.

Measurement of Construct Volume, Speed of Sound, and Acoustic Attenuation

As shown in Figure 4, a significant decrease in construct volume to about 25–30% of the original volume occurred between days 1 and 7, and the construct volume then stabilized between days 7 and 21. No statistically significant differences in construct volumes were detected between control and osteogenic media at any of the time-points. There was a slight increase and then plateau in the speed of sound over development time for constructs in both control and osteogenic media. The acoustic attenuation parameter increased almost linearly over development time in culture, with no significant differences between constructs in control and osteogenic media. Since the acoustic attenuation is typically an

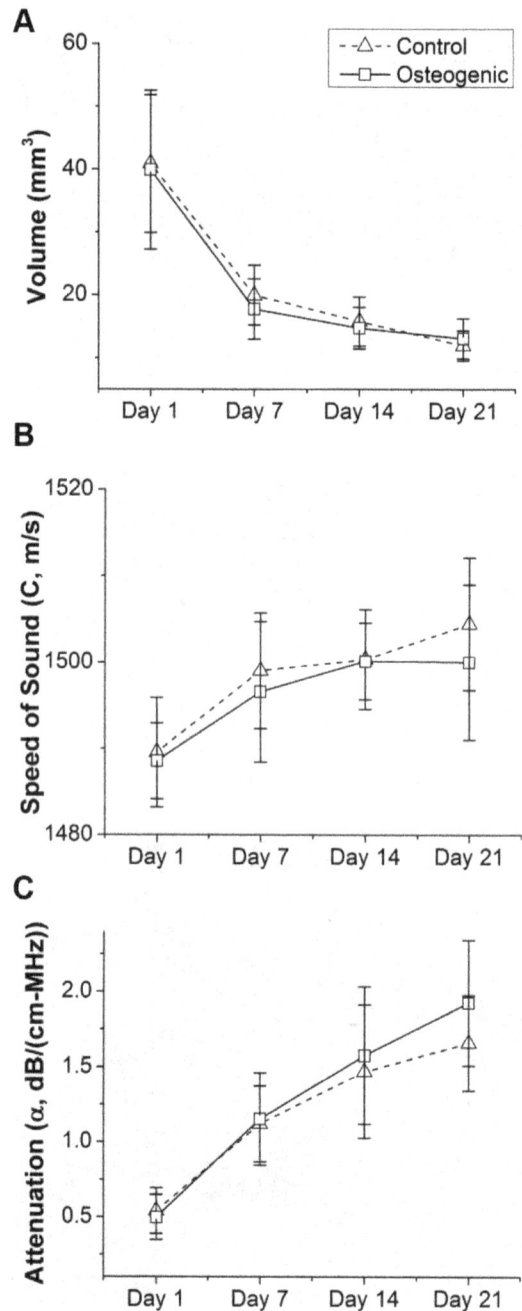

Figure 4. Backscatter analysis of MC3T3-seeded constructs in control and osteogenic media over time in culture. Quantification of (A) construct volume, (B) speed of sound, and (C) attenuation at each time-point.

indicator of increased acoustic impedance and/or scatterer concentration, this increase may indicate cell proliferation and/or mineral deposition. These results of construct volume, speed of sound, and acoustic attenuation were used in further analysis of the constructs including estimation of spectral parameters and calcium deposits by the differentiated cell constructs in osteogenic media.

SUSI analysis of Size of Cells or Scatterers in Constructs

Figure 5A–D shows the seeded MC3T3 cells in the collagen hydrogels with the F-actin filaments stained in green and cell

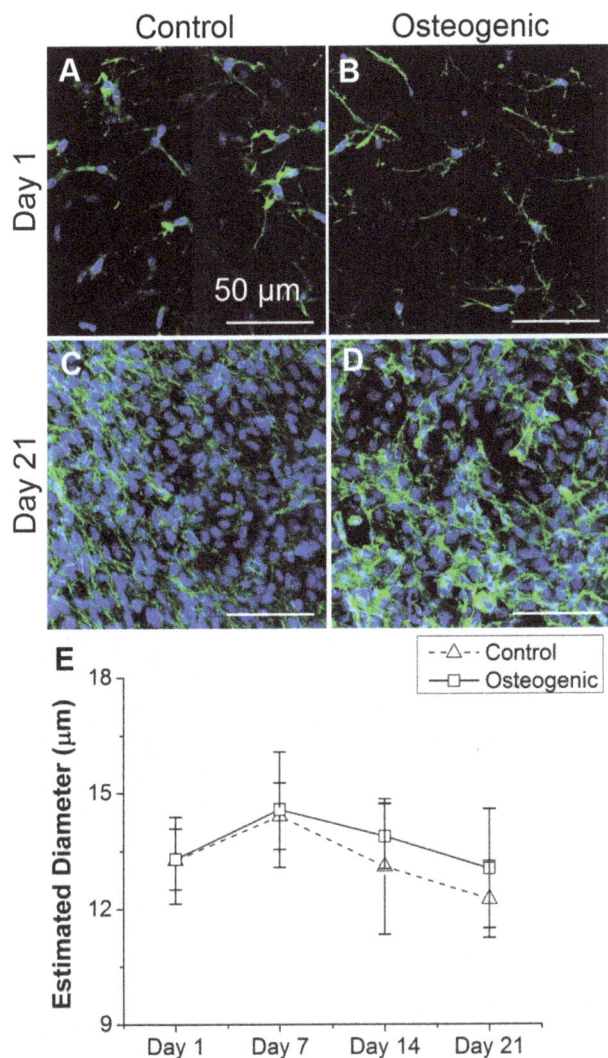

Figure 5. Developmental changes in sizes of MC3T3 cells seeded in collagen constructs. (A) –(D) Fluorescence staining of MC3T3 cells embedded in collagen constructs in control and osteogenic media on day 1 and day 21 (actin cytoskeleton is stained green, nuclei are stained blue). (E) Estimated diameter of cells from SUSI analysis over time in culture. Scale bar = 50 μm. Best viewed in color.

Figure 6. Quantified development of MC3T3 cells seeded in collagen constructs (A) Total number of cells as assessed by DNA quantification of MC3T3-seeded collagen constructs in control and osteogenic media over time in culture. (B) Equivalent number of acoustic scatterers as estimated from SUSI analysis. (C) Relative acoustic impedance estimated from (A) and (B).

nuclei stained in blue. Fluorescent imaging was performed on multiple slices of the hydrogel and was observed to be consistent across slices. The central region of the construct was chosen to represent the images in Figure 5A–D. From these images, diameters of the cell nuclei were estimated using a customized MATLAB script to be 6.0 ± 1.0 μm ($n = 4$) on day 1 and 7.6 ± 1.9 μm ($n = 4$) on day 21; these values are not statistically different. The effective size of the cells over the three week period was also estimated using the slope parameter obtained from SUSI analysis (Figure 5E). The average diameter for the cells, was approximately 14 μm with a range from about 12–15 μm, and remained essentially unchanged during the three week culture period for constructs in both control and osteogenic media. The significantly higher estimate value for scatterer diameter from SUSI compared to nucleus diameter measurement from DAPI stain suggest that the entire cell may have involved in the scattering of ultrasound, not merely cell nucleus.

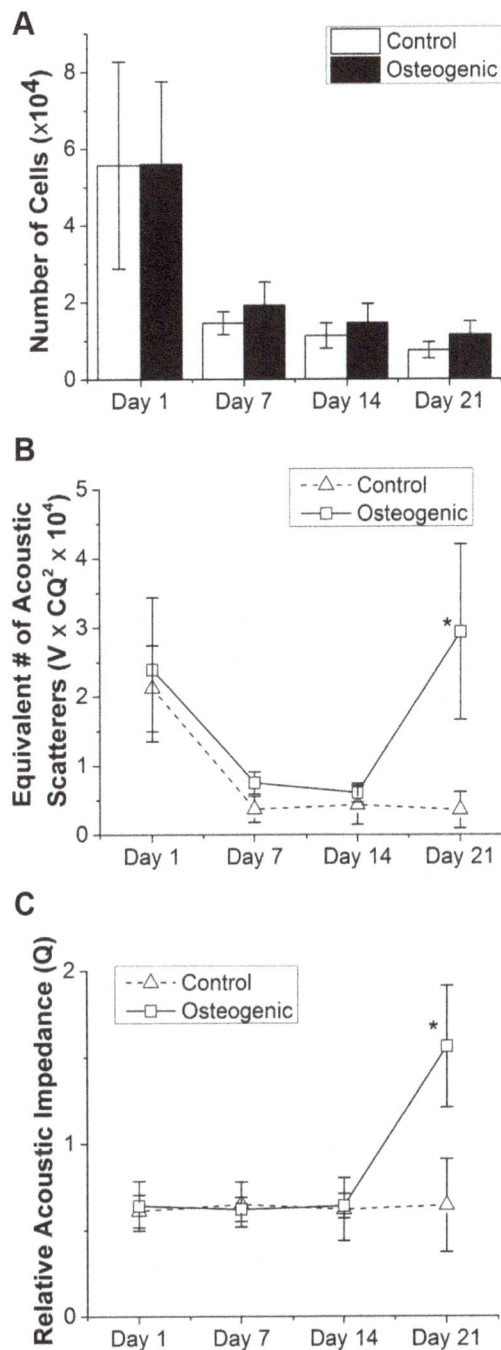

Acoustic Concentration and Calcium Deposition in Constructs

The total amount of DNA in a construct was measured biochemically and converted to the total number of cells by determining the average amount of DNA per cell on a construct at day 0, when the number of cells was known (0.5×10^6 cells/construct). As shown in Figure 6A, the number of cells decreased

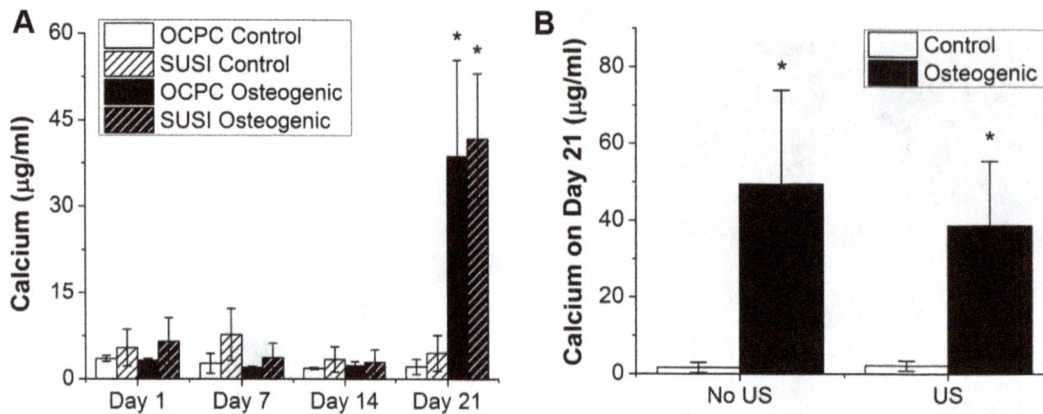

Figure 7. Amount of calcium mineral secreted by MC3T3-seeded collagen constructs in control and osteogenic media. (A) Calcium content as determined by OCPC assay and SUSI estimation. (B) Comparison of calcium content between constructs cultured in control and osteogenic media with and without exposure to ultrasound imaging.

by about 60% from day 1 to day 7 in constructs in both the control and osteogenic groups, indicating probable cell death or migration out of the constructs. Thereafter the number of cells remained constant from days 7 to 21 with no significant differences between or within media groups. These cell numbers correlated well with the equivalent number of acoustic scatterers, which is the acoustic concentration (CQ^2) estimated by SUSI analysis multiplied by the construct volume, with the exception of the day 21 measurement in the osteogenic group (Figure 6B). The acoustic concentration from SUSI depends on both the relative acoustic impedance of the scatterers (Q) and the actual number of scatterers (C). Thus assuming the actual number of scatterers or cells remained constant, the increased equivalent number of acoustic scatterers (or increased acoustic concentration) can be attributed to an increase of the relative acoustic impedance of the scatterers during the last days of incubation of the constructs (Figure 6C). The significantly increased acoustic impedance on day 21 may therefore be indicative of changes due to the differentiation process in the constructs. Since acoustic impedance depends on the mass density and the speed of sound in scatterers, the increased acoustic impedance may reflect an increase in mass density due to calcium deposition associated with cell differentiation. As calcium is much denser than water, its presence is expected to significantly increase the relative acoustic impedance of the scatterers.

As a quantitative marker to identify the extent of osteogenic differentiation of seeded MC3T3 cells [36], we estimated the calcium content using both standard biochemical assays (OCPC method) and SUSI. The estimated calcium concentration from SUSI on day 21 was 41.7 ± 11.4 μg/ml ($n = 9$) and was comparable with the measured values of 38.7 ± 16.7 μg/ml ($n = 10$) from the OCPC method. No statistically significant difference in calcium deposition at day 21 was detected in constructs subjected to ultrasound imaging and those without ultrasound imaging performed (Figure 7B).

Spatiotemporal Evolution of Constructs by Parametric Ultrasound Imaging

The estimated microstructural properties (acoustic scatterer size and secreted calcium concentration) from SUSI analysis were used to generate parametric, color coded images overlaid on B-mode images, allowing visual assessment of the spatiotemporal evolution of constructs during development. As an example, Figure 8 shows the estimated microstructural parameters in a representative region of interest for constructs in control and osteogenic groups throughout the 21 day culture period. The representative region was chosen based on the bandwidth of the ultrasound focus (1.5–2.0 mm). Although, local variations in scatterer diameter estimation were observed, the average scatterer diameter of the construct

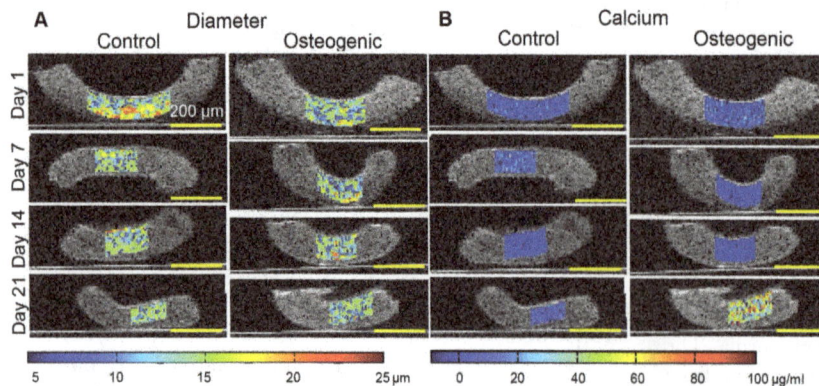

Figure 8. Overlaid B-mode (grayscale) and color maps of SUSI parameters. (A) Cell diameter, and (B) mass of calcium deposition of MC3T3 cells embedded in collagen constructs in control and osteogenic media. Best viewed in color.

did not vary significantly during the development process in either control or osteogenic medium (Figure 8A). The estimated calcium concentration was relatively constant at low values at all time-points to day 14, but exhibited a significantly higher value in the constructs in osteogentic medium on day 21. The increase in calcium content is indicative of osteogenic differentiation of MC3T3 cells in the constructs.

Discussion

In this study, we demonstrated that high resolution ultrasound imaging provided non-destructive monitoring of MC3T3-seeded collagen constructs over 3 weeks. Physical parameters including the volume of each individual construct, speed of sound and acoustic attenuation in the constructs were obtained from simple analysis of the ultrasound RF signals. Notably, SUSI analysis provided estimation and assessment of key microstructural characteristics of the constructs, beyond what can be generated by conventional ultrasound images. These parameters included cell size, acoustic scatterer concentration, cell number, and mineral deposition. Since system-dependent factors are removed from SUSI analysis by calibration, the parameters provided are objective and instrument-independent. Therefore such data have broad utility and are particularly useful for inter-study compar-isons. These features make ultrasound imaging and particularly SUSI a very attractive tool for biomaterials and tissue engineering research, and as a tool in quality assurance as engineered tissues approach the market. Below we discuss implications of our results and limitation of this study.

When embedded in 3D collagen hydrogels, many cell types including fibroblasts, smooth muscle cells, cardiomyocytes and osteoblasts, will remodel the collagen by exerting contractile forces that can align and compact the matrix [37]. These forces are significantly higher than the forces required for cell locomotion and it has been proposed that this force generation is targeted at matrix remodeling, rather than cell migration [38,39]. This morphogenic phenomenon has been studied widely, and the mechanisms are still not fully understood. Assessment of tissue construct morphology in 3D in a non-invasive method is important to study these changes, and to quantitatively charac-terize the degree of remodeling. In this study, we showed that high resolution 3D ultrasound imaging could be used to noninvasively track morphological changes in tissue constructs longitudinally, revealing the significant compaction of unconstrained MC3T3-seeded constructs from day 1 to day 7.

We showed that ultrasound imaging provides non-destructive monitoring without affecting the structure or function of the constructs. Cell diameter as determined by SUSI was in the range of 12.5–15.5 µm, which matched the size determined by parallel fluorescent confocal imaging. SUSI analysis also revealed that the total number of acoustic scatterers in unconstrained MC3T3-seeded collagen constructs decreased by approximately 80% over the 21 day culture period in control medium. This result was in agreement with destructive biochemical DNA measurement performed in parallel. The decrease in cell number may have been a result of cell death or migration from the construct, possibly as a result of the decrease in construct volume that resulted from gel compaction. A similar pattern in cell number was observed in a previous study of unconstrained constructs seeded with undiffer-entiated mesenchymal stem cells (MSC) [40].

The acoustic concentration is defined as CQ^2, where C is the number concentration of the scatterers and Q the acoustic impedance of the scatterers. The parameter Q depends on the physical and acoustic properties of the scatterer, particularly the mass density and speed of sound. Therefore assessment of the total number of acoustic scatterers (CQ^2 multiplied by construct volume) can provide information regarding changes in the construct microstructure. MC3T3 cells secrete mineral into the surrounding matrix as they undergo osteogenic differentiation [36]. They thereby modify the properties of the acoustic scatterers in the construct by increasing their mass density and thus the relative acoustic impedance. Therefore, relative acoustic impedance can be used as an indicator of changes in cellular state during osteogenic differentiation. In the current study, we qualitatively characterized the differentiation process by monitoring changes in the relative acoustic impedance, and also generated quantitative values of the mass of calcium deposited based on the relative acoustic impedance values. These data matched well with parallel measurement from conventional destructive biochemical tests for calcium content. We therefore have demonstrated that ultrasound imaging can be used to estimate the mass of calcium mineral in 3D collagen constructs as MC3T3 cells differentiate in osteogenic medium.

The current study highlights the advantages of high frequency ultrasound imaging for monitoring of tissue construct develop-ment. The use of a high frequency in our study provided high spatial resolution, which allowed detailed characterization of the constructs in vitro. However, most in vivo and eventual clinical applications will require the use of lower frequency imaging (e.g. 10 MHz) to allow deeper tissue penetration, which will reduce the resolution of the images. Another limitation of our current method is that our estimation of the relative acoustic impedance in this study required knowledge of the number of cells, which may not be readily available non-destructively. Relative changes to the acoustic concentration, a composite parameter that is obtained directly from SUSI analysis, can be used to noninvasively assess relative changes in the constructs. However, at this stage we are not able to use this parameter to generative an absolute quantitative estimation of calcium deposition without knowledge of the cell number. We assumed that cells and the calcium they deposited were a single scatterer, however future work will examine the ability to resolve different cell types and matrix components using SUSI.

Conclusions

High resolution spectral ultrasound imaging (SUSI) was shown to provide non-destructive 3D imaging of in-vitro osteogenic differentiation of MC3T3 pre-osteoblasts seeded within collagen hydrogels. SUSI analysis enabled accurate measurement of construct volume and estimation of cell size, acoustic cell concentration, cell acoustic impedance, and calcium deposition within the construct. The technique was used to compare constructs cultured in maintenance and osteogenic media formu-lations, and SUSI data were validated by independent biochemical methods. Overlay of spectral parameter data on B-mode images allowed clear visualization of the spatiotemporal changes in construct composition. This study demonstrates that spectral ultrasound imaging can be a useful noninvasive tool to quantita-tively characterize the development of orthopedic engineered tissues over time.

Supporting Information

Figure S1   Experiment for estimation of relative acous-tic impedance of MC3T3 cells on day 0 and validation of estimated cell concentration from SUSI analysis. (A)–(D) Ultrasound B-mode (grayscale) images of MC3T3-seeded collagen constructs on Day 0 at 0.5, 1, 2 and 5×10^6 cells/ml cell concentrations, respectively. (E) Cell diameter, and (F) cell concentration estimated from SUSI analysis and compared to seeded cell concentration at day 0.

Table S1 Tabular values of estimated Polybead® polystyrene microsphere (Polysciences Inc.) bead size and concentration and comparison with their true values.

Author Contributions

Conceived and designed the experiments: MSRG RRR JPS CXD. Performed the experiments: MSRG RRR AWP DJC. Analyzed the data: MSRG RRR AWP DJC JPS CXD. Contributed reagents/materials/analysis tools: JPS, CXD. Wrote the paper: MSRG RRR JPS CXD.

References

1. Langer R, Vacanti JP (1993) Tissue Engineering. Science 260: 920–926.
2. Yang SY, Hsiung SK, Hung YC, Chang CM, Liao TL, et al. (2006) A cell counting/sorting system incorporated with a microfabricated flow cytometer chip. Measurement Science & Technology 17: 2001–2009.
3. Oliver MH, Harrison NK, Bishop JE, Cole PJ, Laurent GJ (1989) A Rapid and Convenient Assay for Counting Cells Cultured in Microwell Plates - Application for Assessment of Growth-Factors. Journal of Cell Science 92: 513–518.
4. Butler WB (1984) Preparing Nuclei from Cells in Monolayer-Cultures Suitable for Counting and for Following Synchronized Cells through the Cell-Cycle. Analytical Biochemistry 141: 70–73.
5. Han JW, Breckon TP, Randell DA, Landini G (2012) The application of support vector machine classification to detect cell nuclei for automated microscopy. Machine Vision and Applications 23: 15–24.
6. Ho KY, Hu HCH, Keyak JH, Colletti PM, Powers CM (2013) Measuring bone mineral density with fat-water MRI: comparison with computed tomography. Journal of Magnetic Resonance Imaging 37: 237–242.
7. Nazarian A, Snyder BD, Zurakowski D, Muller R (2008) Quantitative micro-computed tomography: A non-invasive method to assess equivalent bone mineral density. Bone 43: 302–311.
8. Jones AC, Milthorpe B, Averdunk H, Limaye A, Senden TJ, et al. (2004) Analysis of 3D bone ingrowth into polymer scaffolds via micro-computed tomography imaging. Biomaterials 25: 4947–4954.
9. Oe K, Miwa M, Nagamune K, Sakai Y, Lee SY, et al. (2010) Nondestructive evaluation of cell numbers in bone marrow stromal cell/beta-tricalcium phosphate composites using ultrasound. Tissue Engineering Part C-Methods 16: 347–353.
10. Fite BZ, Decaris M, Sun YH, Sun Y, Lam A, et al. (2011) Noninvasive multimodal evaluation of bioengineered cartilage constructs combining time-resolved fluorescence and ultrasound imaging. Tissue Engineering Part C-Methods 17: 495–504.
11. Kreitz S, Dohmen G, Hasken S, Schmitz-Rode T, Mela P, et al. (2011) Nondestructive method to evaluate the collagen content of fibrin-based tissue engineered structures via ultrasound. Tissue Engineering Part C-Methods 17: 1021–1026.
12. Walker JM, Myers AM, Schluchter MD, Goldberg VM, Caplan AI, et al. (2011) Nondestructive evaluation of hydrogel mechanical properties using ultrasound. Annals of Biomedical Engineering 39: 2521–2530.
13. Tunis AS, Czarnota GJ, Giles A, Sherar MD, Hunt JW, et al. (2005) Monitoring structural changes in cells with high-frequency ultrasound signal statistics. Ultrasound in Medicine and Biology 31: 1041–1049.
14. Lizzi FL, Ostromogilsky M, Feleppa EJ, Rorke MC, Yaremko MM (1987) Relationship of ultrasonic spectral parameters to features of tissue microstructure. Ieee Transactions on Ultrasonics Ferroelectrics and Frequency Control 34: 319–329.
15. Lizzi FL, Astor M, Liu T, Deng C, Coleman DJ, et al. (1997) Ultrasonic spectrum analysis for tissue assays and therapy evaluation. International Journal of Imaging Systems and Technology 8: 3–10.
16. Nair A, Kuban BD, Tuzcu EM, Schoenhagen P, Nissen SE, et al. (2002) Coronary plaque classification with intravascular ultrasound radiofrequency data analysis. Circulation 106: 2200–2206.
17. Qian J, Maehara A, Mintz GS, Margolis MP, Lerman A, et al. (2009) Impact of Gender and Age on In Vivo Virtual Histology-Intravascular Ultrasound Imaging Plaque Characterization (from the global Virtual Histology Intravascular Ultrasound [VH-IVUS] Registry). American Journal of Cardiology 103: 1210–1214.
18. Gudur MSR, Kumon RE, Zhou Y, Deng CX (2012) High-Frequency Rapid B-Mode Ultrasound Imaging for Real-Time Monitoring of Lesion Formation and Gas Body Activity During High-Intensity Focused Ultrasound Ablation. Ieee Transactions on Ultrasonics Ferroelectrics and Frequency Control 59: 1687–1699.
19. Kumon RE, Gudur MSR, Zhou Y, Deng CX (2012) High-Frequency Ultrasound M-Mode Imaging for Identifying Lesion and Bubble Activity during High-Intensity Focused Ultrasound Ablation. Ultrasound in Medicine and Biology 38: 626–641.
20. Siebers S, Schwabe M, Scheipers U, Welp C, Werner J, et al. (2004) Evaluation of ultrasonic texture and spectral parameters for coagulated tissue characterization. 2004 IEEE Ultrasonics Symposium, Vols 1-3: 1804–1807.
21. Feleppa EJ (2008) Ultrasonic tissue-type imaging of the prostate: Implications for biopsy and treatment guidance. Cancer Biomarkers 4: 201–212.
22. Oelze ML, O'Brien WD, Blue JP, Zachary JF (2004) Differentiation and characterization of rat mammary fibroadenomas and 4T1 mouse carcinomas using quantitative ultrasound imaging. Ieee Transactions on Medical Imaging 23: 764–771.
23. Vlad RM, Alajez NM, Giles A, Kolios MC, Czarnota GJ (2008) Quantitative Ultrasound Characterization of Cancer Radiotherapy Effects in Vitro. International Journal of Radiation Oncology Biology Physics 72: 1236–1243.
24. Lizzi FL (1997) Ultrasonic scatterer property images of the eye and prostate. 1997 Ieee Ultrasonics Symposium Proceedings, Vols 1 & 2: 1109–1117.
25. Lizzi FL, Feleppa EJ, Alam SK, Deng CX (2003) Ultrasonic spectrum analysis for tissue evaluation. Pattern Recognition Letters 24: 637–658.
26. Kumon RE, Repaka A, Atkinson M, Faulx AL, Wong RCK, et al. (2012) Characterization of the pancreas in vivo using EUS spectrum analysis with electronic array echoendoscopes. Gastrointestinal Endoscopy 75: 1175–1183.
27. Kumon RE, Deng CX, Wang XD (2011) Frequency-Domain Analysis of Photoacoustic Imaging Data from Prostate Adenocarcinoma Tumors in a Murine Model. Ultrasound in Medicine and Biology 37: 834–839.
28. Insana MF, Wagner RF, Brown DG, Hall TJ (1990) Describing Small-Scale Structure in Random-Media Using Pulse-Echo Ultrasound. Journal of the Acoustical Society of America 87: 179–192.
29. Kolios MC, Czarnota GJ, Lee M, Hunt JW, Sherar MD (2002) Ultrasonic spectral parameter characterization of apoptosis. Ultrasound in Medicine and Biology 28: 589–597.
30. Czarnota GJ, Kolios MC, Abraham J, Portnoy M, Ottensmeyer FP, et al. (1999) Ultrasound imaging of apoptosis: high-resolution non-invasive monitoring of programmed cell death in vitro, in situ and in vivo. British Journal of Cancer 81: 520–527.
31. Gudur M, Rao RR, Hsiao YS, Peterson AW, Deng CX, et al. (2012) Noninvasive, Quantitative, Spatiotemporal Characterization of Mineralization in Three-Dimensional Collagen Hydrogels Using High-Resolution Spectral Ultrasound Imaging. Tissue Engineering Part C-Methods 18: 935–946.
32. Czekanska EM, Stoddart MJ, Richards RG, Hayes JS (2012) In Search of an Osteoblast Cell Model for in Vitro Research. European Cells & Materials 24: 1–17.
33. Al-Munajjed AA, Plunkett NA, Gleeson JP, Weber T, Jungreuthmayer C, et al. (2009) Development of a Biomimetic Collagen-Hydroxyapatite Scaffold for Bone Tissue Engineering Using a SBF Immersion Technique. Journal of Biomedical Materials Research Part B-Applied Biomaterials 90B: 584–591.
34. Rao RR, Peterson AW, Ceccarelli J, Putnam AJ, Stegemann JP (2012) Matrix composition regulates three-dimensional network formation by endothelial cells and mesenchymal stem cells in collagen/fibrin materials. Angiogenesis 15: 253–264.
35. Rao RR, Peterson AW, Stegemann JP (2013) Winner for outstanding research in the Ph.D. category for the 2013 Society for Biomaterials meeting and exposition, April 10–13, 2013, Boston, Massachusetts: Osteogenic differentiation of adipose-derived and marrow-derived mesenchymal stem cells in modular protein/ceramic microbeads. J Biomed Mater Res A 101: 1531–1538.
36. Declerq HA, Verbeeck RMH, De Ridder LIFJM, Schacht EH, Cornelissen MJ (2005) Calcification as an indicator of osteoinductive capacity of biomaterials in osteoblastic cell cultures. Biomaterials 26: 4964–4974.
37. Wakatsuki T, Elson EL (2003) Reciprocal interactions between cells and extracellular matrix during remodeling of tissue constructs. Biophysical Chemistry 100: 593–605.
38. Dembo M, Wang YL (1999) Stresses at the cell-to-substrate interface during locomotion of fibroblasts. Biophysical Journal 76: 2307–2316.
39. Galbraith CG, Sheetz MP (1997) A micromachined device provides a new bend on fibroblast traction forces. Proceedings of the National Academy of Sciences of the United States of America 94: 9114–9118.
40. Zscharnack M, Hepp P, Richter R, Aigner T, Schulz R, et al. (2010) Repair of Chronic Osteochondral Defects Using Predifferentiated Mesenchymal Stem Cells in an Ovine Model. American Journal of Sports Medicine 38: 1857–1869.

Similar Properties of Chondrocytes from Osteoarthritis Joints and Mesenchymal Stem Cells from Healthy Donors for Tissue Engineering of Articular Cartilage

Amilton M. Fernandes[1,2], Sarah R. Herlofsen[1,2], Tommy A. Karlsen[1,2], Axel M. Küchler[1,2], Yngvar Fløisand[3], Jan E. Brinchmann[1,2]*

1 The Norwegian Center for Stem Cell Research, University of Oslo, Oslo, Norway, 2 Institute of Immunology, Oslo University Hospital Rikshospitalet, Oslo, Norway, 3 Department of Hematology, Oslo University Hospital Rikshospitalet, Oslo, Norway

Abstract

Lesions of hyaline cartilage do not heal spontaneously, and represent a therapeutic challenge. *In vitro* engineering of articular cartilage using cells and biomaterials may prove to be the best solution. Patients with osteoarthritis (OA) may require tissue engineered cartilage therapy. Chondrocytes obtained from OA joints are thought to be involved in the disease process, and thus to be of insufficient quality to be used for repair strategies. Bone marrow (BM) derived mesenchymal stem cells (MSCs) from healthy donors may represent an alternative cell source. We have isolated chondrocytes from OA joints, performed cell culture expansion and tissue engineering of cartilage using a disc-shaped alginate scaffold and chondrogenic differentiation medium. We performed real-time reverse transcriptase quantitative PCR and fluorescence immunohistochemistry to evaluate mRNA and protein expression for a range of molecules involved in chondrogenesis and OA pathogenesis. Results were compared with those obtained by using BM-MSCs in an identical tissue engineering strategy. Finally the two populations were compared using genome-wide mRNA arrays. At three weeks of chondrogenic differentiation we found high and similar levels of hyaline cartilage-specific type II collagen and fibrocartilage-specific type I collagen mRNA and protein in discs containing OA and BM-MSC derived chondrocytes. Aggrecan, the dominant proteoglycan in hyaline cartilage, was more abundantly distributed in the OA chondrocyte extracellular matrix. OA chondrocytes expressed higher mRNA levels also of other hyaline extracellular matrix components. Surprisingly BM-MSC derived chondrocytes expressed higher mRNA levels of OA markers such as *COL10A1*, *SSP1* (osteopontin), *ALPL*, *BMP2*, *VEGFA*, *PTGES*, *IHH*, and *WNT* genes, but lower levels of *MMP3* and *S100A4*. Based on the results presented here, OA chondrocytes may be suitable for tissue engineering of articular cartilage.

Editor: Jorge Sans Burns, University Hospital of Modena and Reggio Emilia, Italy

Funding: This work was supported by a grant from South-Eastern Norway Regional Health Authority, Storforsk and Stamceller grants from the Research Council of Norway and the Gidske and Peter Jacob Sørensens Foundation for the Promotion of Science. The funders had no role in study design, data collection and analysis, decision to publish, or preparation of the manuscript.

Competing Interests: The authors have declared that no competing interests exist.

* E-mail: jan.brinchmann@rr-research.no

Introduction

Articular cartilage defects have been reported in up to 20% of all patients undergoing arthroscopy of the knee [1–4]. Lesions of hyaline articular cartilage heal poorly. Untreated defects will usually give rise to pain, functional impairment, and eventually osteoarthritis (OA) [5,6]. However, OA may also arise as a primary degenerative lesion of the hyaline cartilage [7]. Several treatment modalities have been developed over the past decades in attempts to reestablish fully functional cartilage tissue [8]. Cartilage cell therapy was introduced by using autologous chondrocyte implantation (ACI) [9]. However, the repair tissue generated following ACI frequently contains fibrocartilage, and ACI has not proven to be superior to other surgical techniques [10–12]. Unfortunately, OA patients frequently end up having total joint replacement surgery.

Tissue engineering combines the use of cells and biomaterials for repair of damaged tissues [13]. This approach has emerged as a promising strategy for cartilage repair [14]. In order to optimize this strategy, we need to determine which cell population to use, the best biomaterial, and whether to use growth- and differentiation factors or other differentiation promoting stimuli. The cells most likely to produce hyaline extracellular matrix (ECM) are articular chondrocytes and mesenchymal stem cells (MSCs) [15–17]. Chondrocytes are obtained from biopsies taken from the periphery of the articular cartilage in the course of arthroscopy. This creates a small lesion which also heals poorly, and may occasionally induce donor-site morbidity [8,18]. However, the chondrocytes have already produced perfect hyaline ECM within the joint, and should in theory be able to do so again. In order to be useful for treatment of large cartilage defects, they need to be expanded *in vitro*. It is known that chondrocytes dedifferentiate when cultivated conventionally as monolayer cells [19]. This is the reason why ACI tends to produce fibrocartilage repair tissue. For chondrocytes obtained from OA joints, an additional problem arises. Here, all the chondrocytes within the joint are thought to be involved in the disease process [20]. Although all the molecular events involved in OA pathogenesis are not known, many of the markers describing chondrocyte hypertrophy during embryogenesis are also seen in OA chondrocytes [21]. During embryogenesis, chondrocyte hypertrophy is thought to be followed by apoptosis, vasculogenesis, and bone formation [22]. If any of these events should occur in chondrocytes used for tissue engineering of cartilage, this would affect the usefulness even of chondrocytes harvested away from the visible lesion in OA joints for engineering of hyaline cartilage for transplantation therapy.

MSCs are thought to resemble the progenitor cells responsible for the cartilage anlagen during embryological bone and cartilage formation [23]. Given the right differentiation clues they should be able to produce perfect hyaline ECM. MSCs may be obtained from BM with minimal discomfort and no residual morbidity, and are easily expanded *in vitro* [24]. Importantly, MSCs have been described as being immunoprivileged cells with immunosuppressive properties [25], which suggest that allogeneic BM-MSCs from young, healthy donors may represent an off-the-shelf choice of cells for tissue engineering of repair cartilage for OA patients. We have recently published a detailed description of gene and protein expression of human BM-MSCs in alginate hydrogel discs undergoing *in vitro* chondrogenic differentiation during exposure to differentiation medium [17]. Here we showed that mRNA encoding type II collagen (COL2) and other chondrogenic proteins, such as aggrecan (ACAN) and SOX5, 6 and 9, were quickly and highly expressed. However, mRNA coding for COL10 was also expressed. COL10 has been described in the literature as a marker for chondrocyte hypertrophy in limb development and endochondral bone formation [26]. Thus, both OA chondrocytes and BM-MSCs may be able to unleash genetic programs similar to those involved in embryonic chondrocyte hypertrophy.

Despite the obvious roles of chondrocytes and MSCs as the cell candidates for tissue engineering of hyaline cartilage for OA patients, to the best of our knowledge, no direct comparison has been made between human populations of these cells under identical differentiation conditions in biomaterial/cell cultures. Thus, in the present study we used our recently established culture system to expand OA chondrocytes *in vitro* [27]. These cells were subsequently differentiated in alginate discs under conditions that were identical to those used in our study of BM-MSCs which was performed at the same time [17]. Using material from the two cell populations in parallel analyses, we compared the kinetics of expression of hyaline cartilage-specific ECM genes between the two cell populations, and demonstrated synthesis of ECM proteins and glycosaminoglycans. Finally, we performed a microarray comparison of the global gene expression between OA chondrocytes and BM-MSCs differentiated in alginate hydrogel discs under identical conditions. The results show that the OA chondrocytes produced more hyaline ECM components, expressed lower levels of many of the markers for chondrocyte hypertrophy, bone differentiation, and vasculogenesis. Thus, OA chondrocytes may be suitable for tissue engineering of articular cartilage.

Materials and Methods

Chemicals

All the chemicals were purchased from Sigma-Aldrich (St. Louis, MO) unless otherwise stated.

Isolation and Culture of the Chondrocytes

Biopsies of articular cartilage were provided from OA patients with primary OA undergoing knee replacement surgery. All the cartilage biopsies were taken from a part of the surface of the femoral condyle considered by the surgeon to look like intact and healthy cartilage. The three donors were 60–65 years of age, and all provided written informed consent. The study was approved by the Regional Committee for Medical Research Ethics, Southern Norway, Section A. The biopsies were cut into tiny pieces and digested with Collagenase type XI (1.2 mg/mL) at 37°C for 90–120 minutes. The digested cartilage pieces were then washed and resuspended in Dulbecco's modified Eagle's medium (DMEM)-F12 (Gibco, Paisley, UK) supplemented with 20% fetal bovine serum (FBS, Cambrex, East Rutherford, NJ), 50 µg/mL ascorbic acid, 1% penicillin and streptomycin (P/S), and 1.5 µg/mL amphotericin B. The culture medium was changed every 3–4 days. At 70–80% confluence, cells were detached with trypsin/EDTA and seeded into new culture flasks. 25 cm^2 culture flasks (Nunc, Roskilde, Denmark) were used for the first passage, before proceeding with 175 cm^2 flasks. Amphotericin B was discontinued and FBS was reduced to 10% after the first passage.

Isolation and Culture of Mesenchymal Stem Cells from Bone Marrow

The mononuclear cell fractions were isolated from human bone marrows of three healthy, voluntary donors aged 24–50 years by density gradient centrifugation (Lymphoprep, Fresenius Kabi, Oslo, Norway) as described [17]. The cells were seeded in 175-cm^2 tissue culturing flasks and cultured in expansion medium containing DMEM-F12, 1% PS, 2.5 µg/ml amphotericin B, and 10% fetal bovine serum. After 48 hours the non-adherent cells were removed by medium exchange. The adhering cells were expanded in monolayer culture with medium change twice a week until colonies reached 70% confluence. Cultures were passaged using Trypsin/EDTA and reseeded at 5000 cells/cm^2. After the first passage amphotericin B was removed from the culture medium. For practical purposes the cells were frozen at passage 2 in DMEM-F12 containing 20% FBS and 5% DMSO. After thawing, cells were reseeded at a density of 5000 cells/cm^2 and passaged when 70% confluent.

Chondrogenic Differentiation in 3D Alginate Scaffold

At passage 2–3, after 18–20 days of cultivation for the OA chondrocytes, and at passage 3–4 for the BM-MSCs, the cells were embedded in equal amounts of 1% Pronova LVG-alginate and 1% calcium alginate in a 4.6% mannitol solution. This scaffold is a self-gelling system provided by FMC BioPolymer AS/NovaMatrix (Sandvika, Norway) [17,28,29]. The alginate gels were shaped into 0.8 mL discs containing 5×10^6 cells/mL in 12 well culture plates (Nonstick, Nunc). After washing with DMEM, the discs were washed with a 50 mM SrCl$_2$ solution while solidifying. After solidification the scaffolds were again washed with DMEM medium and moved to 6 well culture plates (Nonstick, Nunc). After establishment of the cells in the alginate discs, chondrogenic differentiation was induced by high-glucose DMEM (4.5 g/l) supplemented with 1 mM sodium pyruvate (Gibco), 0.1 mM ascorbic acid-2-phosphate, 0.1 µM dexamethasone, 1% ITS (insulin 25 µg/ml, transferrin 25 µg/ml, and sodium selenite 25 ng/ml), 1.25 mg/ml human serum albumin (Octapharma, Hurdal, Norway), 500 ng/ml bone morphogenic protein-2 (BMP-2, InductOs, Wyeth, Taplow, UK), and 10 ng/ml recombinant human transforming growth factor-β1 (TGF-β1, R&D Systems, Minneapolis, MN). Medium was changed every day for the first three days, then every 2–4 days. During expansion and differentiation cultures the same batches of all supplements, FBS, differentiation reagents, and alginate were used for the two cell populations.

Total RNA Isolation

The alginate scaffolds were depolymerized on a thermo shaker 37°C for 30 minutes using 5 U/mL of the enzyme GLyase (kindly donated by Professor Gudmund Skjåk-Bræk). 1 mL GLyase solution was used to treat 400 μL of the alginate discs. RNA was isolated from the cells using the TRIzol method following the company's protocol (Invitrogen, Carsbad, CA). The cell samples were dissolved in TRIzol, frozen in liquid nitrogen, and stored at −80°C.

Real-time Reverse Transcription-polymerase Chain Reaction Quantification

Isolated total RNA was treated with DNase I (Ambion, Austin, TX) followed by quantification with NanoDrop ND-1000 Spectrophotometer (Nanodrop Technologies, Wilmington, DE). 200 ng of total RNA from each sample was reverse transcribed into cDNA (total volume of 20 μL) using High Capacity cDNA Reverse Transcription kit (Applied Biosystems, Abingdon, UK) following the manufacturer's protocol. The cDNA samples were analyzed by relative quantification using the 7300 Real-Time reverse transcriptase quantitative PCR (RT-qPCR) System (Applied Biosystems) and TaqMan® Gene Expression assay following manufacture's protocol. Primers from Applied Biosystems are listed in Table 1. B2M proved to be the most stably expressed gene within the differentiated OA chondrocytes and BM-MSCs, and was therefore used as the endogenous control. All the expression levels were normalized to the expression of the endogenous control.

Fluorescence Immunohistochemistry

All antibodies and concentrations are listed in Table 1. The following secondary antibodies were used: Alexa 488-conjugated goat anti-rabbit antibody (used at 5 μg/mL), purchased from Invitrogen, and Cy3-conjugated donkey anti-rat IgG (used at 2 μg/mL) and Cy3-conjugated donkey anti-mouse IgG (used at 1.4 μg/mL), both purchased from Jackson Immuno Research Europe (Newmarket, UK). Formalin-fixed, paraffin-embedded 3D culture samples from day 21 were sectioned and deparaffinised, using standard laboratory procedures, and postfixed for 10 minutes in 4% paraformaldehyde in phosphate-buffered saline (PBS; Electron Microscopy Sciences, Hatfield, PA). Tissue sections were boiled for 12 minutes in antigen retrieval buffer. TrisEDTA buffer (pH 9.0) was used for antibody ab26041, staining for SOX5, and 0.05% citraconic anhydride in ddH2O (pH 7.4) was used for all the other antibodies. One section per slide was incubated with primary antibodies diluted in 1.25% bovine serum albumin with 0.1% saponin in PBS. The other section served as negative control and was incubated with the same buffer without antibody. Sections were incubated overnight at 4°C. Subsequently, secondary fluorochrome-conjugated antibodies were applied to both sections for 1.5 hours at room temperature. Sections were mounted with ProLong Gold antifading reagent with DAPI (Invitrogen), two sections per slide, one with and one without

Table 1. Taqman assay primers used in real-time reverse transcription polymerase chain reaction and antibodies used in immunohistochemistry.

Gene symbol	Protein	Primers Taqman assay no.	Antibodies Designation (concentration)	Type and Ig class	Source
ACAN	Aggrecan	Hs00202971_m1	969D4D11 (2.28 μg/mL)	Mouse monoclonal IgG1	BioSource
ALPL	Alkaline phosphatase	Hs00758162_m1	–	–	–
B2M	Beta-2 microglobulin	Hs99999907_m1	–	–	–
COL1A1	Collagen type I	Hs00164004_m1	I-8H5 (1.00 μg/mL)	Mouse monoclonal IgG2a	MP Biomedicals
COL2A1	Collagen type II	Hs00156568_m1	II-4C11 (0.83 μg/mL)	Mouse monoclonal IgG1	MP Biomedicals
COL3A1	Collagen type III	Hs00943809_m1	–	–	–
COL9A1	Collagen type IX	Hs00156680_m1	–	–	–
COL10A1	Collagen type X	Hs00166657_m1	X53 (1:200)	Mouse monoclonal IgG1	Prof. Klaus von der Mark
COMP	Cartilage oligomeric matrix protein	Hs00164359_m1	–	–	–
FN1	Fibronectin 1	Hs01549976_m1	–	–	–
IBSP	Integrin-binding sialoprotein/bone sialoprotein	Hs00173720_m1	–	–	–
MMP13	Matrix metalloproteinase 13	Hs00233992_m1	–	–	–
MMP3	Matrix metalloproteinase 3	Hs00867308_m1	–	–	–
RUNX2	Runt-related transcription factor 2	Hs00231692_m1	–	–	–
SOX5	SRY (sex determining region Y)-box containing gene 5	Hs00374709_m1	ab26041 (1.42 μg/mL)	Rabbit polyclonal	Abcam
SOX6	SRY (sex determining region Y)-box containing gene 6	Hs00264525_m1	HPA001923 (1.67 μg/mL)	Rabbit polyclonal	Sigma
SOX9	SRY (sex determining region Y)-box containing gene 9	Hs00165814_m1	AB5535 (0.20 μg/mL)	Rabbit polyclonal	Millipore
SSP1	Secreted phosphoprotein 1/bone sialoprotein 1/osteopontin	Hs00959010_m1	–	–	–
VCAN	Versican	Hs01007941_m1	255915 (2.50 μg/mL)	Rat monoclonal IgG1	R&D

primary antibody. Microscopy was performed with a Nikon Eclipse E-600 fluorescence microscope equipped with Nikon Plan-Fluor objective lenses and Color View III digital camera controlled by Cell-B software (Olympus; www.olympusglobal.com/en/). No fluorescence signal was detectable in negative control sections. Analysis of signal, specific for the primary antibodies used, was performed with automatic camera settings, to enable optimal visualization of structures.

Microarray Analysis

RNA samples from chondrogenically differentiated OA chondrocytes and BM-MSCs at day 21 were analyzed at the Norwegian Microarray Consortium according to the manufacturer's protocol. Biotin labeled cRNA was transcribed using Illumina® TotalPrep RNA Amplification Kit (Ambion). cRNA hybridized onto Illumina HumanWG-6 v3 Expression BeadChips was subsequently stained with streptavidin-Cy3. The chips were then scanned using Illumina® BeadArray™ Reader. Results were imported and quantile normalized in Illumina GenomeStudio v. 2009.1 Gene Expression v. 1.1.1. for data extraction and initial quality control by using the array annotation file "HumanWG-6_V3_0_R3_11282955_A.bgx". Further quality control, reprocessing, log(2) transformation and expression analysis were performed in the microarray analysis program J-Express 2009 [30]. The quality control consisted of box plot, correspondence analysis plot, and hierarchical clustering with distance matrix. The statistical analysis was performed using Rank Product (RP) [31] for calculating the differential expression between the two groups. A cut-off value was set at >2 fold expression change. RP's own q-value (adjusted p-value) <0.02 for both positive and negative scores was used to produce two lists representing the upregulation-values for the two cell groups. Gene Ontology [32] (GO) overrepresentation analysis, using Bonferroni correction (p<0.05), was then performed on the genes that were found differently expressed by RP.

Statistical Analysis

Differences in gene expression measured by RT-qPCR in OA chondrocytes and BM-MSCs were evaluated using unpaired Student's t-test, assuming Gaussian distribution. p<0.05 was considered significant.

Results

Gene Expression Analysis

RT-qPCR analysis for a number of selected genes expressed in chondrogenically differentiated OA chondrocytes and BM-MSCs, which had been cultivated under identical conditions in alginate scaffolds, are presented in Figure 1. For the collagen genes, COL1A1 remained quite stable, while the expression of COL2A1 and COL10A1 increased slightly from day 7 to day 21 in the OA chondrocytes. The expression of these genes in the differentiating BM-MSCs, however, increased considerably in this period. The difference between the BM-MSCs and the OA chondrocytes for mean expression of COL1A1, COL2A1, and COL10A1 at day 21 were approximately 6 fold, 3 fold and 85 fold, respectively. The expression of COL2A1 on day 7 was considerably higher in OA chondrocytes than in BM-MSC derived chondrocytes, consistent with retained expression of COL2A1 in the chondrocytes in the course of in vitro culture.

The SOX transcription factors SOX5 and SOX6 mRNA had similar expression levels when comparing the OA and BM-MSC derived chondrocytes, and with relatively little change in expression over the measured time points. SOX9, however, was

consistently expressed at higher levels in the BM-MSC derived chondrocytes compared to OA chondrocytes. The expression of the non-collagenous matrix protein genes ACAN, COMP, and VCAN were grossly unaltered over time in the chondrocytes. In the BM-MSC derived chondrocytes, however, ACAN expression showed approximately 5-fold increase between days 7 and day 21, starting at much lower levels than in the OA chondrocytes. The gene markers of hypertrophic chondrocytes ALPL, MMP13, and RUNX2 were also investigated and compared between the cell types. The differentiating BM-MSCs had a 10-fold increase in ALPL gene expression between days 7 and 21, with a considerably higher expression level in the BM-MSC derived chondrocytes than in the OA chondrocytes at day 21. MMP13 had a rather similar trend in both OA and BM-MSC derived chondrocytes with a 10-fold decrease in expression levels between day 7 and 21, with a trend towards lower expression in the OA chondrocytes. The expression level of MMP13 at the end of the culture period was very low. RUNX2 showed a higher level of expression in the BM-MSC derived chondrocytes over the measured time points, though both cell types showed a rather stable expression profile.

Protein Expression Analysis

Fluorescence immunohistochemistry images of the chondrogenically differentiated OA chondrocytes and BM-MSCs embedded in alginate scaffold on day 21 are shown in Figure 2A and B, respectively. COL2 was found to surround nearly all the cells of both cell types. SOX9 was detected as a nuclear signal in more than half of the cells. We found no correlation between the nuclear SOX9 expression signal and the amount of extracellular COL2 signal in either of the cell types. We found both SOX9 positive and negative cells with and without extracellular COL2 signal. COL1 was detected as an extracellular signal around more than half of the cells in both cell populations. Interestingly, a major difference was observed between the two cell populations for the synthesis of COL10. Among the OA chondrocytes extremely few cells were positive for COL10, which stained only in the cytoplasm. For the BM-MSC derived chondrocytes, however, COL10 was found to be synthesized by the vast majority of the cells, and was stained both in the cytoplasm and in the immediate pericellular region at this time point. The expression of the transcription factor proteins SOX5 and SOX6 was similar for the two cell populations, and appeared to be higher than the fraction of SOX9 positive cells. ACAN was found to be synthesized and secreted by almost all the cells within both populations, with a trend towards a more generalized intercellular distribution in the OA chondrocytes. Surprisingly, VCAN demonstrated a strong cytoplasmic distribution in many cells, despite low expression on the RT-qPCR. Little VCAN was detected in the ECM.

Comparison of mRNA Microarray Analysis

Evaluation of differential gene expression by RP analysis showed that 427 genes were upregulated in the OA chondrocytes and 388 genes in the BM-MSC derived chondrocytes after 21 days of differentiation (Table S1 and S2). The GO distribution hierarchy trees for these genes are shown in Table S3. Following Bonferroni correction, the Extracellular Region category (GO: 0005576) came out as most significantly overrepresented GO term among the upregulated genes in the OA chondrocytes. This category was also significantly overrepresented in the BM-MSC derived chondrocytes. As this category was likely to include genes involved in the synthesis and regulation of the ECM, we chose to identify the 82 genes upregulated in OA chondrocytes and the 64 genes upregulated in BM-MSC derived chondrocytes in this category. These genes, subcategorized by ourselves, are presented

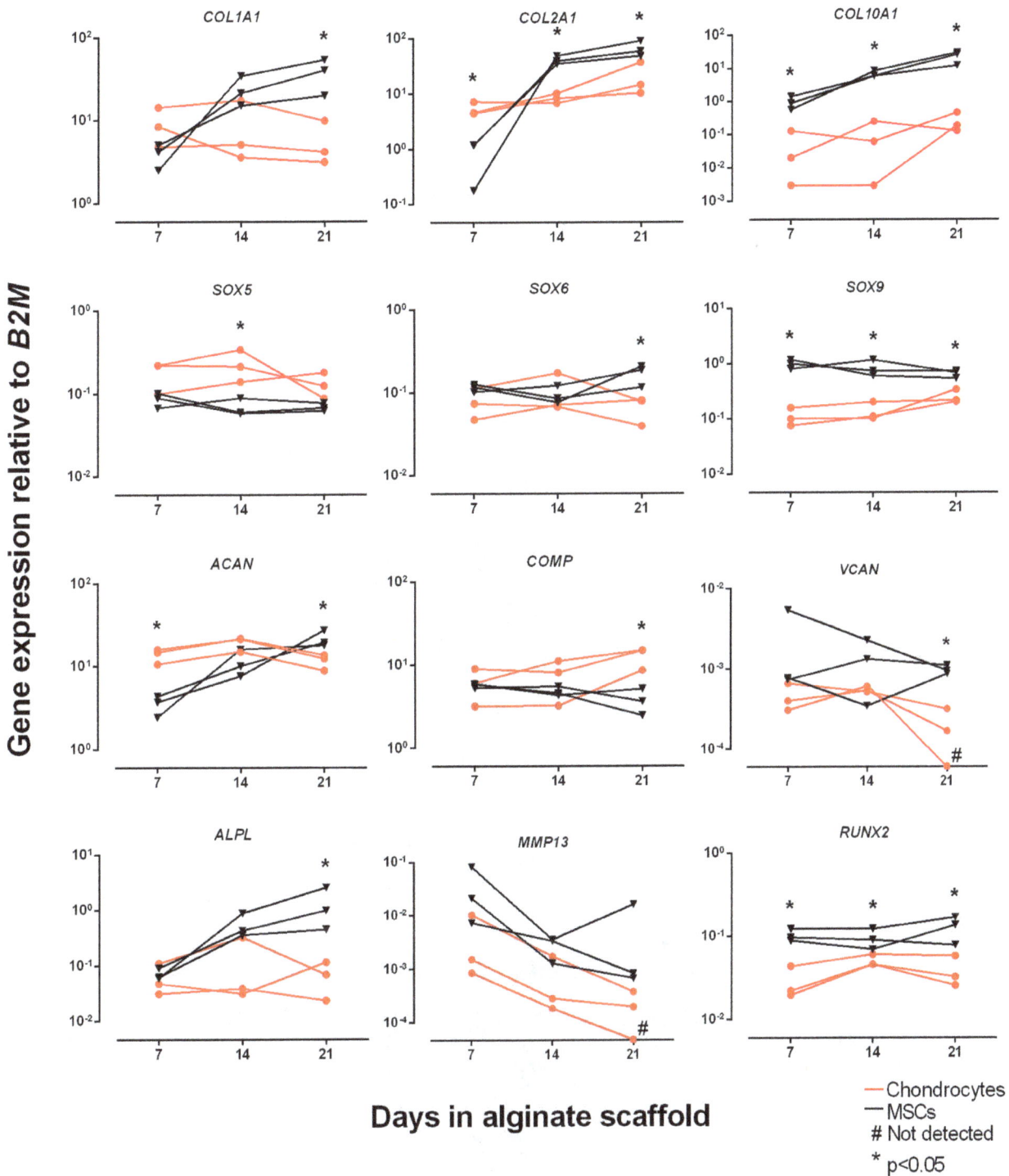

Figure 1. Real-time reverse transcription-polymerase chain reaction quantification (RT-qPCR) of genes expressed in human osteoarthritis (OA) chondrocytes and human bone-marrow (BM) mesenchymal stem cells (MSCs) embedded in self-gelling alginate scaffold with chondrogenic medium. Quantification was made on days 7, 14, and 21. The gene expression was normalized to *Beta-2 microglobulin (B2M)* expression levels. The mRNA gene expression levels for the OA chondrocytes are represented by the red lines and circular symbols, and the BM-MSC derived chondrocytes by the black lines and triangular symbols. Each data point represents the mean of technical triplicates, and each line represents one of three donors run in parallel. "#" signifies non-detectable levels of expression. * signifies statistically significant difference between cell populations (p<0.05).

in Table 2. Most surprisingly, a number of genes associated with hypertrophic chondrocytes and OA were found to be expressed at higher levels in differentiated BM-MSCs than in OA chondro-

cytes: *COL10A1, SSP1, ALPL, BMP2, VEGFA, IHH* and several *WNT* genes [21,33–36]. The differences for the OA markers *MMP13* and *RUNX2*, observed using the more sensitive RT-

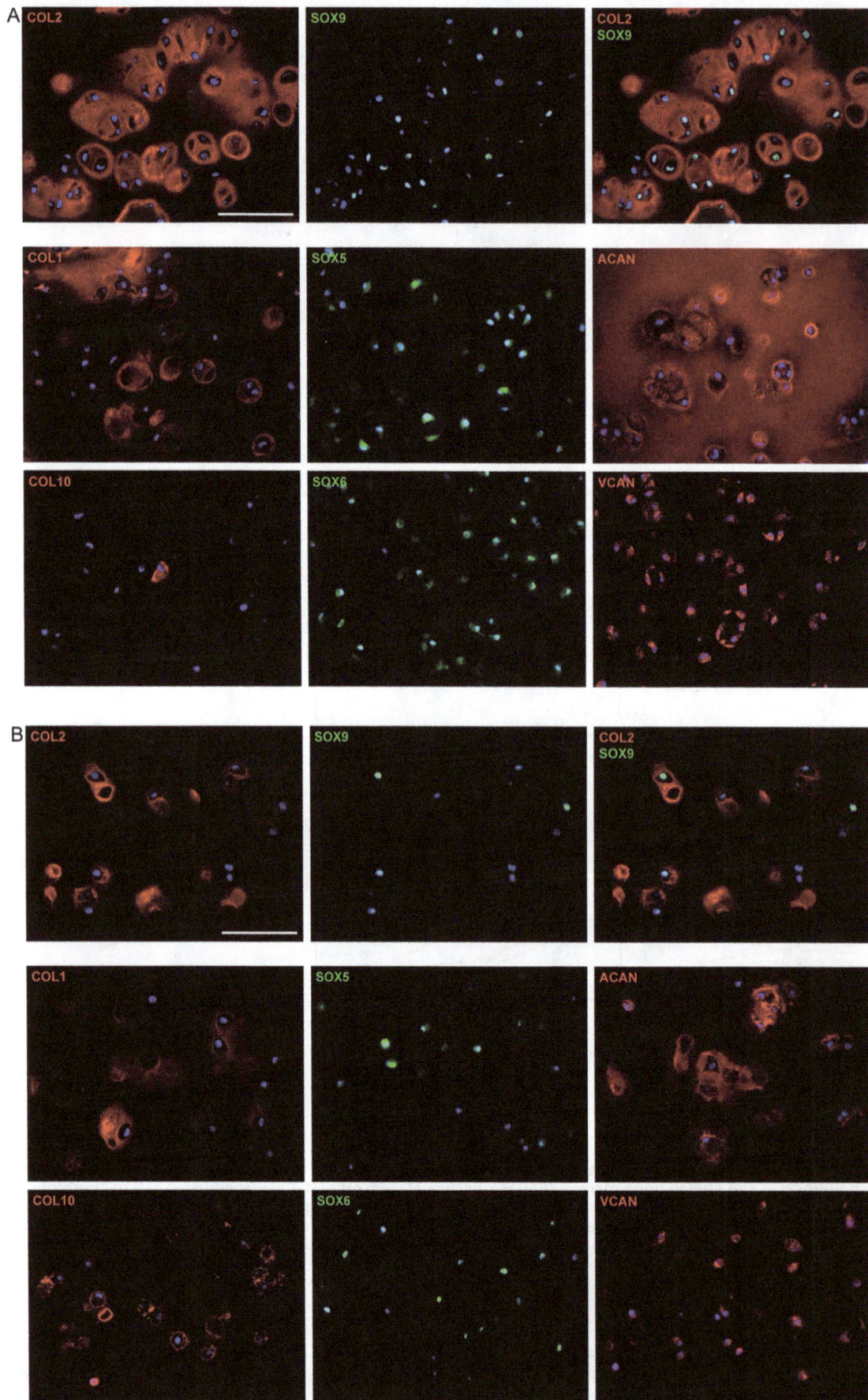

Figure 2. Fluorescence immunohistochemical analysis of protein expression in OA chondrocytes (A) and BM-MSC derived chondrocytes (B) embedded in alginate scaffold with chondrogenic medium for 21 days. COL2 (red color) and SOX9 (green color) were co-stained and is shown both as separate images and in overlay. COL1, COL10, ACAN and VCAN are shown in red color, whereas SOX5, SOX6, and SOX9 are shown in green color. Cellular nuclei were counterstained with DAPI (blue). Scale bar = 100 μm, applies to all images.

qPCR assay, could not be verified here. Similarly, the differences observed for *COL2A1* and *COL1A1* by RT-qPCR could not be verified by the microarray method. Among genes associated with hyaline cartilage, *COL9A1*, *CLIP*, and *DCN* were expressed at higher levels in the OA chondrocytes, while *FN1* and *FBLN1* were expressed at higher levels in the BM-MSC derived chondrocytes. Genes encoding matrix metallopeptidases and ADAM metallo-peptidases were upregulated in the OA chondrocytes, while a number of WNT pathway genes were upregulated in the differentiated BM-MSCs. In general, more genes encoding matrix molecules, and particularly hyaline matrix molecules were found to be upregulated in the OA chondrocytes on day 21 of chondrogenic differentiation in alginate. However, a number of genes encoding molecules important for hyaline cartilage, such as COL11, matrillin 3, biglycan, fibromodulin, hyaluronan, and proteoglycan link protein 1, were expressed at variable and similar levels in the two cell populations (data not shown).

We went on to verify the upregulation of some of the genes by RT-qPCR. *B2M* was again used as endogenous control. We analyzed the expression of *COL3A1*, *COL9A1*, *MMP3*, *IBSP*, *SSP1*, and *FN1*, and found that the changes observed in the microarray analysis could in all cases be confirmed, and that in most cases the fold change differences were greater in the RT-qPCR analysis. The results are shown in Figure 3.

As the chondrocytes were harvested from OA joints, and thus could be suspected of expressing genes known to be involved in OA pathogenesis, we identified a large number of genes known to be involved in this disease from the microarray comparison. These were genes encoding enzymes known to be involved in ECM remodeling (*MMPs 1, 3, 8, 9, 13* and *14*, and *ADAMTS4* and *5*), genes encoding enzymes involved in prostaglandin E2 synthesis (*PTGS2*, *PTGES*, *PTGES2*, and *PLA2G4A*), genes encoding some other enzymes (*NOS2A*, *HTRA1*, *DDR2*, and *CASP3*), genes encoding inflammatory substances (*ILs 1A, 1B, 6, 17A, 17B, 17C, 17D, 17F, 18, TNF*, and *IL1R1*), some genes in the S100A series (*S100A4, A8, A9*, and *A11*), the secreted signaling molecule *IHH* and the damage-associated molecular pattern protein *HMGB1*, all of these genes that have been described in recent reviews on the pathogenesis of OA [21,37,38]. The results are shown in Figure 4. Surprisingly, most of these genes were either not expressed or only expressed at levels close to background in any of these cell populations. However, *MMP3*, *S100A4*, *ADAMTS4*, and *IL17D* were all expressed at higher levels in the OA chondrocytes, while *PTGES* and *IHH* were expressed at higher levels in the differentiated BM-MSCs. *HTRA1*, *PTGS2*, *PTGES2*, *DDR*, *CASP3*, *IL18*, and *S100A11* were expressed at moderate to high levels in both populations.

Discussion

Tissue engineering of hyaline cartilage by *in vitro* culture of cells in a biomaterial may provide the best treatment option for lesions of articular cartilage. However, in order to tolerate the functional demands put on the cartilage within weight bearing joints over decades, the implant should be as close to native hyaline cartilage as possible, both for the composition and structure of the ECM. OA patients are candidates for treatment with tissue engineered cartilage. Autologous cells, which may be used to produce cartilage

in vitro for these patients, are articular chondrocytes and MSCs. However, as OA may be a generalized disease within the affected joint, and perhaps also a systemic disease [36,38], these cell populations may both be affected by the OA disease process. An alternative may be to use allogeneic BM-MSCs from healthy donors. Such cells are easily obtained, as the bone marrow aspiration process carries very limited pain and morbidity, and the cells are easily expanded *in vitro*. Transplantation of allogeneic cells would normally lead to immune rejection of the transplanted cells. However, several observations and facts suggest that this might not occur if allogeneic MSCs are used in scaffolds for treatment of cartilage lesions. First, MSCs have been described to be, at least in part, immunoprivileged cells [39,40]. Second, MSCs are known to have immunosuppressive properties [25], and this could stop an allogeneic immune response in its early stages. And finally, hyaline cartilage has no blood supply and is, to a minimal degree, in contact with the immune system. A tissue engineered implant entirely within the articular cartilage may, therefore, in effect be within an immunosequestered site. Thus, allogeneic MSCs differentiated to chondrocytes and used in tissue engineering strategies for patients with early OA may represent a viable treatment strategy. On this background, we decided to compare chondrocytes harvested from joints with very advanced OA with BM-MSCs from healthy donors for their ability to act as a cellular substrate for *in vitro* tissue engineering of hyaline cartilage.

COL2 is the most abundant and important of the ECM molecules in cartilage. Using our recently established *in vitro* culture procedure for articular chondrocytes, allowing these cells to expand in their own matrix [27,41], *COL2A1* mRNA expression was maintained in OA chondrocytes after the *in vitro* expansion, and remained at high levels during chondrogenic differentiation in 3D culture. At the protein level, COL2 was synthesized into the extracellular space around most of the chondrocytes at three weeks. BM-MSCs do not express *COL2A1* mRNA during 2D expansion, but following chondrogenic differentiation in 3D, the *COL2A1* mRNA level increased quickly, and was slightly higher than that observed for the OA chondrocytes at two and three weeks. This difference was not discernible in the less sensitive microarray assay, and there was no obvious difference in COL2 protein synthesis at three weeks. Of other molecules known to be involved in hyaline chondrogenesis, mRNAs for COL9, cartilage intermediate layer protein 1, and decorin were increased in OA chondrocytes by microarray analysis. However, some hyaline cartilage mRNAs, such as *FN1*, *FBLN1*, and *ACAN* were slightly higher in the BM-MSC derived chondrocytes. Notably; mRNAs for most of the other molecules required for synthesis of hyaline cartilage ECM were expressed in both cell types, at similar levels. This suggests that both cell types, embedded in alginate disc scaffolds and exposed to chondrogenic differentiation medium are able to produce the molecules required for synthesis of hyaline cartilage ECM.

Certain molecules, not native to hyaline cartilage, were also expressed in these cells. *COL1A1* encoding the collagen predominantly responsible for the fibrous component of fibrocartilage. This mRNA was highly expressed in both cell types, but more in BM-MSC derived chondrocytes. *COL3A1*, which encodes a protein involved with COL1 in fibrocartilage, but also has a place in hyaline cartilage, was slightly higher in OA chondrocytes.

Table 2. List of genes within the GO term Extracellular Region which were differentially expressed in OA chondrocytes and chondrogenically differentiated human BM-MSCs.

Upregulated in OA chondrocytes			Upregulated in differentiated BM-MSCs		
Symbol	Name	Fold increase	Symbol	Name	Fold increase
Extracellular matrix proteins			**Extracellular matrix proteins**		
COL3A1	Collagen type 3	5.7	COL27A1	Collagen type 27	2.6
COL9A1	Collagen type 9	2.8	COL10A1	Collagen type 10	6.6
COL16A1	Collagen type 16	2.5			
COL24A1	Collagen 24	2.5			
CLIP	Cartilage intermediate layer protein	2.9	IBSP	Integrin-binding sialoprotein/bone sialoprotein	9.6
CILP2	Cartilage intermediate layer protein 2	5.1	SPP1	Secreted phosphoprotein 1/bone sialoprotein 1/osteopontin	10.2
CRTAC1	Cartilage acidic protein 1	8.6	FN1	Fibronectin 1	2.6
MFAP4	Microfibrillar-associated protein 4	6.0	MATN4	Matrilin 4	8.7
CHI3L2	Chitinase 3-like 2	2.5			
FBLN5	Fibulin 5	4.3			
AMTN	Amelotin	6.4			
ELN	Elastin	4.3			
DPT	Dermatopontin	12.5			
MGP	Matrix Gla protein	3.8			
EFEMP1	EGF-containing fibulin-like ECM protein 1	4.1			
FNDC1	Fibronectin type III domain containing 1	3.6			
SMOC1	SPARC related modular calcium binding 1	4.3	SPARCL1	SPARC-like 1/hevin	3.3
SMOC2	SPARC related modular calcium binding 2	2.8			
ECM2	Extracellular matrix protein 2	4.0			
LAMA2	Laminin alpha 2	2.3	LAMA4	Laminin alpha 4	4.2
Glycoproteins, proteoglycans, and PG modifiers			**Glycoproteins, proteoglycans, and PG modifiers**		
OGN	Osteoglycin	9.8	FBLN1	Fibulin	3.0
MAMDC2	MAM domain containing 2/mamcan	5.5	GPC3	Glypican 3	5.8
SULF1	Sulfatase 1	2.6	GPC1	Glypican 1	3.3
OMD	Osteomodulin	4.9	ENG	Endoglin/CD105	3.0
DCN	Decorin	3.4	FSTL3	Follistatin-like 3	2.6
Matrix remodeling			**Matrix remodeling**		
ADAMTS1	ADAM metallopeptidase with thrombospondin 1 motif	2.3	A2M	Alpha 2 macroglobulin	5.6
ADAMTSL3	ADAMTS-like 3	5.4	LOX	Lysyl oxidase	3.8
MXRA5	Matrix-remodeling associated 5	4.3	LOXL4	Lysyl oxidase-like 4	3.3
SERPINA5	Serpin peptidase inhibitor, clade A, 5	5.2	LOXL3	Lysyl oxidase-like 3	2.6
SERPINA1	Serpin peptidase inhibitor, clade A, 1	7.0	SERPINE2	Serpin peptidase inhibitor, clade E, 2	5.6
SPINT2	Serine peptidase inhibitor, Kunitz type, 2	6.3	KAZALD1	Kazal-type serine peptidase inhibitor domain 1	2.8
SLPI	Secretory leukocyte peptidase inhibitor	2.8	P4HB	Prolyl 4-hydroxylase, beta	2.6
MMP7	Matrix metallopeptidase 7	25.0	ALPL	Alkaline phosphatase	6.1

Table 2. Cont.

Upregulated in OA chondrocytes			Upregulated in differentiated BM-MSCs		
Symbol	Name	Fold increase	Symbol	Name	Fold increase
MMP3	Matrix metallopeptidase 3	21.1	MMP2	Matrix metallopeptidase 2	2.5
MMP23A	Matrix metallopeptidase 23A	2.8	PAMR1	Peptidase associated with muscle regeneration 1	14.1
			CTSD	Cathepsin D	8.4
Growth factors, GF receptors, and GF antagonists			**Growth factors and GF receptors**		
TNFRSF11B	Tumor necrosis factor receptor 11B/ osteoprotegerin	11.0	IGFBP4	Insulin-like growth factor binding protein 4	4.0
IGFBP6	Insulin-like growth factor binding protein 6	6.4	IGFBP5	Insulin-like growth factor binding protein 5	14.6
GDF10	Growth differentiation factor 10	13.1	GDF15	Growth differentiation factor 15	6.6
PDGFD	Platelet derived growth factor D	3.3	VEGFA	Vascular endothelial growth factor A	4.4
GREM1	Gremlin 1	4.2	TGFBI	Transforming growth factor, beta-induced	5.1
TGFBR3	Transforming growth factor beta receptor 3	2.8	MDK	Midikine	2.5
BMP4	Bone morphogenetic protein 4	2.7	BMP2	Bone morphogenetic protein 2	3.6
			PGF	Placental growth factor	5.6
Signaling pathways			**Signaling pathways**		
CD47	CD47 molecule	2.4	WNT4	Wingless-type MMTV integration site family 4	4.9
WISP2	WNT1 inducible signaling protein 2	4.6	WNT5A	Wingless-type MMTV integration site family 5A	2.4
SFRP5	Secreted frizzled-related protein 5	4.0	WNT5B	Wingless-type MMTV integration site family 5B	4.5
JAG1	Jagged 1	2.6	WNT 11	Wingless-type MMTV integration site family 11	3.9
SCUBE2	Signal peptide, CUB domain, EGF-like 2	2.6	SFRP1	Secreted frizzled-related protein 1	3.1
			DKK1	Dickkopf 1	3.7
			IHH	Indian hedgehog	4.2
			PTGDS	Prostaglandin D2 synthase	3.4
			SCUBE3	Signal peptide, CUB domain, EGF-like 3	3.9
Cytokines			**Cytokines**		
CYTL1	Cytokine-like 1	36.8	MIF	Macrophage migration inhibitory factor	2.8
IL17RB	Interleukin 17 receptor B	2.4	CXCL14	Chemokine (C-X-C motif) ligand 14	14.0
CCL2	Chemokine (C-C motif) ligand 2	3.3			
Cell adhesion			**Cell adhesion**		
MSLN	Mesothelin	5.1	FLRT3	Fibronectin leucine rich transmembrane protein 3	2.8
SVEP1	Sushi, von Willebrand factor type A, EGF and pentraxin domain containing 1	7.6	CLEC3A	C-type lectin domain family 3 A	3.7
TNC	Tenascin C	7.7	CLEC11A	C-type lectin domain family 11 A	2.5
CHAD	Chondroadherin	3.1	LGALS3BP	Lectin, galactoside-binding, soluble, 3 binding protein	4.0
CDH13	Cadherin 13	2.3	THBS2	Thrombospondin 2	3.7
THBS3	Thrombospondin 3	4.6			
Miscellaneous			**Miscellaneous**		
CFH	Complement factor H	3.3	CFD	Complement factor D	3.8

Table 2. Cont.

Upregulated in OA chondrocytes			Upregulated in differentiated BM-MSCs		
Symbol	Name	Fold increase	Symbol	Name	Fold increase
CFB	Complement factor B	2.4	PZP	Pregnancy-zone protein	2.8
C5	Complement factor 5	3.6	PCSK5	Proprotein convertase subtilisin/kexin 5	2.5
ABI3BP	ABI gene family, member 3 binding protein	13.3	ALDOA	Aldolase A	2.7
RPESP	RPE-spondin	4.0	F12	Coagulation factor XII	2.7
WFDC1	WAP four-disulfide core domain 1	2.7	TUBA4A	Tubulin alpha 4A	2.7
DMKN	Dermokine	2.9	DHRS11	Dehydrogenase/reductase (SDR family) 11	3.0
TMSB4X	Thymosin, beta 4, X-linked	2.9	CCDC80	Coiled-coil domain containing 80	2.4
CA2	Carbonic anhydrase 2	3.7	SLIT3	Slit homolog 3	2.9
GPX3	Glutathione peroxidase 3	8.2	TF	Transferrin	3.6
ISM1	Isthmin 1 homolog	2.7	ACTN4	Actinin, alpha 4	2.6
PLA2G2A	Phospholipase A2, group 2A	8.3	ADM	Adrenomodulin	3.0
PON3	Paraoxonase 3	3.3	GPI	Glucose phosphate isomerase	3.1
POSTN	Periostin	4.6	HLA-C	Human leukocyte antigen class I, C	2.4
CPXM2	Carboxypeptidase X (M14 family) 2	2.3	MICA	MHC class I polypeptide-related sequence A	2.9
C2orf40	Chromosome 2 open reading frame 40	12.0			
OLFM1	Olfactomedin	5.4			
FST	Follistatin	2.5			
PRRG4	Proline rich Gla 4	2.6			
CNPY4	Canopy 4	2.3			
RAMP1	Receptor activity modifying protein 1	2.5			
ANGPT1	Angiopoietin 1	3.2			
ANGPTL7	Angiopoietin-like 7	4.7			
ANGPTL5	Angiopoietin-like 5	13.7			

Interestingly, several molecules that are thought to be markers of OA pathogenesis showed higher expression in the differentiated BM-MSCs than in the differentiated OA chondrocytes. The molecule most frequently mentioned in this context is *COL10A1* [42]. *COL10A1* was one of the most differentially expressed molecules between the two cell populations, 85-fold higher in the differentiated BM-MSCs by RT-qPCR. This difference seemed to be reflected at the protein level also, not predominantly by the amount of COL10 synthesized from each cell, but by the proportion of cells synthesizing COL10. Most of the differentiated BM-MSCs synthesized COL10, while <1% of the OA chondrocytes had synthesized this collagen. For the OA chondrocytes, this may represent the cells known to produce COL10 *in vivo* [43,44]. This could mean that the differentiation cocktail was unable to induce COL10 production in the chondrocytes that did not already produce COL10 *in vivo*. Undifferentiated BM-MSCs also express negligible or no COL10 mRNA [17]. Here, evidently, the epigenetic restrictions to COL10 mRNA and protein synthesis could be overcome by the differentiation cocktail and 3D culture, at least in the majority of the cells. Other molecules known to be markers of chondrocyte hypertrophy during embryological chon-

drogenesis are *RUNX2, ALPL,* and *MMP13.* Of these, RUNX2 and ALPL were expressed at higher levels in BM-MSC derived chondrocytes, while the expression of MMP13 was very low in both cell populations by RT-qPCR, and not discernible in the microarray assay. Many of the processes associated with chondrocyte hypertrophy, including chondrocyte apoptosis and ECM mineralization have been suggested to also be involved in OA pathogenesis [21,23,35,36,45]. We have seen no evidence of apoptosis or mineralization occurring during 6–8 week cultures of BM-MSCs exposed to chondrogenic differentiation medium in alginate (A. Küchler, unpublished results), which may suggest that these genetic markers of chondrocyte hypertrophy and OA pathogenesis have no functional correlate during *in vitro* chondrogenesis. However, COL10 is distributed within the extracellular space in these longer term cultures. COL10 is a short, sheet forming collagen [46], unlikely to contribute to tensile strength, and as such not desirable in hyaline ECM.

A number of other genes known to encode proteins involved in OA pathogenesis but not associated with chondrocyte hypertrophy were also investigated. These proteins have been detected in the synovial fluid and in areas of ECM depletion in the cartilage of

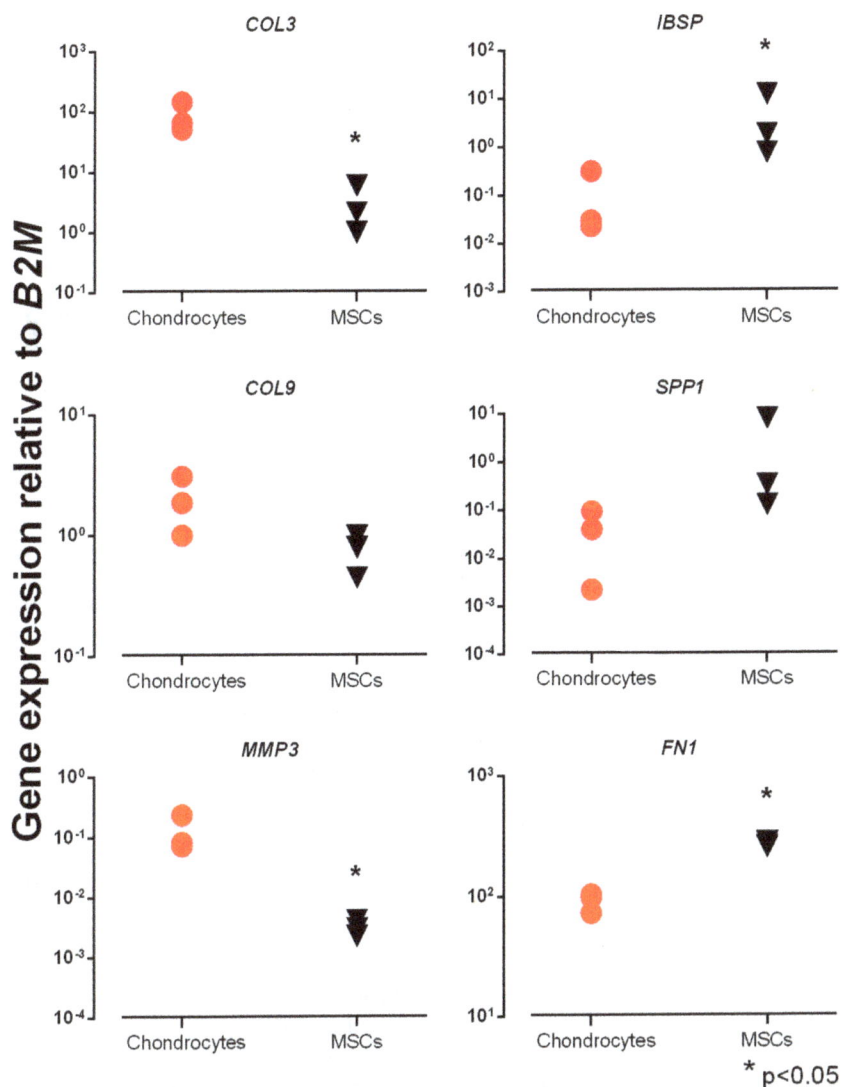

Figure 3. RT-qPCR verification of selected genes from the microarray comparison of the OA chondrocytes (red circles) and BM-MSC derived chondrocytes (black triangles) following 21 days of chondrogenic differentiation in alginate discs. Gene expression was normalized to *B2M* expression levels. Shown are mean values of the technical triplicates for each donor. * signifies statistically significant difference between cell populations (p<0.05).

OA joints. Interestingly, many of these genes were expressed neither in OA chondrocytes nor in BM-MSC derived chondrocytes. This was true particularly for most of the mediators of inflammation known to be associated with OA, but also for most of the enzymes known to degrade hyaline cartilage. One possible explanation for this unexpected finding could be that the OA chondrocytes expressed the OA associated genes at the time of their harvest from the OA cartilage, and subsequently lost expression in the course of *in vitro* expansion and chondrogenic redifferentiation. If so, the fact that the OA genes are no longer expressed may mean that these chondrocytes will not contribute to OA progression when the tissue engineered cartilage is implanted, although this issue can really only be ascertained following *in vivo* implantation. Another possibility could be that the chondrocytes that were used for this *in vitro* chondrogenesis assay, harvested from a healthy-looking part of the OA cartilage, never expressed these genes. This may be the case if the OA-associated proteins were produced by other cells in the vicinity such as the

chondrocytes located within the lesion, by cells in the synovial tissue or both. In this situation the cultured OA chondrocytes also most probably would not contribute to progression towards OA after implantation.

Four OA-associated genes were clearly more highly expressed in the cultured OA chondrocytes than in the BM-MSC derived chondrocytes: *MMP3, ADAMTS4, IL17D,* and *S100A4. MMP3* encodes a matrix metalloproteinase which degrades fibronectin, laminin, COL3, 4, 9, and 10, and cartilage proteoglycans. Of these, fibronectin, COL9, and proteoglycans, such as ACAN, are important for the normal functioning of articular cartilage. Of these again, only ACAN was investigated at the protein level in the present study and, paradoxically, found to be more widely distributed around the OA chondrocytes. This may suggest that the upregulation of the *MMP3* mRNA in the OA chondrocytes may not have a detrimental effect on cartilage ECM, at least not on ACAN. Complete insight into the effect of MMP3 on articular cartilage would, however, require a more targeted study.

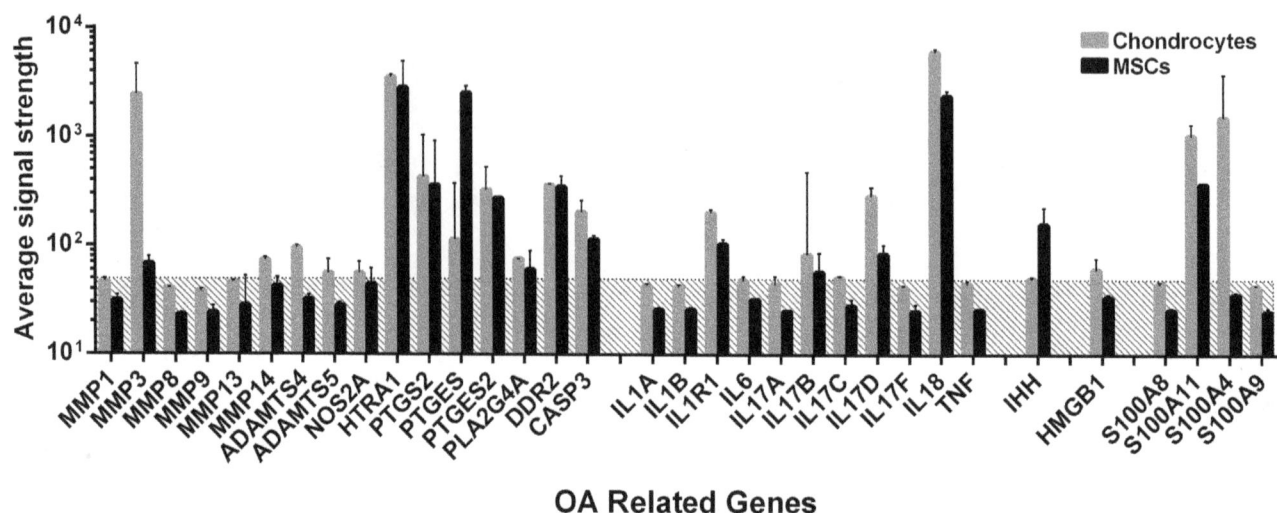

OA Related Genes

Figure 4. mRNA concentration of selected OA related genes expressed by OA chondrocytes (grey columns) and BM-MSC derived chondrocytes (black columns) on day 21 of differentiation. Expression values are extracted from the whole-genome BeadChip analysis. Each donor data point is the mean value of the Illumina® BeadArray™ Reader scan for that gene and that donor. Columns represent median values for the three donors, and variability whiskers represent range.

ADAMTS4 was expressed at low levels in OA chondrocytes, but was not expressed at all in the differentiated BM-MSCs. ADAMTS4 is also a metalloproteinase which degrades ACAN. Thus, as with *MMP3*, the observation that more ACAN seemed to be observed around the OA chondrocytes than the BM-MSC derived chondrocytes would suggest that the low-level upregulation of *ADAMTS4* gene expression did not significantly influence ECM synthesis of the OA chondrocytes in alginate discs. *S100A4* encodes a calcium-binding protein. It is best known for its role in cancer metastasis, but has also been found to be upregulated in the synovium of rheumatoid arthritis to a much greater extent than in OA. It is thought to be involved in regulation of several matrix-degrading enzymes, but its exact role in OA remains to be determined [47]. Finally, *IL17D* was also expressed at a higher level in OA chondrocytes than in differentiated BM-MSCs. The encoded cytokine, also known as interleukin (IL) 27, is best known for its ability to induce other inflammatory cytokines from endothelial cells and immune cells. In OA it is thought to be induced by IL-1β and TNFα [37]. However, genes encoding these cytokines were not expressed in either cell subset. Thus, the mechanism responsible for the expression of *IL17D* mRNA and the importance of this expression for the use of OA chondrocytes in tissue engineering of cartilage remains unknown at this time.

Two genes associated with OA, but not directly involved in chondrogenesis, *PTGES* and *IHH*, were expressed at higher levels in differentiated BM-MSCs than in OA chondrocytes. *PTGES* is involved in the synthesis of prostaglandin E2 (PGE2). The gene encoding the first enzyme in this pathway, cytosolic phospholipase A2s (*PLA2G4A*) was expressed at very low levels in both cell types. The gene encoding the subsequent enzyme, cycloxygenase 2 (*PTGS2*), which is normally the rate limiting enzyme, was expressed at moderate levels in both cell types. Finally, prostaglandin E synthase (*PTGES*) was very much more expressed in the differentiated BM-MSCs [48]. PGE2 could potentially be an unwanted molecule produced by implanted cartilage made from BM-MSCs. However, since *PLA2G4A* is only barely expressed in the two cell types, which most likely makes this the rate limiting step, it seems unlikely that the upregulation of *PTGES* will transfer a propensity for OA development through the implanted cartilage.

IHH has only recently been implicated in the pathogenesis of OA [49]. The mechanism is not fully elucidated, but stimulation of RUNX2 and ADAMTS5 was suggested. In the cells discussed here, ADAMTS5 was expressed at extremely low levels or not at all, while there was a trend for RUNX2 to be expressed at higher levels in the BM-MSC derived chondrocytes by RT-qPCR, perhaps consistent with a stimulatory effect by upregulated IHH.

In conclusion, we show here that both OA chondrocytes and BM-MSCs produce the vast majority of the mRNA molecules required for synthesis of hyaline cartilage when placed in an alginate biomaterial and exposed to chondrogenic differentiation cocktail. For those molecules tested, mRNA expression was translated into protein synthesis. Unfortunately, both cell populations also produced molecules responsible for fibrocartilage. Surprisingly, many of the markers thought to be associated with OA pathogenesis were expressed at higher levels in the differentiated BM-MSCs than in the OA chondrocytes, and the differential synthesis of COL10 may have functional consequences. Karlsson et al, using pellet mass cultures and a restricted number of gene assays, have made similar observations [50]. To make a final decision between these cells, more studies are required to examine other differentiation strategies, other biomaterials, more parallel donors, and longer duration of cell/scaffold constructs *in vitro*. Subsequently, the chosen cell/scaffold constructs should be tested in animal models where OA chondrocytes may be compared with healthy BM-MSCs within the microenvironment, and under the dynamic weight and strain conditions to which native cartilage is exposed *in vivo* [51–53]. Eventually, these studies should result in the generation of tissue engineered implants which will provide patients with lesions of articular cartilage with a treatment option that will last for life.

Supporting Information

Table S1 Genes upregulated in osteoarthritis (OA) chondrocytes embedded in alginate scaffold with chondrogenic medium after 21 days.

Table S2　Genes upregulated in human bone marrow (BM)-MSCs embedded in alginate scaffold with chondrogenic medium after 21 days.

Table S3　Gene Ontology comparison results on OA chondrocytes (A) and chondrogenic differentiated human BM-MSCs (B).

Acknowledgments

We are grateful to dr Stephan Röhrl, Department of Orthopedic Surgery, Oslo University Hospital and drs. Øystein Høvik and Bjørn Ødegaard, Lovisenberg Diakonale Hospital, Oslo for providing us with OA cartilage from knee replacement surgery. We would like to thank professor Gudmund Skjåk-Bræk, Department of Biotechnology, Norwegian University of Science and Technology, Trondheim, Norway for providing us with G-Lyase; professor Klaus von der Mark, Department of Experimental Medicine I, Nikolaus-Fiebiger-Center of Molecular Medicine, Friedrich-Alexander-University of Erlangen-Nuremberg, Erlangen, Germany for the generous gift of X53 supernatant; Jon Halvor F. Lunde and Linda T. Dorg, Department of Pathology, University of Oslo for technical help.

Author Contributions

Conceived and designed the experiments: AMF SRH TAK AMK JEB. Performed the experiments: AMF SRH TAK YF AMK. Analyzed the data: AMF SRH TAK AMK JEB. Contributed reagents/materials/analysis tools: AMF SRH TAK AMK YF JEB. Wrote the paper: AMF SRH TAK AMK YF JEB.

References

1. Curl WW, Krome J, Gordon ES, Rushing J, Smith BP, et al. (1997) Cartilage injuries: a review of 31,516 knee arthroscopies. Arthroscopy 13: 456–460.
2. Hjelle K, Solheim E, Strand T, Muri R, Brittberg M (2002) Articular cartilage defects in 1,000 knee arthroscopies. Arthroscopy 18: 730–734.
3. Aroen A, Loken S, Heir S, Alvik E, Ekeland A, et al. (2004) Articular cartilage lesions in 993 consecutive knee arthroscopies. Am J Sports Med 32: 211–215.
4. Widuchowski W, Widuchowski J, Trzaska T (2007) Articular cartilage defects: Study of 25,124 knee arthroscopies. Knee 14: 177–182.
5. Roos H, Adalberth T, Dahlberg L, Lohmander LS (1995) Osteoarthritis of the knee after injury to the anterior cruciate ligament or meniscus: The influence of time and age. Osteoarthritis Cartilage 3: 261–267.
6. Lohmander LS, Ostenberg A, Englund M, Roos H (2004) High prevalence of knee osteoarthritis, pain, and functional limitations in female soccer players twelve years after anterior cruciate ligament injury. Arthritis Rheum 50: 3145–3152.
7. Dieppe PA, Lohmander LS (2005) Pathogenesis and management of pain in osteoarthritis. Lancet 365: 965–973.
8. Gomoll AH, Farr J, Gillogly SD, Kercher J, Minas T (2010) Surgical management of articular cartilage defects of the knee. J Bone Joint Surg Am 92: 2470–2490.
9. Brittberg M, Lindahl A, Nilsson A, Ohlsson C, Isaksson O, et al. (1994) Treatment of deep cartilage defects in the knee with autologous chondrocyte transplantation. N Engl J Med 331: 889–895.
10. Grigolo B, Roseti L, De FL, Piacentini A, Cattini L, et al. (2005) Molecular and immunohistological characterization of human cartilage two years following autologous cell transplantation. J Bone Joint Surg Am 87: 46–57.
11. Knutsen G, Drogset JO, Engebretsen L, Grontvedt T, Isaksen V, et al. (2007) A randomized trial comparing autologous chondrocyte implantation with microfracture. Findings at five years. J Bone Joint Surg Am 89: 2105–2112.
12. Vasiliadis HS, Wasiak J (2010) Autologous chondrocyte implantation for full thickness articular cartilage defects of the knee. Cochrane Database Syst Rev CD003323.
13. Ahmed TA, Hincke MT (2010) Strategies for articular cartilage lesion repair and functional restoration. Tissue Eng Part B Rev 16: 305–329.
14. Marlovits S, Zeller P, Singer P, Resinger C, Vecsei V (2006) Cartilage repair: generations of autologous chondrocyte transplantation. Eur J Radiol 57: 24–31.
15. Jakobsen RB, Shahdadfar A, Reinholt FP, Brinchmann JE (2010) Chondrogenesis in a hyaluronic acid scaffold: comparison between chondrocytes and MSC from bone marrow and adipose tissue. Knee Surg Sports Traumatol Arthrosc 18: 1407–1416.
16. Boeuf S, Richter W (2010) Chondrogenesis of mesenchymal stem cells: role of tissue source and inducing factors. Stem Cell Res Ther 1: 31.
17. Herlofsen SR, Kuchler AM, Melvik JE, Brinchmann JE (2011) Chondrogenic differentiation of human bone marrow-derived mesenchymal stem cells in self-gelling alginate discs reveals novel chondrogenic signature gene clusters. Tissue Eng Part A 17: 1003–1013.
18. Matricali GA, Dereymaeker GP, Luyten FP (2010) Donor site morbidity after articular cartilage repair procedures: a review. Acta Orthop Belg 76: 669–674.
19. Benya PD, Shaffer JD (1982) Dedifferentiated chondrocytes reexpress the differentiated collagen phenotype when cultured in agarose gels. Cell 30: 215–224.
20. van der Kraan PM, van den Berg WB (2012) Chondrocyte hypertrophy and osteoarthritis: role in initiation and progression of cartilage degeneration? Osteoarthritis Cartilage 20: 223–232.
21. Pitsillides AA, Beier F (2011) Cartilage biology in osteoarthritis-lessons from developmental biology. Nat Rev Rheumatol 7: 654–663.
22. Mackie EJ, Tatarczuch L, Mirams M (2011) The skeleton: a multi-functional complex organ: the growth plate chondrocyte and endochondral ossification. J Endocrinol 211: 109–121.
23. Goldring MB, Tsuchimochi K, Ijiri K (2006) The control of chondrogenesis. J Cell Biochem 97: 33–44.
24. Pittenger MF, Mackay AM, Beck SC, Jaiswal RK, Douglas R, et al. (1999) Multilineage potential of adult human mesenchymal stem cells. Science 284: 143–147.
25. Uccelli A, Moretta L, Pistoia V (2008) Mesenchymal stem cells in health and disease. Nat Rev Immunol 8: 726–736.
26. Ballock RT, O'Keefe RJ (2003) The biology of the growth plate. J Bone Joint Surg Am 85A: 715–726.
27. Shahdadfar A, Loken S, Dahl JA, Tunheim SH, Collas P, et al. (2008) Persistence of collagen type II synthesis and secretion in rapidly proliferating human articular chondrocytes in vitro. Tissue Eng Part A 14: 1999–2007.
28. Melvik JE, Dornish M, Onsoyen E, Berge AB, Svendsen T (2006) Self-gelling alginate systems and uses thereof. United States Patent Application #20060159823.
29. Kristiansen A, Andersen T, Melvik JE (2007) Cells, gels and alginates. Eur Biopharm Rev 34–38.
30. Dysvik B, Jonassen I (2001) J-Express: exploring gene expression data using Java. Bioinformatics 17: 369–370.
31. Breitling R, Armengaud P, Amtmann A, Herzyk P (2004) Rank products: a simple, yet powerful, new method to detect differentially regulated genes in replicated microarray experiments. FEBS Lett 573: 83–92.
32. Consortium GO (2010) The Gene Ontology in 2010: extensions and refinements. Nucleic Acids Res 38: D331–D335.
33. Tchetina EV, Squires G, Poole AR (2005) Increased type II collagen degradation and very early focal cartilage degeneration is associated with upregulation of chondrocyte differentiation related genes in early human articular cartilage lesions. J Rheumatol 32: 876–886.
34. Pullig O, Weseloh G, Gauer S, Swoboda B (2000) Osteopontin is expressed by adult human osteoarthritic chondrocytes: protein and mRNA analysis of normal and osteoarthritic cartilage. Matrix Biol 19: 245–255.
35. Drissi H, Zuscik M, Rosier R, O'Keefe R (2005) Transcriptional regulation of chondrocyte maturation: potential involvement of transcription factors in OA pathogenesis. Mol Aspects Med 26: 169–179.
36. Corr M (2008) Wnt-beta-catenin signaling in the pathogenesis of osteoarthritis. Nat Clin Pract Rheumatol 4: 550–556.
37. Goldring MB, Otero M (2011) Inflammation in osteoarthritis. Curr Opin Rheumatol 23: 471–478.
38. Aspden RM (2008) Osteoarthritis: a problem of growth not decay? Rheumatology (Oxford) 47: 1452–1460.
39. Salem HK, Thiemermann C (2010) Mesenchymal stromal cells: current understanding and clinical status. Stem Cells 28: 585–596.
40. Schu S, Nosov M, O'Flynn L, Shaw G, Treacy O, et al. (2011) Immunogenicity of allogeneic mesenchymal stem cells. J Cell Mol Med.
41. Karlsen TA, Shahdadfar A, Brinchmann JE (2011) Human primary articular chondrocytes, chondroblasts-like cells, and dedifferentiated chondrocytes: differences in gene, microRNA, and protein expression and phenotype. Tissue Eng Part C Methods 17: 219–227.
42. Shen G (2005) The role of type X collagen in facilitating and regulating endochondral ossification of articular cartilage. Orthod Craniofac Res 8: 11–17.
43. Kielty CM, Kwan AP, Holmes DF, Schor SL, Grant ME (1985) Type X collagen, a product of hypertrophic chondrocytes. Biochem J 227: 545–554.
44. Schmid TM, Popp RG, Linsenmayer TF (1990) Hypertrophic cartilage matrix. Type X collagen, supramolecular assembly, and calcification. Ann N Y Acad Sci 580: 64–73.
45. Schmid TM, Linsenmayer TF (1990) Immunoelectron microscopy of type X collagen: supramolecular forms within embryonic chick cartilage. Dev Biol 138: 53–62.
46. Schmid TM, Linsenmayer TF (1985) Immunohistochemical localization of short chain cartilage collagen (type X) in avian tissues. J Cell Biol 100: 598–605.
47. Oslejskova L, Grigorian M, Gay S, Neidhart M, Senolt L (2008) The metastasis associated protein S100A4: a potential novel link to inflammation and

consequent aggressive behaviour of rheumatoid arthritis synovial fibroblasts. Ann Rheum Dis 67: 1499–1504.

48. Crofford LJ (2001) Prostaglandin biology. Gastroenterol Clin North Am 30: 863–876.

49. Lin AC, Seeto BL, Bartoszko JM, Khoury MA, Whetstone H et al. (2009) Modulating hedgehog signaling can attenuate the severity of osteoarthritis. Nat Med 15: 1421–1425.

50. Karlsson C, Brantsing C, Svensson T, Brisby H, Asp J, et al. (2007) Differentiation of human mesenchymal stem cells and articular chondrocytes:

analysis of chondrogenic potential and expression pattern of differentiation-related transcription factors. J Orthop Res 25: 152–163.

51. Chu CR, Szczodry M, Bruno S (2010) Animal models for cartilage regeneration and repair. Tissue Eng Part B Rev 16: 105–115.

52. Grigolo B, Lisignoli G, Desando G, Cavallo C, Marconi E, et al. (2009) Osteoarthritis treated with mesenchymal stem cells on hyaluronan-based scaffold in rabbit. Tissue Eng Part C Methods 15: 647–658.

53. Mueller-Rath R, Gavenis K, Gravius S, Andereya S, Mumme T, et al. (2007) In vivo cultivation of human articular chondrocytes in a nude mouse-based contained defect organ culture model. Biomed Mater Eng 17: 357–366.

Functional Characterization of Detergent-Decellularized Equine Tendon Extracellular Matrix for Tissue Engineering Applications

Daniel W. Youngstrom[1], Jennifer G. Barrett[2]*, Rod R. Jose[3], David L. Kaplan[3]

1 Department of Biomedical and Veterinary Sciences, Virginia-Maryland Regional College of Veterinary Medicine, Virginia Tech, Leesburg, Virginia, United States of America, 2 Department of Large Animal Clinical Sciences, Marion duPont Scott Equine Medical Center, Virginia-Maryland Regional College of Veterinary Medicine, Virginia Tech, Leesburg, Virginia, United States of America, 3 Department of Biomedical Engineering, Tissue Engineering Resource Center, Tufts University, Medford, Massachusetts, United States of America

Abstract

Natural extracellular matrix provides a number of distinct advantages for engineering replacement orthopedic tissue due to its intrinsic functional properties. The goal of this study was to optimize a biologically derived scaffold for tendon tissue engineering using equine flexor digitorum superficialis tendons. We investigated changes in scaffold composition and ultrastructure in response to several mechanical, detergent and enzymatic decellularization protocols using microscopic techniques and a panel of biochemical assays to evaluate total protein, collagen, glycosaminoglycan, and deoxyribonucleic acid content. Biocompatibility was also assessed with static mesenchymal stem cell (MSC) culture. Implementation of a combination of freeze/thaw cycles, incubation in 2% sodium dodecyl sulfate (SDS), trypsinization, treatment with DNase-I, and ethanol sterilization produced a non-cytotoxic biomaterial free of appreciable residual cellular debris with no significant modification of biomechanical properties. These decellularized tendon scaffolds (DTS) are suitable for complex tissue engineering applications, as they provide a clean slate for cell culture while maintaining native three-dimensional architecture.

Editor: Nuno M. Neves, University of Minho, Portugal

Funding: The authors would like to acknowledge partial funding by the Virginia Tech Institute for Critical Technology and Applied Science (ICTAS), the Morris Animal Foundation, and the Tissue Engineering Resource Center (TERC) at Tufts University [NIH P41 EB002520]. The funders had no role in study design, data collection and analysis, decision to publish, or preparation of the manuscript.

Competing Interests: The authors have declared that no competing interests exist.

* E-mail: jgbarrett@vt.edu

Introduction

Extracellular matrix (ECM) has emerged as a fundamental tool for developing regenerative tissue prostheses [1]. Simultaneously, bioengineered scaffolds (either natural or synthetic) are critical to improving our understanding of the complex relationship between three-dimensional topographical and biomechanical environments and stem cell growth and differentiation [2,3]. Regardless of the intended application, decellularization protocols are designed to remove cells and debris while preserving the three-dimensional organization and ultrastructure of the extracellular matrix. Biomaterials based on stem cells seeded on decellularized tissue scaffolds are emerging as exciting options for clinical therapy, by obviating the need for traditional organ transplant or autologous donation techniques which are associated with significant morbidity [4–7]. Moreover, the study of stem cell/matrix interactions in a native scaffold environment furthers our understanding of basic cellular behavior and stem cell differentiation.

Tendon is an important target for tissue engineering due to its frequent involvement in musculoskeletal pathology and its comparatively simple organization. Tendon tissue repairs slowly, has a poor functional endpoint after healing, and often suffers re-injury [8]. Furthermore, there is an unmet need for functional and

readily integrated graft material for human patients that suffer from traumatic tendon rupture or loss [9,10]. The horse is an excellent model for tendon research due to its pathophysiological similarities with human degenerative orthopedic disease [11]. Additionally, there is a relatively large quantity of donor tissue available compared with other model animals, as well as significant clinical demand. Equine athletes routinely function close to the mechanical threshold for tendon damage [12] in a mildly hyperthermic environment [13], resulting in cumulative cellular and extracellular breakdown as well as changes in tissue biochemistry [14]. Tendinopathy results when this deterioration exceeds the capacity for restorative remodeling [15,16]. Since ECM components are highly conserved, potential immunogenic reactivity is minimal [17]. Furthermore, due to the high *in* vivo tensile load experienced by the equine flexor digitorum superficialis tendon (5% average strain at a speed of 7 m/s) [18], a maximal load on the order of 10 kN [12], and the low cellularity and vascularity of the tissue, it is a strong and homogeneous starting material as a source of xenogeneic scaffold material.

It has long been recognized that cells demonstrate a complex array of behaviors in response to culture on natural collagenous matrices [19]. This characteristic has been exploited in biotechnology, as natural matrices inherently possess bioactive factors that induce host-mediated healing [20,21]. MSC differentiation pathways are dependent on the structural and chemical composition of ECM [22], as well as its mechanical properties [23]. This is particularly true for tendon progenitor cells, whose niche is dominated by extracellular proteins [24]. Therefore, successful engineering of scaffolds for orthopedic research requires the correct physical and chemical environment to induce *de novo* tissue generation. Preparation of these tissues ideally removes cellular material, but leaves behind the critical fibrillar collagen ultrastructure, as well as the majority of glycosaminoglycans (GAGs), to aid in tissue regeneration.

Successful decellularization is a tissue-dependent procedure, and optimization of protocols has until now been lacking for dense strong tendon such as equine flexor digitorum superficialis. Nevertheless, testing protocols published for other fibrous tissues could translate into an optimized protocol for this novel tissue. Low concentrations of tri(*n*-butyl)phosphate (TnBP) and SDS have shown utility in a preliminary decellularization study of rat tail tendon [25]. An analogous study of porcine diaphragm tendon revealed that TnBP caused adequate loss of cellularity and preserved ECM architecture, while other commonly implemented detergents, including Triton X-100, did not [26]. Conversely, Triton X-100, when used in combination with other methods, reportedly aided in decellularization of flexor digitorum superficialis tendons in a chicken model [27], and has also been used in combination with SDS [28,29]. SDS has effectively removed cellular debris from connective tissues when other methods have failed [30], and has been used successfully in whole-organ [31,32] and even multi-organ [33] decellularization. Enzymatic decellularization without detergent exposure has also been documented for certain tissues [34–36], and it is common practice to combine physical agitation, chemical manipulation, and enzymatic digestion to achieve a finished product [17]. It is apparent that decellularization procedures must be optimized by tissue type and donor species, and tailored to the needs of the application.

Our aim was to test several different decellularization protocols and compare their ability to decellularize equine flexor digitorum superficialis tendon while maintaining collagen ultrastructure and minimizing loss of GAG content. Our hypothesis was that a treatment protocol using SDS would remove the majority of cellular debris without reducing collagen content or compromising structural organization of the DTS, preserving scaffold topography and mechanical properties. We anticipated higher detergent concentrations would result in GAG loss, so multiple concentrations were compared against each other and against other decellularization procedures referenced in the literature. We additionally hypothesized that our DTS would be compatible with allogeneic MSC culture, resulting in no significant loss of viability or proliferative potential.

Materials and Methods

Experimental Design

Equine flexor digitorum superficialis tendons were aseptically harvested from the forelimbs of eight castrated male sporting horses (mean age 12.8 ± 1.4 years) euthanized for conditions unrelated to musculoskeletal disease. All procedures were approved by the Institutional Animal Care and Use Committee of Virginia Tech. Tendons were longitudinally sectioned into 400 µm-thick ribbons using a Padgett Model B electric dermatome (Integra Lifesciences, USA). Samples were sectioned into squares approximately 3 cm^2 in area and were randomly assigned to treatment groups. With three replicates of each control and treated scaffold per horse, a total of 144 tendon samples were

required for initial characterization. Each outcome variable was performed in triplicate on samples from each of the eight horses.

Two small representative sections were removed from every scaffold sample for biochemical characterization. These sections were dehydrated overnight in an 80°C oven, weighed, and placed into low binding affinity microcentrifuge tubes (Eppendorf, USA). One portion from each sample (4.7 mg mean dry mass) was digested in 1 mL of 1 mg/mL papain [37] (Sigma, USA) for 36 hours in a heated water bath at 65°C, while the other (2.8 mg mean dry mass) was digested with 0.1 mg/mL pepsin (Sigma, USA) in 0.5 M acetic acid for 48 hours at 4°C. All concentrations were normalized to scaffold dry mass. The remaining portions of the original scaffold samples were reserved for microscopy.

Following biochemical and microscopic characterization, the optimal protocol (2% SDS) was selected for additional comparison versus the untreated tendon control, including analysis of ultrastructure, biocompatibility and cellular integration. This was performed in triplicate on material from four horses.

Scaffold Preparation

Tendon samples were assigned to either the untreated control group and immediately frozen at −80°C, or to a treatment group consisting of one of the following detergent types: (1) phosphate buffered saline [PBS] (Lonza, USA) control; (2) 1% tri(n-butyl)phosphate [TnBP] (Aldrich, USA); (3) 1% SDS (Sigma, USA) plus 0.5% Triton X-100 (Sigma, USA) [SDS/TX100]; (4) 1% SDS; (5) 2% SDS. All detergents were buffered in 1 M Tris-HCl (Fisher Scientific, USA), pH 7.8. In a previous study, we determined that no live cells and no residual mRNA were present in tendon samples after four freeze/thaw cycles [37]. Following that procedure, samples were individually suspended in 2 mL of their assigned detergent types in six-well plates in a refrigerated shaker at 4°C for 48 hours to allow adequate perfusion of tendon matrices. Tendon samples were then rinsed in PBS six times to remove residual detergent [38] before incubation with 0.05% trypsin-EDTA (Gibco, USA) for 10 minutes and washed with water. Samples were then stored at 4°C for 24 hours with the addition of 1 µl/mL of a mammalian tissue protease inhibitor cocktail (Sigma, USA). Additional treatment steps included incubation in DNase-I (STEMCELL Technologies, Canada) for 30 minutes and 95% ethanol (Sigma, USA) for two hours at 4°C, separated with and followed by three 10-minute washes in water. Treatment steps were conducted in a gyratory shaker (New Brunswick Scientific, USA). Samples were divided for each assay as described above.

Biochemical Characterization

DNA content, as a marker for cell debris, was quantitatively measured in the papain scaffold digests following a single ethanol-based extraction technique using a fluorometric dye, Quant-iT PicoGreen (Molecular Probes, USA) in a ratio of 170 µL working solution to 30 µL samples/standards in a 96-well plate.

Protein content in 1:10 dilution papain digests was measured via a standard Bradford absorbance assay with Coomassie Brilliant Blue G-250 (Fisher Scientific, USA). 50 µL of each sample was combined with 200 µL reagent solution in a 96-well plate, referencing type-I rat tail collagen (Gibco, USA) as a standard.

Solubilized collagen following digestion was assessed in 100 µL aliquots of acid/salt-washed pepsin scaffold digests using a Sircol kit (Biocolor Ltd., UK), which is based on the specific binding of Sirius red to the $[Gly-x-y]_n$ triple helix motif characteristic of collagen.

Sulfated GAG content in the papain digests was quantified through a spectrophotometric assay based on 1,9-dimethylmethylene blue (Sigma, USA) and compared to a standard curve of chondroitin sulfate A from bovine trachea (Sigma, USA), with a combination of 50 μL samples/standards and 200 μL DMMB solution. Data was in all cases normalized to scaffold dry weight, obtained prior to digestion.

Microscopic Evaluation

Samples for microscopy underwent fixation in 4% paraformaldehyde and were submitted for paraffin embedding, sectioning at 5 μm, and routine histological staining (Histoserv, Inc., USA). Longitudinal cross sections were stained with hematoxylin and eosin (H&E) or Masson's trichrome. Images were acquired using standard brightfield techniques on an Olympus IM inverted microscope.

DNA was microscopically observed in tendon samples with ethidium homodimer-1 (EthD-1) (Invitrogen, USA) [528EX, 617EM] under an AMG EVOS$_{FL}$ digital inverted fluorescence microscope (AMG, USA). Identical brightness and exposure settings were used for each image. A similar procedure was replicated for cell culture analysis, with the addition of calcein, AM [494EX, 517EM] via an Invitrogen Live/Dead cytotoxicity assay. Cell viability over a period of four days was assessed by a combination of live/dead fluorescence and the CellTiter 96 assay (Promega, USA).

Scanning Electron Microscopy

Samples for SEM were dehydrated in a graded ethanol series (15%, 30%, 50%, 70%, 95%, and 100%), critical-point dried in CO_2, and sputter coated with gold. Samples were visualized in an FEI Quanta 600 FEG scanning electron microscope and representative images of scaffold ultrastructure were acquired.

Biomechanics

Mechanical testing was conducted on duplicate samples from native tendon and DTS from four horses, with a mean gauge length of 16.15 ± 0.29 mm, 5.79 ± 0.14 mm width, and 439 ± 17 μm thickness. Scaffolds underwent failure testing at a 1 mm/s extension rate parallel to the fiber orientation while submerged in a 37°C PBS bath on an Instron 3366 (Instron, USA) using a 100 N load cell and pneumatic clamps with 100-grit sandpaper. A tangent modulus of 0.003 MPa was set for 0% strain following conversion from the raw load/extension relationship. Ultimate tensile stress was measured empirically, and its corresponding strain was calculated using a fourth-order polynomial fit to the stress/strain curve. Elastic modulus was reported at the maximum first derivative of this relationship, and yield point was determined by the corresponding maximum of the second derivative.

Biocompatibility

Tendon samples (0.4 cm^2) from the freeze/thaw group and from the 2% SDS experimental group were dehydrated and sterilized in the graded ethanol series as above, rehydrated in sterile PBS, and tested for biocompatibility with equine MSCs according to the procedures below. For longer-term cell culture, 1×4.5 cm DTS sections were prepared in accordance with the 2% SDS-based decellularization procedure. Additionally, a section of each scaffold was soaked in 0.05% Tween-20 (Fisher Scientific, USA), and tested with a commercial limulus amoebocyte lysate-based chromogenic endotoxin assay (GenScript, USA). Values were compared to positive and negative controls, and referenced to a broad-range standard curve of endotoxin.

Cell Culture

Bone marrow aspirate was obtained from the sternum of a 3 year old male horse and processed using routine stem cell separation procedures [37] that were approved by the Institutional Animal Care and Use Committee. Bone marrow-derived MSCs were cultured in low-glucose GlutaMAX DMEM with 110 μg/mL sodium pyruvate (Gibco, USA) supplemented with 10% MSC FBS (Gibco, USA) and 100 U/mL sodium penicillin, 100 μg/mL streptomycin sulfate (Sigma, USA) at 37°C, 5% CO_2, and 90% humidity. Cells were expanded and passaged twice at 80% confluence, then 250,000-cell aliquots were directly seeded over the longitudinal surface of the scaffolds in a 200 μL meniscus of the same media. Samples were then incubated for 48 hours, after which scaffolds were transferred to fresh containers. Adhesion efficiency was assessed by performing a manual count of non-adherent cells in a hemocytometer following scaffold transfer and trypsinization of the culture wells. A CellTiter 96 assay was performed for to quantify viable cells after four days. A live/dead cell staining kit was also used to visualize scaffolds using fluorescence microscopy on day four as described above.

Cellular integration within engineered tissue constructs was examined after an 11-day culture period. An autologous MSC/DTS pair from a 9 year old female horse was cultured at low density (20,000 cells/cm^2) in the same media as previously described with the addition of 35.7 μg/mL ascorbic acid and clamped on both ends. The sample was incubated for a seeding period of 3 days followed by 8 days of immersion in media, with the solution changed every 3 days. At the experimental end point, the sample was collected and analyzed histologically to assess cellularity deep in the scaffold.

Statistical Analysis

Quantitative data is presented as mean ± standard error for all reported assays. Differences across treatment means for each of the eight research horses were assessed using one-way repeated measures multivariate analysis of variance (MANOVA). Data was grouped by significance with $P \leq 0.05$ via alphabetical notation in applicable figures; data points that share a letter are not statistically different from one another. Additionally, decellularized groups were individually tested against the untreated controls for each of the biochemical, mechanical, and cytocompatibility outcome variables with a one-way t-test. Statistical significance was declared for those values that were found to meet the criteria of $\alpha \leq 0.01$ and $\beta \leq 0.02$, noted with an asterisk. All analyses were performed using the commercial software programs JMP 9 (SAS Institute Inc., USA), Excel 12 (Microsoft, USA), and Mathematica 8 (Wolfram, USA).

Results

2% SDS Removes Cellular Debris from Tendon Explants

Histological sections stained with H&E are displayed alongside equivalent EthD-1 labeled scaffolds in the first figure. Untreated tendon histology shows tenocyte nuclei exhibiting a characteristic elongated morphology, in parallel alignment with collagen fibrils in lacunae. Nuclear remains of resident tenocytes are evident, as blue staining in the H&E sections, and as red fluorescence following EthD-1 hybridization due to membrane disruption from repetitive freeze/thaw cycles. A marked reduction in DNA content was demonstrated in all three SDS experimental groups (Figure 1). Quantitative analysis of these data was performed using Image J software (National Institutes of Health, USA), and this difference was statistically significant (data not shown). Most notably, the 2% SDS decellularized group retained only 0.11 ± 0.05 μg/mg resid-

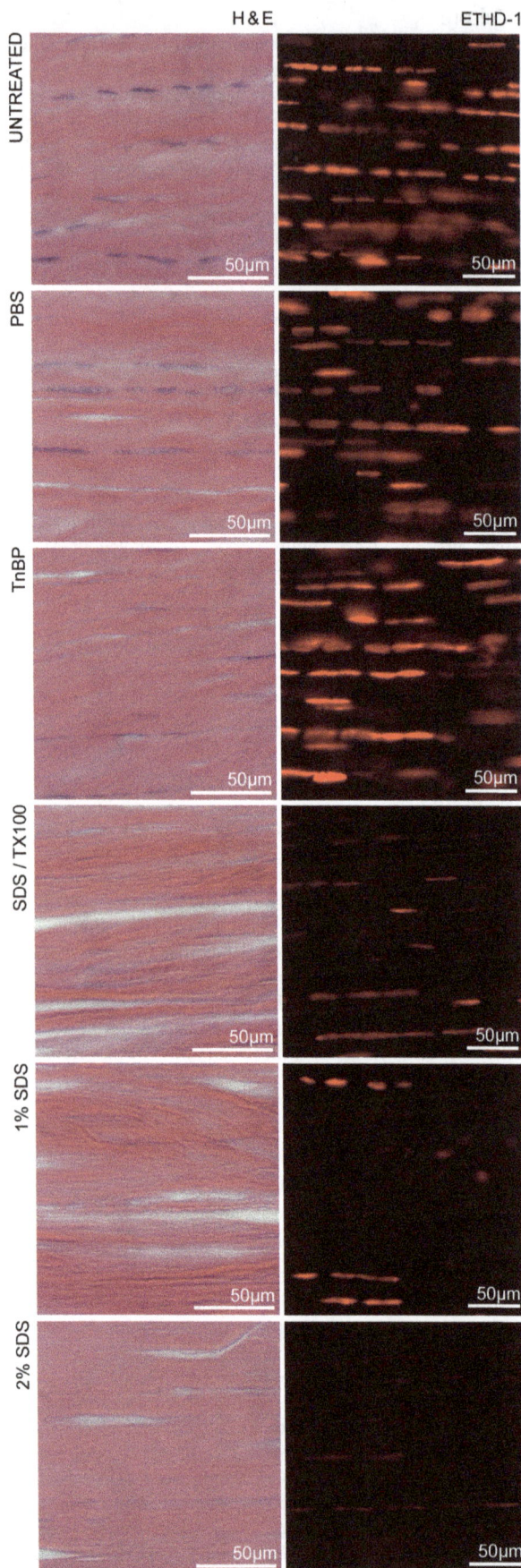

Figure 1. Loss of nuclear content following decellularization. Brightfield images of H&E-stained histological cross-sections (5 µm-thick) and fluorescence micrographs of scaffolds imaged with EthD-1 (400 µm-thick). Marked loss of DNA content is evident in SDS-decellularized experimental groups, with 2% SDS resulting in the most dramatic decellularization. All images are representative and were acquired from the same horse for ease of comparison.

ual DNA compared with 0.67 ± 0.12 µg/mg in untreated tendon. Subjectively, no major alterations in scaffold architecture were noted in histological samples.

Tendon Composition is not Significantly Altered by Decellularization

Tendon samples were assessed for the following biochemical outcome variables: DNA, protein, soluble collagen, and GAG levels (Figure 2). All detergent-treated groups exhibited a statistically significant reduction in DNA versus untreated tendon. The 2% SDS treatment reduced DNA content to $16.8 \pm 6.9\%$ of untreated tendon, and this reduction was significant ($p = 0.0081$ with 0.0023 sphericity, power of >0.999). There was some protein loss in the SDS-treated samples, reaching significance ($p < 0.001$) in the 1% concentration group and approaching significance ($p = 0.0052$) in the 2% group. There was no significant difference between SDS treatment groups ($p = 0.30$). The majority of GAG content was preserved. The 2% SDS decellularized tendon resulted in a small reduction in GAG content (2.83 ± 1.64 µg/mg decrease) with $p = 0.042$ and a power of 0.86.

Decellularized Tendon Maintains Desired Ultrastructure

Because the 2% SDS treatment maintained collagen content and resulted in minimal GAG loss and maximal DNA removal, this treatment was chosen for further characterization. Untreated tendon (Figure 3A) underwent structural comparison with scaffolds decellularized with 2% SDS. The characteristic deep blue coloration of collagen after staining with Masson's trichrome was apparent even in decellularized tendon, consistent with maintenance of collagen content (Figure 3B). Samples were also evaluated via scanning electron microscopy (Figure 3C), revealing no microscale topographical differences in tendon structure aside from a subjective increase in porosity as observed in the transverse plane. No statistically significant alterations in tensile properties were observed between native equine tendon and 2% SDS decellularized scaffolds. A summary of biomechanical outcome variables is included in Table 1.

Scaffolds are Biocompatible and Free of Endotoxin

Scaffolds decellularized with 2% SDS were compared to untreated tendon for assessment of biocompatibility (Figure 4). Prior to testing, all samples were screened and tested negative for endotoxin (data not shown). There was no significant difference in seeding efficiency between decellularized tendon and untreated tendon scaffolds; $45.9 \pm 7.0\%$ of plated cells adhered to decellularized scaffolds compared with $38.9 \pm 3.3\%$ for untreated tendon ($p = 0.77$). Cell counts on day four post-seeding indicated a global mean cell density of $420,000 \pm 17,600$ cells per square centimeter of scaffold with no statistical significance between untreated and 2% SDS treated tendon. Fluorescence micrographs on day four using a live/dead cell staining kit demonstrated a healthy population of MSCs with no detectable cell death (Figure 4C). The untreated tendon had background staining of DNA, as expected. Subjectively, seeded cells exhibited an elongated morphology in parallel with the aligned scaffolds.

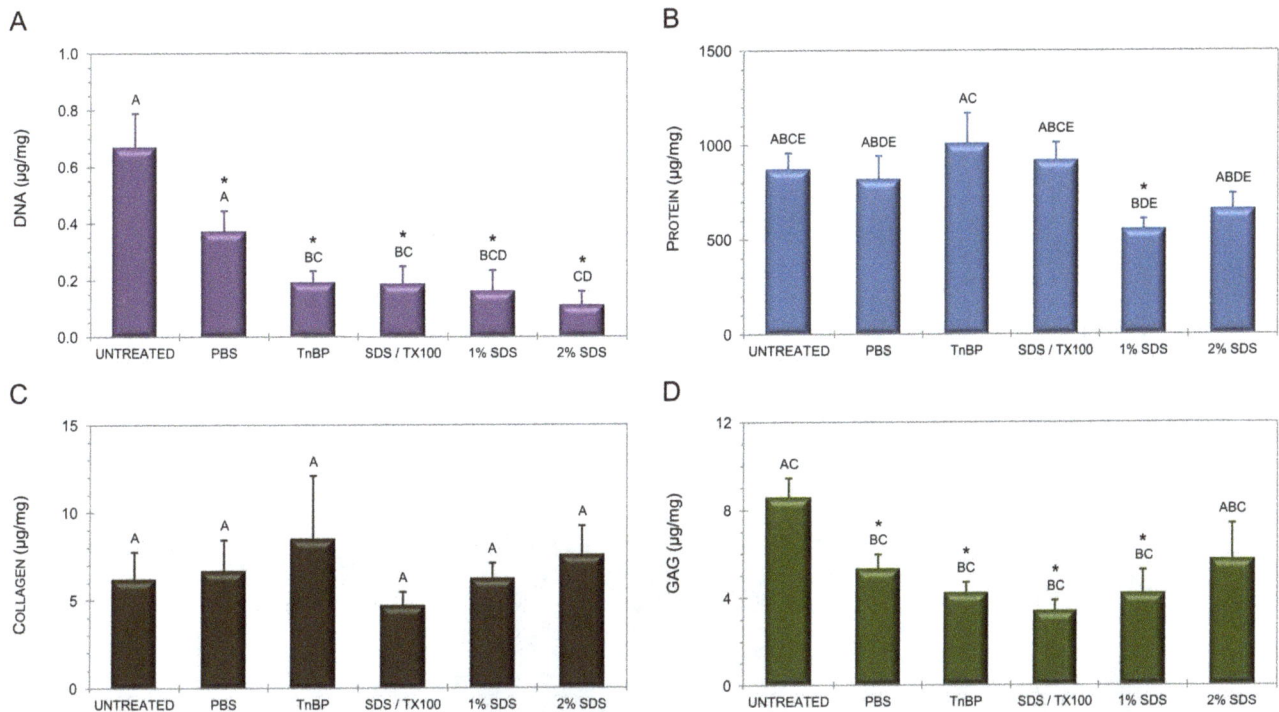

Figure 2. Treatment alters tendon matrix composition. Biochemical analysis of tendon scaffolds, including DNA (A), total protein (B), soluble collagen (C), and GAG content (D). Statistical significance between untreated control and treatment is annotated by use of an asterisk. Statistical differences between treatments are indicated by different letters (repeated measures MANOVA).

MSCs Proliferate on and Penetrate Deep into DTS

Culturing autologous MSCs on DTS over 11 days demonstrated marked cellular proliferation and integration, as visualized histologically (Figure 5). This strongly supports the suitability for DTS as a platform for tissue engineering applications.

Discussion

DTS provides a structurally similar physical environment to native tendon with minimal residual cellular debris. Engineered scaffolds including DTS are geared to provide a tunable microenvironment in which we may study fundamental development as well as cell-mediated mechanisms of tissue regeneration. Additionally, advanced scaffolds such as DTS may prove valuable in extending the phenotypic stability of tenocytes [39] and tendon stem cells [40], which experience drift over extended culture periods [41]. Our novel protocol resulted in formation of biocompatible, acellular constructs that will prove valuable in this pursuit. There are several ECM-based scaffolds on the market for clinical use, but most have demonstrated limited clinical efficacy as standalone products [7] and no tissue-specific commercial options for tendon and ligament engineering or augmentation are available from animal sources [42]. Moving forward, optimization of tendon ECM decellularization focusing on structural stability and accessibility to cell seeding are crucial to the development of useful reconstructive scaffold materials [43]. Equine DTS represents an improvement to current competing technologies in that regard, as it provides a unique combination of long graft length, mechanical strength, and inexpensive small-scale production.

We evaluated the effect of several candidate detergents in combination with physical and enzymatic cell disruption, followed by a series of wash steps and sterilization in ethanol, on longitudinal tendon ribbons. These sections were acquired from

the core regions of tendon explants to take maximum advantage of intrinsic properties including comparatively low cell density and vascularity as well as uniform ultrastructure and mechanics compared with other regions such as epitenon [44–46]. As 300 µm is reported as the minimum desirable tendon scaffold thickness from a biomechanical standpoint [47], we selected a thickness of 400 µm to compensate for potential loss of surface ultrastructure following decellularization. This thickness allows complex tissue construction techniques such as stacking [48], rolling [49], and forming composites [50], yet is strong enough to facilitate early range of motion. And, it would allow for repair of partial tears, and replacement of small tendons such as flexor tendons of the hand via a rolling or stacking technique. Scaffolds underwent rigorous comparison with validated methods for efficacy of removal of cellular debris, maintenance of ECM composition and functional properties, and compatibility for allogeneic MSC culture. 2% SDS treatment, in combination with trypsinization, treatment with DNase-I, and ethanol sterilization, resulted in optimal characteristics.

DNA was selected as a sensitive indicator of cell debris due to its high stability, and common use as a proxy for other cellular debris. While DNA is not cytotoxic, it is a useful index that strongly correlates with adverse host immunogenicity [51]. EthD-1 was selected as a marker for nuclear content following confirmation that no live cells or mRNA remained after four freeze/thaw cycles [37]. The data indicates that 2% SDS removed a mean 83% of DNA from the tendon scaffolds, and 1% SDS resulted in a 76% reduction. This represents an improvement over other effective published decellularization protocols for tendon (67% [27]). SDS in combination with Triton X-100 removed 72% of DNA, and TnBP was the least effective detergent-based method with 71%. Detergent-free decellularization (PBS group) removed only 44% of nuclear content. These numbers correlate with fluorescence

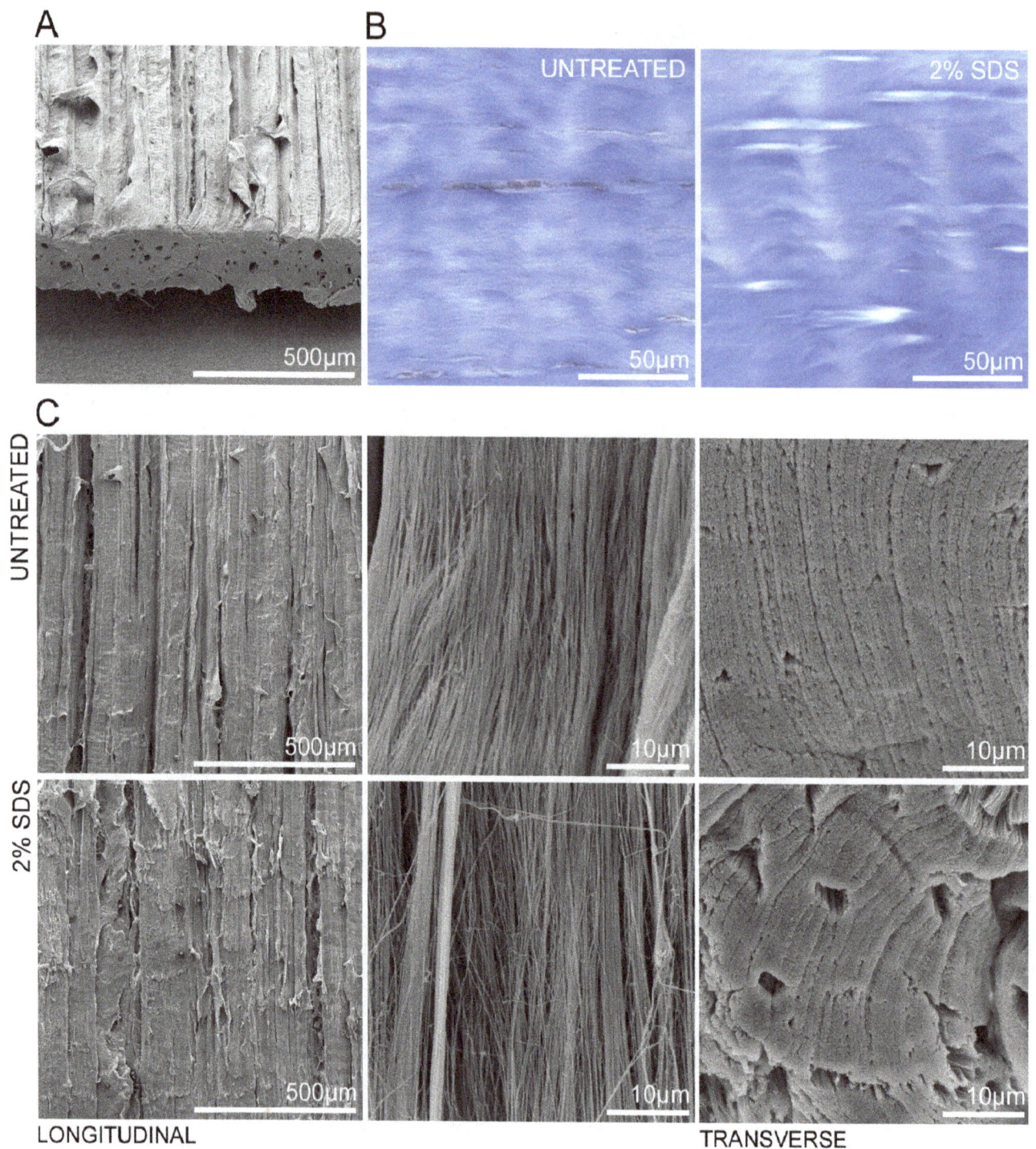

Figure 3. Scaffold ultrastructure. (A) Scanning electron micrograph of untreated tendon strip, angled to show both longitudinal and transverse section architecture. (B) Histological sections stained with Masson's trichrome indicate maintenance of collagen content with a slight increase in porosity. (C) Comparison of SEM images obtained from untreated and 2% SDS decellularized tendon samples, shown at 100x and 5000x magnifications longitudinally, and 5000x transversely. Collagen ultrastructure is not adversely altered by detergent treatment.

images of intact scaffolds. However, it is interesting to note that residual TnBP may enhance EthD-1 fluorescence, as the intensity displayed on the representative micrograph does not correspond to results obtained with complimentary techniques, including H&E staining.

The Bradford assay demonstrated a mean 24% reduction of total protein, which may correspond to loss of soluble intracellular proteins, chromosomal proteins such as histones, organelle proteins such as lysosomal proteases and/or noncollagenous extracellular matrix proteins. The statistical significance of the protein loss following 1% SDS decellularization most likely

Table 1. DTS biomechanics.

	Native	DTS
Yield Strain (%)	8.337±0.142	7.829±0.442
Failure Strain (%)	8.712±0.510	8.851±0.216
Yield Stress (MPa)	5.774±0.440	5.782±0.775
Failure Stress (MPa)	6.029±0.414	6.045±0.759
Elastic Modulus (MPa)	76.13±4.12	70.31±5.91

Tensile testing indicated no statistically significant alterations in scaffold mechanics following decellularization. Data represents mean ± standard error values.

It is interesting to note that our procedure leaves biologically relevant residual GAG content, as experiments in porcine ACL demonstrated that SDS decellularization dramatically reduced sulfated GAG content compared with untreated tissue [52]. Maintenance of GAG content is desirable due to the role of the molecules in signal modulation, water composition, and tissue biomechanics [53]. The DMMB assay demonstrated a 33% reduction in GAG content which approached statistical significance (as it reached in other experimental groups), but residual GAG content was much higher than anticipated. Statistical differences between the 1% and 2% SDS-treated groups in protein and GAG outcomes may be attributed to the relatively small sample size. A canine tendon decellularization protocol that did not use detergent has been shown to maintain native proteoglycan content [36], but our data indicates that equine tendon requires detergent decellularization, which may fortunately still be accomplished while preserving the integrity of functional macromolecules. This highlights the necessity for tissue-specific optimization of decellularization protocols.

corresponds to intracellular proteins or non-collagenous matrix proteins, as soluble collagen content following DTS digestion was statistically unchanged with all experimental groups failing to approach any of the statistical significance criteria.

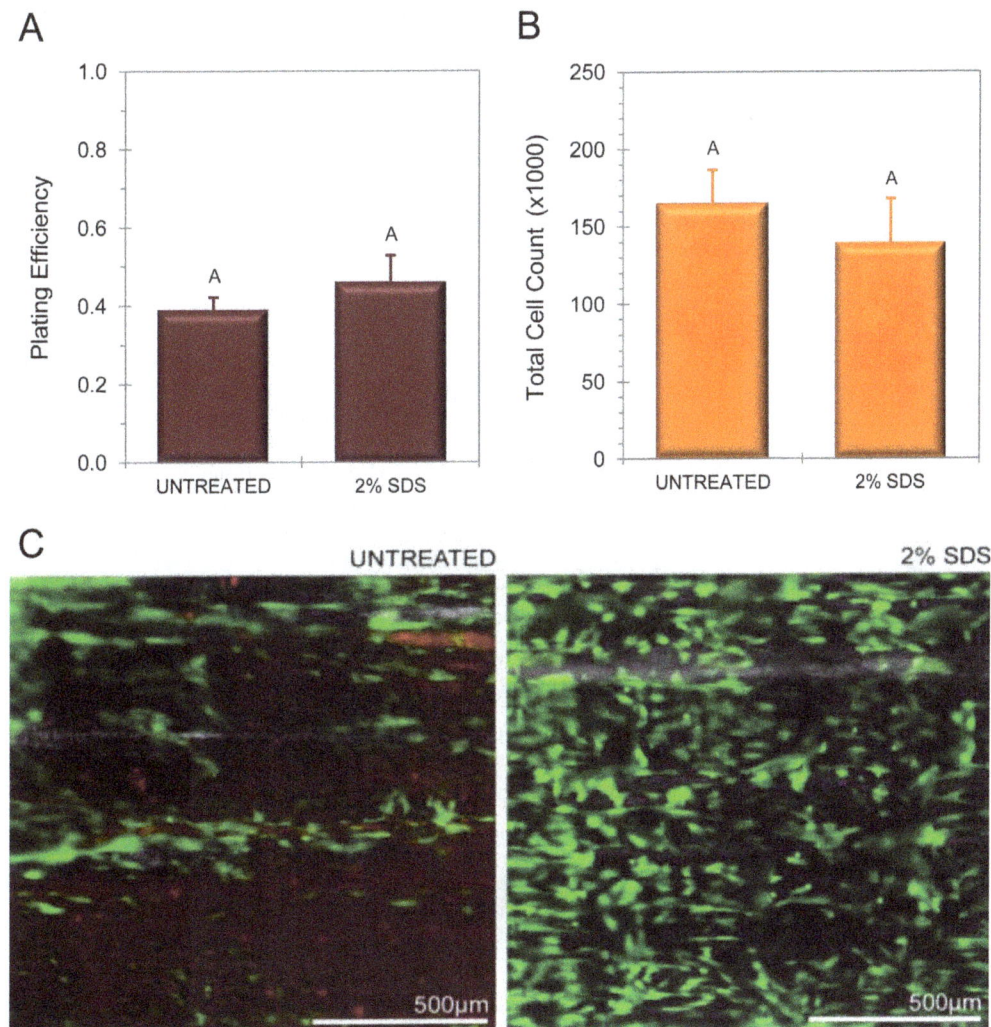

Figure 4. Biocompatibility as assessed with MSC culture. (A) Plating efficiency calculated as a ratio of MSCs adherent to scaffolds following a 48-hour incubation period. (B) Total cell count at 96 hours obtained via MTS assay, indicating no significant reduction in proliferation or metabolic activity on DTS. (C) Fluorescence micrographs portraying representative samples of untreated and decellularized tendon, with live cells stained green (calcein) and dead cells stained red (EthD-1).

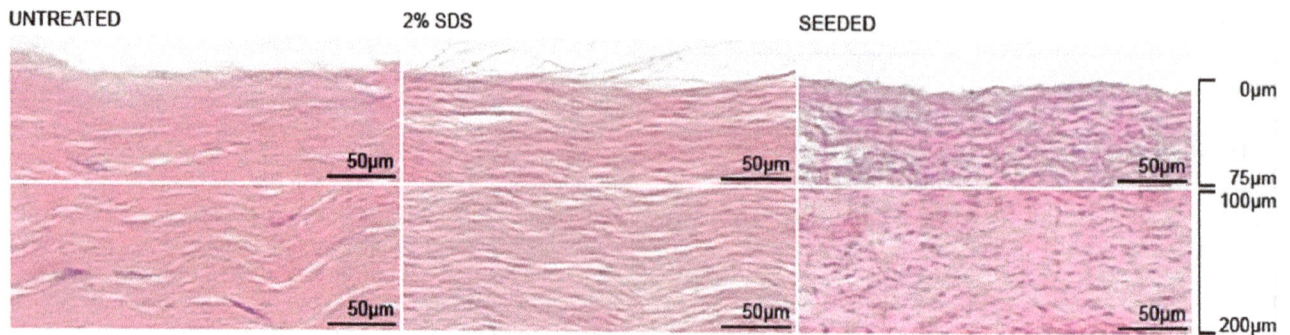

Figure 5. Cellular integration. H&E staining demonstrates infiltration of MSCs deep into DTS following 11 days of static tissue culture.

Scanning electron microscopy allowed analysis of potential ultrastructural modifications to decellularized scaffolds. Subjectively, SEM images did not demonstrate disruption of collagen alignment or fiber thickness at the micrometer level, though a slight reduction in fibril density was evident in the SDS-treated group that may prove beneficial for seeding scaffolds with cells. This may correspond with the 18% increase in seeding efficacy noted in subsequent cell culture, as the increase in porosity and surface area was hospitable toward MSCs. Histological sections stained with Masson's trichrome revealed similar staining of ECM, indicative of native collagen composition, while also suggesting a subtle increase in porosity, supporting the electron micrographs.

Biomechanical parameters of DTS did not significantly deviate from control tissue, contrary to results reported in other decellularization protocols [54]. An insignificant extension of the toe region and a decrease in yield strain was observed (p = 0.16) with a slight increase in elastic modulus (p = 0.11), indicating trends that may emerge with a larger sample size. Interestingly, these data suggest the potential for safe elongation protocols up to 7% strain before plastic deformation is observed, with an ideal linear extension region centered on 4.1±0.2% strain (data not shown). This characterization is necessary for experiments involving mechanical manipulation of cell-seeded constructs.

While endotoxin contamination has not been associated with long-term derogative effects in tissue graft integration or remodeling [55], testing is essential for evaluation of biomaterials destined for bioreactor cell culture and medical device applications [56]. Our DTS contained no detectable levels of endotoxin. In order to assess the initial feasibility of using DTS as a raw material for tissue engineering, we seeded our scaffolds with MSCs for four days due to their frequent use in cell-based orthopedic tissue engineering studies, and the possibility that autologous MSCs could be used on DTS as a therapeutic graft material [57]. MSC culture was successful in terms of plating efficiency, cell proliferation, and viability. As stated, DTS had a higher seeding efficiency than identically sectioned control tissue, perhaps due to increased porosity resulting from the decellularization process. For viability, no dead cells were seen adherent to the scaffolds among 2% SDS DTS samples. A statistical comparison of cell viability on

DTS versus control tissue could not be performed due to the presence of high background staining of endogenous dead cells in non-decellularized control tendon. The influence of long-term culture on the growth and differentiation of MSC-seeded DTS constructs has only begun to be explored, and will be the subject of further investigation. However, the results of an 11-day study of autologous MSC/DTS constructs showed excellent cellular integration throughout the depth of the scaffold and further supports its lack of cytotoxicity as a scaffold material.

Acellular ECM represents a versatile, biocompatible scaffold for three-dimensional cell culture. This study validated a novel natural scaffold for tendon tissue engineering purposes. Equine flexor digitorum superficialis tendon is well suited for use in tissue engineering applications, bioreactor cell culture studies and graft material as a result of its robust mechanical properties, homogeneous low cellularity, low vascularity, and long length and width. Our protocol, based on 2% SDS detergent decellularization in conjunction with free/thaw lysis, trypsinization, DNase-I digestion, and ethanol sterilization induces practical acellularity without compromising functionality. The remaining extracellular matrix material maintains the biochemical composition, ultrastructure, and mechanics of native tendon, yet has minimal residual cellular debris. This provides a clean slate for subsequent cell culture, allowing exploitation of the features intrinsic to physiological tendon matrix without concern over functional or immunological interference from the original resident cells.

Acknowledgments

The authors would like to acknowledge the ICTAS Nanoscale Characterization and Fabrication Laboratory (NCFL) and the Morphology Laboratory at the Virginia-Maryland Regional College of Veterinary Medicine for their technical assistance with this study.

Author Contributions

Conceived and designed the experiments: DWY JGB RRJ DLK. Performed the experiments: DWY JGB RRJ. Analyzed the data: DWY JGB RRJ DLK. Contributed reagents/materials/analysis tools: DWY JGB RRJ DLK. Wrote the paper: DWY JGB RRJ DLK.

References

1. Weber B, Emmert MY, Schoenauer R, Brokopp C, Baumgartner L, et al. (2011) Tissue engineering on matrix: future of autologous tissue replacement. Semin Immunopathol 33: 307–315.
2. Kolf CM, Cho E, Tuan RS (2007) Mesenchymal stromal cells. Biology of adult mesenchymal stem cells: regulation of niche, self-renewal and differentiation. Arthritis Res Ther 9: 204.
3. Guilak F, Cohen DM, Estes BT, Gimble JM, Liedtke W, et al. (2009) Control of stem cell fate by physical Interactions with the extracellular matrix. Cell Stem Cell 5: 17–26.
4. Zhang J, Li B, Wang JH (2011) The role of engineered tendon matrix in the stemness of tendon stem cells in vitro and the promotion of tendon-like tissue formation in vivo. Biomaterials 32: 6972–6981.

5. Ahmad Z, Wardale J, Brooks R, Henson F, Noorani A, et al. (2012) Exploring the application of stem cells in tendon repair and regeneration. Arthroscopy 28: 1018–1029.

6. Caplan AI (2007) Adult mesenchymal stem cells for tissue engineering versus regenerative medicine. J Cell Physiol 213: 341–347.

7. Chen J, Xu J, Wang A, Zheng M (2009) Scaffolds for tendon and ligament repair: review of the efficacy of commercial products. Expert Rev Med Devices 6: 61–73.

8. Blevins FT, Djurasovic M, Flatow EL, Vogel KG (1997) Biology of the rotator cuff tendon. Orthop Clin North Am 28: 1–16.

9. McIntosh JK, Jablons DM, Mule JJ, Nordan RP, Rudikoff S, et al. (1989) In vivo induction of IL-6 by administration of exogenous cytokines and detection of de novo serum levels of IL-6 in tumor-bearing mice. J Immunol 143: 162–167.

10. Loppnow H, Flad HD, Durrbaum I, Musehold J, Fetting R, et al. (1989) Detection of interleukin 1 with human dermal fibroblasts. Immunobiology 179: 283–291.

11. Patterson-Kane JC, Becker DL, Rich T (2012) The pathogenesis of tendon microdamage in athletes: the horse as a natural model for basic cellular research. J Comp Pathol 147: 227–247.

12. Dowling BA, Dart AJ (2005) Mechanical and functional properties of the equine superficial digital flexor tendon. Vet J 170: 184–192.

13. Birch HL, Wilson AM, Goodship AE (1997) The effect of exercise-induced localised hyperthermia on tendon cell survival. J Exp Biol 200: 1703–1708.

14. Birch HL, Bailey AJ, Goodship AE (1998) Macroscopic 'degeneration' of equine superficial digital flexor tendon is accompanied by a change in extracellular matrix composition. Equine Vet J 30: 534–539.

15. Thorpe C, Clegg P, Birch H (2010) A review of tendon injury: why is the equine superficial digital flexor tendon most at risk? Equine Vet J 42: 174–180.

16. Thornton G, Hart D (2011) The interface of mechanical loading and biological variables as they pertain to the development of tendinosis. J Musculoskelet Neuronal Interact 11: 94–105.

17. Gilbert TW, Sellaro TL, Badylak SF (2006) Decellularization of tissues and organs. Biomaterials 27: 3675–3683.

18. Butcher MT, Hermanson JW, Ducharme NG, Mitchell LM, Soderholm LV, et al. (2009) Contractile behavior of the forelimb digital flexors during steady-state locomotion in horses (Equus caballus): an initial test of muscle architectural hypotheses about in vivo function. Comp Biochem Physiol A Mol Integr Physiol 152: 100–114.

19. Kleinman HK, Klebe RJ, Martin GR (1981) Role of collagenous matrices in the adhesion and growth of cells. J Cell Biol 88: 473–485.

20. Hodde J (2002) Naturally occurring scaffolds for soft tissue repair and regeneration. Tissue Eng 8: 295–308.

21. Kleinman HK, Philp D, Hoffman MP (2003) Role of the extracellular matrix in morphogenesis. Curr Opin Biotechnol 14: 526–532.

22. Santiago JA, Pogemiller R, Ogle BM (2009) Heterogeneous differentiation of human mesenchymal stem cells in response to extended culture in extracellular matrices. Tissue Eng Part A 15: 3911–3922.

23. Engler AJ, Sen S, Sweeney HL, Discher DE (2006) Matrix elasticity directs stem cell lineage specification. Cell 126: 677–689.

24. Bi Y, Ehirchiou D, Kilts TM, Inkson CA, Embree MC, et al. (2007) Identification of tendon stem/progenitor cells and the role of the extracellular matrix in their niche. Nat Med 13: 1219–1227.

25. Cartmell JS, Dunn MG (2000) Effect of chemical treatments on tendon cellularity and mechanical properties. J Biomed Mater Res 49: 134–140.

26. Deeken CR, White AK, Bachman SL, Ramshaw BJ, Cleveland DS, et al. (2011) Method of preparing a decellularized porcine tissue using tributyl phosphate. J Biomed Mater Res B Appl Biomater 96: 199–206.

27. Whitlock PW, Smith TL, Poehling GG, Shilt JS, Van Dyke M (2007) A naturally derived, cytocompatible, and architecturally optimized scaffold for tendon and ligament regeneration. Biomaterials 28: 4321–4329.

28. Woods T, Gratzer PF (2005) Effectiveness of three extraction techniques in the development of a decellularized bone-anterior cruciate ligament-bone graft. Biomaterials 26: 7339–7349.

29. Rieder E, Kasimir MT, Silberhumer G, Seebacher G, Wolner E, et al. (2004) Decellularization protocols of porcine heart valves differ importantly in efficiency of cell removal and susceptibility of the matrix to recellularization with human vascular cells. J Thorac Cardiovasc Surg 127: 399–405.

30. Elder BD, Eleswarapu SV, Athanasiou KA (2009) Extraction techniques for the decellularization of tissue engineered articular cartilage constructs. Biomaterials 30: 3749–3756.

31. Ott HC, Matthiesen TS, Goh SK, Black LD, Kren SM, et al. (2008) Perfusion-decellularized matrix: using nature's platform to engineer a bioartificial heart. Nature Med 14: 213–221.

32. Uygun BE, Soto-Gutierrez A, Yagi H, Izamis M-L, Guzzardi MA, et al. (2010) Organ reengineering through development of a transplantable recellularized liver graft using decellularized liver matrix. Nat Med 16: 814–820.

33. Park KM, Woo HM (2012) Systemic decellularization for multi-organ scaffolds in rats. Transplant Proc 44: 1151–1154.

34. Schenke-Layland K, Vasilevski O, Opitz F, Konig K, Riemann I, et al. (2003) Impact of decellularization of xenogeneic tissue on extracellular matrix integrity for tissue engineering of heart valves. J Struct Biol 143: 201–208.

35. Omae H, Zhao C, Sun YL, An KN, Amadio PC (2009) Multilayer tendon slices seeded with bone marrow stromal cells: a novel composite for tendon engineering. J Orthop Res 27: 937–942.

36. Ning LJ, Zhang Y, Chen XH, Luo JC, Li XQ, et al. (2012) Preparation and characterization of decellularized tendon slices for tendon tissue engineering. J Biomed Mater Res A 100: 1448–1456.

37. Stewart AA, Barrett JG, Byron CR, Yates AC, Durgam SS, et al. (2009) Comparison of equine tendon-, muscle-, and bone marrow-derived cells cultured on tendon matrix. Am J Vet Res 70: 750–757.

38. Cebotari S, Tudorache I, Jaekel T, Hilfiker A, Dorfman S, et al. (2010) Detergent decellularization of heart valves for tissue engineering: toxicological effects of residual detergents on human endothelial cells. Artif Organs 34: 206–210.

39. Yao L, Bestwick CS, Bestwick LA, Maffulli N, Aspden RM (2006) Phenotypic drift in human tenocyte culture. Tissue Eng 12: 1843–1849.

40. Lui PPY (2012) Identity of tendon stem cells – how much do we know? J Cell Mol Med 20: 1–10.

41. Halfon S, Abramov N, Grinblat B, Ginis I (2011) Markers distinguishing mesenchymal stem cells from fibroblasts are downregulated with passaging. Stem Cells Dev 20: 53–66.

42. Place ES, Evans ND, Stevens MM (2009) Complexity in biomaterials for tissue engineering. Nat Mater 8: 457–470.

43. Schulze-Tanzil G, Al-Sadi O, Ertel W, Lohan A (2012) Decellularized tendon extracellular matrix - a valuable approach for tendon reconstruction? Cells 1: 1010–1028.

44. Abrahamsson SO, Lundborg G, Lohmander LS (1989) Segmental variation in microstructure, matrix synthesis and cell proliferation in rabbit flexor tendon. Scand J Plast Reconstr Surg Hand Surg 23: 191–198.

45. Kraus-Hansen AE, Fackelman GE, Becker C, Williams RM, Pipers FS (1992) Preliminary studies on the vascular anatomy of the equine superficial digital flexor tendon. Equine Vet J 24: 46–51.

46. Birch HL, Smith TJ, Poulton C, Peiffer D, Goodship AE (2002) Do regional variations in flexor tendons predispose to site-specific injuries? Equine Vet J 34: 288–292.

47. Qin TW, Chen Q, Sun YL, Steinmann SP, Amadio PC, et al. (2012) Mechanical characteristics of native tendon slices for tissue engineering scaffold. J Biomed Mater Res B Appl Biomater 100: 752–758.

48. Tran RT, Thevenot P, Zhang Y, Gyawali D, Tang L, et al. (2010) Scaffold sheet design strategy for soft tissue engineering. Nat Mater 3: 1375–1389.

49. Alberti KA, Xu Q (2012) Slicing, stacking and rolling: fabrication of nanostructured collagen constructs from tendon sections. Adv Healthc Mater, Dec 12 [Epub ahead of print].

50. Godier-Furnemont AF, Martens TP, Koeckert MS, Wan L, Parks J, et al. (2011) Composite scaffold provides a cell delivery platform for cardiovascular repair. Proc Natl Acad Sci U S A 108: 7974–7979.

51. Crapo PM, Gilbert TW, Badylak SF (2011) An overview of tissue and whole organ decellularization processes. Biomaterials 32: 3233–3243.

52. Gratzer PF, Harrison RD, Woods T (2006) Matrix alteration and not residual sodium dodecyl sulfate cytotoxicity affects the cellular repopulation of a decellularized matrix. Tissue Eng 12: 2975–2983.

53. Badylak SF (2002) The extracellular matrix as a scaffold for tissue reconstruction. Semin Cell Dev Biol 13: 377–383.

54. Pridgen BC, Woon CY, Kim M, Thorfinn J, Lindsey D, et al. (2011) Flexor tendon tissue engineering: acellularization of human flexor tendons with preservation of biomechanical properties and biocompatibility. Tissue Eng Part C Methods 17: 819–828.

55. Daly KA, Liu S, Agrawal V, Brown BN, Huber A, et al. (2012) The host response to endotoxin-contaminated dermal matrix. Tissue Eng Part A 18: 1293–1303.

56. Gorbet MB, Sefton MV (2005) Endotoxin: the uninvited guest. Biomaterials 26: 6811–6817.

57. Caplan AI (2005) Review: mesenchymal stem cells: cell-based reconstructive therapy in orthopedics. Tissue Eng 11: 1198–1211.

3D Non-Woven Polyvinylidene Fluoride Scaffolds: Fibre Cross Section and Texturizing Patterns Have Impact on Growth of Mesenchymal Stromal Cells

Anne Schellenberg[1], Robin Ross[2], Giulio Abagnale[1], Sylvia Joussen[1], Philipp Schuster[2], Annahit Arshi[2], Norbert Pallua[3], Stefan Jockenhoevel[2,4], Thomas Gries[2], Wolfgang Wagner[1]*

1 Stem Cell Biology and Cellular Engineering, Helmholtz-Institute for Biomedical Engineering, RWTH Aachen University Medical School, Aachen, Germany, 2 Institute for Textile Technology RWTH Aachen University, Aachen, Germany, 3 Department of Plastic and Reconstructive Surgery, Hand Surgery, Burn Center, RWTH Aachen University, Aachen, Germany, 4 Department of Applied Medical Engineering, Helmholtz-Institute for Biomedical Engineering, RWTH Aachen University, Aachen, Germany

Abstract

Several applications in tissue engineering require transplantation of cells embedded in appropriate biomaterial scaffolds. Such structures may consist of 3D non-woven fibrous materials whereas little is known about the impact of mesh size, pore architecture and fibre morphology on cellular behavior. In this study, we have developed polyvinylidene fluoride (PVDF) non-woven scaffolds with round, trilobal, or snowflake fibre cross section and different fibre crimp patterns (10, 16, or 28 needles per inch). Human mesenchymal stromal cells (MSCs) from adipose tissue were seeded in parallel on these scaffolds and their growth was compared. Initial cell adhesion during the seeding procedure was higher on non-wovens with round fibres than on those with snowflake or trilobal cross sections. All PVDF non-woven fabrics facilitated cell growth over a time course of 15 days. Interestingly, proliferation was significantly higher on non-wovens with round or trilobal fibres as compared to those with snowflake profile. Furthermore, proliferation increased in a wider, less dense network. Scanning electron microscopy (SEM) revealed that the MSCs aligned along the fibres and formed cellular layers spanning over the pores. 3D PVDF non-woven scaffolds support growth of MSCs, however fibre morphology and mesh size are relevant: proliferation is enhanced by round fibre cross sections and in rather wide-meshed scaffolds.

Editor: Masaya Yamamoto, Institute for Frontier Medical Sciences, Kyoto University, Japan

Funding: This work was supported by the Stem Cell Network North Rhine Westphalia, by the Else-Kröner-Fresenius Stiftung, and by the German Research Foundation (WA 1706/3-2 within SPP1327). The funders had no role in study design, data collection and analysis, decision to publish, or preparation of the manuscript.

* E-mail: wwagner@ukaachen.de

Introduction

Mesenchymal stromal cells (MSCs) raise high expectations in regenerative medicine. They can easily be expanded *in vitro*, comprise a subset with multilineage differentiation potential often referred to as "mesenchymal stem cells", and they have immunomodulatory properties [1–4]. Usually, MSCs are culture expanded on tissue culture plastic (TCP) – particularly on 2D polystyrene surfaces. For therapeutic applications the cells are then harvested and injected in suspension. However, tissue engineering of complex lesions or interventions aimed at repairing hierarchically organized tissues requires the implantation of cells in a suitable scaffold which facilitates 3D cell integration [5–7].

A multitude of biomaterials has been used in tissue engineering. Hydrogels and fibrous scaffolds based on synthetic or natural polymers were shown to be suitable for MSC expansion [8–10]. Fibre based structures represent a promising approach for tissue engineering due to their close resemblance to native extracellular matrix (ECM)[11]. Such fibre based structures have large surface areas and porosity that can be adjusted to the specific cellular requirements [12]. Hence, these biomaterials are intensively studied for numerous applications in tissue engineering, including ligament repair [13,14], bone and cartilage regeneration [15,16] and soft tissue replacement [17,18]. Depending on the application it is advantageous to either use a biodegradable material which is resorbed in the course of tissue regeneration and restructuring, or to rather use a non-biodegradable material if long-term stability and functionality is required. Non-degradable meshes of polymers can be used to coat a broad range of implants. Vascular stents, for example, have to remain attached and integrated into the surrounding tissue to keep the lumen permanently opened. Various biostable polymers such as polyethylene terepthalate (PET), have been used to coat stents [19].

Polyvinylidene Fluoride (PVDF) has been used as suture material [20,21], for the construction of hernia meshes [22,23] and for mechanical supporting meshes in vascular tissue engineering [24]. It provides good biocompatibility, is biologically inert, non-toxic, non-degradable and resistant to bacterial infections [25]. In contrast to other polymers such as polypropylene (PP) or PET, PVDF shows less inflammatory reactions, less fibrotic tissue formation and it is more resistant to hydrolysis and degradation [20–23]. Furthermore, monofilament sutures out of

PVDF reveal very good long-term stability under tension and can be sterilized by beta or gamma radiation [26].

Previous studies demonstrated that surface patterning and 3D composition of tissue engineering scaffolds have major impact on cellular behavior [27]. Under *in vivo* conditions, the extracellular microenvironment (e.g. basement membrane and ECM) is not flat, but rather arranged in semi-aligned sheets with grooves, ridges and pores [28,29]. It has been shown that mechanical cues such as micro-patterns and substrate elasticity influence cell growth and differentiation [30,31]. In fact, various cell types such as MSCs, osteoblasts, fibroblasts and endothelial cells align, elongate and migrate along structured surfaces [32]. Non-woven structures which are mechanically bonded together by entangling fibres resemble some of the characteristics of ECM and may therefore be promising scaffolds for tissue engineering [33,34]. So far little is known about the effect of fibre cross section geometry or pore architecture on MSCs growth and integration.

In the present study, highly porous PVDF non-wovens were produced with varying fibre cross section and crimp and subsequently used as scaffold for MSCs. We demonstrate that MSCs adhere and proliferate better on non-wovens with round fibre cross section, although the surface area on trilobal and snowflake cross section is significantly larger. Electron microscopy revealed that MSCs form small layers spanning over the non-woven pores. Overall, PVDF non-wovens support MSCs growth and differentiation and therefore represent suitable alternatives for tissue engineering.

Material and Methods

Ethics statement

MSCs from adipose tissue were isolated after patient's written consent using guidelines approved by the Ethic Committee of the University of Aachen (Permit number: EK163/07).

Manufacturing of PVDF non-woven scaffolds

Polymeric PVDF granules (PVDF Solef 1006 by Solvay Solexis S. A., Tavaux, France) were processed into multifilament fibres with 24 single fibres using the coextrusion spinning plant (Fourné Polymertechnik GmbH, Germany) in a single extrusion mode. Besides round shaped filaments also trilobal and snowflake-shaped fibres were produced. The respective spinneret geometries are shown in Figure 1A–C. A Fully-Drawn-Yarn (FDY) Take Up with one godet-duo and two heatable single godets were used. The temperature profile of the spin line was set as follows: 1st extruder zone 235°C, 2nd extruder zone 240°C, 3rd extruder zone 240°C, melt pipe 245°C, melt pump 250°C, spinning head 245°C. As spin finish Dryfi PP I (Schill+Seilacher GmbH, Germany) was used. For subsequent processing steps Silastol R641 (Schill+Seilacher GmbH, Germany) was applied to the PVDF fibres.

The production of non-woven fabrics consists of three main processing steps: texturing, web formation, and bonding (Figure 1B). Starting point for the non-woven process is the multifilament yarn mentioned above. To obtain a stable non-woven structure the yarn must be textured. A texturing-method which produces a homogeneous and reproducible wavelike texture is called "knit-deknit". Therefore, we used a circular knitting machine (TK 83 of Harry Lucas Textilmaschinenfabrik, Neu-münster, Germany) with various knitting parameters to adjust the amplitude and the wavelength of the texture. A permanent shape was facilitated by thermal fixation the yarn at 120°C for 5 minutes. After cutting the multifilament into staple fibres, the cut fibre bundles were separated into single fibres. This "opening process" was carried out by the lab-scaled card MDTA3 (Zellweger, Uster,

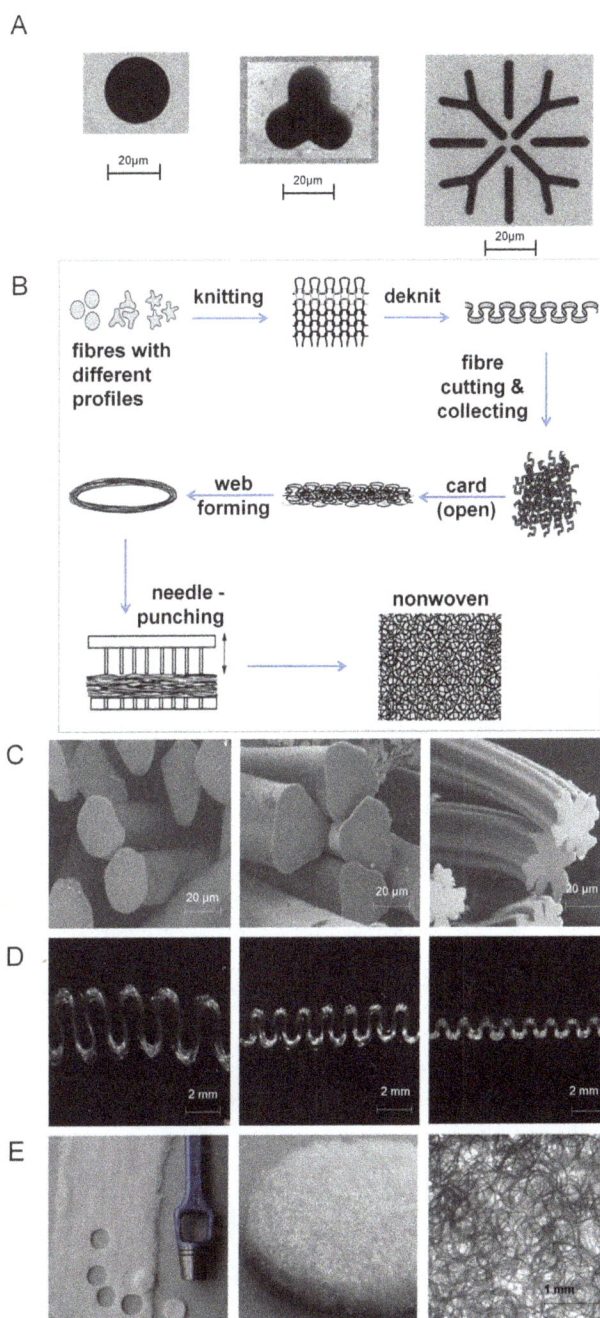

Figure 1. Manufacturing steps of three-dimensional PVDF non-wovens. Round shaped (f24 0.4 L/D 2, left), trilobal (Y24 250×552 L/D 2, middle), and snowflake (f24 L/D 2, right) spinnerets (A). Schematic overview of the fabrication process of non-wovens made of PVDF fibres (B). SEM pictures of round (left), trilobal (middle) and snowflake (right) non-wovens cross-sections (C). Fibre texturizing of fibres knitted with 10 (left), 16 (middle), or 28 (right) needles/inch (D). Round scaffolds are punched out of the non-woven fabrics (E).

Swiss) and repeated three times to obtain a homogenous web. Subsequently, this web was mechanically bonded by using a lab-scaled needle-punching machine designed at the Institute for Textile Technology Aachen (ITA). For the needle-punching, 15×18×40×3 FBD6 needles (Groz-Beckert, Albstadt, Germany) were selected. Finally, round scaffolds with a diameter of 15 mm were manually punched out of the non-woven fabrics (Figure 1E).

For production of non-structured, even PVDF substrates 1.5 g PVDF granules were melted at 200°C for 5 min and subsequently pressed for 1 min at 150°C. The PVDF were subjected to hydrogen peroxide gas plasma treatment in a plasma sterilizer (Sterrad 100 s, Johnson & Johnson, Brunswick, NJ, USA). The duration of the plasma treatment was 7 min (505 mTorr).

Isolation and characterisation of MSCs

MSCs were isolated from lipoaspirates from human healthy donors. In brief, lipoaspirates were washed in 9 g/L NaCl, centrifuged at 300 g for 10 minutes and the middle layer was subsequently digested with 2 g/L collagenase type I (Biochrom, Berlin, Germany) and with 15 g/L BSA (PAA, Pasching, Austria) for 45 min at 37°C under constant shaking. The digested tissue was passed through a 100 µm cell strainer and seeded in culture flasks (Nunc Thermo Fisher Scientific, Langenselbold, Germany). Culture medium consisted of Dulbecco's Modified Eagles Medium-Low Glucose (DMEM-Low Glucose; PAA, Pasching, Austria) with 2 mM L-glutamine (Gibco/Invitrogen, Eugene, OR, USA), 100 U/mL penicillin/streptomycin (Gibco/Invitrogen, Eugene, OR, USA) and 10% human platelet lysate (HPL) which was pooled from five platelet units of healthy donors as described before [35–37]. Cultures were maintained at 37°C in a humidified atmosphere containing 5% CO_2 with medium changes twice per week. MSCs were always harvested by trypsinization upon 80% confluent growth, counted with a Neubauer counting chamber (Brand, Wertheim, Germany) and re-seeded in a density of 10,000 cells/cm^2 in 75 cm^2 culture flasks. Immunophenotypic analysis of various surface markers (CD14, CD29, CD31, CD34, CD45, CD73, CD90, and CD105) and *in vitro* differentiation potential towards adipogenic and osteogenic lineages was validated for cell preparations as described in our previous work [38].

Seeding of MSCs on non-wovens

Non-wovens were placed in 6-well plates and disinfected by ethanol treatment. MSCs at passage 3 were then harvested by trypsinisation and about 19,000 cells were seeded with a pipette on PVDF non-wovens with a surface of 1.9 cm^2. Culture conditions were then used as described above with medium changes twice per week.

Proliferation analysis

Proliferation of MSCs on flat PVDF surface was quantified by cell counting: MSCs were seeded at a density of 10,000/cm^2 and cells were counted at day 5 with a Neubauer counting chamber with Trypan blue exclusion (3 technical replicas per condition). Cell population doublings (PDs) within 5 days were then estimated by the ratio of seeded *versus* harvested cells. Since not all cells can be harvested from the 3D non-wovens we have estimated proliferation within these scaffolds using the Alamar Blue assay according to manufacturer's instructions (Invitrogen, Eugene, OR,

USA) [39]. Cells were analysed at day 1, 5, 10 and 15 after seeding on the non-wovens. To exclude cells from the underlying TCP, the nonwovens were always transferred into a fresh well and then incubated with 1 ml culture medium containing 1×10^{-4} M resazurin (Invitrogen, Eugene, OR, USA) for 5 h at 37°C. The medium was removed from the scaffolds and fluorescence was measured with a Tecan Infinite 2000 plate reader at excitation and emission wavelengths of 560 nm and 590 nm, respectively.

Adipogenic and osteogenic differentiation

Adipogenic differentiation was induced by culture medium consisting of DMEM (PAA) containing 10% HPL, 0.5 mM isobutylmethylxanthine (IBMX; Sigma, St. Louis, MO, USA), 1 µM dexamethasone (Sigma) and 10 µM insuline (Sigma) as described before [40,41]. Osteogenic differentiation medium consists of DMEM-low glucose (PAA) with 2 mM L-glutamine (Sigma), 100 U/mL pen/strep (Lonza), 100 nM dexamethasone (Sigma), 200 µM L-ascorbic acid-2-PO$_4$ (Sigma), 10 mM β-glycerophosphate (Sigma). For staining of lipid droplets we initially tried the conventional staining method with Oil red but this dye has high affinity for the PVDF-substrates. Therefore, we stained fat droplets with the green fluorescent dye BODIPY (4,4-difluoro-1,2,5,7,8-pentamethyl-4-bora-3a,4a-diaza-s-indacene) counter-stained with DAPI (4',6-Diamidin-2-phenylindol; both Molecular Probes, Eugene, Oregon, USA). Fluorescence microscopy pictures were taken from representative areas. Images acquisition was performed using a Leica DM IL LED microscope (Leica, Wetzlar, Germany) with a 10x dry objective (numerical aperture: 0.3; Leica) and a camera (Leica DFC420C) equipped with Leica application suite 3.3.1 software. For analysis of calcium deposits upon osteogenic differentiation we tested Alizarin Red as described before [42].

Quantitative RT-PCR analysis

In addition, we validated adipogenic differentiation by increased gene expression of adiponectin (*ADIPOQ*), fatty acid binding protein 4 (*FABP4*) and peroxisome proliferator-activated receptor γ (*PPARγ*) after two weeks as described before [43]. Osteogenic differentiation was also addressed by the expression of alkaline phosphatase (*ALP*), runt-related transcription factor 2 (*RUNX2*), osteocalcin (*BGLAP*), and osterix (*OSX*). For each PVDF non-woven we have also analyzed MSCs without induction of differentiation. Expression levels were calculated for each sample in relation to *GAPDH* (ΔCT) [44]. To determine up-regulation of differentiation markers the expression levels were subsequently normalized to the corresponding undifferentiated controls (ΔΔCT).

Scanning electron microscopy

MCSs seeded on PVDF non-wovens were fixed in 3% glutaraldehyde for at least 1 hour at room temperature, rinsed

Table 1. Fibre Characteristics.

	draw-ratio	fineness of the multifilament yarn [dtex]	single fibre fineness [dtex]	single fibre fineness [µm]*	tensile strength [cN/dtex]	elongation at break [%]
round	2	180.9	7.5	23	2.3	61.1
trilobal	2.5	257.6	10.7	41	2.1	90.9
snowflake	2	175.7	7.3	29	2.1	39.9

*Mean single fibre fineness [µm] was calculated for the different fibre profiles by deviation of the ideally round fibre profile.

Table 2. Characterisation of non-wovens properties.

	pore size [µm]	standard deviation [µm]	pore number	standard deviation
round 10	246	19	107	95
round 16	192	84	93	151
round 28	203	15	80	31
trilobal 10	213	14	208	35
trilobal 16	229	13	82	94
trilobal 28	224	12	140	170
snow flake 10	211	7	226	107
snow flake 16	246	22	125	155
snow flake 28	254	26	83	38

with PBS and dehydrated by serial incubations in 30%, 50%, 70%, 90%, and 100% ethanol for 10 minutes at room temperature. The compounds were then critical point dried in liquid CO_2 and sputter-coated with a 30 nm gold layer using an ion sputter coater (LEICA EM SCD 500). Samples were analysed using an environmental scanning electron microscope at the electron microscope facility, RWTH Aachen University (ESEM XL 30 FEG, FEI, Philips, Eindhoven, Netherlands).

Statistics

Results are expressed as mean ± standard deviation of at least three independent experiments. To estimate the probability of differences we have adopted the paired two-sided Student's T-test.

Results

PVDF is a suitable biomaterial for MSCs

Before analysing MSC growth in 3D non-woven PVDF scaffolds we performed preliminary experiments with 2D PVDF-surfaces in comparison to normal polystyrene tissue culture plastic (TCP). MSCs did proliferate on PVDF substrates although the proliferation rate – estimated by measuring the population doublings (PDs) within five days – was significantly higher on conventional TCP. We compared two different PVDF preparations, PVDF1006 and PVDF1008 – the latter characterized by

lower viscosity and higher molecular weight (Solvay, Solexis 2006). No differences in MSC growth were observed between PVDF1006 and PVDF1008 (6.2 PDs, 4.7 PDs, and 4.9 PDs within 5 days on TCP, PVDF1006 and PVDF1008, respectively). Therefore we have used PVDF1006 for all subsequent experiments (Figure S1A). Processing of multifilament PVDF fibres requires a thin oil film coating step to prevent static charging. We considered that residual traces of oil might interfere with MSC growth on PVDF non-wovens, thus we analysed MSC growth on PVDF substrates treated with two distinct oil preparations (R641 and PP1) followed by three additional washing steps with PBS and either with or without additional washing steps with ultrasonic treatment. Overall, MSCs displayed similar proliferation rates on PVDF with different oil preparations and washing procedures (ranging from 4.5 PDs to 4.9 PDs within 5 days) indicating that in our experimental setting residual oil traces are either not present or do not have major impact on cell growth (Figure S1B). Alternatively, the PVDF substrates were subjected to plasma treatment to promote cell attachment and proliferation [45,46] but this did not increase MSC proliferation and therefore this approach was not adopted for the following experiments (Figure S1C). Taken together, 2D-PVDF is a suitable biomaterial to support growth of MSCs, but the proliferation rates are significantly lower than on conventional TCP.

Seeding efficiency of MSCs into PVDF non-wovens

Afterwards, we generated PVDF fibres with different cross-sectional shape (round, trilobal and snowflake). The draw-ratio (ratio between draw-off godet and feed-godet), fineness of the multifilament yarn, single fibre fineness, mechanical characteristics of tensile strength and elongation at break point of the as-spun fibres are shown in Table 1. Each of these fibres was further texturized by knit-de-knit procedures using 10, 16 or 28 needles/inch resulting in nine different well characterized PVDF non-wovens with different values of pore size and pore number (Table 2)(Figure 1D). The average fibre circumference varied significantly according to the different fibre profiles: 79.4 µm for round fibres, 96.6 µm for trilobal fibres and 150.8 µm for snow flake fibres. The average non-woven weight ranged from 61.3 mg to 74.7 mg with an average non-woven thickness of 2.8 mm (±0.2 mm).

These parameters might affect the seeding efficiency and the capability of MSCs to adhere to the scaffolds. To address this question, MSCs were seeded on the non-wovens by pipetting: after 15 minutes we determined the percentage of cells which failed to attach on the biomaterials because they passed through the pores

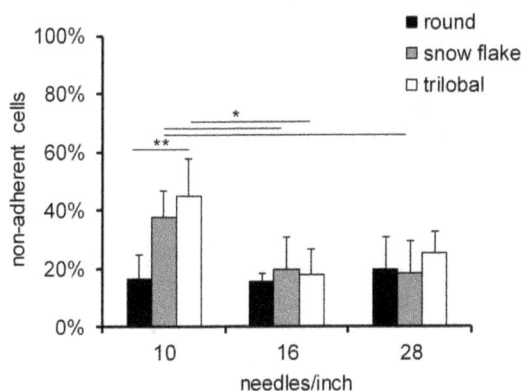

Figure 2. Cell loss upon initial seeding in PVDF non-wovens with different fibre texturizing. Percentage of non-adherent cells was determined by cell counting 15 min after seeding (n = 3). *p<0.05; **p<0.005.

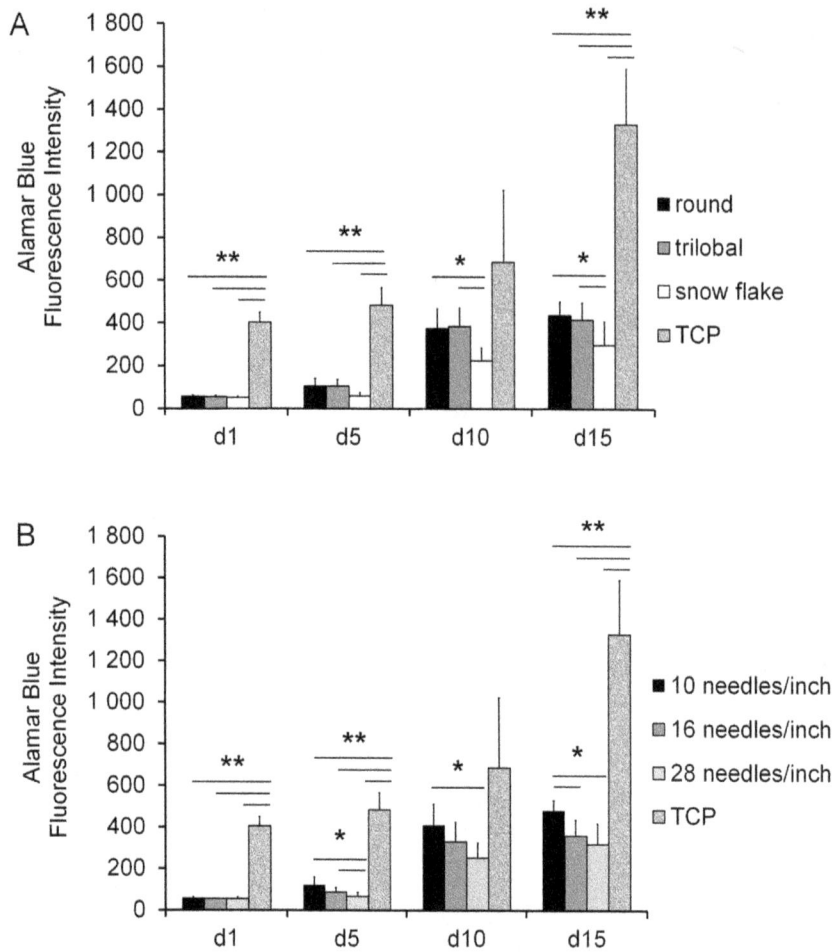

Figure 3. Proliferation rates of MSCs in 3D-PVDF non-wovens. Effect of fibre shape on cell growth was assessed with the Alamar Blue assay 1, 5, 10, and 15 days after seeding. Cells in non-wovens with round or trilobal fibres displayed higher proliferations rates at day 10 and 15. Diagram bars represent the mean value of non-wowens with 10, 16 and 28 needles/inch texturizing (**A**). Effect of different texturizing on cell growth: cells in non-wovens with 10 needles/inch texturizing resulted in the highest proliferation rates at day 5, 10 and 15. Diagram bars represent the mean value of non-wowens with round, trilobal or snowflake fibre shape (**B**; n = 3; *p<0.05; **p<0.005; ***p<0.0005).

of the scaffold and adhered to the underlying TCP. Adherent cells were harvested and counted with a Neubauer chamber. As expected, more cells attached to PVDF non-wovens with a denser meshwork. Notably, considerably more cells adhered to non-woven scaffolds with round fibre shape compared to trilobal fibres (n = 3; p = 0.038) indicating that the cross-sectional shape of fibres has impact on MSCs adhesion (Figure 2).

Proliferation of MSCs in different non-woven scaffolds

To determine whether MSCs are capable of proliferating within PVDF non-wovens we have first tried to harvest the MSCs from the scaffolds after one week of culture by trypsinisation. However, fluorescent microscopic analysis revealed that many MSCs were retained in the PVDF non-wovens despite this procedure. Therefore, we have estimated the proliferation rate using the Alamar Blue assay after 1, 5, 10, and 15 days. Overall, the proliferation rate was significantly higher for MSCs seeded on 2D TCP. Nevertheless, we observed a continuous increase of fluorescence intensity in all PVDF non-wovens indicating that MSCs do proliferate within these scaffolds. Interestingly, at day 10 and 15 the highest proliferation rate was observed on fibres with a round or trilobal cross section rather than on those with snowflake structures (Figure 3A). Furthermore, the proliferation rate was

significantly higher in wide-meshed non-wovens (knitted with 10 needles per inch) (Figure 3B).

MSCs span across the non-woven pores

MSCs growth within the 3D scaffolds was then analysed using scanning electron microscopy (Figure 4). As a general trend, cells aligned along the fibres during their growth and particularly on fibres with snow flake cross section they were arranged in individual fibres. In all non-woven scaffolds cells were preferentially located at the intersection of neighboring fibres. Notably, cell layers span across the non-woven pores displaying cellular sheets with tight cell-cell contacts; thus, in these cellular sheets, single cells are hardly distinguishable in contrast to the well-defined cell layers seen on 2D PVDF controls. Furthermore, we observed that cells seemed to be embedded into some kind of fibrous network, possibly extra-cellular matrix (ECM) proteins. This was observed on TCP and in 3D PVDF non-wovens indicating that the cells contribute to functionalize the scaffold by themselves.

Differentiation of MSCs within non-woven scaffolds

Lastly, we analyzed whether MSCs embedded in non-woven structures maintain their *in vitro* differentiation potential. For adipogenic differentiation formation of fat droplets can be

TCP	Round	Trilobal	Snow flake

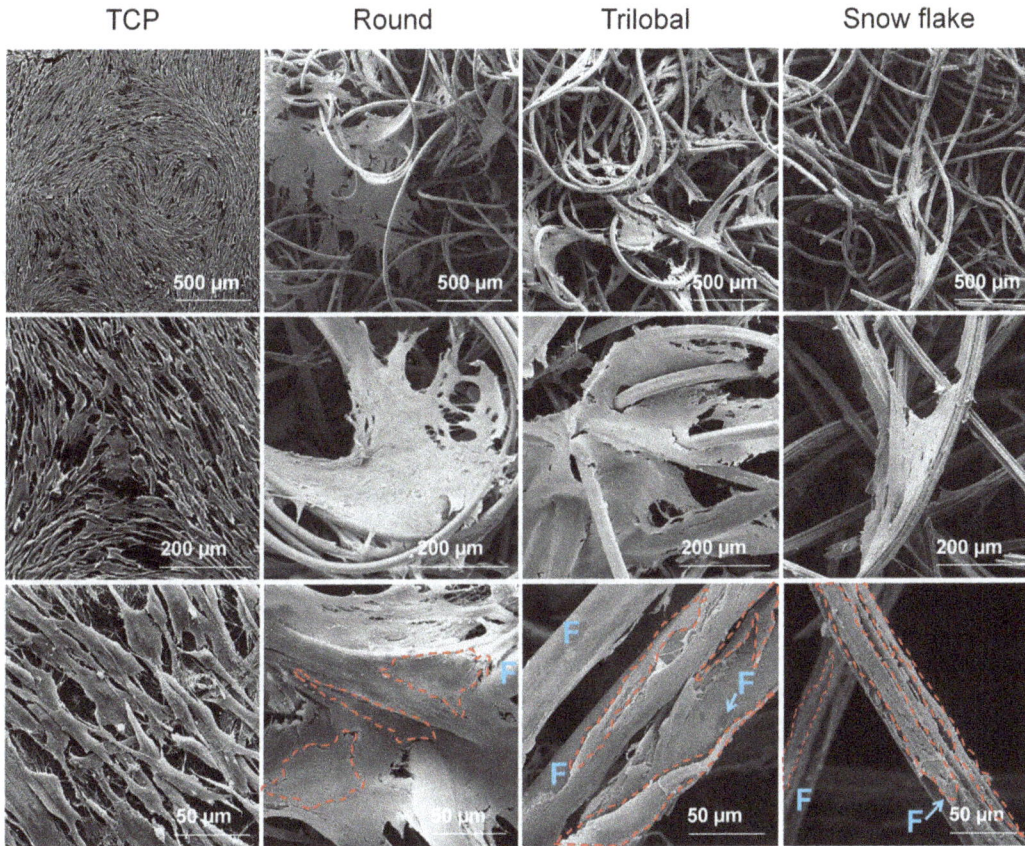

Figure 4. MSC growth on TCP and in 3D-PVDF non-wovens. SEM pictures showing MSC morphology on TCP and in PVDF non-wovens: MSCs span over non-woven pores forming large confluent cell layers. Photos at higher magnifications show that cells align along the fibres and accumulate at fibre intersections. Some individual cells are exemplarily depicted by red dotted lines and PVDF fibres are marked with F.

observed at single cell level. Such BODIPY positive cells were mainly found on the fibre intersections (Figure 5A). The frequency of differentiated *versus* non-differentiated cells appeared to be similar as compared to TCP control. Expression of adipogenic markers *FABP4*, *ADIPOQ* and *PPARγ* was highly up-regulated as compared to non-differentiated controls and there was no

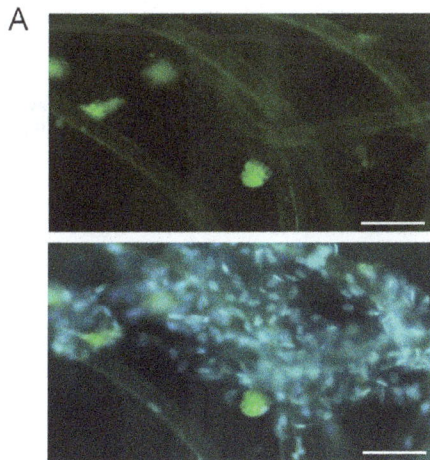

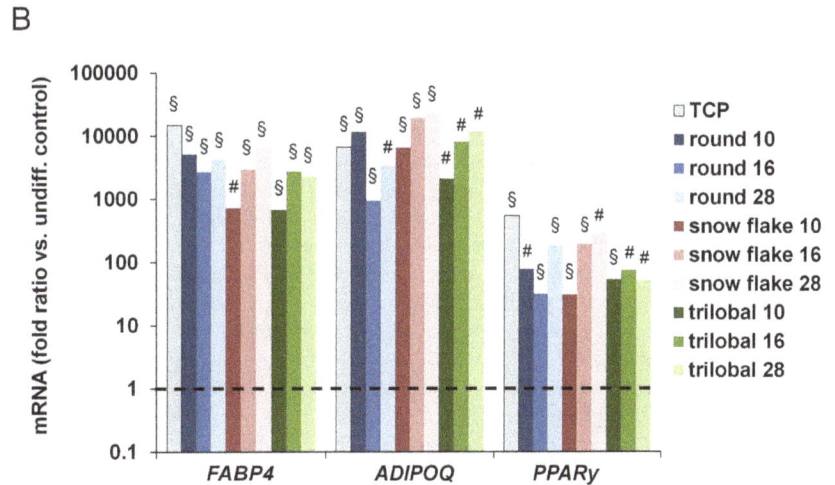

Figure 5. Adipogenic differentiation of MSCs in 3D-PVDF non-wovens. Fluorescence microscopy pictures of MSCs in PVDF non-wovens stained for lipid droplets with BODIPY (green) and nuclei with DAPI (blue) staining (**A**). Adipogenic differentiation was also validated on gene expression level for the adipogenic markers *FABP4, ADIPOQ* and *PPARγ*. Gene expression changes are demonstrated in relation to non-differentiated MSCs cultured on the corresponding substrate, indicated by the dotted line (**B**; * P<0.05; # P<0.01; § P<0.001; scale bars =100 μm).

difference between MSC cultured either on TCP or non-wovens (Figure 5B). However, analysis of osteogenic differentiation was hampered by the fact that Alizarin Red staining has high affinity to the PVDF-substrates (Figure S2A) and up-regulation of osteogenic markers was only moderate and not reliable (Figure S2B). This exemplifies, that analysis of in vitro differentiation is not trivial in 3D scaffolds. At least adipogenic differentiation potential of MSCs seems to be maintained when cultured on 3D non-woven scaffolds.

Discussion

The three-dimensional scaffold architecture of a biomaterial has major impact on cellular behavior and cell fate. Non-wovens represent a new concept of biomaterial. Due to their complex 3D conformation they can provide interesting perspectives for some surgical interventions. For example, PVDF non-woven scaffolds might be used as meshes in hernia repair. MSCs are known to mediate the wound healing process by supporting vascularization and by differentiating into many cell types. Studying in vitro the interactions between PVDF non-wovens and MSCs may be used to estimate cellularization of the implant also in vivo [47].

Polyvinylidene fluoride has been shown to be biocompatible with MSC growth before [48] and this finding was confirmed by our observations. Nonetheless MSCs gained significantly less PDs on PVDF substrates compared to TCP control. Commercially available TCP is usually treated with plasma gas to obtain more hydrophilic surfaces carrying functional groups to facilitate cell attachment. Previous studies reported that argon plasma treated surfaces promoted adhesion and proliferation of various cell types [45,46,49,50]. The results obtained in the present study indicate that the hydrogen peroxide plasma gas used here to treat the PVDF surface does not improve cell adhesion and proliferation on the substrate, but it is conceivable that alternative plasma treatment with argon gas would even further increase cell growth.

Upon initial seeding we observed the highest cell loss in scaffolds with lowest fibre texturizing density (10 needles/inch). These results are in line with the concept that less fibre crimp implies larger pores in the final scaffold and thereby increases the amount of cells falling through it upon seeding [51]. Moreover less fiber crimp leads to fewer fibre-fibre contacts and our results support the notion that cells preferably attach and localise at zones of fibre intersections [52,53]. The fibre shape hardly affected cellular attachment in scaffolds with high fibre texturizing. However, in scaffolds with larger pores, the round fibre cross sections displayed significantly higher attachment than trilobal and snowflake fibres. Initially, we expected that cells stick better to snowflake structures since they have the largest area and provide a structured surface made of parallel grooves. Considering our results we now hypothesize that surface patterning reduces the area to establish focal adhesions and therefore reduces cellular attachment.

Our results demonstrate that MSCs seeded in 3D-PVDF scaffolds remain viable and proliferate over an extended period of time. The highest proliferation rates were observed between day 5 and day 10, even though they were still much lower than on TCP. It has been reported that MSCs grown in 3D culture systems show a prolonged lag phase of about 5 days before they enter linear growth [54]. Furthermore we observed a correlation between different scaffold porosities and the degree of cell proliferation: fibre texturizing with less needles/inch results in a lower crimp wave length and the larger pores therefore facilitate higher permeability. Hence it is conceivable that the increased cell growth in scaffolds with fibre texturizing of 10 needles/inch result of higher nutrient and oxygen supply.

Scanning electron microscopy pictures revealed that MSCs on 3D-PVDF structures displayed a typical flat, fibroblast like morphology aligned along the fibres. Furthermore, they spanned sheet-like across pores - similar to 2D growth on conventional TCP – which renders it difficult to distinguish single cells. In analogy, previous studies reported that cells seeded in 3D scaffolds form multilayer sheets of cells embedded within a secreted extracellular matrix [55].

Matrix elasticity, surface chemistry and topography are parameters that can direct stem cells differentiation. McBeath et al. described how cellular morphology alone regulated commitment of MSCs [56]. Our osteogenic and adipogenic differentiation assays were initially performed using traditional staining methods with Alizarin Red and Oil Red, respectively. However, quantification of colour intensity is hard to assess inside of a complex 3D multi-layered fibrillar biomaterial. Moreover, PVDF showed high affinity for the Alizarin Red and Oil Red. Therefore, we analysed adipogenic differentiation using the green fluorescent BODIPY dye that allowed us to analyse cells at a single cell level. The number of cells with many fat droplets was relatively low but in a similar range as described in previous studies [38]. Overall, adipogenic differentiation of MSCs appeared to be very similar in 3D non-wovens as compared to TCP. Moreover, we did not observed spontaneous adipogenic differentiation of MSCs in non-woven scaffolds in normal culture medium.

Conclusions

Three-dimensional-PVDF non-woven structures represent a new kind of biomaterial that can be seeded with MSCs. Seeding affinity is higher on round cross sections and smaller pores. MSCs cultivated in 3D-PVDF non-woven scaffolds showed an impaired proliferation capacity but form coherent cell layers bridging the scaffold pores. These results support the notion that PVDF scaffolds are suitable biomaterials for implants in tissue engineering which support MSC growth.

Supporting Information

Figure S1 MSC growth on 2D-PVDF surfaces. MSC were seeded on TCP, PVDF1006 and PVDF1008 (1,000 cells/cm^2) and population doublings (PDs) within 5 days were estimated by cell counting (**A**). Likewise MSC proliferation was assessed on PVDF1006 substrates after treatment with the oil preparations R641 or PPI and subsequent removal of oil residues with ultrasonic treatment (**B**). Treatment of PVDF substrates with hydrogen peroxide plasma showed no increase in MSC proliferation (**C**).

Figure S2 Attempts to analyse osteogenic differentiation of MSCs in 3D-PVDF non-wovens. Upon Alizarin Red staining the PVDF non-wovens were rigorously washed. The image depicts non-specific staining even in non-differentiated controls which hampers reliable quantification (A). Osteogenic differentiation was alternatively estimated by gene expression of ALP, RUNX2, osteocalcin, and osterix. However, up-regulation of these markers was not reliable and consistent, too (B).

Acknowledgments

The authors would like to thank Manfred Bovi for excellent technical assistance in electron microscopy. We are grateful to Dr. Xiaomin Zhu, German Wool Research Institute (DWI; RWTH Aachen) for helping with the biomaterial press.

Author Contributions

Conceived and designed the experiments: AA PS S. Jockenhoevel TG WW. Performed the experiments: AS S. Joussen RR. Analyzed the data: AS RR WW. Contributed reagents/materials/analysis tools: NP. Wrote the paper: AS GA PS RR WW.

References

1. Dominici M, Le Blanc K, Mueller I, Slaper-Cortenbach I, Marini F, et al. (2006) Minimal criteria for defining multipotent mesenchymal stromal cells. The International Society for Cellular Therapy position statement. Cytotherapy 8: 315–317.
2. Salem HK, Thiemermann C (2010) Mesenchymal stromal cells: current understanding and clinical status. Stem Cells 28: 585–596.
3. Sensebe L, Krampera M, Schrezenmeier H, Bourin P, Giordano R (2009) Mesenchymal stem cells for clinical application. Vox Sang 98: 93–107.
4. de GL, Lucarelli E, Alessandri G, Avanzini MA, Bernardo ME, et al. (2013) Mesenchymal stem/stromal cells: a new "cells as drugs" paradigm. Efficacy and critical aspects in cell therapy. Curr Pharm Des 19: 2459–2473.
5. Karam JP, Muscari C, Montero-Menei CN (2012) Combining adult stem cells and polymeric devices for tissue engineering in infarcted myocardium. Biomaterials 33: 5683–5695.
6. Takahashi Y, Tabata Y (2004) Effect of the fiber diameter and porosity of non-woven PET fabrics on the osteogenic differentiation of mesenchymal stem cells. J Biomater Sci Polym Ed 15: 41–57.
7. Walenda G, Hemeda H, Schneider RK, Merkel R, Hoffmann B, et al. (2012) Human platelet lysate gel provides a novel three dimensional-matrix for enhanced culture expansion of mesenchymal stromal cells. Tissue Eng Part C Methods 18: 924–934.
8. Leisten I, Kramann R, Ventura Ferreira MS, Bovi M, Neuss S, et al. (2012) 3D co-culture of hematopoietic stem and progenitor cells and mesenchymal stem cells in collagen scaffolds as a model of the hematopoietic niche. Biomaterials 33: 1736–1747.
9. Neuss S, Apel C, Buttler P, Denecke B, Dhanasingh A, et al. (2008) Assessment of stem cell/biomaterial combinations for stem cell-based tissue engineering. Biomaterials 29: 302–313.
10. Pasquinelli G, Orrico C, Foroni L, Bonafe F, Carboni M, et al. (2008) Mesenchymal stem cell interaction with a non-woven hyaluronan-based scaffold suitable for tissue repair. J Anat 213: 520–530.
11. Tuzlakoglu K, Reis RL (2009) Biodegradable polymeric fiber structures in tissue engineering. Tissue Eng Part B Rev 15: 17–27.
12. Zong X, Bien H, Chung CY, Yin L, Fang D, et al. (2005) Electrospun fine-textured scaffolds for heart tissue constructs. Biomaterials 26: 5330–5338.
13. Cooper JA, Lu HH, Ko FK, Freeman JW, Laurencin CT (2005) Fiber-based tissue-engineered scaffold for ligament replacement: design considerations and in vitro evaluation. Biomaterials 26: 1523–1532.
14. Cardwell RD, Dahlgren LA, Goldstein AS (2012) Electrospun fibre diameter, not alignment, affects mesenchymal stem cell differentiation into the tendon/ligament lineage. J Tissue Eng Regen Med (Epub ahead of print).
15. Puppi D, Mota C, Gazzarri M, Dinucci D, Gloria A, et al. (2012) Additive manufacturing of wet-spun polymeric scaffolds for bone tissue engineering. Biomed Microdevices 14: 1115–1127.
16. Emans PJ, Jansen EJ, van Iersel D, Welting TJ, Woodfield TB, et al. (2013) Tissue-engineered constructs: the effect of scaffold architecture in osteochondral repair. J Tissue Eng Regen Med 7: 751–756.
17. Morgan SM, Ainsworth BJ, Kanczler JM, Babister JC, Chaudhuri JB, et al. (2009) Formation of a human-derived fat tissue layer in P(DL)LGA hollow fibre scaffolds for adipocyte tissue engineering. Biomaterials 30: 1910–1917.
18. Chandrasekaran AR, Venugopal J, Sundarrajan S, Ramakrishna S (2011) Fabrication of a nanofibrous scaffold with improved bioactivity for culture of human dermal fibroblasts for skin regeneration. Biomed Mater 6: 015001.
19. Mani G, Feldman MD, Patel D, Agrawal CM (2007) Coronary stents: a materials perspective. Biomaterials 28: 1689–1710.
20. Mary C, Marois Y, King MW, Laroche G, Douville Y, et al. (1998) Comparison of the in vivo behavior of polyvinylidene fluoride and polypropylene sutures used in vascular surgery. ASAIO J 44: 199–206.
21. Laroche G, Marois Y, Guidoin R, King MW, Martin L, et al. (1995) Polyvinylidene fluoride (PVDF) as a biomaterial: from polymeric raw material to monofilament vascular suture. J Biomed Mater Res 29: 1525–1536.
22. Klinge U, Klosterhalfen B, Ottinger AP, Junge K, Schumpelick V (2002) PVDF as a new polymer for the construction of surgical meshes. Biomaterials 23: 3487–3493.
23. Klink CD, Junge K, Binnebosel M, Alizai HP, Otto J, et al. (2011) Comparison of long-term biocompability of PVDF and PP meshes. J Invest Surg 24: 292–299.
24. Tschoeke B, Flanagan TC, Cornelissen A, Koch S, Roehl A, et al. (2008) Development of a composite degradable/nondegradable tissue-engineered vascular graft. Artif Organs 32: 800–809.
25. Berger D, Bientzle M (2009) Polyvinylidene fluoride: a suitable mesh material for laparoscopic incisional and parastomal hernia repair! A prospective, observational study with 344 patients. Hernia 13: 167–172.
26. Urban E, King MW, Guidoin R, Laroche G, Marois Y, et al. (1994) Why make monofilament sutures out of polyvinylidene fluoride? ASAIO J 40: 145–156.
27. Sarkar S, Dadhania M, Rourke P, Desai TA, Wong JY (2005) Vascular tissue engineering: microtextured scaffold templates to control organization of vascular smooth muscle cells and extracellular matrix. Acta Biomater 1: 93–100.
28. Chai C, Leong KW (2007) Biomaterials approach to expand and direct differentiation of stem cells. Mol Ther 15: 467–480.
29. Lim JY, Donahue HJ (2007) Cell sensing and response to micro- and nanostructured surfaces produced by chemical and topographic patterning. Tissue Eng 13: 1879–1891.
30. Engler AJ, Sen S, Sweeney HL, Discher DE (2006) Matrix elasticity directs stem cell lineage specification. Cell 126: 677–689.
31. Oh S, Brammer KS, Li YS, Teng D, Engler AJ, et al. (2009) Stem cell fate dictated solely by altered nanotube dimension. Proc Natl Acad Sci U S A 106: 2130–2135.
32. Flemming RG, Murphy CJ, Abrams GA, Goodman SL, Nealey PF (1999) Effects of synthetic micro- and nano-structured surfaces on cell behavior. Biomaterials 20: 573–588.
33. Roessger A, Denk L, Minuth WW (2009) Potential of stem/progenitor cell cultures within polyester fleeces to regenerate renal tubules. Biomaterials 30: 3723–3732.
34. Min BM, Lee G, Kim SH, Nam YS, Lee TS, et al. (2004) Electrospinning of silk fibroin nanofibers and its effect on the adhesion and spreading of normal human keratinocytes and fibroblasts in vitro. Biomaterials 25: 1289–1297.
35. Horn P, Bork S, Wagner W (2011) Standardized Isolation of Human Mesenchymal Stromal Cells with Red Blood Cell Lysis. Methods Mol Biol 698: 23–35.
36. Schallmoser K, Bartmann C, Rohde E, Reinisch A, Kashofer K, et al. (2007) Human platelet lysate can replace fetal bovine serum for clinical-scale expansion of functional mesenchymal stromal cells. Transfusion 47: 1436–1446.
37. Lohmann M, Walenda G, Hemeda H, Joussen S, Drescher W, et al. (2012) Donor age of human platelet lysate affects proliferation and differentiation of mesenchymal stem cells. PLoS ONE 7: e37839.
38. Schellenberg A, Stiehl T, Horn P, Joussen S, Pallua N, et al. (2012) Population Dynamics of Mesenchymal Stromal Cells during Culture Expansion. Cytotherapy 14: 401–411.
39. O'Brien J, Wilson I, Orton T, Pognan F (2000) Investigation of the Alamar Blue (resazurin) fluorescent dye for the assessment of mammalian cell cytotoxicity. Eur J Biochem 267: 5421–5426.
40. Pittenger MF, Mackay AM, Beck SC, Jaiswal RK, Douglas R, et al. (1999) Multilineage potential of adult human mesenchymal stem cells. Science 284: 143–147.
41. Schellenberg A, Lin Q, Schuler H, Koch CM, Joussen S, et al. (2011) Replicative senescence of mesenchymal stem cells causes DNA-methylation changes which correlate with repressive histone marks. Aging (Albany NY) 3: 873–888.
42. Gregory CA, Gunn WG, Peister A, Prockop DJ (2004) An Alizarin red-based assay of mineralization by adherent cells in culture: comparison with cetylpyridinium chloride extraction. Anal Biochem 329: 77–84.
43. Cholewa D, Stiehl T, Schellenberg A, Bokermann G, Joussen S, et al. (2011) Expansion of adipose mesenchymal stromal cells is affected by human platelet lysate and plating density. Cell Transplant 20: 1409–1922.
44. Pfaffl MW (2001) A new mathematical model for relative quantification in real-time RT-PCR. Nucleic Acids Res 29: e45.
45. Garcia JL, Asadinezhad A, Pachernik J, Lehocky M, Junkar I, et al. (2010) Cell proliferation of HaCaT keratinocytes on collagen films modified by argon plasma treatment. Molecules 15: 2845–2856.
46. Luna SM, Silva SS, Gomes ME, Mano JF, Reis RL (2011) Cell adhesion and proliferation onto chitosan-based membranes treated by plasma surface modification. J Biomater Appl 26: 101–116.
47. Hegewald AA, Medved F, Feng D, Tsagogiorgas C, Beierfuss A, et al. (2013) Enhancing tissue repair in annulus fibrosus defects of the intervertebral disc: analysis of a bio-integrative annulus implant in an in-vivo ovine model. J Tissue Eng Regen Med (Epub ahead of print).
48. Neuss S, Stainforth R, Salber J, Schenck P, Bovi M, et al. (2008) Long-term survival and bipotent terminal differentiation of human mesenchymal stem cells (hMSC) in combination with a commercially available three-dimensional collagen scaffold. Cell Transplant 17: 977–986.
49. Engelmayr GC Jr, Papworth GD, Watkins SC, Mayer JE Jr, Sacks MS (2006) Guidance of engineered tissue collagen orientation by large-scale scaffold microstructures. J Biomech 39: 1819–1831.
50. Sun T, Smallwood R, MacNeil S (2009) Development of a mini 3D cell culture system using well defined nickel grids for the investigation of cell scaffold interactions. J Mater Sci Mater Med 20: 1483–1493.
51. O'Brien FJ, Harley BA, Yannas IV, Gibson LJ (2005) The effect of pore size on cell adhesion in collagen-GAG scaffolds. Biomaterials 26: 433–441.

52. Hutmacher DW, Schantz T, Zein I, Ng KW, Teoh SH, et al. (2001) Mechanical properties and cell cultural response of polycaprolactone scaffolds designed and fabricated via fused deposition modeling. J Biomed Mater Res 55: 203–216.

53. Baumchen F, Smeets R, Koch D, Graber HG (2009) The impact of defined polyglycolide scaffold structure on the proliferation of gingival fibroblasts in vitro: a pilot study. Oral Surg Oral Med Oral Pathol Oral Radiol Endod 108: 505–513.

54. Grayson WL, Ma T, Bunnell B (2004) Human mesenchymal stem cells tissue development in 3D PET matrices. Biotechnol Prog 20: 905–912.

55. Edwards SL, Church JS, Alexander DL, Russell SJ, Ingham E, et al. (2011) Modeling tissue growth within nonwoven scaffolds pores. Tissue Eng Part C Methods 17: 123–130.

56. McBeath R, Pirone DM, Nelson CM, Bhadriraju K, Chen CS (2004) Cell shape, cytoskeletal tension, and RhoA regulate stem cell lineage commitment. Dev Cell 6: 483–495.

Tissue Engineering in Animal Models for Urinary Diversion

Marije Sloff[1]*, Rob de Vries[2], Paul Geutjes[1], Joanna IntHout[3], Merel Ritskes-Hoitinga[2], Egbert Oosterwijk[1ᵔ], Wout Feitz[1ᵔ]

1 Department of Urology, Nijmegen Centre for Molecular Life Sciences, Radboud University Medical Center, Nijmegen, The Netherlands, **2** SYRCLE (SYstematic Review Centre for Laboratory animal experimentation), Central Animal Laboratory, Radboud University Medical Center, Nijmegen, The Netherlands, **3** Department for Health Evidence, Radboud University Medical Center, Nijmegen, The Netherlands

Abstract

Tissue engineering and regenerative medicine (TERM) approaches may provide alternatives for gastrointestinal tissue in urinary diversion. To continue to clinically translatable studies, TERM alternatives need to be evaluated in (large) controlled and standardized animal studies. Here, we investigated all evidence for the efficacy of tissue engineered constructs in animal models for urinary diversion. Studies investigating this subject were identified through a systematic search of three different databases (PubMed, Embase and Web of Science). From each study, animal characteristics, study characteristics and experimental outcomes for meta-analyses were tabulated. Furthermore, the reporting of items vital for study replication was assessed. The retrieved studies (8 in total) showed extreme heterogeneity in study design, including animal models, biomaterials and type of urinary diversion. All studies were feasibility studies, indicating the novelty of this field. None of the studies included appropriate control groups, i.e. a comparison with the classical treatment using GI tissue. The meta-analysis showed a trend towards successful experimentation in larger animals although no specific animal species could be identified as the most suitable model. Larger animals appear to allow a better translation to the human situation, with respect to anatomy and surgical approaches. It was unclear whether the use of cells benefits the formation of a neo urinary conduit. The reporting of the methodology and data according to standardized guidelines was insufficient and should be improved to increase the value of such publications. In conclusion, animal models in the field of TERM for urinary diversion have probably been chosen for reasons other than their predictive value. Controlled and comparative long term animal studies, with adequate methodological reporting are needed to proceed to clinical translatable studies. This will aid in good quality research with the reduction in the use of animals and an increase in empirical evidence of biomedical research.

Editor: Wei-Chun Chin, University of California, Merced, United States of America

Funding: This work is funded by Fonds NutsOhra: Kunststoma 1102-56(http://www.stichtingnutsohra.nl/). The funders had no role in study design, data collection and analysis, decision to publish or preparation of the manuscript.

Competing Interests: The authors have declared that no competing interests exist.

* Email: marije.sloff@radboudumc.nl

ᵔ These authors contributed equally to this work.

Introduction

Urinary diversion with gastrointestinal (GI) tissue remains the gold standard treatment for patients suffering from end-stage bladder disease caused by bladder cancer or congenital malformations, e.g. bladder exstrophy or spina bifida [1,2]. There are three approaches to create a urinary diversion in these patients. The first and most commonly used type among surgeons is the incontinent ileocutaneostomy; a urinary conduit with a skin-outlet. Alternatively, continent diversions can be formed non-orthotopically with a skin-outlet or orthotopically as a neobladder [3,4]. Although the use of GI tissue provides a satisfactory outcome in most cases, it can be associated with severe complications. These can be either related to the bowel surgery (obstruction, infections, fistulas, etc.) or to the urostomy implantation (metabolic disorders, stone formations, infections, etc.) [5,6].

A tissue engineering and regenerative medicine (TERM) approach may provide new possibilities by creating a man-made construct to replace GI tissue for urinary diversions. The implementation of such constructs could prevent invasive bowel surgery and the potentially life-threatening complications, therefore reducing health care costs. Several investigators have focused on the development of new materials for this purpose, including naturally derived materials, synthetic polymers and decellularized scaffolds [7–9]. These biomaterials can be applied with and without autologous cells, using the regenerative capacity of the body [10,11].

In the field of urogenital reconstruction, bladder domes for cystoplasty and uretheral reconstruction with man-made constructs have already been used in patients [12,13]. However, despite the progress in *in vitro* research and animal experimentation, clinical translation of TERM approaches for urinary diversion has been negligible. Translation from bench to bedside for these tissue engineered constructs starts with the analysis of biodegradability, biocompatibility and foreign body response, which is usually performed in small rodents. To engineer and regenerate specific tissues, evaluation should preferably be performed at relevant anatomical sites with appropriately sized constructs to permit easy clinical translation. Large animal models closely mimicking the human body are therefore desirable, but their use might be ethically debatable [14]. In general, the choice of an animal model is dependent on financial considerations, the investigators experience, ethical sensitivity and practical limitations. Even though other and better translatable models might be available [15]. To our knowledge a superior animal for tissue engineering and urinary diversion has not been identified yet.

To decide on the most suitable type of animal model an evidenced-based systematic review is essential, since it potentially increases the chance of successful clinical translation [16,17]. We therefore systematically searched the current literature for all types of studies on the efficacy of tissue engineered constructs in animal models for urinary diversion. The results were analyzed with respect to survival, side effects, functionality and urothelium formation, to investigate whether there was sufficient evidence to decide if any animal model was superior for evaluating tissue engineered constructs for urinary diversion applications.

Materials and Methods

1. Search strategy

We identified relevant studies, including peer reviewed articles and (congress) abstracts, through a systematic search of PubMed, EMBASE (OvidSP) and Web of Science up until the 23rd of January 2013, following the approach as described by de Vries et al., and Leenaars et al. [16,17]. In all three databases synonyms for tissue engineering (e.g. tissue engineering, tissue engineered, regenerative medicine or biomaterials) were combined with synonyms for urinary diversion (e.g. orthotopic diversion, neobladders, continent or incontinent stomas). MeSH terms (PubMed) and EMTREE terms (EMBASE) were used when available and were combined with additional free-text words from titles or abstracts ([tiab] or/ti,ab.). For the complete strategy, see Table S1. In PubMed and Embase (OvidSP), the results were filtered for animal studies, using previously designed 'animal filters' [18,19]. The included primary studies and relevant reviews on the subject were screened for additional relevant references.

2. Study selection

Only primary studies that evaluated tissue engineered construct for urinary diversion in animal models were included. From the retrieved set of papers, duplicates and triplicates were manually deleted from EndNote, considering the preference PubMed> EMBASE>Web of Science. Based on title/abstract, primary screening was performed by a single review author (MS), deleting articles that clearly did not involve tissue engineering or urinary diversion. In case of any doubt, articles were included for further screening. Secondary screening of title/abstract was independently performed in Early Review Organizing Software (EROS, IECS, Buenos Aires, Argentina, www.eros-systematic-review.org) by two review authors (MS and RdV). The following inclusion criteria were used: 1) urinary diversion, 2) tissue engineering, 3) (living) animals of any species, and 4) primary articles. In this step, a procedure was considered to be a urinary diversion if it involved a total/radical cystectomy or implantation of a stoma or pouch connected to at least one ureter. Articles that described a partial cystectomy, hemi-cystectomy or bladder augmentation were excluded, because they do not relate to urostomy. Tissue engineered constructs were defined as biomaterials or polymers that aided the (re)construction of tissues. Articles were either categorized as 'included', 'excluded' or 'more information

necessary' if important details were not included in the abstract. Any discrepancies were discussed and re-evaluated until consensus was reached. Full-text articles were retrieved and evaluated for definite inclusion/exclusion, based on the same criteria used for the secondary screening. The reference lists of the included studies and manually identified reviews on the subject were screened for any missed references. Unfortunately, one of the included studies was published in Korean and we did not have the resources to have it translated [20]. The article was therefore excluded from this review.

3. Data extraction

From every included study, basic information (author, year of publication, etc.), animal characteristics (species, sex, etc) and study characteristics (biomaterial, follow-up, etc.) were extracted and tabulated by MS and RdV after reaching consensus (Table 1). The outcome of the studies for the meta-analysis was assessed using extracted data on mortality, adverse effects, occlusion (blockade of urinary flow) and the formation of urothelium on the implanted construct (Table 2).

4. Methodological quality assessment

All included studies were feasibility studies only, i.e. no comparison was made between the new (tissue engineering) and classical treatment (GI tissue) or any other relevant control group. Therefore, performing a risk of bias-assessment was not possible, and we consequently focused on the quality of the reporting of data and outcomes of the studies (Table 3).

5. Meta-analyses

Meta-analyses were performed for the outcome measures functionality (absence of occlusion) and formation of urothelium in seven studies. One study did not describe the animal species and was therefore excluded [26]. Since appropriate control groups were not included in any of the studies, it was not possible to perform a standard meta-analysis using, for example, odds ratios. We therefore performed a meta-analysis of proportions, more specifically of the number of animals in which a functional construct or urothelium was formed as a proportion of the total number of treated animals. First, exact binomial confidence intervals were calculated for the individual studies. To circumvent continuity corrections (some studies had 0 events), an arcsine transformation of the proportions was carried out for the meta-analyses [21]. Because high heterogeneity was expected, the individual proportions were pooled using a random effects model. Given the low number of studies, a Hartung-Knapp adjustment for random effects models was applied [22]. I^2 was used as a measure of heterogeneity. The analyses were conducted in R (version 3.0.1; R Core Team 2012), using the metafor package [23,24]. To explore the potential influence of animal size on the effect, the studies were ranked according to the subgroups "large" (rabbits and larger) and "small" (rats and mice) models in the forest plots. The small group sizes prevented calculation of an overall effect per subgroup and therefore only visually derived tendencies are presented.

Results and Discussion

1. Study inclusion

Database searches yielded 573 references for PubMed, 855 references for Embase and 315 references for Web of Science (Figure 1). After removal of duplicates and triplicates 1157 references remained. During the primary screening in EndNote, 883 references that did not meet our inclusion criteria were

Table 1. Study characteristics.

Reference	Publication	Species (strain)	Sex	weight/age	Group size	Type of intervention	Biomaterial	Size scaffold	Cell type/amount	Culture period	Study Design	Evaluation time point	outcome measures
1	Basu 2012	Research paper	Minipigs (Göttingen swine)	M/F	12–16 kg	gr 1: 8 gr 2: 8 gr 3: 8 gr 4: 8	Urinary conduit 2 ureters	PLGA-coated PGA	*	1 cm2 bladder biopsy, 2 cm2 adipose biopsy 50 mL peripheral blood All: 30–40×10^6 SMC	6 days	gr 1: bladder SMC gr 2: adipose SMC gr 3: blood SMC gr 4: unseeded	84+/−5 days	Occlusion? Histology Immunohistochemistry
2	Bertram 2009	Poster	Canines	*	*	gr 1: 8 gr 2: 8 gr 3: 8 gr 4: 8	Neo-bladder	Tengion Autologous Neo-bladder (PLGA)	*	SMC	*	gr 1: 4×10^6 SMC gr 2: 12×10^6 SMC gr 3: 25×10^6 SMC gr 4: reimplanted bladder	9 months	Histology Contractile response Electrical field stimulation
3	De Filippo 2009	Abstract	*	*	*	*	Neo-bladder	PLGA-based scaffold	*	a	*	gr 1: cystecomized animals gr 2: weight/age matched human patients	6 months	Occlusion? Histology Immunohistochemstry Voiding intervals Bladder capacity
4	Dorflinger 1985	Research paper	Dogs (Mongrel)	F	gr 1: 17–21 kg gr 2: 22–27 kg b	gr 1: 5 gr 2: 5	Neo-bladder	Silicone prosthesis with PGA mesh	65 cc	NA	NA	gr 1: thin prosthesis 2 months preimplant gr 2: thick prosthesis 1 month preimplant	6 months	Occlusion Histology Urogram Mactroscopic evaluation Colony-forming assay
5	Drewa 2007	Research paper	Rats (Wistar)	M	300 g, 6 months	gr 1: 3 gr 2: 3	Urinary conduit 1 ureter	Small intestinal submucosa (SIS), porcine	3 cm 3-layered c	Fibroblast 3T3 2×10^8 cells in alginate gel	*	gr 1: 2×10^8 3T3 cells gr 2: unseeded	2 and 4 weeks	Occlusion Histology Pyelogram Macroscopic evaluation
6	Geutjes 2012	Research article	Pigs (Landrace)	F	50 kg	gr 1: 6 gr 2: 4	Urinary conduit 1 ureter	Collagen and Vypro polymer	l=12 cm, d=15 mm	UC from 4 cm2 bladder biopsy: 10×10^6 cells	6 days	gr 1: bladder UC gr 2: unseeded	1 month	Occlusion Histology Immunohistochemistry Loopogram Macroscopic evaluation
7	Kloskowski 2012	Abstract	Rats (Wistar)	*	*	gr 1: 12 gr 2: 2	Urinary conduit 1 ureter	Decellularized aortic arch or PCL scaffold	*	NA	NA	gr 1: aortic arch or gr 2: PCL	3 weeks	Occlusion Histology
8	Liao 2013	Research paper	Rabbits (New Zealand White)	M	2,0–2,5 kg	gr 1: 24 gr 2: 6	Urinary conduit 2 ureters	Rabbit bladder acellular matrix (BAM)	l=4 cm d=0,8 cm	UC from 4 cm2 bladder biopsy: 80×10^7 cells	7 days	gr 1: bladder UC gr 2: unseeded	1, 2, 4 and 8 weeks d	Occlusion? Histology Immunohistochemistry

SMC = smooth muscle cells, PLGA = poly(lactic-co-glycolic acid), PCL = polycaprolactone, PGA = polyglycolic acid, UC = urothelial cells, * = not mentioned, N.A. = not applicable, ? = is implied in the text, but not specifically mentioned, a = cells are used, type and amount are not mentioned, b = group 3 included only partial cystetomies, as well as 1 dog in group 1 (not included), c = diameter of a 12-Fr catheter, d = evaluation time point of control group is unclear.

Table 2. Scoring of the included studies.

	Reference	Follow-up	Mortality	Adverse effects	Formation of UD	Urothelium formation
1	Basu 2012	84+/−5 days	0%[a]	*	32/32[a]	32/32[b]
2	Bertram 2009	9 months	0%?	none	24/24?[c]	24/24?[c]
3	De Filippo 2009	6 months	0%?	*	*	*
4	Dorflinger 1985	6 months	60%	hydronephrosis (5#), hydroureter (4#), pyonephrosis (2#), inflammation (3#), leakage (2#), infection (1#), ulcers (1#), reflux (2#)	2/10	2/10
5	Drewa 2007	2 or 4 weeks	0%	adhesion (3#), inflammation (4#), leakage (1#), pseudocyst (1#), hydronephrosis (4#), hydroureter (3#)	3/6	1/6
6	Geutjes 2012	1 month	11%	stenosis (3#), leakage (2#), hydroureteronephrosis (all), hydroureter (all)	8/9	6/9
7	Kloskowski 2012	3 weeks	28%[d]	Inflammation (all),	0/14	1/14[e]
8	Liao 2013	1, 2, 4 or 8 weeks	13%[f]	scarring (4#), atresia (4#), hydronephrosis (4#), fistulas (2#), inflammation(2#)[g]	26/30[h]	26/30[h]

* = not mentioned, ? = it is implied that all animals survived and formed a functional conduit with urothelial layers, a = all animals were euthanized at indicated time points, animals remained healthy, no explicit mentioning of occlusions, b = no mentioning of place of sampling. Unclear whether it covers the entire conduit, c = group 4 does not include TE, leaving 24 animals for UD, d = deaths were only in aortic arch group, e = mentioning of formation of cell layers, not specific on type of cell layer, f = all animals died in the experimental group, g = complications were observed in the control group only, h = unclear if this accounts for all animals.

removed. Secondary screening of the remaining 274 references in EROS led to the removal of 206 references. Full text analyses of the remaining 68 references resulted in the inclusion of only 8 studies: two abstracts, one poster and five full-text papers (Figure S1). Screening of the reference lists of these papers and manually identified relevant reviews on the subject did not yield any new references. Thus, the final set included: Geutjes et al. (2012), Basu et al. (2012), Kloskowski et al. (2012), Liao et al. (2013), De Filippo et al. (2009), Bertram et al. (2009), Drewa et al. (2007), and Dørflinger et al. (1985) [11,25–31]. Although the number of published studies on tissue engineering for urinary diversion in animal models appeared to be substantial, only these selected studies applied tissue engineered constructs for urinary diversion at relevant clinical sites in animal models. The recent increase in published papers on this subject is remarkable. Although all relevant databases were explored, we cannot exclude that some data reside within company protected domains. Moreover, studies tend to be published only when results are positive and statistically significant [32]. These two factors may have resulted in an incomplete data set and they may have introduced a publication bias in this systematic review. Due to the limited number of included studies, we were not able to estimate the risk and effect of this publication bias.

2. Study characteristics

Analysis of study characteristics revealed extraordinary diversity (Table 1). Various animal species, including pigs, minipigs, dogs, rats and rabbits, were used. They were implanted with constructs of either biodegradable polymers (poly(lactic-*co*-glycolic acid) (PLGA), polyglycolic acid (PGA) or collagen) or decellularized material (small intestinal submucosa (SIS) and bladder acellular matrix (BAM)). Kloskowski et al., used a biodegradable polymer polycaprolactone (PCL) and decellularized aortic arches. Although three different approaches for urinary diversion are known (urinary conduit, abdominal pouch and neobladder), the included studies only created urinary conduits (5 studies) or neobladders (3

studies). An animal model for an abdominal pouch was not described.

There are three different study designs within the pool of included studies: studies that use cellular or acellular scaffolds and studies that compare these two constructs (Table 1 and 4). The most frequently used study design was the comparison between cellular and acellular constructs (4 studies), investigating whether (pre-)seeding of the biomaterial resulted in a superior outcome. These studies generally isolated their cells from a bladder biopsy, although alternative sources like adipose tissue, peripheral blood and cell lines were investigated (Basu et al., Drewa et al., Geutjes et al. and Liao et al.). The other studies compared different materials (Kloskowski et al.), cell concentrations (Bertram et al.) or different experimental designs (Dørflinger et al.). Filippo et al. compared the behavior of a cell-seeded PLGA-based scaffold in cystectomized animals with its behavior in children enrolled in a Phase II trial for bladder augmentation. This diversity in animal species, biomaterial or study design complicates the interpretation of the results of this systematic review.

Since tissue engineering for urinary diversion is a relatively new area of research, the main focus of the studies was to determine the feasibility of implantation and subsequent behavior of the designed constructs. Follow-up time was less than a year in all cases, whereas constructs need to be functional throughout the remainder of a patients' life time. Stoma complications can occur after several years in patients, and short follow-up in animals will not provide evidence on late complications. This indicates the necessity of appropriate control groups with the classical techniques and longer follow-up for at least 1 year.

3. Quality assessment

The lack of control groups precluded a risk of bias-analysis. We therefore focused on the reporting quality of the studies (Table 3). Our specific interests were the animal characteristics, the composition, dimensions and preparations of the construct, the

Table 3. Quality assessment.

	Abstracts/Poster			Research Articles				
	Bertram 2009	De Filippo 2009	Kloskowski 2012	Basu 2012	Dorflinger 1985	Drewa 2007	Geutjes 2012	Liao 2013
Study Design								
Q1: Is the animal species described?	yes	no	yes	yes	yes	yes	yes	yes
Q2: Is the specific strain described?	no	no	yes	yes	?	yes	yes	yes
Q3: Is the sex of the animal specified?	no	no	no	yes	yes	yes	yes	yes
Q4: Is the age or weight of the animals specified?	no	no	no	yes	yes	yes	yes	yes
Q5: Is the number of animals specified?	yes	no	yes	yes	yes	yes	yes	yes
Q6: Is the creation of the urinary diversion clearly described?	yes	?	yes	yes	yes	yes	yes	yes
Q7: Is the composition of the tissue-engineered construct clearly described?	no	no	yes	yes	yes	yes	yes	yes
Q8: Are the dimensions of the implanted construct clearly described?	no	no	no	no	yes	yes	yes	yes
Q9: Is the preparation or culture period clearly described?	no	no	no	yes	N.A.	?	yes	yes
Outcomes								
Q10: Are the used outcome measures clearly described?	yes	yes	yes	yes	yes	yes	yes	yes
Q11: Is the follow-up time after implantation clearly described?	yes	yes	yes	yes	yes	yes	yes	?a
Q12: Are there any comments on the representativeness of the results?	no	no	no	no	yes	yes	?	no
Q13: Is the location of sampling for histology clearly described?	no	no	no	no	N.A.	yes	yes	no
Q14: Is it explicitly indicated that there were any drop-outs?	no	yes	yes	?	yes	yes	yes	yes
Q15: Is the number of drop-outs described?	N.A.	N.A.	yes	N.A.	yes	N.A.	yes	yes
Q16: Are the reasons for dropping out specified?	N.A.	N.A.	no	N.A.	yes	N.A.	yes	yes
Q17: Is it stated that there were any adverse effects?	yes	?	?	no	yes	yes	yes	yes

? = not clearly described in the text, N.A. = not applicable, a = unknown for the control group.

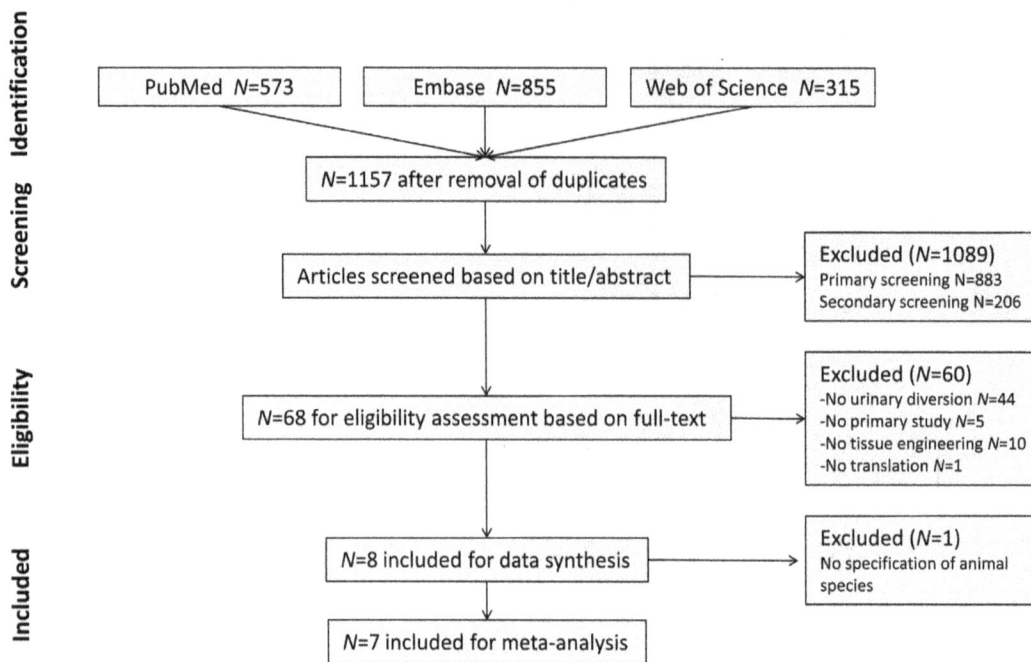

Figure 1. Flow-chart of search and screening process. Primary screening exclusion was performed in End-Note. Criteria included: no urinary diversion, no tissue engineering or reconstruction of ureter or urethra. Secondary inclusion was performed in EROS. Criteria included: no urinary diversion, no tissue engineering, no animal study or no primary study.

representativeness of the results and the adequate reporting of drop-outs.

The reporting was relatively poor in the abstracts compared to the full-text papers. Many abstracts omitted important details that are essential to compare different studies, including animal species, strain, number and sex, type of tissue engineered scaffolds, description of the composition, dimensions and preparations of the construct (Q1-Q9). This might have been partly the consequence of the word and space limitation for abstracts.

All included studies appropriately described the predefined outcome measures and stated at which time evaluation took place. Only two studies commented on the representativeness of the figures for the overall study outcomes (Dørflinger et al. and Drewa et al.) (Q10-Q13). In total, three papers were complete in their reporting of the study design (Drewa et al., Geutjes et al., and Liao et al.).

Although guidelines for standardized reporting of animal experimentation have been described, these have not yet been generally accepted [33,34]. We observed that in some studies, especially in the abstracts, these guidelines were not implemented.

It is crucial to further improve methodological reporting to aid future research, even in abstracts.

4. Data synthesis

To determine the efficacy of the tissue engineered constructs, we focused on the outcome measures: formation of a functional conduit or reservoir and the formation of urothelium in the regenerated tissue, evidenced by histology (Table 2). Some studies performed additional analyses, including immunohistochemistry, urograms or pyelograms (Table 1), but these were not considered here. Because constructs can only be functional in the absence of major complications (including mortality), we first looked at the survival of the animals and adverse effects. Survival of the animals was regarded as the first indication for the safety of a construct and represented a condition for the efficacy of a construct. Secondly, since urine needs to exit the body adequately, the formation of a reservoir or conduit with a urothelial lining was deemed essential.

4.1 Adverse effects and mortality. Experience shows that even in some situations animals might die of unrelated causes, with more likelihood in larger experimental groups. A clear and detailed description of the drop-outs will increase the credibility of

Table 4. Cellular vs. Acellular.

	Acellular	Cellular
Basu et al 2012	8/8	24/24
Drewa et al 2007	2/3	1/3
Geutjes et al 2012*	4/4	5/6
Liao et al 2013	4/6	24/24

Amount of functional conduits formed in comparative studies with cellular and acellular groups. * was tabulated after correspondence with the first author.

the study. Surprisingly, two studies did not explicitly report on the mortality rate, but the studies suggest that all animals survived the procedure in good health (Basu et al, and Bertram et al.) They imply that a functional urinary conduit with urothelial linings is formed in all animals without any adverse effects (Table 2 and Table 3, Q14-Q17). Such a successful score was not described in any of the other studies, which raises the possibility that the success rate was overestimated. The reported mortality of the other studies ranged between 60% in the pioneering study in 1985 (Dørflinger et al.) to 13% in the most recently published paper (Liao et al.), suggesting the application of improved constructs or improved surgical techniques over the years. Geutjes et al. is the only study that reports on both related and unrelated deaths.

All studies report on inflammation of the construct and the surrounding tissues, although this does not necessarily constitute a negative effect. Some degree of inflammation may guide tissue regeneration, remodeling and the formation of blood vessels. The formation of a vasculature structure in large constructs still remains a major problem in tissue engineering [35,36]. Other common adverse effects found in the majority of the studies were hydronephrosis and hydroureters, prevention of which remains the biggest challenge for tissue engineers. The designed constructs were not able to control urinary pressure, leading to reflux to the kidney and the aforementioned conditions.

The formation of stones in urinary diversions is one of the complications in using GI tissue. Remarkably, only one study, performed in pigs reported the formation of calcifications (Geutjes et al.). The absence of stones, particularly in the included rabbit study, is unexpected (Liao et al.). Implantation of a tissue engineered patch in the bladder wall of rabbits resulted in a high incidence of stone formation and encrustation, indicating that this animal model in particular is prone to develop stones [37,38]. Although a different biomaterial was used in a different setting, no stones were formed up until two months in the study performed by Liao et al. In humans, it takes months to years to develop urinary stones and perhaps the follow-up time in these studies is too short (<1 year) to detect stones [39]. Obviously, due to the difference in diet composition, flow speed, composition and pH of urine in different species, urinary stones might not develop in some animals. Long-term follow-up is necessary to exclude encrustation or stone formation.

4.2 Efficacy. We conducted meta-analyses for the outcome measures functionality and urothelium formation (Figures 2A and B). The aim of these meta-analyses was not to obtain a precise point estimate, but rather to get an impression of the quantitative relations between the results of the individual studies and to detect trends [40]. The heterogeneity between the studies was very high ($I^2 = 96\%$ for functionality and $I^2 = 95\%$ for urothelium formation). Although this did not justify pooling of the results, nevertheless, overall proportions were calculated for functionality and urothelium formation (69% [CI: 18–100%] and 65% [CI: 17–98%], respectively). Due to the large heterogeneity and small number of studies, these overall proportions should be interpreted with caution.To prevent misinterpretations, overall proportions are only shown as dotted vertical lines in the forest plots.

Successful formation of functional conduits with full urothelial linings, the primary goal of all studies, varied between the studies. Only small numbers of functional conduits and urothelial coverage were observed by Dørflinger et al., Drewa et al. and Kloskowski et al. In contrast, Liao et al. showed that in 4/5 cases a functional conduit was formed with urothelial linings. The large heterogeneity and small number of studies complicates identification of the underlying cause of the different results. We can only speculate

whether these were the consequence of the biomaterial, the animal model or confounding factors.

4.3 Animal models. The small number and heterogeneity of the studies did not permit subgroup analysis for different animal species. However, the forest plots (Figures 2A and B) from the meta-analyses suggested a tendency towards better results in large animal models compared to small animals. Although we cannot exclude the effect of confounding factors, this supports the idea that future research should focus on larger animals. Larger animals have a more similar anatomy to the human body than, for example, rodents, and allow for evaluation at clinical relevant sites with constructs of a comparable size. Moreover, surgical techniques and materials are more comparable to the human setting, therefore mimicking the human situation as closely as possible. In 1985, the first report on the use of a construct to tissue engineer a urinary diversion as a replacement for GI tissue in large animals was published (Dørflinger et al.). Even though the mortality rate was high, a functional reservoir could be formed showing the feasibility of this approach. In later studies small animals were used (Kloskowski et al. and Drewa et al.) but the success rate in these studies was low. This might be explained by the surgical challenge with a higher risk of complications in smaller animals.

The extraordinary heterogeneity in animal studies has previously been reported by Roosen et al. [41], who reviewed different animal models for the classical types of urinary diversion using GI tissue. A uniform conclusion regarding the most suitable animal model could not be drawn and the author advised to view the results with caution, when translating these for clinical implementation.

4.4 Cellular and acellular scaffolds. The studies that compared acellular and cellular scaffolds investigated the effect of (pre-) seeding on tissue regeneration (Table 1 and 4). Basu et al., Liao et al. and Geutjes et al. expanded autologous cells from a biopsy. Basu et al. stated that (pre-)seeding of the scaffolds provided an additional advantage, although this was not substantiated. The same conclusion was drawn by Liao et al., but here the authors clearly report that the drop-outs and adverse effects in the acellular group were higher. Geutjes et al. observed no difference between cellular and acellular constructs.

In the study by Drewa et al., scaffolds were seeded with 3T3 fibroblast from mouse origin, followed by implantation in rats. Not surprisingly, this resulted in an excessive inflammatory response and consequently a better outcome for the unseeded group.

The presence of a cellular component can trigger M1 macrophages, resulting in fibroblast deposition. In contrast, acellular scaffolds activate the M2 macrophages, which leads to reconstruction and regeneration of tissues [35]. Some researchers have reported that scaffold (pre-)seeding provides an additional advantage for tissue regeneration, since regeneration is less dependent on cellular in-growth [42,43]. Since the effect pre-seeding is outside the scope of this systematic review, a meta-analyses was not performed. Therefore, based on the available information, it is unclear whether (pre-)seeding of the scaffolds provides an advantage for tissue reconstruction in urinary diversions.

4.5 Concluding remarks. Based on this systematic review the most adequate animal model for urinary diversion is still undefined. Only a limited amount of studies could be identified despite the comprehensive search strategy, all showing large heterogeneity. It appears that the predictive value of a particular animal model was not a decisive factor in the studies performed. Nevertheless, the forest-plots suggested a trend towards successful experimentation in larger animal models, supporting the idea that

A Forestplot functionality

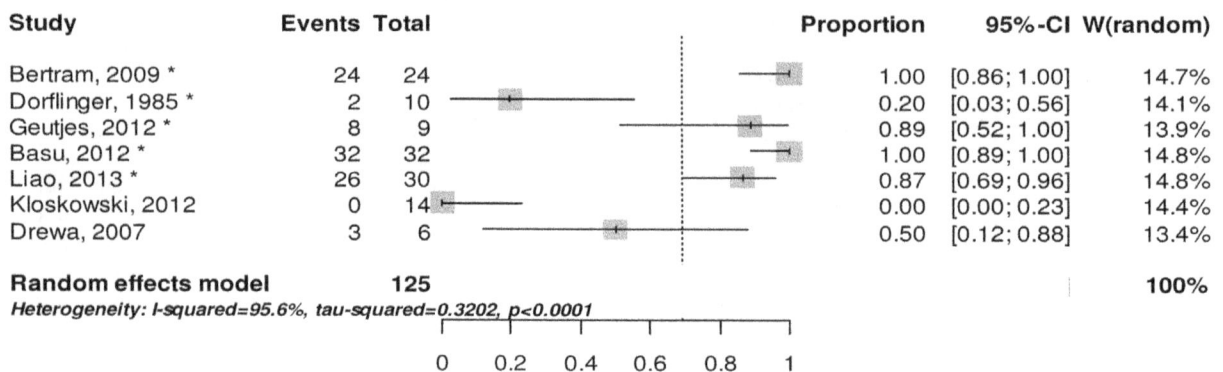

Study	Events	Total		Proportion	95%-CI	W(random)
Bertram, 2009 *	24	24		1.00	[0.86; 1.00]	14.7%
Dorflinger, 1985 *	2	10		0.20	[0.03; 0.56]	14.1%
Geutjes, 2012 *	8	9		0.89	[0.52; 1.00]	13.9%
Basu, 2012 *	32	32		1.00	[0.89; 1.00]	14.8%
Liao, 2013 *	26	30		0.87	[0.69; 0.96]	14.8%
Kloskowski, 2012	0	14		0.00	[0.00; 0.23]	14.4%
Drewa, 2007	3	6		0.50	[0.12; 0.88]	13.4%
Random effects model		**125**				**100%**

Heterogeneity: I-squared=95.6%, tau-squared=0.3202, p<0.0001

0   0.2   0.4   0.6   0.8   1

B Forestplot urothelium formation

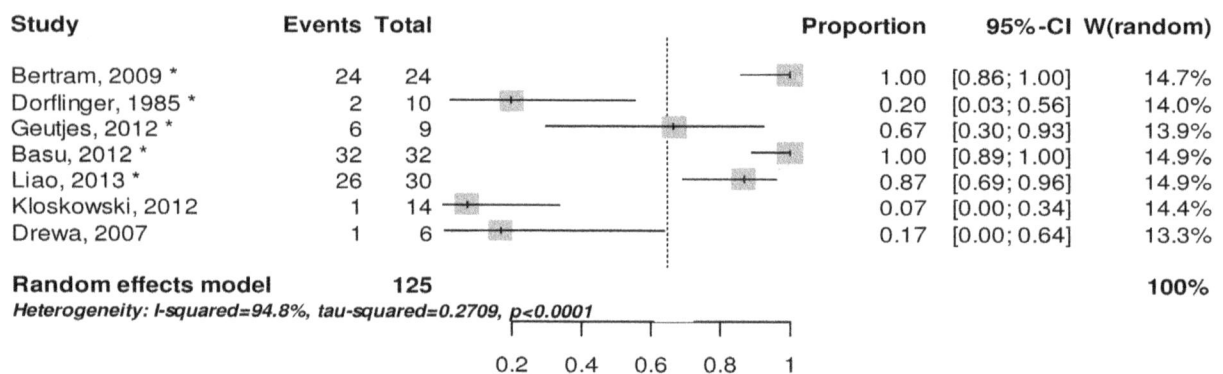

Study	Events	Total		Proportion	95%-CI	W(random)
Bertram, 2009 *	24	24		1.00	[0.86; 1.00]	14.7%
Dorflinger, 1985 *	2	10		0.20	[0.03; 0.56]	14.0%
Geutjes, 2012 *	6	9		0.67	[0.30; 0.93]	13.9%
Basu, 2012 *	32	32		1.00	[0.89; 1.00]	14.9%
Liao, 2013 *	26	30		0.87	[0.69; 0.96]	14.9%
Kloskowski, 2012	1	14		0.07	[0.00; 0.34]	14.4%
Drewa, 2007	1	6		0.17	[0.00; 0.64]	13.3%
Random effects model		**125**				**100%**

Heterogeneity: I-squared=94.8%, tau-squared=0.2709, p<0.0001

0.2   0.4   0.6   0.8   1

Figure 2. Forestplots for functionality (A) and urothelium formation (B). Forest plot of 7 studies. Filippo et al. was excluded for both meta-analyses since the study does not report on the type of animal, functionality and urothelium formation. * indicates studies with large animal species.

future research should focus on the evaluation of constructs in larger animal models which more closely mimic the human body than small (rodent) animal models. Bladder augmentations and urethral replacements have been successful in dogs [44–47]. Although the results of the earliest study were disappointing (Dorflinger) this might be explained by the tissue engineering strategy. Use of a more sophisticated tissue engineering approach did lead to satisfactory results (Bertram). In view of the former, evaluation of urinary diversion constructs in dogs might be a valid alternative.

Despite the limited amount of data available on this subject, a phase I clinical trial was initiated for implantation of a (pre-)seeded tissue-engineered urinary conduit in cystectomized patients [48]. Whether it is acceptable to expose patients to a new device which has not been investigated in a long term and multicenter animal study remains a matter of debate. Critics of animal experimentation might say that it is difficult to find a representative animal model to improve clinical translation. In order to improve translational practice, a re-evaluation of preclinical practice is warranted, in which systematic reviews and meta-analyses of animal studies can provide a valuable tool [49].

This systematic review focused on the identification of an appropriate animal model to investigate tissue engineering for urinary diversion. Remarkably, only feasibility studies were identified, which are necessary to evaluate potentially valuable

techniques. To obtain a more thorough estimation of the feasibility and applicability, a large variety of approaches should be investigated and should include different biomaterials, growth factors and large animal models. Evaluation of the functionality and advantages of newly developed constructs should include appropriate controls and evaluate long-term effects and outcomes to investigate safety and efficacy of these constructs. Moreover, standardized assessment methods are desirable. The ultimate control would be the use of GI tissue, the golden standard technique currently used in the clinic. Assessment of these preclinical experiments should focus on both functionality and tissue regeneration, using urodynamic measurements and histological evaluations. To continue to clinically translatable studies, standardization of (large) controlled and comparative studies with adequate methodological reporting is required as laid down in existing guidelines [33,34].

Animal experimentation remains subject to ethical debate and - among others- for animal welfare reasons it is therefore essential that research is properly conducted and reported. Nevertheless, the reporting of the methodology and representativeness of results was rather poor in the reviewed studies. Improvements in this area are urgently needed for ethical, scientific and economic reasons, and will allow researchers to repeat studies reliably and make an unbiased decision on the most correct animal model for their future experiments. The scientific community should adopt the existing guidelines for standardized reporting of animal experimentation and implement them in journal author guidelines, which should be enforced by editors, to improve standardized reporting in full-text articles. This will aid in good quality research with more relevant output for the clinic and reduction of the use of animals in biomedical research.

Supporting Information

Figure S1 Inclusion and exclusion of papers during secondary and full-text screening in EROS. After primary screening 274 papers were analyzed in EROS, resulting in the inclusion of 8 papers. Papers were excluded which were without urinary diversion, tissue engineering or animals or when they were reviews. For one study, we did not have the resources for translation.

Table S1 Complete search strategy for PubMed, EMBASE and Web of Science. MeSH and EMTREE terms were used when possible and combined with free-text words from title or abstract ([tiab] or.ti,ab.).

Author Contributions

Conceived and designed the experiments: MS RdV PG EO MRH WF. Performed the experiments: MS RdV. Analyzed the data: MS RdV JH. Contributed reagents/materials/analysis tools: JH. Wrote the paper: MS RdV PG JH MRH EO WF.

References

1. National Cancer Institute (2010) What you need to know about bladder cancer. NIH publiciation No. 10–1559.
2. Pannek J, Senge T (1998) History of urinary diversion. Urol Int 60:1–10
3. Hautmann RE (2003) Urinary diversion: ileal conduit to neobladder. J Urol 169: 834–842.
4. Kaefer M, Hendren WH, Bauer SB, Goldenblatt P, Peters CA, et al. (1998) Reservoir calculi: a comparison of reservoirs constructed from stomach and other enteric segments. J Urol 160: 2187–2190.
5. Hautmann RE, Hautmann SH, Hautmann O (2011) Complications associated with urinary diversion. Nat Rev Urol 8: 667–677.
6. Nieuwenhuizen JA, de Vries RR, Bex A, van der Poel HG, Meinhardt W, et al. (2008) Urinary diversions after cystectomy: the association of clinical factors, complications and functional results of four different diversions. Eur Urol 53: 834–842.
7. Drewa T, Adamowicz J, Sharma A (2012) Tissue Engineering for the oncologic urinary bladder. Nat Rev Urol 9: 561–572.
8. Kim BS, Baez CE, Atala A (2000) Biomaterials for tissue engineering. World J Urol 18: 2–9.
9. Beiko DT, Knudsen BE, Watterson JD, Cadieux PA, Reid G, et al. (2004) Urinary tract biomaterials. J Urol 171: 2438–2444.
10. Drewa T, Sir J, Czajkowski R, Wozniak A (2006) Scaffold seeded with cells is essential in urothelium regeneration and tissue remodeling in vivo after bladder augmentation using in vitro engineered graft. Transplant Proc 38: 133–135.
11. Geutjes P, Roelofs L, Hoogenkamp H, Walraven M, Kortmann B, et al. (2012) Tissue engineered tubular construct for urinary diversion in a preclinical porcine model. J Urol 188: 653–660.
12. Raya-Rivera A, Esquiliano DR, Yoo JJ, Lopez-Bayghen E, Soker S, et al. (2011) Tissue-engineered autologous urethras for patients who need reconstruction: an observational study. Lancet 377: 1175–1182.
13. Atala A, Bauer SB, Soker S, Yoo JJ, Retik AB (2006) Tissue-engineered autologous bladders for patients needing cystoplasty. Lancet 367:1241–1246.
14. Wood MW, Hart LA (2007) Selectin appropriate animal models and strains: making the best use of research, information and outreach. AATEX 14: 303–306.
15. Nordgren A (2004) Moral imagination in tissue engineering research on animal models. Biomaterials 25: 1723–1734.
16. De Vries RB, Buma P, Leenaars M, Ritskes-Hoitenga M, Gordijn B (2012) Reducing the number of laboratory animals used in tissue engineering research by restricting the variety of animal models. Articular cartilage tissue engineering as a case study. Tissue Eng Part B Rev 18: 427–435.
17. Leenaars M, Hooijmans CR, van Veggel N, ter Riet G, Leeflang M, et al. (2012) A step-by-step guide to systematically identify all relevant animal studies. Lab Anim 46: 24–31.
18. Hooijmans CR, Tillema A, Leenaars M, Ritskes-Hoitenga M (2010) Enhancing search efficiency by means of a search filter for finding all studies on animal experimentation in PubMed. Lab Anim 44: 170–175.
19. De Vries RB, Hooijmans CR, Tillema A, Leenaars M, Ritskes-Hoitinga M (2011) A search filter for increasing the retrieval of animal studies in Embase. Lab Anim 45:168–270.
20. Lee YS, Cho SY, Kim HW, Kang SH, Kim HY, et al. (2009) Preliminary study of tissue-engineered ileal conduit using poly (e-caprolactone) (PCL) nano-sheet seeded with muscle-derived stem cells. Korean J Urol. 50: 282–287.
21. Rücker G, Schwarzer G, Carpenter J, Olkin I (2009) Why add anything to nothing? The arcsine difference as a measure of treatment effect in meta-analysis with zero cells. Stat Med 28: 721–728.
22. Knapp G, Hartung J (2003) Improved tests for a random effects meta-regression with a single covariate. Stat Med 22: 2693–2710.
23. R Core Team (2012) R: a language and environment for statistical computing. R foundation for statistical computing. ISBN 3-900051-07-0.
24. Viechtbauer W (2010) Conducting meta-analyses in R with the metafor package. J Stat Soft 36: 1–48.
25. Liao W, Yang S, Song C, Li Y, Meng L, et al. (2013) Tissue-engineered tubular graft for urinary diversion after radical cystectomy in rabbits. J Surg Res 182: 185–191.
26. De Filippo R, Bertram TA, Jayo MJ, Seltzer E (2009) Adaptive regulation of regenerated bladder size after implantation with tengion neo-bladder augment™ early clinical outcomes and preclinical evidence. J Urol 181: 267.
27. Drewa T (2007) The artificial conduit for urinary diversion in rats: a preliminary study. Transplant Proc 39: 1647–1651.
28. Dörflinger T, Frimodt-Møller PC, England DM, Madsen PO, Bruskewitz R (1985) Prosthetic urinary bladder implantation to facilitate bladder regeneration. Neurourol Urodyn 4: 47–59.
29. Basu J, Jayo MJ, Ilagan RM, Guthrie KI, Sangha N, et al. (2010) Regeneration of native-like neo-urinary tissue from non-bladder cell sources. Tissue Eng Part A 18: 1025–1034.
30. Kloskowski T, Jundzill A, Gurtowska N, Olkowska J, Kowalczuk T, et al. (2012) Urinary conduit construction using tissue engineering. J Regen Med Tissue Eng 6: 23.
31. Bertram TA, Christ GJ, Andersson KE, Aboushwareb T, Fuellhase C, et al. (2009) Pharmacologic response of regenerated bladders in a preclinical model. FASEB J 23: 291.
32. Knight J (2003) Negative results: Null and void. Nature 422: 554–555.
33. Kilkenny C, Browne WJ, Cuthill IC, Emerson M, Altman DG (2010) Improving bioscience research reporting: the ARRIVE guidelines for reporting on animal research. PLoS Biol 8: e1000412.
34. Hooijmans C, Leenaars M, Ritskes-Hoitenga M (2010) A gold standard publication checklist to improve the quality of animal studies, to fully integrate the Three Rs, and to make systematic reviews more feasible. Altern Lab Anim 38: 167–182.
35. Brown BN, Valentin JE, Stewart-Akers AM, McCabe GP, Badylak SF (2009) Macrophage phenotype and remodeling outcomes in response to biologic scaffolds with and without cellular component. Biomaterials 30: 1482–1491.
36. Badylak SF, Valentin JE, Ravindra AK, McCabe GP, Stewars-Akers AM (2008) Macrophage phenotype as a determinant of biologic scaffold remodeling. Tissue Eng Part A 14: 1835–1842.
37. Nuininga JE, van Moerkerk H, Hanssen A, Hulsbergen CA, Oosterwijk-Wakka J, et al. (2004) A rabbit model to tissue engineer the bladder. Biomaterials 25: 1657–1661.
38. Grover PK, Ryall RL (1994) Urate and calcium oxalate stones: from repute to rhetoric to reality. Miner Electrolyte Metab 20: 361–370.
39. Evans K, Costabile RA (2005) Time to development of symptomatic urinary calculi in a high risk environment. J.Urol 173: 858–861.
40. Vesterinen HM, Sena ES, Egan KJ, Hirst TC, Churolov L, et al. (2014) Meta-analyis of data from animal studies: a practical guide. J Neurosci Methods 221: 92–102.
41. Roosen A, Woodhouse CR, Wood DN, Stief CG, McDougal WS, et al. (2011) Animal models in urinary diversion. BJU Int 109: 6–23.
42. De Filippo RE, Yoo JJ, Atala A (2002) Urethral replacement using cell seeded tabularized collagen matrices. J Urol 168: 1789–1792.
43. Zhang Y, Kropp BP, Lin HK, Cowan R, Cheng EY (2004) Bladder regeneration with cell-seeded small intestinal submucosa. Tissue Eng 10:181–187.
44. Sievert KD, Fandel T, Wefer J, Gleason CA, Nunes L, et al. (2006) Collagen I:III ratio in canine heterologous bladder acellular matrix grafts. World J Urol 24:101–109.
45. Probst M, Piechota HJ, Dahiya R, Tanagho EA (2000) Homologous bladder augmentation in dog with the bladder acellular matrix graft. BJU Int 85: 362–371.

46. Orabi H, Aboushwareb T, Zhang Y, Yoo JJ, Atala A (2013) Cell-seeded tabularized scaffolds for reconstruction of long urethral defects: a preclinical study. Eur Urol 63: 531–538.

47. Roth CC, Mondalek FG, Kibar Y, Ashley RA, Bell CH, et al. (2011) Bladder regeneration in a canine model using hyaluronic acid-poly(lactic-co-glycolic-acid) nanoparticle modified porcine small intestinal submucosa. BJU Int 108: 148–155.

48. ClinicalTrials.gov (2014) Incontinent urinary diversion using an autologous neo-urinary conduit. Identifier NCT 01087697.

49. Hooijmans CR, Ritskes-Hoitinga M (2013) Progress in using systematic reviews of animal studies to improve translational research. PLoS Med 10: e1001482.

A Glycosaminoglycan Based, Modular Tissue Scaffold System for Rapid Assembly of Perfusable, High Cell Density, Engineered Tissues

Ramkumar Tiruvannamalai-Annamalai[1], David Randall Armant[2,3], Howard W. T. Matthew[1,4]*

1 Department of Biomedical Engineering, Wayne State University, Detroit, Michigan, United States of America, **2** Departments of Obstetrics & Gynecology, Wayne State University, Detroit, Michigan, United States of America, **3** Program in Reproductive & Adult Endocrinology, National Institute of Child Health & Human Development, National Institutes of Health, Bethesda, Maryland, United States of America, **4** Department of Chemical Engineering and Materials Science, Wayne State University, Detroit, Michigan, United States of America

Abstract

The limited ability to vascularize and perfuse thick, cell-laden tissue constructs has hindered efforts to engineer complex tissues and organs, including liver, heart and kidney. The emerging field of modular tissue engineering aims to address this limitation by fabricating constructs from the bottom up, with the objective of recreating native tissue architecture and promoting extensive vascularization. In this paper, we report the elements of a simple yet efficient method for fabricating vascularized tissue constructs by fusing biodegradable microcapsules with tunable interior environments. Parenchymal cells of various types, (i.e. trophoblasts, vascular smooth muscle cells, hepatocytes) were suspended in glycosaminoglycan (GAG) solutions (4%/1.5% chondroitin sulfate/carboxymethyl cellulose, or 1.5 wt% hyaluronan) and encapsulated by forming chitosan-GAG polyelectrolyte complex membranes around droplets of the cell suspension. The interior capsule environment could be further tuned by blending collagen with or suspending microcarriers in the GAG solution These capsule modules were seeded externally with vascular endothelial cells (VEC), and subsequently fused into tissue constructs possessing VEC-lined, inter-capsule channels. The microcapsules supported high density growth achieving clinically significant cell densities. Fusion of the endothelialized, capsules generated three dimensional constructs with an embedded network of interconnected channels that enabled long-term perfusion culture of the construct. A prototype, engineered liver tissue, formed by fusion of hepatocyte-containing capsules exhibited urea synthesis rates and albumin synthesis rates comparable to standard collagen sandwich hepatocyte cultures. The capsule based, modular approach described here has the potential to allow rapid assembly of tissue constructs with clinically significant cell densities, uniform cell distribution, and endothelialized, perfusable channels.

Editor: Christina Chan, Michigan State University, United States of America

Funding: Funding for these studies was provided by a DRICTR award from Wayne State University, and grant awards from the National Science Foundation (CBET-1067323) and the National Institutes of Health (HD067629). The funders had no role in study design, data collection and analysis, decision to publish, or preparation of the manuscript.

Competing Interests: The authors have declared that no competing interests exist.

* E-mail: hmatthew@wayne.edu

Introduction

Fabrication of 3D constructs that promote cell-cell interaction, extra cellular matrix (ECM) deposition and tissue level organization is a primary goal of tissue engineering [1]. Accomplishing these prerequisites with the currently available conventional scaffolds and fabrication techniques still remains a challenge. Some of the tissue types that have been successfully engineered include skin [2], bone [3–5] and cartilage [4,6,7]. Significant success has also been achieved in nerve regeneration [8], corneal construction [9–11] and vascular tissue engineering [12]; However-er, the success rate has been relatively low in engineering complex tissue types such as liver, lung, and kidney due to their complex architectures and metabolic activities.

In conventional preformed scaffolds, the cell viability depends on diffusion of oxygen, nutrients and growth factors from the surrounding host tissues, and it is limited to 100–200 microns thickness at cell densities comparable to that of normal tissues [13]. Hence in constructs with larger dimensions, efficient mass transfer

and subsequent cell survival can be achieved only by significantly reducing cell densities or by tolerating hypoxic conditions. Moreover, in a porous scaffold, uniform distribution throughout the construct is difficult to achieve, and the seeded cells will stay on the peripheral surface of the construct forming a thin peripheral layer. In addition, these scaffolds cannot facilitate incorporation of multiple cell types in a controlled manner. Hence the slow vascularization, mass transfer limitation, low cell density and non-uniform cell distribution limits conventional methods from engineering large and more complex organs. Therefore, an innate structure that supports functional vascularization is imperative for engineering large tissues grafts. Many strategies have been proposed to incorporate vascular structure that includes creating endothelial microchannels inside scaffolds [14,15], surface mod-ification and/or controlled releasing of pro-vasculogenic growth factor and cytokines [16–18], coculturing vascular cell types for microvessel formation [19] etc. Despite their limited success, none of these approaches is able to incorporate an extensive vasculature as seen in natural organs.

The bioinspired modular tissue engineering approach has emerged in recent years as a promising fabrication strategy to address the common shortcomings of a preformed scaffold by assembling tissue constructs from the bottom up [20,21]. Using this principle, complex tissues and organs can be engineered efficiently from microscale modules as opposed to the top down approach of conventional scaffolds [21]. This approach is increasingly becoming a promising tool to study and recreate vascular physiology in tissue engineering applications [22,23]. Some of the proposed modular TE strategies include 3D tissue printing [24–26], cell sheets technology [27] and assembly of cell laden hydrogels [20,28] (Figure 1).

Here we report the development of a simple yet efficient method for assembling tissue prototypes with embedded, endothelialized channels by fusing microscale capsules. Our method of cell encapsulation was previously developed for perfusion culture of highly metabolic cells [29–33].We are now extending its use to fabricate modular tissue constructs. Capsules were seeded internally with various test cell types and externally with vascular endothelial cells to authenticate our proof of principle and growth and metabolic performance were studied. To enhance the versatility of these microcapsules, the effects of tuning the capsule interiors with collagen gels or microcarriers were explored.

Materials and Methods

Cell culture conditions

All chemical and culture reagents were purchased from Sigma-Aldrich unless mentioned otherwise. The human trophoblast cell line HTR-8/SVneo[34] (HTBs) was used as the model cell type for some studies due to their high proliferative capacity and ability to form dense, tissue-like aggregates. The cells were cultured in 10 cm tissue culture dishes, using F12/DMEM supplemented with 5% fetal bovine serum (FBS), 50 mg/ml gentamycin and 2.5 mg/L Amphotericin-B.

For co-culture studies, vascular smooth muscle cells (SMCs) were isolated from rat aorta and endothelial cells (AECs) were isolated from sheep aorta using established enzymatic procedures [2,35,36]. Sheep aortas were procured from a slaughterhouse under an educational license (Wolverine Packing Company, Detroit, MI). Aortas were obtained within 2 hours of slaughter and used for AEC isolation immediately. Human umbilical vein endothelial cells (HUVECs) obtained from ATCC (Manassas, VA) were also used as vascular component. Primary cells were used from passages 3 to 6. SMCs and AECs were maintained in DMEM supplemented with 10% FBS, 50 mg/ml gentamycin, and 2.5 mg/L Amphotericin-B. In addition, SMC cultures were supplemented with 2 ng/ml fibroblast growth factor 2 and AECs with 50 ng/ml of epidermal growth factor. For HUVECs, MCDB 131 medium supplemented with Endothelial Cell Growth Kit-VEGF (ATCC) was used. During co-cultures of parenchymal and vascular components, a 50–50 mixture of the respective culture media was used. Primary hepatocytes were isolated from Sprague dawley rats weighing 250–450 g by the two-step collagenase perfusion technique described by Seglen [37] and modified by Dunn [38,39]. Cell viability averaged 90–95%, as assessed by trypan blue exclusion, and the average yield was 4×10^8 viable cells per liver. Type I collagen was isolated from Sprague dawley rat tail

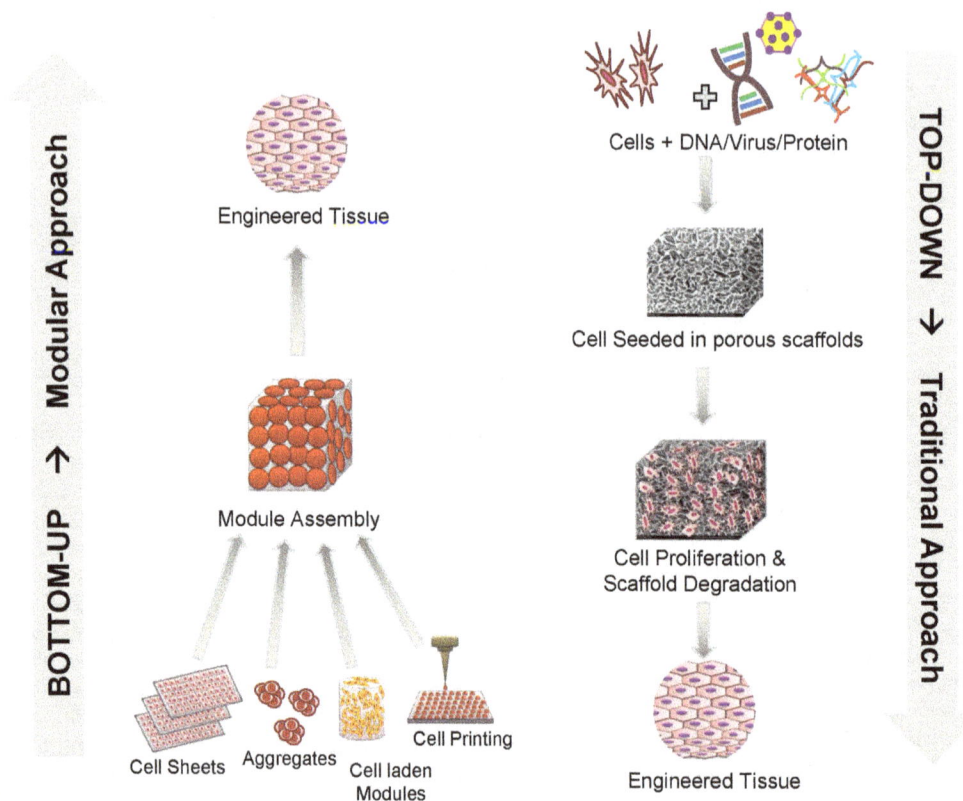

Figure 1. Bottom-up vs. top-down approaches in tissue engineering. The traditional, top-down approach (right) involves seeding cells into full sized porous scaffolds to form tissue constructs. This approach poses many limitations such as slow vascularization, diffusion limitations, low cell density and non-uniform cell distribution. In contrast, the modular or bottom-up approach (left) involves assembling small, non-diffusion limited, cell-laden modules to form larger structures and has the potential to eliminate the shortcomings of the traditional approach.

tendons as previous described [38] and used for hepatocyte collagen sandwich cultures. Hepatocyte culture medium consisted of high glucose DMEM medium supplemented with 10% fetal bovine serum (FBS), 0.5 U/mL insulin, 7 ng/mL glucagon, 20 ng/mL epidermal growth factor, 7.5 μg/mL hydrocortisone, 100 mg/L gentamycin and 2.5 mg/L amphotericin B. Culture medium was collected and analyzed for albumin and urea synthesis using established methods [33,40]. All dish cell cultures were maintained at 37°C in a 5% CO2/95% air humidified incubator.

Ethics statement

Harvesting of rat hepatocytes and aortic smooth muscle cells for culture was carried out in strict accordance with the recommendations in the Guide for the Care and Use of Laboratory Animals of the National Institutes of Health. The cell isolation protocol was approved by the Animal Investigation Committee of Wayne State University (Protocol Number: A-07-16-10). Surgery and liver perfusion were performed under ketamine/xylazine anesthesia, and all efforts were made to minimize suffering.

Biopolymer materials

The materials used in preparing our microcapsules and modular scaffolds were: chitosan from crab shells, molecular weight ~600 kDa (Sigma); chondroitin 4-sulfate sodium salt from bovine trachea, molecular weight ~50–100 kDa (Sigma); hyaluronic acid sodium salt from *Streptococcus equii*, molecular weight 1500–1800 kDa (Sigma); dextran sulfate sodium salt, molecular weight ~500 kDa (SCBT); heparin sodium salt from porcine intestinal mucosa, molecular weight 17–19 kDa (Celsus); carboxymethylcellulose sodium salt, molecular weight 250 kDa (Sigma); polygalacturonic acid sodium salt (Sigma) and collagen type-I isolated from Sprague Dawley rat tail tendons (Invitrogen).

Aqueous solutions of the polyanions (chondroitin 4-sulfate (CSA), carboxymethylcellulose (CMC), hyaluronic acid (HA), polygalacturonic acid (PGA)) were prepared in a HEPES-sorbitol buffer containing: 0.4 g/L KCl, 0.5 g/L NaCl, 3.0 g/L HEPES-sodium salt, and 36 g/L sorbitol, pH 7.3. Polyanion solutions were sterilized by autoclaving at 121°C. The two formulations of polyanionic solutions studied for capsule formation were: (a) 4 wt% CSA/1.5 wt% CMC and (b) 1.0 or 1.5 wt% HA. To prepare the polycationic solution, chitosan powder was suspended in water (3 g in 250 ml) and autoclaved at 121°C. Under sterile conditions, 0.6 ml of glacial acetic acid was added to the aqueous suspension and stirred for 4 hours to partially dissolve the chitosan. Likewise, 19 g of sorbitol was autoclaved in 250 ml of water and then mixed with the chitosan solution. Undissolved chitosan was removed by centrifugation at 500 G. PGA (0.1 wt%) in HEPES-sorbitol buffer was used for surface stabilization of capsules. For capsule experiments employing collagen, cold collagen-I solution was diluted to 2 mg/ml in 1 mM HCl, and then neutralized with 10X DMEM (9:1 ratio). Normal saline (0.9 wt% NaCl) was used for capsule washing immediately after formation.

Cell encapsulation

Cells were encapsulated in microcapsules produced by polyelectrolyte complexation between cationic chitosan and polyanions as described in detail previously [29,30]. In brief, the 5–10 million cells were suspended in 1 ml of a polyanionic solution (either 4 wt% CSA/1.5 wt% CMC, or 1.5 wt% HA). Droplets of the cell suspension (~0.8 mm diameter) were dispensed into 30 ml of stirred chitosan solution containing 2–3 drops of Tween 20. A 24 gauge Teflon catheter was used to generate droplets and filtered

air was blown coaxially to shear away the droplets at a suitable size. Care was taken during encapsulation process to ensure uniform droplet size. Capsule membranes were formed almost instantaneously by ionic complexation between the oppositely charged polymers. Capsules were allowed to mature for ~1 min in the stirred chitosan, followed by two washes with normal saline and surface stabilization by washing with 0.1% PGA solution. Microcapsules were subsequently equilibrated with culture medium for ~60 min and then transferred to suitable culture conditions (Figure 2A).

The interior environment of the capsules could be enhanced with collagen gel or adhesion surface-providing microcarriers when desired. For capsules with an internal collagen matrix, chilled Type I collagen solution (1 mg/ml in 1 mM HCL) was neutralized with 10X DMEM in a 9:1 ratio, and mixed with an equal volume of double strength polyanionic solution (e.g. 8% CSA/3% CMC). Cells were then suspended in this mixture instead of the regular polyanion solution, and capsules were made as described previously. For microcarrier co-encapsulation, PBS swelled microcarriers were suspended along with cells in normal strength polyanionic solution at a volume ratio of 0.5:1 (packed cells+microcarriers:polyanion solution). The suspension was then dispensed as droplets to generate capsules as described above. Capsules enhanced with interior collagen or microcarriers were subjected to similar washing and surface stabilization steps as described above prior to culture.

Endothelial cell seeding on capsule surfaces

Capsules were coated with an adsorbed layer of Type I collagen prior to externally seeding endothelial cells. For coating collagen on the outer surface, non-surface stabilized capsules (i.e. capsules without a PGA final wash) were washed in dilute acidic collagen solution (0.2 mg/ml of collagen in 1 mM acetic acid) for 1–2 min and then equilibrated with culture medium for 30 mins. Capsules that had been previously surface stabilized with PGA, were first washed with dilute chitosan solution (0.06% chitosan) prior to the dilute collagen wash. The equilibration culture medium was then removed and an endothelial cell suspension (HUVECs or AECs) in medium was added to settled capsules in a 50 ml centrifuge tube. Cells were seeded at a density of 10^6 cells per ml of capsules and incubated at 37°C for 60 minutes with gentle resuspension every 10 min. After incubation, the seeded capsules were transferred to bioreactor chambers or tissue culture dishes for further experiments.

Evaluation of capsule wall permeability

The permeability of capsule walls was studied fluorometrically by measuring the rate of diffusion of tetramethylrhodamine-labelled bovine serum albumin (BSA-TMR) from the capsules, as detailed before [29,30,41]. The rate was used to calculate an overall mass transfer coefficient for the capsule wall membrane under mixing conditions.

A precise number of capsules (100–150 per sample, n = 3) of similar size from each formulation were counted out and equilibrated in HBSS (pH 7.4) containing 2.5 mg/ml BSA (13% TMR-labeled. The equilibration saturated all the BSA binding sites of the capsule wall and efficiently loaded BSA into the capsules. After washing and resuspension in fresh HBSS, the capsules were redistributed into three fluorescence cuvettes (4 ml volume, 1 cm light path) at 35–50 capsules per cuvette. The HBSS volume was made up to 3 ml and the cuvettes were sealed and mixed horizontally on a linear shaker at 100 rpm. The outward diffusion of BSA was followed by measuring the fluorescence of the external HBSS at exitation/emission wavelengths of 541/572 nm

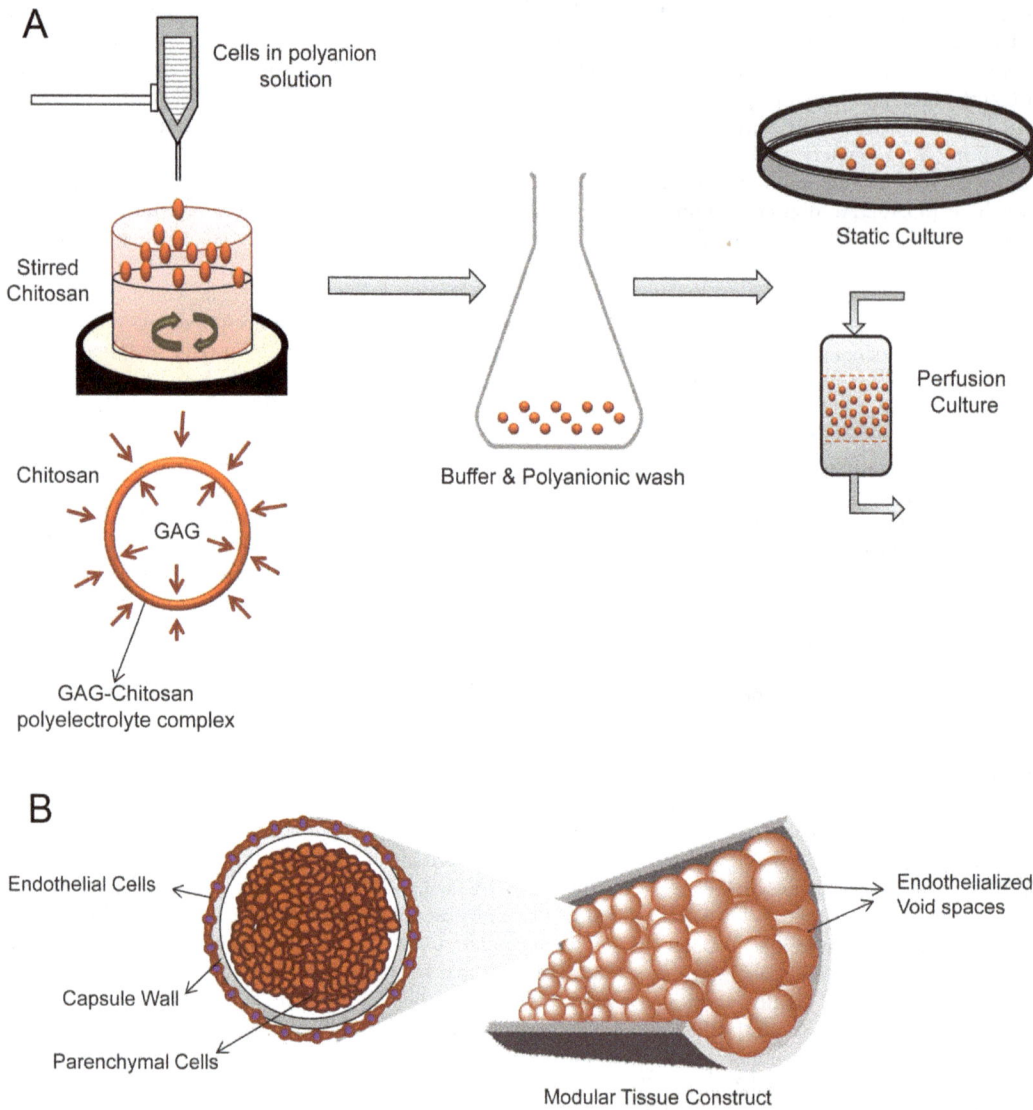

Figure 2. Microencapsulation through complex coacervation and modular assembly. (A) Droplets of cells suspended in a polyanionic solution were dispensed into a stirred chitosan solution. Ionic interactions between the oppositely charged polymers formed an insoluble ionic complex membrane at the droplet-solution interface, thus encapsulating the suspended cells. Capsule were washed surface-stabilized with a suitable anionic polymer solution, and transferred to culture. (B) Cell laden capsules can be assembled in a packed bed fashion with interconnected endothelialized channels that may enable perfusion of fluids such as blood with limited adverse reactions.

at regular time intervals for 3 h. The BSA concentration was determined using a standard curve covering the range of 0–50 μg/mL total albumin. The overall mass-transfer coefficient for diffusion across the capsule wall (K) was calculated by solving the differential equation obtained through an unsteady state mass balance on the external solution [29].

$$V \frac{dC}{dt} = KA(C_C - C) \tag{1}$$

$$NC_C V_C + VC = NC_\infty V_C + VC_0 = M \tag{2}$$

Where: K is the overall mass transfer coefficient for membrane diffusion; M is the total mass of solute present in the cuvette (M =

C_r ($V+NV_c$)); V and V_c are the volume of external solution and volume of capsules, respectively; N is the number of capsules; A is the total surface area (A = N*surface area of single capsule); C is the concentration of solute in external solution; C_o is the initial extracapsular concentration; C_c is the concentration of solute in the capsules; C_∞ is the initial intracapsule concentration; C_r is the final concentration after end of 3 hours; and t is time. The solution to the equations 1 and 2 yields,

$$\ln(Q) = \left(\frac{KA(V + NV_C)}{NVV_C} \right) t \tag{3}$$

Where Q is a dimensionless concentration-dependent parameter defined as

$$Q = \frac{M - (V + NV_C)C_0}{M - (V + NV_C)C}$$

The overall mass-transfer coefficient for transmembrane diffusion, K, was determined by plotting ln(Q) vs. time and determining the slope of the linear portion of the curve by linear regression. The intrinsic permeability, P, of each capsule wall was determined from the relation:

$$K = \frac{P}{\delta}$$

Where δ is the thickness of the capsule wall.

Assembly of modular constructs

Individual capsules were fused into 3D constructs in one of two ways. In the first method, freshly formed capsules were fused by allowing them to sit in contact with each other after the second saline wash, but before the surface stabilizing PGA wash. Freshly formed capsules were washed once with normal saline and then transferred in saline to a cylindrical mold with a 250 micron mesh at the base. Capsules were allowed to settle within the mold and held stationary for 2–3 minutes to allow inter-capsule adhesion. The excess saline was then drained and the capsule surfaces in the fused construct were then stabilized by briefly rinsing with saline, followed by a diluted polyanion solution (i.e. 0.1% PGA or 0.4% CSA/0.15% CMC), followed by a final PBS rinse.

In the second fusion method, previously stabilized and cultured capsules were first reloaded with a polyanion by incubation in a diluted polyanion solution (0.1% heparin or 0.4% CSA/0.15% CMC). The capsules were then transferred to a cylindrical mold with the mesh base. After draining excess polyanion solution, the mold with reloaded capsules was perfused with 0.06 wt% chitosan solution to ionically fuse the capsules. Excess chitosan solution was drained, the capsules were rinsed with normal saline and surface stabilized by a brief perfusion with a dilute polyanion solution. The fused modular construct was then removed from the mold for further culture or analysis.

Evaluation of cell proliferation inside capsules

Cell proliferation inside capsules was characterized using either a Hoechst DNA quantification assay [42] or an MTT assay. Briefly, 30 capsules were distributed into each well of a 24 well plate. Capsules were maintained under standard culture conditions, and one well was sacrificed at each time point. The capsules were gently ruptured using a fire-polished Pasteur pipette, and the cell aggregates within were lysed using cell lysis buffer (0.1% SDS, 10 mM Tris-HCl, 1 mM EDTA) to extract whole DNA. To an aliquot of this extract was added an equal volume of Hoechst 33258 dye dissolved at 1 mg/ml in TNE buffer (50 mM Tris-HCL, 100 mM NaCl, 0.1 mM EDTA). Fluorescence of the mixture was then measured (EX/EM 350/450 nm). A calf thymus DNA standard curve was used to determine the total DNA concentration. For the MTT proliferation assay, capsules were washed in PBS and suspended in phenol red free DMEM containing 2 mg/ml MTT. After incubation for 4 hours at 37°C, the solution was aspirated and 150 μL of DMSO was added to extract the formazan crystals. After 10 mins of rotary agitation, the absorbance of the DMSO extract was measured at 540 nm using a spectrophotometer. Exponential cell growth was assumed and the

specific growth rate was determined by fitting the following equation to the absorbance reading:

$$\ln\left(\frac{A}{A_0}\right) = \mu(t - t_0))$$

Where A_0 and A are initial and final absorbance or fluorescence readings respectively, t_0 and t are initial and final time points, and μ is the specific growth rate in $time^{-1}$.

Cell viability imaging and histology

Cell viability was assessed using Calcein-AM and ethidium homodimer (Cytotoxicity Kit L3224, Invitrogen). The cell laden capsules were washed with PBS and incubated in serum free DMEM containing 4 μM Calcein-AM and 4 μM ethidium homodimer for 20 min at 37°C. For long-term tracking of HUVECs on capsules and fused capsule constructs, CellTracker™ Green CMFDA (Invitrogen) was used. Briefly, adherent cells were rinsed with PBS and incubated in a serum free culture medium containing 5 μM CellTracker Green probe for 45–60 min. After the incubation the medium was replaced with pre-warmed normal medium and incubated for another 30 min for the dye to undergo modification due to intracellular esterases. The cells were then trypsinized and seeded onto capsule outer surface. Cell fluorescence was then observed using wide-field fluorescence microscopy and laser scanning confocal microscopy (Zeiss LSM-410).

The distribution and organization of cells and matrix inside the encapsulated cultures were investigated by histology. Cell laden individual capsules and fused capsule constructs were washed in PBS, fixed in 10% buffered formalin, dehydrated in an ethanol series, paraffin embedded, sectioned (4–6 μm) and stained using Hematoxylin and Eosin (H&E) or Masson's trichrome stains (Sigma-Aldrich). The stained sections were observed using bright field microscopy.

Perfusion culture of encapsulated hepatocytes

Encapsulated primary rat hepatocytes (encapsulation density: 20×10^6 cells/mL of CSA/CMC) were maintained in perfusion cultures under both packed bed and fluidized bed conditions as previously described [29,33]. For fluidized perfusion, non-fused capsules were fluidized by a continuous upward flow of the culture medium in a cylindrical chamber within a continuous circulation flow circuit. For the packed bed cultures, the capsules were fused in a cylindrical flow chamber as described above and subjected to a downward flow of the medium in a continuous circulation flow circuit. Medium exiting the culture chamber was oxygenated using a silicone tubing oxygenator (supplied with 95% air/5% CO_2) and recirculated using a peristaltic pump. The flow rates were adjusted to maintain physiological pressure differences (<100 mmHg) across the chamber (4–5 mL per minute). The perfusion system was maintained at 37°C for 1–2 weeks and medium was changed every 2–3 days. Medium samples were collected daily for evaluation of urea and albumin synthesis by the hepatocytes.

Analysis of albumin and urea synthesis

Standard methods for measuring albumin and urea production rates were used to assess hepatocyte function. Culture medium collected from collagen sandwich cultures and perfusion bioreactor cultures at regular intervals was analyzed for rat serum albumin by ELISA with purified rat albumin (Sigma) and a peroxidase conjugated anti-rat albumin antibody (Bethyl). Urea production was quantified using the diacetylmonoxime method as previously described [43]. Standard curves for both quantification techniques were generated using purified rat albumin or urea dissolved in culture medium. Absorbances were measured with a Spectramax microplate reader.

Statistical analyses

Measurements were performed in triplicate (n = 3). Data are plotted as means with error bars representing standard deviation. Statistical comparisons were done using Student's t-test with a 95% confidence limit. Differences with $p < 0.05$ were considered statistically significant.

Results

High density encapsulated cell cultures

We investigated the effects of different polyanions and polyanion blends on the capsule characteristics and encapsulated cell growth patterns. The 4%CSA/1.5%CMC capsules (Figure 3 A–F) and the 1.5% HA capsules (Figure 3H) were sturdy and could be easily handled with forceps. In contrast, the 1.5%HA/1.5%CMC mixture formed very thin walled capsules (Figure 3G), many of which ruptured after a week of culture. Capsules made with 4% DXS or 3% CMC were thin walled and ruptured within 2–3 hours due to osmotic swelling (Figure 3I).

Encapsulated HTBs grew rapidly, eventually filling the capsules, and most cells appeared viable with a distinct nucleus up to at least day 30 (Figure 3 A, B, C, D, E, F). This indicated that the capsule wall was sufficiently permeable to nutrients to allow maintenance of a dense, tissue-like cell mass. The estimated capsule cell density ($\sim 6 \times 10^7$ cells/cm^3, assessed via image analysis) at day-30 was high enough to replicate the cell density in many tissues. By the end of week-3, HTBs had invaded the capsule wall as seen in Figure 3C. This in vitro invasion suggests that the capsule materials may be degraded within a relatively short time frame upon implantation in vivo. No necrotic core was observed within the encapsulated cell mass at least until 45 days of static culture.

Endothelial cell growth on capsule surfaces

The growth of sheep aortic endothelial cells (AEC) and HUVECs was investigated by seeding these cells onto the outside surfaces of CSA/CMC capsules. Endothelial cells attached poorly to surface stabilized capsule surfaces. However, their attachment and growth greatly improved when type-I collagen was coated onto the outer surface of the CSA/CMC capsules. HUVECs seeded on the collagen coated CSA/CMC capsules attached well and formed a viable monolayer within 24 hours of seeding (Figure 4A, B, C). SEM images of capsules fixed 1 hour post-seeding (Figure 4E) showed a continuous, but irregular monolayer of cells in varying stages of spreading. SEM images 24 hours after seeding showed a well spread and smooth endothelial monolayer, with few areas of exposed capsule surface (Figure 4D). This morphology was maintained for at least 14 days under static culture conditions.

Growth rates of encapsulated smooth muscle cells

Sheep aortic smooth muscle cells were encapsulated for purposes of evaluating the performance of a normal parenchymal cell type. Use of these cells also allowed indirect evaluation of their interaction with endothelial cells in a subsequent study. SMC specific growth rate data showed that the cells proliferated significantly better in HA than in CSA/CMC capsules ($p < 0.05$) during the first 36 hours of culture (Figure 5). However, differences between the formulations were less pronounced after

36 hours ($p < 0.10$). The difference in cell proliferation might be attributable to hyaluronan-specific signaling through CD44 cell surface receptors [44]. Alternatively, CSA, a sulfated GAG, may have bound and partially sequestered growth factors necessary for SMC growth. In addition, the HA capsules appeared to support slightly better cell attachment to the internal surface than CSA/CMC, thereby promoting formation of several small aggregates rather than the single large spheroid typically seen in CSA/CMC capsules. Smaller aggregates are less likely to be adversely affected by diffusion limitations and may thus exhibit higher growth rates in the early stages than larger aggregates.

Capsule membrane permeability

Results of the diffusion studies on encapsulated BSA are shown in Figure 6. A typical plot of the dimensionless concentration factor ln (Q) vs. time, for three replicates runs of the CSA/CMC capsule formulation is shown in Figure 6A. The higher slope of the curve observed at early time points, is likely due to the rapid desorption of weakly bound albumin from the capsule wall. For our diffusion calculations, only the slope of the later, linear portion of the curve was used. Figure 6B compares the values of the overall mass-transfer coefficient (K), permeability coefficient (P), and wall thickness (δ) for the two most stable capsule formulations (HA and CSA/CMC). As expected, the HA capsules exhibited ~ 3 fold higher permeability than CSA/CMC capsules due to the higher molecular mass of HA and its expected formation of a looser polyelectrolyte complex network with chitosan. However, the overall mass transfer coefficient which correlates directly with the overall rate of BSA diffusion from capsules was higher in the CSA/CMC capsules, mainly due to their thinner walls. Significant post-formation swelling was also observed in CSA/CMC capsules, which nearly doubled their initial diameter. No such swelling was observed with HA capsules. Overall, the results indicate that both capsule types are permeable to globular proteins of moderate size, and suggest that the permeability of the capsule wall might be further tunable via the molecular weight of the capsule materials.

Tuning the interior capsule microenvironment

The hollow nature of the GAG-based microcapsules allowed us to tune the inner microenvironment by co-encapsulating additional materials. We investigated the effect of incorporating a collagen gel matrix and the inclusion of microcarriers by co-encapsulating them in microcapsules along with cells. When a collagen matrix was included along with the HTBs, the aggregates formed were smaller, more numerous, loosely organized and distributed inside the CSA/CMC capsules (Figure 7(A, B)). The collagen matrix did not appear to affect the invasiveness of the HTBs, and overall capsule integrity remained unaltered. The looser organization of HTBs in the presence of collagen was likely due to integrin-mediated cell-matrix adhesion which competed with cadherin-mediated cell-cell adhesion. This modification may be particularly useful for reducing the effective sizes of cell aggregates in larger capsules, and thereby reducing intra-aggregate diffusion limitations.

To further modulate the intracapsule environment, we investigated the effect of gelatin coated dextran microcarriers (Cytodex-3, Sigma) encapsulated in HA and CSA/CMC capsules, on aortic smooth muscle cell proliferation and viability. The inclusion of microcarriers in the capsules reduced the formation of large aggregates and dispersed the cells more evenly across the capsule. Interestingly, the inclusion of the microcarriers in the polyanionic solution resulted in greatly reduced swelling and smaller sized capsules (Figure 7C) compared to capsules formed without microcarriers. Calcein AM and Ethidium Homodimer fluores-

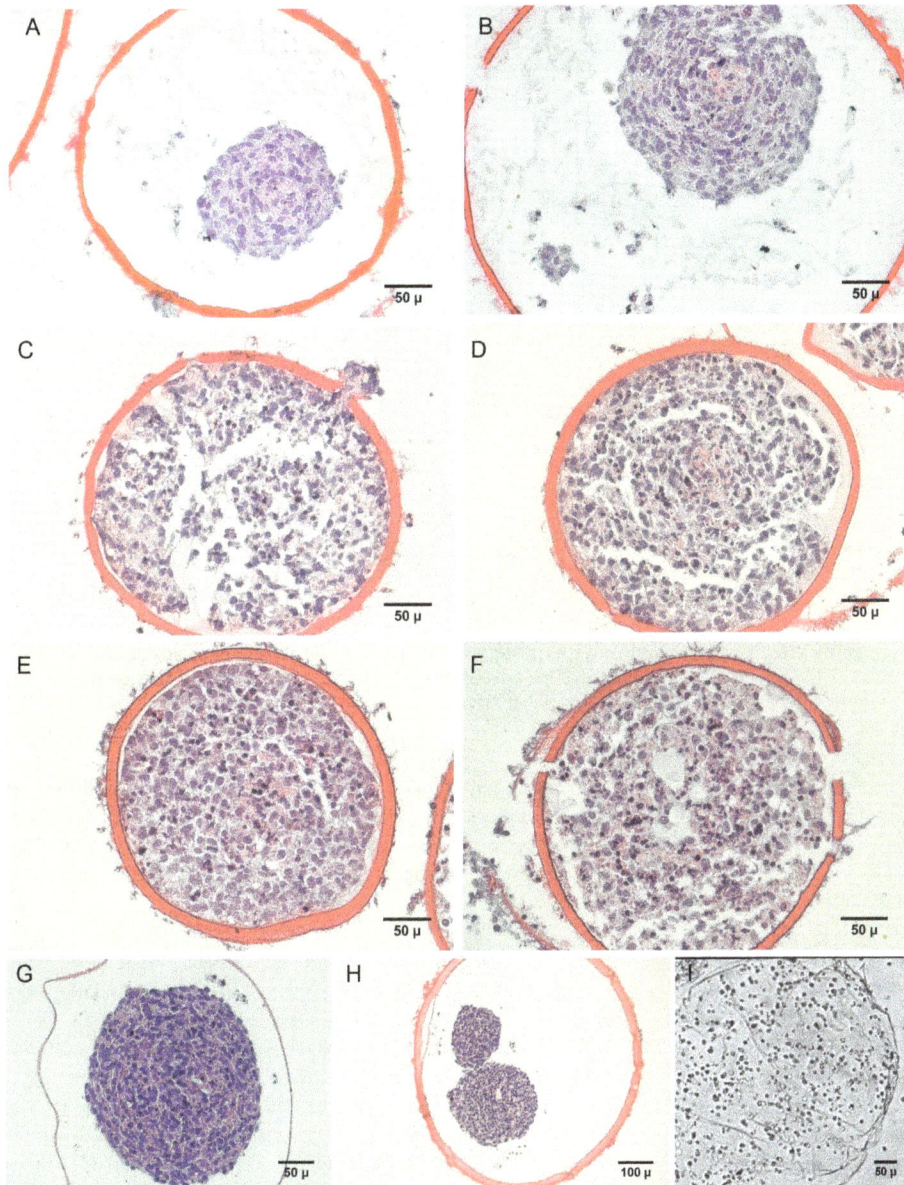

Figure 3. Histology of microencapsulated cultures of human trophoblasts (HTBs) in various GAG-chitosan capsule formulations. HTBs in CSA/CMC capsules on days (A) 5, (B) 10, (C) 15, (D) 20, (E) 25, (F) 30. (G) Hyaluronan/CMC capsules. (H) HA capsules. (I) Dextran sulfate/CMC capsules quickly ruptured due to osmotic swelling.

cence imaging after 12–14 days of encapsulated culture showed that encapsulation of microcarriers along with the cells promoted microcarrier adhesion and proliferation of SMCs as shown in Figure 7(C, D, E). Similar adhesion and growth results were not observed in capsules without microcarriers or with microcarriers made of dextran alone (Cytodex-1, Sigma).

Cell-contractable capsules for higher density cultures

Some cell types may not grow well in a GAG-only ECM environment and hence attaining high cell densities could be a challenge. To address this, we developed a capsule formulation that promoted the cell-mediated contraction of capsules and the rapid elimination of excess capsule volume. We investigated the ability of collagen to promote contraction of the capsules by encapsulating smooth muscle cells with varying volume ratios of

collagen gelling solution to GAG solution. Initially when collagen was encapsulated along with HTBs in CSA/CMC capsules at a final concentration of 1.5 mg/ml, the growing cells formed aggregates which were loosely packed as discussed earlier (Figure 7 A–B). When collagen of the same concentration was encapsulated along with SMCs at a similar cell density in CSA/CMC capsules, the cells were initially uniformly dispersed within the collagen gel inside the capsules. However, within 24 hrs the SMCs contracted the collagen matrix and formed a denser mass of cells and matrix as shown in Figure 8 (A, B, C). Even though the internal matrix was contracted, the walls of the CSA/CMC capsules were unyielding and retained their spherical shape (Figure 8B). Similar experiments were conducted with SMCs in HA capsules using various concentrations of HA and collagen-I to examine the collagen-mediated contraction effect. HA capsules containing SMCs suspended in a collagen gel were found to

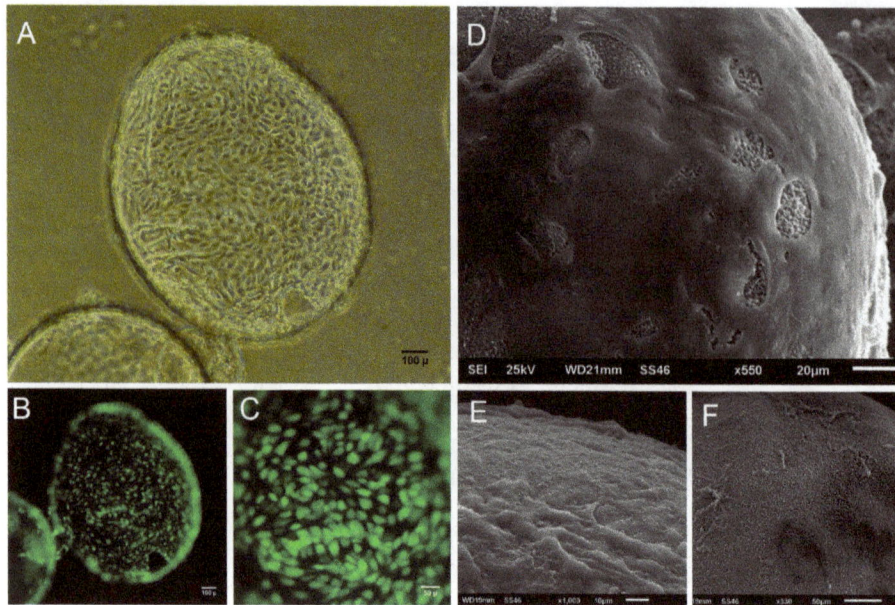

Figure 4. HUVECs seeded on CSA/CMC capsules after surface coating with collagen. (A) Phase contrast image of a capsule coated with a monolayer of HUVECs, 24 hours after cell seeding. (B,C) CellTracker Green fluorescence images of HUVECs seeded on the outer surface shown in A. (D,E) SEM images of HUVEC seeded capsule surfaces after 1 hour (E) and 24 hours (D). (F) Non-seeded capsule surface.

support cell-mediated contraction and crumpling of the entire capsule wall simultaneously with contraction of the interior gel (Figure 8 D, E, F). Sectioning and staining (H&E and Trichrome) revealed a structure with cells embedded within a dense collagen matrix inside the capsule and also large amount of collagen integrated into the capsule walls (Figure 8(G, H, I)). Maximal contraction with ~75% reduction in volume (~37% reduction in diameter) was achieved using a polyanion solution containing 0.33 wt% HA and 1.3–1.4 mg/ml of collagen-I (Figure 9). The capsule contraction was confirmed to be cell mediated based on the insignificant reduction in capsule diameter of cell-free capsules (Figure 9). This collapsible capsule formulation may be a useful tool for preparing reduced diameter capsules with higher density

cell content, for use with cells that do not proliferate well in-vitro. The technology may also be utilized to improve diffusional transport performance in encapsulated culture by reducing the excess capsule volume and effectively increasing the surface to volume ratio.

Cocultures of smooth muscle and endothelial cells

Many studies have demonstrated functional relationships between endothelium and adjacent cell types [45–49,50,51]. Designing tissue assembly approaches that allow critical paracrine interactions is mandatory for achieving in vivo-like performance of engineered tissues. Towards this end, we sought to characterize the growth of encapsulated smooth muscle cells with and without capsule surface-seeded endothelial cells in HA capsules. Phase contrast imaging showed that encapsulated SMCs co-cultured with AECs on the external capsules surfaces exhibited greater proliferation than SMC-only capsules, as indicated by the larger SMC aggregates seen in Figure 10B. This result suggests that the HA chitosan membrane possessed enough permeability to allow stimulatory paracrine signaling between AECs and SMCs.

Assembly and perfusion bioreactor culture of capsule modules

Assembly of capsules into three-dimensional modular constructs is fabrication critical step in our modular tissue engineering approach to generating vascularized tissue. We investigated various methods for assembling larger 3D constructs from pre-cultured individual capsules. The most successful method involved reloading cultured capsules with a polyanion, followed by perfusion with a diluted polycation solution. Outward diffusion of reloaded GAG during the polycation perfusion step deposited a polyelectrolyte complex that effectively fused capsules together around points of contact. This method yielded self-supporting structures with interconnecting, perfusable spaces as shown in Figures 11 and 12.

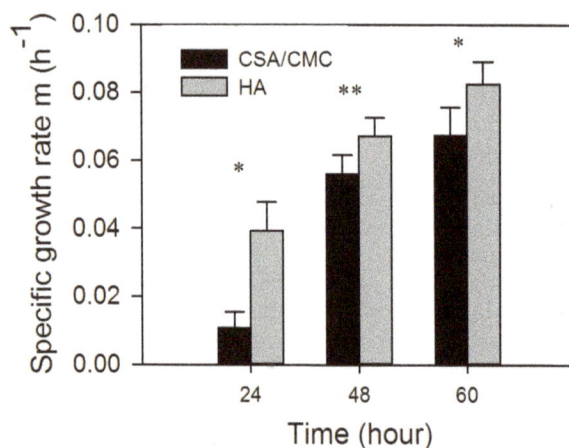

Figure 5. Specific growth rates of aortic smooth muscle cells in HA and CSA/CMC capsules. Specific growth rates were calculated using DNA measurements. Error bars represent standard deviations of 3–5 independent measurements. Significant differences are denoted by single or double asterix (* = p<0.05; ** = p<0.10).

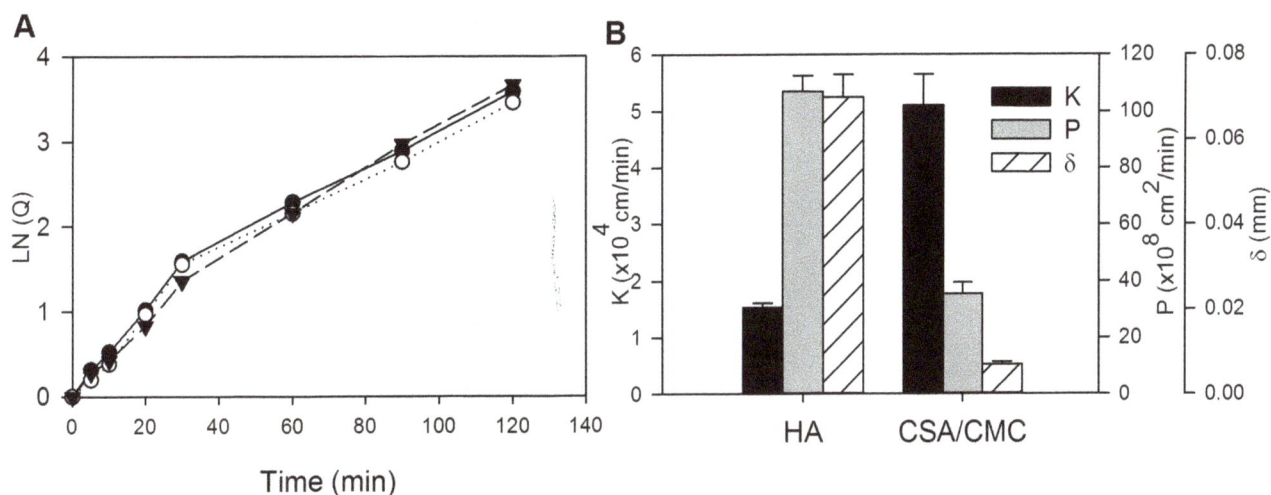

Figure 6. Albumin permeability measurements and mass transfer characteristics of HA and CSA/CMC capsules. (A) Representative plots of the concentration factor ln(Q) vs. time for three replicate runs with the CSA/CMC capsule formulation. (B) Plots of permeability coefficient (P), overall mass transfer coefficient (K) and wall thickness (δ) for the HA and CSA/CMC capsule formulations. Error bars represent the standard deviation of three replicate measurements.

Capsules that were surface-seeded with HUVECs and subsequently fused showed well endothelialized, interconnected channels as seen in phase contrast (Figure 12A) and confocal images (Figure 12B). SEM imaging of an axial section through a fused construct shows the interconnected channels more clearly (Figure 12C).

In a modified procedure, fusion-based assembly of high cell density capsules was explored by encapsulating primary rat hepatocytes in HA-collagen capsules at a density of 10×10^6 cells/ml of the HA/collagen solution, followed by heparin reloading, and centrifugation for 10 seconds at 50G to expel excess intracapsular liquid. H&E stained sections of the resulting construct showed a dense cell mass ($\sim 5 \times 10^7$ cells/cm^3, estimated via image analysis) with reduced, but still significant intercapsule spaces (Figure 13 A–B). Encapsulated hepatocyte constructs assembled without centrifugation showed a much less dense cellular construct (estimated at 9×10^6 cells/cm^3 via image analysis) with more and larger intercapsule spaces (Figure 13 C–

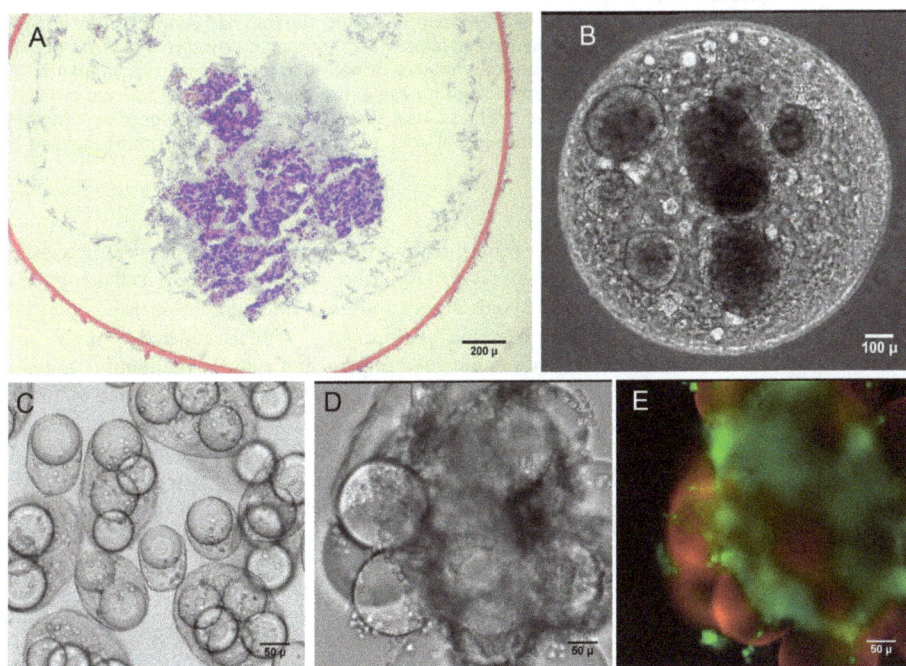

Figure 7. Tuning the inner capsule microenvironment with a collagen gel matrix and microcarriers. (A–B) HTBs in CSA/CMC capsules with a collagen type-I gel after one week of static culture. (A) H&E histology. (B) Phase contrast image. (C–E) SMCs co-encapsulated with gelatin coated dextran (Cytodex-3) microcarriers in HA capsules. (C) 60 min after encapsulation. (D) Day 14 of culture. (E) Calcein-AM stained fluorescence images on day 14 (green = live cells, red = microcarriers).

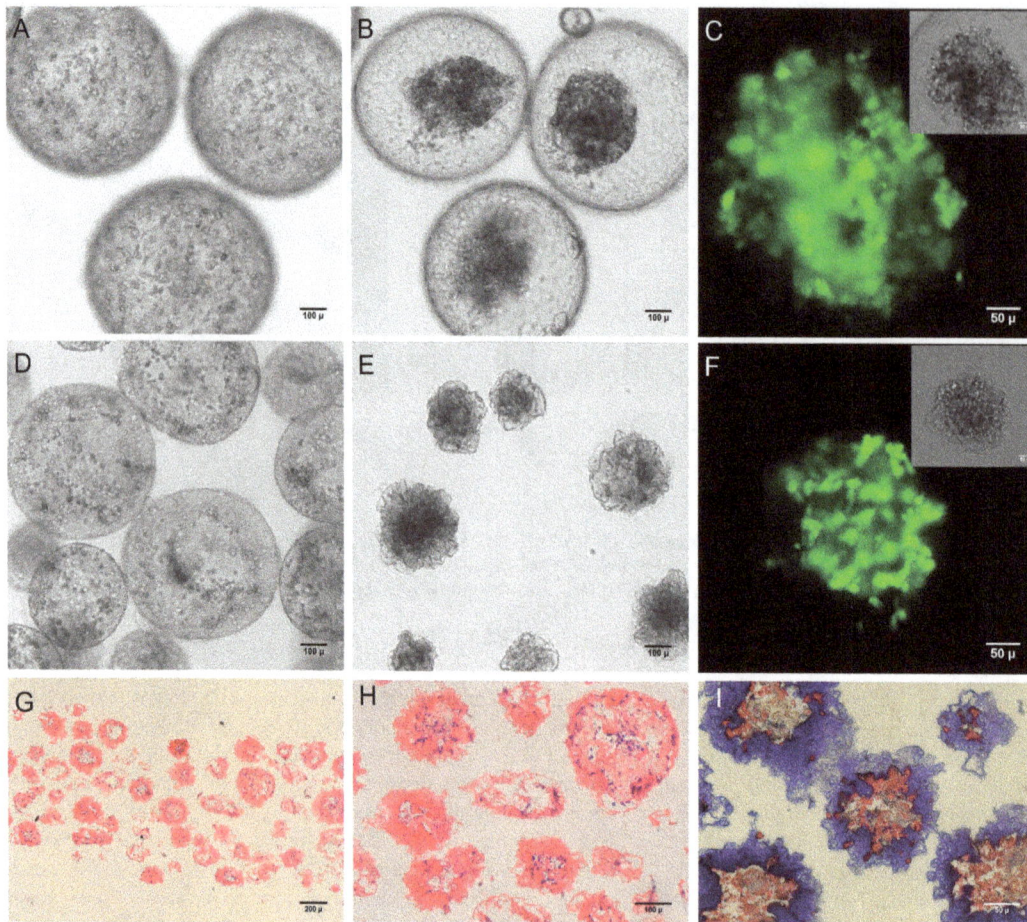

Figure 8. Vascular smooth muscle cells in collagen-containing capsules. (A–C) SMCs in CSA/CMC capsules with a 1 mg/ml collagen gel. (A) 60 min after encapsulation, SMCs are well dispersed in the internal collagen matrix. (B) After 24 hours of culture, the cells had contracted the internal collagen gel and formed a dense cell-matrix mass. (C) Calcein-AM fluorescence of contracted cell mass. Inset shows phase contrast image. (D–F) SMC encapsulated in HA capsules with 1 mg/ml collagen-I gel. (D) 60 min after encapsulation, cells are well dispersed in the internal collagen matrix. (E) After 24 hours of culture, the cells contracted the internal collagen gel, simultaneously collapsing the entire capsule structure to form a denser module with a convoluted surface membrane. (F) Calcein-AM fluorescence of contracted cell mass. Inset shows phase contrast image. (G–I) Histology of contracted capsules. H&E (G,H) staining showing compacted capsule structure with minimal void volume. (I) Masson's Trichrome staining of contracted capsule, showing the distribution of collagen (blue) within the structure.

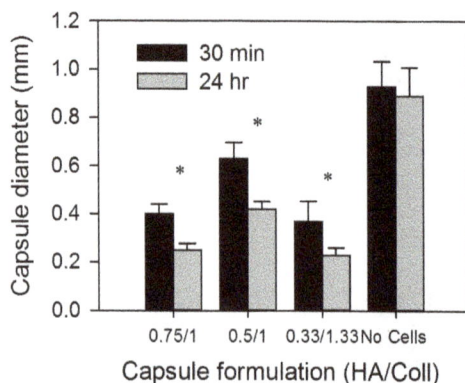

Figure 9. Effects of HA and collagen concentrations on cell-mediated capsule contraction. Maximal cell-mediated contraction was seen in the formulation with a final concentration of 0.33 wt% HA/ 1.33 mg/ml collagen-I. Capsules without cells exhibited an insignificant reduction in capsule diameter. Error bars represent standard deviation from at least 10 capsule measurements. Asterix denote statistically significant differences ($p < 0.05$).

D). These results demonstrate that additional physical processing methods can be used to further adjust the effective cell density and perfusable void space within these modular constructs.

The metabolic performance of encapsulated primary rat hepatocytes maintained in perfusion culture conditions (Figure 14 C–D) was evaluated by measuring urea and albumin synthesis rates and comparing to rates of identical cells in standard collagen sandwich dish cultures (Figure 14A). Perfusion cultures maintained the functionality of encapsulated hepatocytes and healthy spheroids were seen in most capsules (Figure 14B). Albumin and urea synthesis rates in both types of perfusion cultures (Figure 14 G,H) approached those of the collagen sandwich cultures (Figure 14 E–F).

Discussion

Modularity is a phenomenon widely observed in nature, which enables biological systems to achieve precise control over organization and function in very compact spaces. The modularity of the kidney and its component nephrons are excellent examples of this concept. Adopting a similar approach in engineered organs

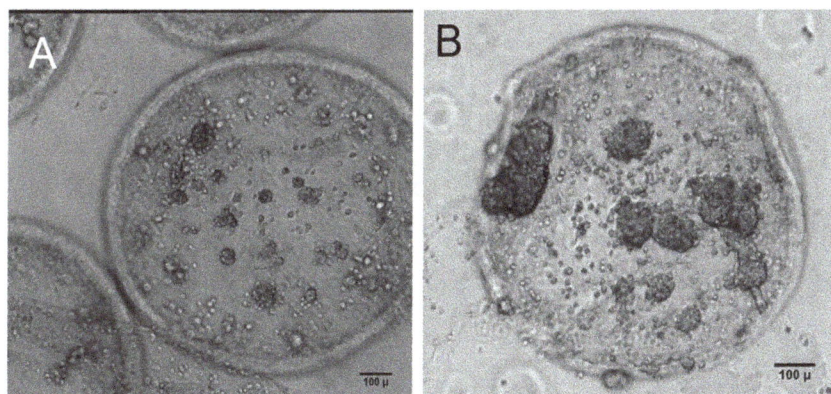

Figure 10. Cell growth in encapsulated cocultures of SMC and AEC. Cocultures of encapsulated SMCs with AECs on the external surfaces of HA/Collagen capsules exhibited increased SMC proliferation compared to encapsulated SMC monocultures. (A) Encapsulated SMCs only, day 7. (B) Encapsulated SMC with AECs, day 7.

has a number of advantages. The scalability of the modular strategy enables rapid fabrication of tissue constructs with greater control over their architecture. The major design challenges of a modular tissue construct include: limiting mass transfer distances, achieving high, tissue-like cell densities, and the ability to form interconnected, vascularizable channels. The GAG-based microcapsules described here allow efficient mass transfer, which is evident from the tissue-density cultures that were maintained for up to 45 days under static culture conditions. The diameter of the capsules can be easily controlled between 0.3 and 2.0 mm, and smaller diameters are achievable using more sophisticated droplet formation methods such as microfluidics. Capsule diameter imposes a natural upper limit on the maximum diffusion distances. The capsule system, in particular the hyaluronan-based capsules, supports direct encapsulation of cells at high, in vivo-like densities. In addition, the cell-contractable capsule formulations provide an additional mechanism for modulating cell density within either the capsules or the fused construct. Under random packing conditions, capsule fusion produced 3D structures with significant void space available for direct perfusion, accessory cell culture, or vascularization. The dimensions and architecture of the intercapsular voids can also be modulated by incorporating additional biomaterial components into fused capsule structures. Such accessory components include fibers, beads, films, tubes, etc. made from chitosan, chitosan-GAG complexes, or other degradable materials. The materials used in our modules are fully degradable, and previous

implantation results with similar materials indicate that in contrast to pure chitosan [52], chitosan-GAG complexes [53] degrade rapidly in vivo and stimulate rapid and extensive neovascularization due to GAG-mediated effects [54–58]. The high density trophoblast cultures were primarily intended to demonstrate the potential of the microcapsules with a highly proliferative human cell type. However, these cultures also provided direct evidence of both the degradability of the GAG-chitosan materials, and the ability of cells to invade the capsule wall. The trophoblast cell line maintains some characteristics of human trophoblasts, in particular the ability to tolerate hypoxic conditions and to invade tissue rapidly. Both characteristics are presumably related to its original, placenta-formation function [59] and may be mediated by focal expression of MMPs, GAG lyases or other matrix degrading enzymes. Wall invasion and cell escape in these trophoblast cultures was evident after week 2 of culture and was clearly captured in histological sections (Figure 3C). This phenomenon strongly suggests that implanted capsules would present only a temporary barrier to integration of encapsulated cells with adjacent tissues. Coupled with the known pro-angiogenic effects of GAG-based materials [60–62], these results further suggest that rapid vascularization is a likely outcome after transplantation of capsule-based constructs

Beyond modular assembly, the ability to incorporate clinically significant cell numbers into an implantable construct of feasible size is an additional challenge. We have shown that cells and

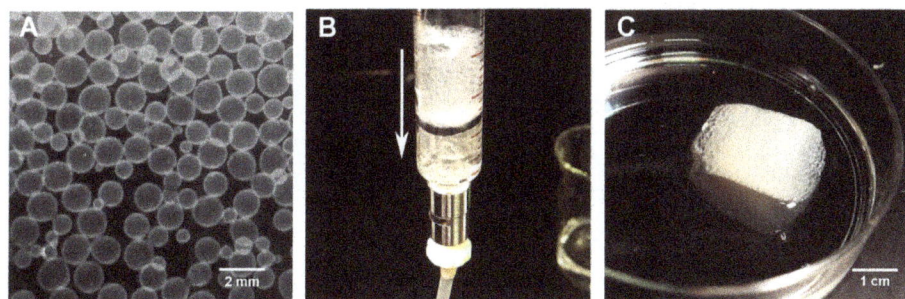

Figure 11. Modular assembly of GAG based microcapsules by fusion. Modular constructs were fabricated by perfusing packed capsules in a chamber of desired dimensions with diluted polymer solutions. This method yielded self-supporting constructs with uniform porosity. (A) Individual capsules in buffer solution before fusion. (B) Capsules being perfused with polymer solution in a perfusion chamber. Arrow indicates direction of flow. (C) Fused construct after removal from perfusion chamber.

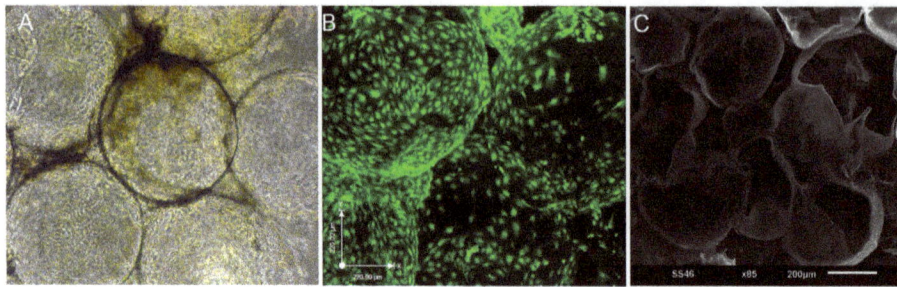

Figure 12. Endothelialized, interconnected channels in a fused modular construct. (A) Phase contrast image of CSA/CMC capsules, seeded externally with HUVECs and fused 48 hours after seeding. (B) Combined confocal image stack of the modular construct shown in A with HUVECS visualized via CellTracker Green staining. (C) SEM image of an axially sectioned, modular construct assembled from fused empty capsules showing interconnected channels.

matrix can be efficiently packed inside capsules of a non-diffusion limited size (Figures 8, 13). Our liver organoid prototype had a cell density of 50×10^6 cells/cm^3 (Figure 13A). This is 40-60% of the hepatocyte cell density of liver tissue. From a practical standpoint, the cell densities achieved in our systems are adequate for liver tissue engineering, as it has been demonstrated that with good blood chemistry, ~10% of total liver mass can support survival in rats [63,64] and humans [65]. It should be noted that maintaining high cell densities inside capsules presents particular diffusion challenges in the case of highly metabolic cells such as primary hepatocytes. Thus, we have also shown that the cell density can be scaled down to compensate for such high metabolic requirements (Figure 13 C–D). In principle, diffusion challenges can be minimized by limiting the maximum capsule diameter to ensure an adequate supply of nutrients and oxygen to all regions of the cell mass. Diffusion inside capsules can be further modulated by controlling the extent of cell distribution and aggregation. In particular, co-encapsulating hydrogel components (e.g. collagen gels) or microcarriers provides a mechanism for tuning the interior microenvironment as well as the architecture of the cell mass. Such

hydrogel materials can benignly interfere or directly compete with large scale cell aggregation, and thus serve to promote formation of multiple smaller or looser cell aggregates.

The encapsulation method also allows incorporation of microcarriers of various biomaterials. As with hydrogels, these microcarriers can produce additional adhesion ligand signaling, organizational barriers or mechanical enhancement. Our results show that gelatin coated dextran microcarriers significantly enhanced the growth and viability of encapsulated smooth muscle cells. These and other cell-adhesive microparticles can also be used to alter the physical properties of the fused capsule construct. It should also be noted that inclusion of microcarriers resulted in capsules with reduced osmotic swelling and substantially reduced internal volumes. This was particularly noteworthy in the case of CSA/CMC capsules which swelled more than HA capsules. We postulate that the increased swelling in the CSA/CMC system was due to higher interior osmotic pressures resulting from combined effects of a higher mass concentration, and lower molecular mass of the interior polymer solution compared to HA capsules. Inclusion of a high volume fraction of microcarriers within a

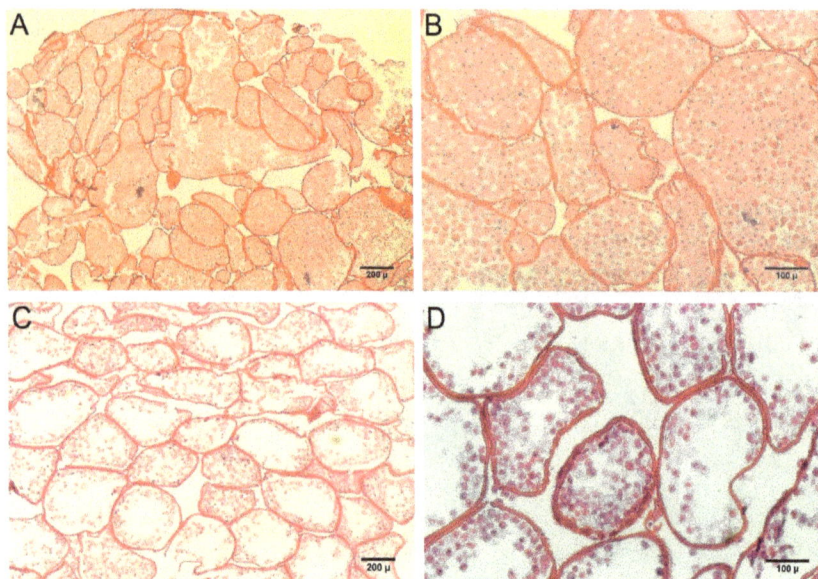

Figure 13. H&E staining of modular constructs based on hepatocytes in HA/collagen capsules. (A, B). Fused construct with reduced fluid volume and porosity due to centrifugation of capsules during the fusion process. (C, D) Construct formed by fusion of capsules settled under unit gravity, resulting in significantly greater fuid volume inside capsules and larger intercapsular spaces suitable for perfusion culture.

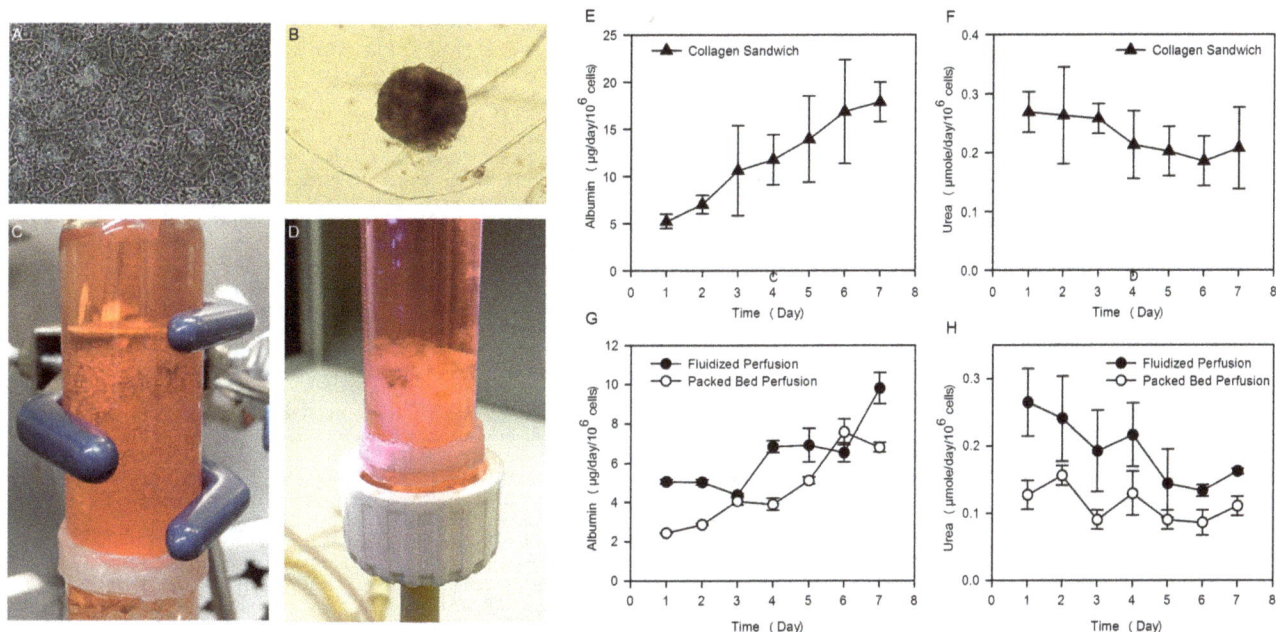

Figure 14. Albumin and urea synthesis rates of hepatocytes in encapsulated perfusion cultures. Primary rat hepatocytes were encapsulated in CSA/CMC capsules with a with 1 mg/ml collagen gel, at a density of 2×10^7 cells/ml of CSA/CMC/collagen solution. (A) Control collagen sandwich dish culture. (B) Encapsulated hepatocytes aggregated into spheroids during culture as either (C) individual capsules in a fluidized bed bioreactor, or as a (D) fused modular construct in a packed bed bioreactor. (E–H) Albumin and urea synthesis rates by the hepatocytes in the three culture conditions. (E,F) Control collagen sandwich cultures. (G, H) Perfusion cultures. Error bars denote standard deviations from 3 replicate measurements.

capsule-forming CSA/CMC droplet reduced both the volume of CSA/CMC solution inside the capsule and also the residual concentration of this solution after capsule membrane formation. This lower final concentration (caused by incorporation of polymer into the capsule membrane) produced a lower osmotic pressure and resulted in contraction of the capsule membrane around the cell+microcarrier mass. In general, the inclusion of microparticles provides a wide range of options for tuning the cellular organization and overall mechanical properties of modular constructs.

Fused capsule modules yielded 3D constructs with inter-capsular spaces that are perfusable in vitro and potentially vascularizable in vivo. The urea and albumin synthesis rates of the perfused cultures indicate that mass transfer rates were sufficient to maintain the encapsulated hepatocytes in our modular constructs. In addition, the interconnected endothelilized channels may provide a foundation for a vascular network and thereby accelerate the process of neovascularization by anastomosing with the host vasculature post-implantation. At the very least, intercapsular endothelial cells are likely to participate in vessel formation between fused capsules. However, the kinetics of this process, and the relative degrees of transplanted vs. host cell organization in the final structure remain to be characterized through animal studies.

Our results also suggest that the capsule membrane can facilitate paracrine signaling as seen by the increase in SMC proliferation during coculture with endothelial cells. This suggests that various other interacting cell types can be cultured in this modular system with a degree of material-based control over cell organization while still allowing substantial paracrine signaling. Several coculture systems have previously been shown to improve morphology and function of engineered tissues including liver [66–

70], bone [71,72,73] and cartilage [74–76]. Our results suggest that similar trophic effects can be achieved with ease in capsule-based modular scaffolds, with added the added benefit of control over cell arrangement and distribution.

Unlike traditional scaffolds, porosity can be either maintained evenly throughout the modular capsule scaffolds or different layers with different capsule sizes and hence different porosity can be easily implemented. GAG-chitosan surfaces can support cell adhesion and proliferation, partly due to GAG-mediated binding of matrix proteins and growth factors [53,77]. External cell adhesion can further be enhanced by directly incorporating cell-adhesive proteins such as collagen into the capsule wall by either blending with the polycationic solution or direct application to external capsule surfaces.

In conclusion, we have demonstrated the formation and use of GAG-based microcapsules to generate a variety of tunable, intracapsular microenvironments. These capsules have been shown suitable for fabrication of porous, 3D constructs that have the potential to mimic native tissue architecture with high cell densities, vascular and parenchymal cell types, and perfusable, endothelium-lined channels. This capsule-based modular tissue assembly approach is a promising strategy that provides a wide range of options for the efficient assembly of three-dimensional, engineered tissues.

Author Contributions

Conceived and designed the experiments: HWTM RTA. Performed the experiments: RTA. Analyzed the data: RTA HWTM. Contributed reagents/materials/analysis tools: DRA HWTM. Wrote the paper: RTA HWTM.

References

1. Langer R, Vacanti JP (1993) Tissue engineering. Science 260: 920–926.
2. MacNeil S (2007) Progress and opportunities for tissue-engineered skin. Nature 445: 874–880.
3. Jones E, Yang X (2011) Mesenchymal stem cells and bone regeneration: Current status. Injury 42: 562–568.
4. Hammouche S, Hammouche D, McNicholas M (2012) Biodegradable bone regeneration synthetic scaffolds: in tissue engineering. Curr Stem Cell Res Ther 7: 134–142.
5. Zhang Y, Venugopal JR, El-Turki A, Ramakrishna S, Su B, et al. (2008) Electrospun biomimetic nanocomposite nanofibers of hydroxyapatite/chitosan for bone tissue engineering. Biomaterials 29: 4314–4322.
6. LaPorta TF, Richter A, Sgaglione NA, Grande DA (2012) Clinical relevance of scaffolds for cartilage engineering. Orthop Clin North Am 43: 245–254, vi.
7. Vinatier C, Bouffi C, Merceron C, Gordeladze J, Brondello JM, et al. (2009) Cartilage tissue engineering: towards a biomaterial-assisted mesenchymal stem cell therapy. Curr Stem Cell Res Ther 4: 318–329.
8. Cunha C, Panseri S, Antonini S (2011) Emerging nanotechnology approaches in tissue engineering for peripheral nerve regeneration. Nanomedicine 7: 50–59.
9. Germain L, Carrier P, Auger FA, Salesse C, Guerin SL (2000) Can we produce a human corneal equivalent by tissue engineering? Prog Retin Eye Res 19: 497–527.
10. Paquet C, Larouche D, Bisson F, Proulx S, Simard-Bisson C, et al. (2010) Tissue engineering of skin and cornea: Development of new models for in vitro studies. Ann N Y Acad Sci 1197: 166–177.
11. Lawrence BD, Marchant JK, Pindrus MA, Omenetto FG, Kaplan DL (2009) Silk film biomaterials for cornea tissue engineering. Biomaterials 30: 1299–1308.
12. Ravi S, Chaikof EL (2010) Biomaterials for vascular tissue engineering. Regen Med 5: 107–120.
13. Carmeliet P, Jain RK (2000) Angiogenesis in cancer and other diseases. Nature 407: 249–257.
14. Hahn MS, Taite LJ, Moon JJ, Rowland MC, Ruffino KA, et al. (2006) Photolithographic patterning of polyethylene glycol hydrogels. Biomaterials 27: 2519–2524.
15. Leslie-Barbick JE, Moon JJ, West JL (2009) Covalently-Immobilized Vascular Endothelial Growth Factor Promotes Endothelial Cell Tubulogenesis in Poly(ethylene glycol) Diacrylate Hydrogels. Journal of Biomaterials Science, Polymer Edition 20: 1763–1779.
16. Chiu LLY, Radisic M (2010) Scaffolds with covalently immobilized VEGF and Angiopoietin-1 for vascularization of engineered tissues. Biomaterials 31: 226–241.
17. Miyagi Y, Chiu LLY, Cimini M, Weisel RD, Radisic M, et al. (2011) Biodegradable collagen patch with covalently immobilized VEGF for myocardial repair. Biomaterials 32: 1280–1290.
18. Nillesen STM, Geutjes PJ, Wismans R, Schalkwijk J, Daamen WF, et al. (2007) Increased angiogenesis and blood vessel maturation in acellular collagen-heparin scaffolds containing both FGF2 and VEGF. Biomaterials 28: 1123–1131.
19. Jain RK (2003) Molecular regulation of vessel maturation. Nat Med 9: 685–693.
20. McGuigan AP, Sefton MV (2006) Vascularized organoid engineered by modular assembly enables blood perfusion. Proc Natl Acad Sci U S A 103: 11461–11466.
21. Nichol JW, Khademhosseini A (2009) Modular Tissue Engineering: Engineering Biological Tissues from the Bottom Up. Soft Matter 5: 1312–1319.
22. Lovett M, Lee K, Edwards A, Kaplan DL (2009) Vascularization strategies for tissue engineering. Tissue Eng Part B Rev 15: 353–370.
23. van der Meer AD, Poot AA, Duits MH, Feijen J, Vermes I (2009) Microfluidic technology in vascular research. J Biomed Biotechnol 2009: 823148.
24. Gaetani R, Doevendans PA, Metz CH, Alblas J, Messina E, et al. (2012) Cardiac tissue engineering using tissue printing technology and human cardiac progenitor cells. Biomaterials 33: 1782–1790.
25. Mironov V, Boland T, Trusk T, Forgacs G, Markwald RR (2003) Organ printing: computer-aided jet-based 3D tissue engineering. Trends Biotechnol 21: 157–161.
26. Fedorovich NE, Wijnberg HM, Dhert WJ, Alblas J (2011) Distinct tissue formation by heterogeneous printing of osteo- and endothelial progenitor cells. Tissue Eng Part A 17: 2113–2121.
27. L'Heureux N, McAllister TN, de la Fuente LM (2007) Tissue-engineered blood vessel for adult arterial revascularization. N Engl J Med 357: 1451–1453.
28. Leung BM, Sefton MV (2010) A modular approach to cardiac tissue engineering. Tissue Eng Part A 16: 3207–3218.
29. Matthew HW, Salley SO, Peterson WD, Klein MD (1993) Complex coacervate microcapsules for mammalian cell culture and artificial organ development. Biotechnol Prog 9: 510–519.
30. Lin SV, Matthew HW (2002) Microencapsulation methods: Glycosaminoglycans and Chitosan. In: Atala A, Lanza R, editors. Methods of Tissue Engineering. pp. 815–823.
31. Matthew HW, Basu S, Peterson WD, Salley SO, Klein MD (1993) Performance of plasma-perfused, microencapsulated hepatocytes: prospects for extracorporeal liver support. J Pediatr Surg 28: 1423–1427; discussion 1427–1428.
32. Matthew HW, Salley SO, Peterson WD Jr, Deshmukh DR, Mukhopadhyay A, et al. (1991) Microencapsulated hepatocytes. Prospects for extracorporeal liver support. ASAIO Trans 37: M328–330.

33. Surapaneni S, Pryor T, Klein MD, Matthew HW (1997) Rapid hepatocyte spheroid formation: optimization and long-term function in perfused microcapsules. ASAIO J 43: M848–853.
34. Graham CH, Hawley TS, Hawley RG, MacDougall JR, Kerbel RS, et al. (1993) Establishment and characterization of first trimester human trophoblast cells with extended lifespan. Exp Cell Res 206: 204–211.
35. Butcher JT, Nerem RM (2004) Porcine aortic valve interstitial cells in three-dimensional culture: comparison of phenotype with aortic smooth muscle cells. J Heart Valve Dis 13: 478–485; discussion 485–476.
36. Christen T, Bochaton-Piallat ML, Neuville P, Rensen S, Redard M, et al. (1999) Cultured porcine coronary artery smooth muscle cells. A new model with advanced differentiation. Circ Res 85: 99–107.
37. Seglen PO (1979) Hepatocyte suspensions and cultures as tools in experimental carcinogenesis. J Toxicol Environ Health 5: 551–560.
38. Dunn JC, Tompkins RG, Yarmush ML (1991) Long-term in vitro function of adult hepatocytes in a collagen sandwich configuration. Biotechnol Prog 7: 237–245.
39. Dunn JC, Yarmush ML, Koebe HG, Tompkins RG (1989) Hepatocyte function and extracellular matrix geometry: long-term culture in a sandwich configuration. FASEB J 3: 174–177.
40. Matthew HWT, Sternberg J, Stefanovich P, Morgan JR, Toner M, et al. (1996) Effects of plasma exposure on cultured hepatocytes: Implications for bioartificial liver support. Biotechnology and Bioengineering 51: 100–111.
41. Crooks CA, Douglas JA, Broughton RL, Sefton MV (1990) Microencapsulation of mammalian cells in a HEMA-MMA copolymer: effects on capsule morphology and permeability. J Biomed Mater Res 24: 1241–1262.
42. Gallagher SR, Desjardins PR (2001) Quantitation of DNA and RNA with Absorption and Fluorescence Spectroscopy. Current Protocols in Protein Science: John Wiley & Sons, Inc. pp. A.4K.1–A.4K.21.
43. Wybenga DR, Di Giorgio J, Pileggi VJ (1971) Manual and Automated Methods for Urea Nitrogen Measurement in Whole Serum. Clinical Chemistry 17: 891–895.
44. Jain M, He Q, Lee WS, Kashiki S, Foster LC, et al. (1996) Role of CD44 in the reaction of vascular smooth muscle cells to arterial wall injury. J Clin Invest 97: 596–603.
45. Armulik A, Abramsson A, Betsholtz C (2005) Endothelial/Pericyte Interactions. Circ Res 97: 512–523.
46. Selwyn AP (2003) Prothrombotic and antithrombotic pathways in acute coronary syndromes. American Journal of Cardiology 91: 3H–11H.
47. Govers R, Bevers L, de Bree P, Rabelink TJ (2002) Endothelial nitric oxide synthase activity is linked to its presence at cell-cell contacts. Biochem J 361: 193–201.
48. Scheppke L, Murphy EA, Zarpellon A, Hofmann JJ, Merkulova A, et al. (2012) Notch promotes vascular maturation by inducing integrin-mediated smooth muscle cell adhesion to the endothelial basement membrane. Blood 119: 2149–2158.
49. Zhao X, Liu L, Wang FK, Zhao DP, Dai XM, et al. (2012) Coculture of vascular endothelial cells and adipose-derived stem cells as a source for bone engineering. Ann Plast Surg 69: 91–98.
50. Fillinger MF, Sampson LN, Cronenwett JL, Powell RJ, Wagner RJ (1997) Coculture of endothelial cells and smooth muscle cells in bilayer and conditioned media models. J Surg Res 67: 169–178.
51. Lavender MD, Pang Z, Wallace CS, Niklason LE, Truskey GA (2005) A system for the direct co-culture of endothelium on smooth muscle cells. Biomaterials 26: 4642–4653.
52. VandeVord PJ, Matthew HW, DeSilva SP, Mayton L, Wu B, et al. (2002) Evaluation of the biocompatibility of a chitosan scaffold in mice. J Biomed Mater Res 59: 585–590.
53. Chupa JM, Foster AM, Sumner SR, Madihally SV, Matthew HWT (2000) Vascular cell responses to polysaccharide materials: in vitro and in vivo evaluations. Biomaterials 21: 2315–2322.
54. Fuster MM, Wang L (2010) Endothelial heparan sulfate in angiogenesis. Prog Mol Biol Transl Sci 93: 179–212.
55. Gaffney J, Matou-Nasri S, Grau-Olivares M, Slevin M (2010) Therapeutic applications of hyaluronan. Mol Biosyst 6: 437–443.
56. Norrby K (2006) Low-molecular-weight heparins and angiogenesis. APMIS 114: 79–102.
57. Stringer SE (2006) The role of heparan sulphate proteoglycans in angiogenesis. Biochem Soc Trans 34: 451–453.
58. West DC, Kumar S (1989) Hyaluronan and angiogenesis. Ciba Found Symp 143: 187–201; discussion 201–185, 281–185.
59. Chang SC, Vivian Yang WC (2013) Hyperglycemia induces altered expressions of angiogenesis associated molecules in the trophoblast. Evid Based Complement Alternat Med 2013: 457971.
60. Mathieu C, Chevrier A, Lascau-Coman V, Rivard GE, Hoemann CD (2013) Stereological analysis of subchondral angiogenesis induced by chitosan and coagulation factors in microdrilled articular cartilage defects. Osteoarthritis Cartilage 21: 849–859.
61. Ferretti A, Boschi E, Stefani A, Spiga S, Romanelli M, et al. (2003) Angiogenesis and nerve regeneration in a model of human skin equivalent transplant. Life Sci 73: 1985–1994.

62. Black AF, Hudon V, Damour O, Germain L, Auger FA (1999) A novel approach for studying angiogenesis: a human skin equivalent with a capillary-like network. Cell Biol Toxicol 15: 81–90.

63. Kobayashi N, Miyazaki M, Fukaya K, Inoue Y, Sakaguchi M, et al. (2000) Transplantation of highly differentiated immortalized human hepatocytes to treat acute liver failure. Transplantation 69: 202–207.

64. Arkadopoulos N, Lilja H, Suh KS, Demetriou AA, Rozga J (1998) Intrasplenic transplantation of allogeneic hepatocytes prolongs survival in anhepatic rats. Hepatology 28: 1365–1370.

65. Bilir BM, Guinette D, Karrer F, Kumpe DA, Krysl J, et al. (2000) Hepatocyte transplantation in acute liver failure. Liver Transpl 6: 32–40.

66. No da Y, Lee SA, Choi YY, Park D, Jang JY, et al. (2012) Functional 3D human primary hepatocyte spheroids made by co-culturing hepatocytes from partial hepatectomy specimens and human adipose-derived stem cells. PLoS One 7: e50723.

67. Kim K, Ohashi K, Utoh R, Kano K, Okano T (2012) Preserved liver-specific functions of hepatocytes in 3D co-culture with endothelial cell sheets. Biomaterials 33: 1406–1413.

68. Kasuya J, Sudo R, Mitaka T, Ikeda M, Tanishita K (2011) Hepatic stellate cell-mediated three-dimensional hepatocyte and endothelial cell triculture model. Tissue Eng Part A 17: 361–370.

69. Yagi H, Parekkadan B, Suganuma K, Soto-Gutierrez A, Tompkins RG, et al. (2009) Long-term superior performance of a stem cell/hepatocyte device for the treatment of acute liver failure. Tissue Eng Part A 15: 3377–3388.

70. Parekkadan B, van Poll D, Megeed Z, Kobayashi N, Tilles AW, et al. (2007) Immunomodulation of activated hepatic stellate cells by mesenchymal stem cells. Biochem Biophys Res Commun 363: 247–252.

71. Sun H, Qu Z, Guo Y, Zang G, Yang B (2007) In vitro and in vivo effects of rat kidney vascular endothelial cells on osteogenesis of rat bone marrow mesenchymal stem cells growing on polylactide-glycoli acid (PLGA) scaffolds. Biomed Eng Online 6: 41.

72. Steiner D, Lampert F, Stark GB, Finkenzeller G (2012) Effects of endothelial cells on proliferation and survival of human mesenchymal stem cells and primary osteoblasts. J Orthop Res 30: 1682–1689.

73. Tao J, Sun Y, Wang QG, Liu CW (2009) Induced endothelial cells enhance osteogenesis and vascularization of mesenchymal stem cells. Cells Tissues Organs 190: 185–193.

74. Qing C, Wei-ding C, Wei-min F (2011) Co-culture of chondrocytes and bone marrow mesenchymal stem cells in vitro enhances the expression of cartilaginous extracellular matrix components. Braz J Med Biol Res 44: 303–310.

75. Meretoja VV, Dahlin RL, Wright S, Kasper FK, Mikos AG (2013) The effect of hypoxia on the chondrogenic differentiation of co-cultured articular chondro-cytes and mesenchymal stem cells in scaffolds. Biomaterials 34: 4266–4273.

76. Bian L, Zhai DY, Mauck RL, Burdick JA (2011) Coculture of human mesenchymal stem cells and articular chondrocytes reduces hypertrophy and enhances functional properties of engineered cartilage. Tissue Eng Part A 17: 1137–1145.

77. Uygun BE, Stojsih SE, Matthew HW (2009) Effects of immobilized glycosaminoglycans on the proliferation and differentiation of mesenchymal stem cells. Tissue Eng Part A 15: 3499–3512.

PERMISSIONS

LIST OF CONTRIBUTORS

Chuanshun Wang, Kai Li, Yaping Ju, Jipeng Li, Yongxing Zhang and Qinghua Zhao
Department of Orthopaedics, Shanghai First People's Hospital, School of Medicine, Shanghai Jiao Tong University, Shanghai, P. R. China

Shige Wang
State Key Laboratory for Modification of Chemical Fibers and Polymer Materials, College of Materials Science and Engineering, Donghua University, Shanghai, P. R. China

Jinhua Li and Xuanyong Liu
State Key Laboratory of High Performance Ceramics and Superfine Microstructure, Shanghai Institute of Ceramics, Chinese Academy of Sciences, Shanghai, P. R. China

Xiangyang Shi
State Key Laboratory for Modification of Chemical Fibers and Polymer Materials, College of Materials Science and Engineering, Donghua University, Shanghai, P. R. China
College of Chemistry, Chemical Engineering and Biotechnology, Donghua University, Shanghai, P. R. China
Department of Chemistry, Chemical Biology, and Biomedical Engineering, Stevens Institute of Technology, Hoboken, New Jersey, United States of America
Department of Physics and Mathematics, School of Biomedical Engineering, Fourth Military Medical University, Xi'an, Shaanxi, People's Republic of China

Wei Chang, Paul Lee, Yuhao Wang, Min Yang, Jun Li and Xiaojun Yu
Department of Chemistry, Chemical Biology, and Biomedical Engineering, Stevens Institute of Technology, Hoboken, New Jersey, United States of America

Sangamesh G. Kumbar
Department of Orthopaedic Surgery, University of Connecticut Health Center, Farmington, Connecticut, United States of America

Alfred Gugerell, Johanna Kober and Maike Keck
Division of Plastic and Reconstructive Surgery, Department of Surgery, Medical University of Vienna, Vienna, Austria

Thorsten Laube, Torsten Walter, Ralf Wyrwa and Matthias Schnabelrauch
Biomaterials Department, INNOVENT e. V., Jena, Germany

Sylvia Nürnberger
Ludwig Boltzmann Institute for Experimental and Clinical Traumatology, Austrian Cluster for Tissue Regeneration, Vienna, Austria
Department of Traumatology, Medical University of Vienna, Vienna, Austria

Elke Grönniger and Simone Brönneke
Research Department Applied Skin Biology, Beiersdorf AG, Hamburg, Germany

Yu Ding
Department of Rehabilitation Medicine and Pain Management Center, Navy General Hospital, Beijing, China

Dike Ruan, Qing He and Chaofeng Wang
Department of Orthopaedics, Navy General Hospital, Beijing, China

Keith D. K. Luk
Department of Orthopaedics and Traumatology, The University of Hong Kong, Pokfulam, Hong Kong, China

Yu-Chieh Chiu and Sevi Kocagöz
Department of Biomedical Engineering, Illinois Institute of Technology, Chicago, Illinois, United States of America

Jeffery C. Larson and Eric M. Brey
Department of Biomedical Engineering, Illinois Institute of Technology, Chicago, Illinois, United States of America
Research Service, Hines Veterans Administration Hospital, Hines, Illinois, United States of America

Tien-En Tan
Yong Loo Lin School of Medicine, National University of Singapore, Singapore
Singapore National Eye Centre, Singapore

Gary S. L. Peh and Benjamin L. George
Tissue Engineering and Stem Cell Group, Singapore Eye Research Institute, Singapore

Howard Y. Cajucom-Uy
Singapore Eye Bank, Singapore

Di Dong
Health Services and Systems Research, Duke-NUS Graduate Medical School, Singapore

Eric A. Finkelstein
Health Services and Systems Research, Duke-NUS Graduate Medical School, Singapore
Lien Centre for Palliative Care, Singapore

Jodhbir S. Mehta
National Eye Centre, Singapore
Tissue Engineering and Stem Cell Group, Singapore Eye Research Institute, Singapore
Department of Clinical Sciences, Duke-NUS Graduate Medical School, Singapore

Talita da Silva Jeremias, Rafaela Grecco Machado, Silvia Beatriz Coutinho Visoni and Andrea Gonçalves Trentin
Departamento de Biologia Celular, Embriologia e Genética, Centro de Ciências Biológicas, Universidade Federal de Santa Catarina, Florianópolis, Santa Catarina, Brasil

Maurício José Pereima
Departamento de Pediatria, Centro de Ciências da Saúde, Universidade Federal de Santa Catarina, Florianópolis, Santa Catarina, Brasil
Hospital Infantil Joana de Gusmão, Florianópolis, Santa Catarina, Brasil

Dilmar Francisco Leonardi
Governador Celso Ramos, Florianópolis, Santa Catarina, Brasil
Departamento de Cirurgia, Universidade do Sul de Santa Catarina, Florianópolis, Santa Catarina, Brasil

Elahe Masaeli
Department of Textile Engineering, Isfahan University of Technology, Isfahan, Iran
Department of Cell and Molecular Biology, Cell Science Research Center, Royan Institute for Biotechnology, ACECR, Isfahan, Iran
Department of Tissue Regeneration, University of Twente, Enschede, The Netherlands

Mohammad Morshed
Department of Textile Engineering, Isfahan University of Technology, Isfahan, Iran

Mohammad Hossein Nasr-Esfahani
Department of Cell and Molecular Biology, Cell Science Research Center, Royan Institute for Biotechnology, ACECR, Isfahan, Iran

Saeid Sadri
Department of Electrical and Computer Engineering, Isfahan University of Technology, Isfahan, Iran

Jun Li and Chen Liu
Department of Orthopaedics, The First Affiliated Hospital of Soochow University, Suzhou, Jiangsu, China

Qianping Guo
Orthopedic Institute, Soochow University, Suzhou, Jiangsu, China

Huilin Yang and Bin Li
Department of Orthopaedics, The First Affiliated Hospital of Soochow University, Suzhou, Jiangsu, China
Orthopedic Institute, Soochow University, Suzhou, Jiangsu, China

Haiwei Xu, Qiang Yang and Yaohong Wu
Department of Spine Surgery, Tianjin Hospital, Tianjin, China
Graduate School, Tianjin Medical University, Tianjin, China

Baoshan Xu, Xinlong Ma and Qun Xia
Department of Spine Surgery, Tianjin Hospital, Tianjin, China

Xiulan Li and Yang Zhang
Cell Engineering Laboratory of Orthopaedic Institute, Tianjin Hospital, Tianjin, China

Chunqiu Zhang
School of Mechanical Engineering, Tianjin University of Technology, Tianjin, China

Yuanyuan Zhang
Wake Forest Institute for Regenerative Medicine, Wake Forest University School of Medicine, Winston-Salem, North Carolina, United States of America

Stefanie Michael, Heiko Sorg, Claas-Tido Peck, Peter M. Vogt and Kerstin Reimers
Department of Plastic, Hand- and Reconstructive Surgery, Hannover Medical School, Hannover, Germany

Lothar Koch, Andrea Deiwick and Boris Chichkov
Laser Zentrum Hannover e.V., Hannover, Germany

Jinhui Shen and Ashwin Nair
Department of Bioengineering, University of Texas at Arlington, Arlington, Texas, United States of America

Ramesh Saxena
Division of Nephrology, University of Texas Southwestern Medical Center at Dallas, Dallas, Texas, United States of America

Cheng Cheng Zhang
Departments of Physiology and Developmental Biology, University of Texas Southwestern Medical Center at Dallas, Dallas, Texas, United States of America

Joseph Borrelli Jr.
Texas Health Physicians Group, Texas Health Arlington Memorial Hospital, Arlington, Texas, United States of America

Liping Tang
Department of Bioengineering, University of Texas at Arlington, Arlington, Texas, United States of America
Department of Biomedical Science and Environmental Biology, Kaohsiung Medical University, Kaohsiung, Taiwan

Eamon J. Sheehy and Conor T. Buckley
Trinity Centre for Bioengineering, Trinity Biomedical Sciences Institute, Trinity College Dublin, Dublin, Ireland
Department of Mechanical and Manufacturing Engineering, School of Engineering, Trinity College Dublin, Dublin, Ireland

Tatiana Vinardell
Trinity Centre for Bioengineering, Trinity Biomedical Sciences Institute, Trinity College Dublin, Dublin, Ireland
Department of Mechanical and Manufacturing Engineering, School of Engineering, Trinity College Dublin, Dublin, Ireland
School of Agriculture and Food Science, University College Dublin, Belfield, Dublin, Ireland

Mary E. Toner
Department of Pathology, School of Dental Science, TrinityCollege Dublin, Dublin, Ireland

Daniel J. Kelly
Trinity Centre for Bioengineering, Trinity Biomedical Sciences Institute, Trinity College Dublin, Dublin, Ireland
Department of Mechanical and Manufacturing Engineering, School of Engineering, Trinity College Dublin, Dublin, Ireland
Advanced Materials and Bioengineering Research Centre (AMBER), Trinity College Dublin, Dublin, Ireland

Aristos A. Athens
Department of Biomedical Engineering, University of California Davis, Davis, California, United States of America
Davis Senior High School, Davis, California, United States of America

Eleftherios A. Makris
Department of Biomedical Engineering, University of California Davis, Davis, California, United States of America
Department of Orthopedic Surgery and Musculoskeletal Trauma, University of Thessaly (BIOMED), Larisa, Greece

Jerry C. Hu
Department of Biomedical Engineering, University of California Davis, Davis, California, United States of America

Stephen D. Thorpe, Thomas Nagel, Simon F. Carroll and Daniel J. Kelly
Trinity Centre for Bioengineering, Trinity Biomedical Sciences Institute, Trinity College Dublin, Dublin, Ireland
Department of Mechanical and Manufacturing Engineering, School of Engineering, Trinity College Dublin, Dublin, Ireland

Madhu Sudhan Reddy Gudur, Rameshwar R. Rao, Alexis W. Peterson, David J. Caldwell, Jan P. Stegemann and Cheri X. Deng
Department of Biomedical Engineering, University of Michigan, Ann Arbor, Michigan, United States of America

Amilton M. Fernandes, Sarah R. Herlofsen, Tommy A. Karlsen, Axel M. Küchler and Jan E. Brinchmann
The Norwegian Center for Stem Cell Research, University of Oslo, Oslo, Norway
Institute of Immunology, Oslo University Hospital Rikshospitalet, Oslo, Norway

Yngvar Fløisand
Department of Hematology, Oslo University Hospital Rikshospitalet, Oslo, Norway

Daniel W. Youngstrom
Department of Biomedical and Veterinary Sciences, Virginia-Maryland Regional College of Veterinary Medicine, Virginia Tech, Leesburg, Virginia, United States of America

Jennifer G. Barrett
Department of Large Animal Clinical Sciences, Marion duPont Scott Equine Medical Center, Virginia-Maryland Regional College of Veterinary Medicine, Virginia Tech, Leesburg, Virginia, United States of America

Rod R. Jose and David L. Kaplan
Department of Biomedical Engineering, Tissue Engineering Resource Center, Tufts University, Medford, Massachusetts, United States of America

Anne Schellenberg, Giulio Abagnale, Sylvia Joussen, Giulio Abagnale, Sylvia Joussen and Wolfgang Wagner
Stem Cell Biology and Cellular Engineering, Helmholtz-Institute for Biomedical Engineering, RWTH Aachen University Medical School, Aachen, Germany

Robin Ross, Philipp Schuster, Annahit Arshi and Thomas Gries
Institute for Textile Technology RWTH Aachen University, Aachen, Germany

Norbert Pallua
Department of Plastic and Reconstructive Surgery, Hand Surgery, Burn Center, RWTH Aachen University, Aachen, Germany

Stefan Jockenhoevel
Institute for Textile Technology RWTH Aachen University, Aachen, Germany
Department of Applied Medical Engineering, Helmholtz-Institute for Biomedical Engineering, RWTH Aachen University, Aachen, Germany

Marije Sloff, Paul Geutjes, Egbert Oosterwijk and Wout Feitz
Department of Urology, Nijmegen Centre for Molecular Life Sciences, Radboud University Medical Center, Nijmegen, The Netherlands

Rob de Vries and Merel Ritskes-Hoitinga
SYRCLE (SYstematic Review Centre for Laboratory animal experimentation), Central Animal Laboratory, Radboud University Medical Center, Nijmegen, The Netherlands

Joanna IntHout
Department for Health Evidence, Radboud University Medical Center, Nijmegen, The Netherlands

Ramkumar Tiruvannamalai-Annamalai
Department of Biomedical Engineering, Wayne State University, Detroit, Michigan, United States of America

David Randall Armant
Departments of Obstetrics & Gynecology, Wayne State University, Detroit, Michigan, United States of America Program in Reproductive & Adult Endocrinology, National Institute of Child Health & Human Development, National Institutes of Health, Bethesda, Maryland, United States of America

Howard W. T. Matthew
Department of Biomedical Engineering, Wayne State University, Detroit, Michigan, United States of America Department of Chemical Engineering and Materials Science, Wayne State University, Detroit, Michigan, United States of America

Index

A

Accelerates Mineralisation, 130, 138
Adipose-derived Stem Cells, 11, 22-23, 34-35, 73, 224-225
Adipose-derived Stem Cells (ascs), 22-23
Adjacent Segment Degeneration (asd), 36
Af Tissue Harvesting, 88
Alkaline Phosphatase (alp) Activity, 3, 12
Annulus Fibrosus (af), 87, 98, 100
Articular Cartilage, 89, 96, 130, 139-141, 144-147, 154-157, 169-170, 175, 179, 181, 191, 209, 224
Articular Cartilage Defects, 181, 224
Autologous Progenitor Cells, 123

B

Beagle Model, 36
Biomechanical Characteristics, 87
Bone Defects, 1, 9-12, 19, 123, 130, 137
Bone Loss, 123

C

Cartilage End Plate (cep), 87
Cartilaginous Tissues Engineered, 130, 139, 157
Cell Culture And Seeding, 15
Cell Traction Forces (ctfs), 87, 92, 96
Cell Transplantation, 21, 36, 44, 65, 181
Cellular Compatibility, 73-74, 85
Characterization Of Scaffolds, 78
Collagen Content, 90, 98-99, 101, 105, 108, 139, 143, 146, 148, 151, 153, 155-157, 168, 184, 186, 188-189
Collagen Cross-links, 140
Compression Testing, 47
Congenital Deformity, 12
Corneal Endothelial Transplantation, 57, 63-65
Cost-minimization Analysis, 57-58, 63-64

D

Decellularization Protocols, 98, 183-184, 187, 189-191
Decellularized Tendon Scaffolds (dts), 183
Deep Dermal Burns, 22
Degenerative Disc Disease (ddd), 36
Degradation, 5, 12, 19-23, 25, 29-35, 46-49, 51-52, 54-56, 75, 86, 98, 130, 137-139, 141, 181, 192
Dermal Substitutes, 66-67, 70-72
Differential Scanning Calorimetry (dsc), 74, 83
Disc Degeneration Disease (ddd), 87

E

Effect Of Gamma Irradiation, 36
Engineer Zonal Articular Cartilage, 139, 146
Extra Cellular Matrix (ecm), 211
Extracellular Matrix (ecm), 7, 23, 74, 95, 99, 144, 155, 160, 169, 183

F

Full-thickness Skin Injuries, 66
Functional Characterization, 183

G

Gastrointestinal (gi) Tissue, 201

H

Hematoxylin And Eosin (h&e)-staining Data, 1
Hemocompatibility Assay, 5
Human Fetal Osteoblasts (hfobs), 12
Hydroxyapatite (ha), 1, 12
Hypertrophic Cartilaginous Grafts, 130

I

Immunofluorescence Staining, 67, 69, 71
Intervertebral Disc Allografts, 36, 44

L

Laponite Bioceramics, 1
Lesions Of Hyaline Cartilage, 169

M

Mesenchymal Stem Cell, 56, 69, 85-86, 127-128, 139, 146, 157-158, 183, 199
Mesenchymal Stem Cells, 1, 11, 21, 35-36, 66, 73-74, 109, 112, 118, 123, 128-130, 138-139, 145, 147, 157-158, 167-170, 173, 181-182, 190-192, 199-200, 224-225
Mesenchymal Stem Cells (mscs), 66, 123, 130, 169, 173
Mesenchymal Stromal Cells, 66-67, 71-73, 129, 181, 190, 192, 199
Multilineage Markers, 66, 71-72

N

Nanofibrous Scaffolds, 11, 21, 23, 73-75, 77, 79-86, 110, 128
Natural Extracellular Matrix, 1, 22, 183
Nerve Tissue Engineering, 73-74, 85
Nucleus Pulposus (np), 37

P

Peritoneum, 123-128

Polymer Scaffolds, 1, 46, 126, 157, 168
Polymer-ceramic Spiral Structured Scaffolds, 12
Porous Hydrogels Of Poly(ethylene Glycol) (peg), 46
Potential Bone Tissue Engineering, 1-2
Progenitor Cell Responses, 123
Promising Tissue Engineering Scaffolds, 22

R
Rat Mesenchymal Stem Cells (rmscs), 1
Regional Variations, 87, 93, 96-97, 106
Regulatory Signals, 139, 146

S
Scanning Electron Microscopy (sem), 2, 25, 70-71, 83, 99, 192
Schwann Cell (scs), 74
Simulated Body Fluid (sbf), 1-2

Skin Regeneration, 66-67, 199
Skin-derived Mesenchymal Stromal Cells (sd-mscs), 66
Skin-like Structures, 111
Sodium Dodecyl Sulfate (sds), 98, 183
Soft Tissue Damage, 22

T
Three-dimensional (3d) Spatial Pattern, 111
Tissue Engineering And Regenerative Medicine (term), 201
Tissue Engineering Bone, 123
Tissue-engineered Constructs, 57-58, 60-64, 199

V
Vascular Endothelial Cells (vec), 211